# Photosynthesis Bibliography

# volume 9 1978

References no. 32555-36587 / AAR-ZVA

Editors Z. Šesták & J. Čatský

Dr W. Junk Publishers – The Hague 1981

Contributors:
Z. Šesták
J. Čatský
I. Tichá
J. Pospíšilová
J. Solárová
D. Hodáňová

ISBN-13:978-90-6193-048-8    e-ISBN-13:978-94-009-8627-5
DOI: 10.1007/978-94-009-8627-5

90 6193 886 4 (series)

# PREFACE

The bibliography includes papers in all fields of photosynthesis research - from studies of model biochemical and biophysical systems of the photosynthesis mechanism to primary production studied by the so-called growth analysis. In addition to papers devoted entirely to photosynthesis, papers on other topics are included if they contain data on photosynthetic activity, photorespiration, chloroplast structure, chlorophyll and carotenoid synthesis and destruction, *etc.*, or if they contain valuable methodological information (measurement of selected environmental factors, leaf area, *etc.*). In many branches it has been difficult to define the limits of interest for photosynthesis researchers. This problem has arisen *e.g.* in topics dealing with the transfer of gases, where - in addition to the papers on carbon dioxide transfer - some papers on water vapour transfer are included, these being of general application or bringing new approaches. On the other hand, many papers dealing with the anatomy and physiology of stomata have been omitted, if the aspect of carbon dioxide or water vapour exchange has not been discussed.

This volume contains references to papers published in the year 1978, and, similarly to Vol.8, also addenda including references published in the preceding period (*i.e.* 1966 to 1977). The numbers of these additional references are labelled with an asterisk (*) in the list of references.

To maximize the value of the bibliography the references are arranged alphabetically by authors' names, and each volume is provided with three indexes. The Authors' Index contains all names of authors, co-authors and editors. The Subject Index covers primary items chosen according to their interest for photosynthesis researchers. Starting with Vol.6, the Subject Index has been newly arranged and enlarged. It contains more details on the electron transport chain, carbon fixation pathways, gas exchange on leaf and canopy level, *etc.*, and also on internal and environmental factors affecting photosynthesis and related processes. In the Plant Index, the most important crop plants and selected plant types and groups are indexed.

Cumulative indexes accompany Volumes 1, 5, and then every fifth volume, *i.e.* Volumes 10, 15, *etc.*

We have tried to cover fully the relevant papers which have appeared in the most important scientific periodicals and books. Articles published in local journals, mimeographed booklets, *etc.*, were chosen mostly from reprints and lists of publications received directly from the authors. Only abstracts published in regular journals were included.

Since some 4000 relevant papers are currently published every year and included in this bibliography, and since almost all citations have been checked with the originals, collecting and preparing for publication of such a large amount of material would have been impossible without the collaboration of the authors of the relevant publications. The courtesy of those authors who have already supplied us with reprints is highly appreciated.

We acknowledge with thanks the cooperation of our colleagues from the Institute of Experimental Botany of the Czechoslovak Academy of Sciences in Prague, especially Mrs. DRAHOMÍRA TĚŽKÁ, Mrs. LUDMILA HÁVOVÁ, Mrs. LENKA KOLČABOVÁ and Mrs. MARIE MANDLOVÁ who helped in preparing the card material. The librarian of our Institute, Mrs. ZORA ZAWOYSKA helped us with checking the references.

Dr. Z. ŠESTÁK and Dr. J. ČATSKÝ

Institute of Experimental Botany
Czechoslovak Academy of Sciences

Flemingovo n. 2
CS-160 00 PRAHA 6
Czechoslovakia

## INSTRUCTIONS FOR USE

All references are arranged alphabetically according to the authors' names and the year of publication. They are numbered and these numbers are used in the indexes. In case of a book title, the number is preceded by B. An asterisk preceding the number denotes the reference published in the preceding period (1966 - 1977).

The references contain the original unshortened title of the paper (book). English, French and German titles are cited in the original language. Titles in other languages are supplemented with a translation in English (sometimes using the title of the respective English abstract or a shortened title with omitted deadweight words). Titles of Japanese, Chinese *etc.* papers are given in English translation only. The journals' names are abbreviated mainly according to the "Style Manual for Biological Journals" (Second Edition, Amer. Institute of Biological Sciences, Washington, D.C. 1964), *e.g.* :

| | | | |
|---|---|---|---|
| Abhandlungen | chinese | Industry | Publishers |
| Abstract | Chromatography | inorganic | quantitative |
| Abteilung | Commission | Institute | Quarterly |
| Academy | Communication | international | Radiation |
| Acta | comparative | Investigation | Radiobiology |
| Africa | Comptes rendus | italian | Rasteniÿ |
| agricultural | Conference | Izvestiya | Recherche |
| Agriculture | Congress | Jahrbuch | Report |
| Agronomy | Contribution | japanese | Research |
| Akademie (-emiya) | Cytochemistry | Japan | Review |
| Algology | Cytology | Journal | royal |
| allgemeine | czechoslovak | Klasse | russian |
| american | Dendrology | Laboratory | russkiÿ |
| America | Department | Landwirtschaft | scandinavicus |
| analytical | Deutschland | Letters | Science |
| Anatomy | Disease | Limnology | Section |
| angewandte | Dissertation | Magazin | Series (-iya) |
| Annals | Doklady | marine | Society |
| annual | Dopovidi | Mathematics | sovetskiÿ |
| anorganisch (-nic) | Ecology | Microbiology | soviet |
| applied | Education | miscellaneous | special |
| Arbeit | Embryology | molecular | SSSR |
| Archiv | Encyclopedia | Monograph | Station |
| Atmosphere | Engineer | moskovskiÿ | Supplement |
| atomic | Enzymology | Mycology | Survey |
| Australia | european | national | Symposium |
| Beiheft | experimental | natural | technical |
| Belgique | Experiment | Naturforschung | Technology |
| Bericht | Faculty | neerlandicus | Tijdschrift |
| biochemical | Federation | Netherland | Transaction |
| Biochemistry | Fizika | New Zealand | Travail (-aux) |
| biokhimicheskiÿ | Fiziologiya | nuclear | tropical |
| Biokhimiya | Forestry | Oceanography | Trudy |
| biological(-ogicheskiÿ) | Forschung | Optics | ukrainian |
| Biology (-ogiya) | Foundation | organic | UK |
| biophysical | France | original | US, USA |
| Biophysics | Gazette | Otdelenie | USSR |
| Bodenkunde | general | Pathology | University |
| bolgarskiÿ | genetical | Pflanzen- | végétal |
| botanical (-anicheskiÿ) | Genetics | Philosophy | Virology |
| Botany | Gesellschaft | physical | Virusforschung |
| british | Giornale | Physics | Volume |
| Bulletin | helveticus | physiological | Weekblad |
| Canada | Histochemistry | Physiology | Wetenschappen |
| cellular (-ulaire) | Histology | Phytopathology | Wissenschaft |
| central | Horticulture | Plant (-arum) | Zeitschrift |
| chemical | hungaricus | polish | Zeitung |
| Chemistry | Husbandry | Proceedings | Zentralblatt |
| chimicus | imperial | Publication | Zhurnal |

The numbers at the end of each reference of a journal article denote : volume (issue) : first page - last page, year of publication. The number of issue is given only in the journal where each issue is paginated separately.

Book titles are cited according to the title page, not to the book jacket or cover (if the names of the editors are not given on the title page, they are not cited in the reference). The publishing house, place and year of publication are included.

Brackets at the end of the reference give bibliographic details and explanations to the contents, not given in the original. The following abbreviations are used most often :

| | | | |
|---|---|---|---|
| ab | abstract | Ital. | Italian |
| Arm. | Armenian | Jap. | Japanese |
| Belorus. | Belorussian | Latv. | Latvian |
| Bil | biliproteins | Lithu. | Lithuanian |
| Bulg. | Bulgarian | Norweg. | Norwegian |
| Car | carotenoids | PC | paper chromatography |
| CC | column chromatography | PhAR | photosynthetically active radiation |
| Chin. | Chinese | | diation |
| Chl | chlorophyll | Pol. | Polish |
| Croat. | Croatian | Ps | photosynthesis |
| Cyt | cytochromes | R | Russian |
| Dan. | Danish | Roum. | Roumanian |
| E | English | Span. | Spanish |
| F | French | Swed. | Swedish |
| G | German | TLC | thin-layer chromatography |
| GC | gas chromatography | Tr | transpiration |
| Georg. | Georgian | Ukr. | Ukrainian |
| Hung. | Hungarian | Uz. | Uzbeg |
| IRGA | infra-red gas analyser | | |

The transliteration of Cyrillic characters is in accordance with the BSI-ASA//SC-Z39 draft table, *i.e.* :

| Translit. | Cyrill. | Translit. | Cyrill. |
|---|---|---|---|
| a | а | p | п |
| b | б | r | р |
| ch | ч | s | с |
| d | д | sh | ш |
| e | е | shch | щ |
| ė | з | t | т |
| f | ф | ts | ц |
| g | г | u | у |
| i | и | v | в |
| ĭ | й | y | ы |
| k | к | ya | я |
| kh | х | yu | ю |
| l | л | z | з |
| m | м | zh | ж |
| n | н | " | ъ |
| o | о | ' | ь |

Several exceptions apply for Ukrainian, Belorussian and Serbian :

| Translit. | Cyrill. | | Translit. | Cyrill. | Translit. | Cyrill. |
|---|---|---|---|---|---|---|
| Ukr. y | и | Serbian : | ć | ħ | č | ч |
| i | і | | dj | ђ | š | ш |
| ï | ï | | dž | џ | c | ц |
| Beloruss. | | | h | х | | |
| ŭ | ў | | j | ј | | |
| | | | lj | љ | | |
| | | | nj | њ | | |

Authors' names are presented in spelling used in the original paper. If this spelling does not correspond to the original spelling used by the author (*e.g.* Russian papers of English authors), one spelling is referred to the other in the Authors' Index.

Printers' errors in the original papers are marked by underlining the respective words (letters).

# ERRATA

| Reference No./page | For | Read |
|---|---|---|

## Volume 6

| | | |
|---|---|---|
| 21932 / p.25 | QUIEROZ, O. | QUEIROZ, O. |
| 22083 / p.33 | CIFFERRI, O. | CIFERRI, O. |
| Authors' index / p.219 | CIFFERI, O. | CIFERRI, O. |
| Authors' index / p.243 | QUEIROZ, O. 24126 | QUEIROZ, O. 21932, 24126 |
| | QUIEROZ, O. 21932 | *delete* |

## Volume 7

| | | |
|---|---|---|
| 26152 / p.55 | GIVAN, A.C.V. | GIVAN, C.V. |
| 27420 / p.128 | PEACH, C. | PAECH, C. |
| 27421 / p.128 | CROSSBIE, T.M. | CROSBIE, T.M. |
| Authors' index / p.198 | CROSSBIE, T.M. | CROSBIE, T.M. |
| Authors' index / p.203 | GIVAN, C.V. 26153 | GIVAN, C.V. 26152-3 |
| | GIVAN, A.C.V. | *delete* |
| Authors' index / p.218 | PEACH, C. | PAECH, C. |
| Subject index / p.257 Hygrometer... | determination | methods |

## Volume 8

| | | |
|---|---|---|
| 29662 / p.67-68 | Fotosynteza | Fotosinteza |
| 30042 / p.89 | GEBBING, M.G.T. | GEBBINK, M.G.T. |
| 30497 / p.117 | in reciprocal | in reciprocal |
| 30648 / p.125 | agr. | Agr. |
| | 6 -   , 1977. | 6 - 10, 1977. |
| 30908 / p.140 | effct | effect |
| 31055 / p.149 | RASHÉTNIKAŬ, U.N. | RASHÉTNIKAŬ, U.M. |
| 31209 / p.158 | Vestsy | Vestsi |
| 31801 / p.193 | strukturnoi | strukturnoĭ |
| | fotosinteziruyushchei | fotosinteziruyushcheĭ |
| 31802 / p.193 | uglekislotnyi | uglekislotnyĭ |
| 32135 / p.212 | diminished proton | diminished proton |
| Authors' index / p.251 | GEBBING, M.G.T. | GEBBINK, M.G.T. |
| Authors' index / p.272 | RASHETNIKAŬ, U.N. | RASHÉTNIKAŬ, U.M. |
| Subject index / p.304 | Decapitation see ... | *delete* |
| | | *add:* Defoliation see |
| | | Decapitation, defoliation... |

| | | |
|---|---|---|
| Subject index / p.306 Ecotypes...gas exchange | | *add:* 31240 |
| Subject index / p.313 Hygrometer ...   determination | | methods |
| Subject index / p.327 Phosphoenolpyruvate carboxylase | | *add:* 31240 |
| Subject index / p.336 Ribulose 1,5-bisphosphate carboxylase | | *add:* 31240 |
| Subject index / p.341 Temperature, physiological, | | |
| | and carbon ... | *add:* 31240 |
| | and electron ... | *add:* 31241 |
| p.342 | and resistances ... | *add:* 31240-1 |
| Plant index / p.346 Algae, green ... | *Chlymadomonas* | *Chlamydomonas* |

*32555 - AARNES, H. : Forskjellige former for fotosyntese hos plantene: sukkulenter, C3 og C4-planter. [Different types of photosynthesis in plants: succulent, $C_3$ and $C_4$ plants.] - Naturen 100 (3) : 115 - 120, 1976. [In Norweg.]

32556 - AASE, J.K. : Relationship between leaf area and dry matter in winter wheat. - Agron J. 70 : 563 - 565, 1978.

*32557 - ABAD-ZAPATERO, C., FOX, J.L., HACKERT, M.L. : The quaternary structure of a unique phycobiliprotein: B-phycoerythrin from Porphyridium cruentum. - Biochem. biophys. Res. Commun. 78 : 266 - 272, 1977.

32558 - ABDULAEV, N.G., FEIGINA, M.Yu., KISELEV, A.V., OVCHINNIKOV, Yu.A., DRACHEV, L.A., KAULEN, A.D., KHITRINA, L.V., SKULACHEV, V.P. : Products of limited proteolysis of bacteriorhodopsin generate a membrane potential. - FEBS Lett. 90 : 190 - 194, 1978.

*32559 - ABDULAEV, N.G., KISELEV, A.V., FEĬGINA, M.Yu., OVCHINNIKOV, Yu.A. : Izuchenie asimmetrichnoĭ orientatsii bakteriorodopsina v purpurnoĭ membrane Halobacterium halobium. [Study on asymmetric arrangement of bacteriorhodopsin in the purple membrane of Halobacterium halobium.] - Bioorg. Khim. 3 : 709 - 710, 1977. [In R, ab : E.]

*32560 - ABDULLAEV, Kh.A., KHASANOV, M.M., NASYROV, Yu.S. : Ul'trastruktura plastid izolirovannykh protoplastov. [Ultrastructure of plastids in isolated protoplasts.] - Dokl. Akad. Nauk tadzh. SSR 18 (6) : 58 - 61, 1975. [In R, ab : Tajik.]

*32561 - ABDULLAEV, Kh.A., MUZAFFAROVA, S.M., NASYROV, Yu.S. : Deĭstvie khloramfenikola na razvitie ul'trastruktury khloroplastov. [Effect of chloramphenicol on development of chloroplast ultrastructure.] - Dokl. Akad. Nauk tadzh. SSR 18 (9) : 59 - 61, 1975. [In R, ab : Tajik.]

32562 - ABDULRAHMAN, F.S., WILLIAMS, G.J. III : The effects of NaCl and temperature on gas exchange and growth in Salicornia fruticosa. - Plant Physiol. 61 (Suppl.) : 6, 1978.

32563 - ABDURAKHMANOV, I.A., GANAGO, A.O., EROKHIN, Yu.E. : Lineĭnyĭ dikhroizm orientirovannykh khromatoforov i pigment-belkovykh kompleksov iz fotosinteziruyushcheĭ bakterii Chromatium minutissimum. [Linear dichroism of oriented chromatophores and pigment-protein complexes from the photosynthesizing bacterium Chromatium minutissimum.] - Dokl. Akad. Nauk SSSR 242 : 1197 - 1199, 1978. [In R.]

32564 - ACKEFORS, H., HERNROTH, L., LINDAHL, O., WULFF, F. : Ecological production studies of the phytoplankton and zooplankton in the Gulf of Bothenia. - Finnish mar. Res. 244 : 116 - 126, 1978. [Chl.]

*32565 - ACKEFORS, H., LINDAHL, O. : Investigations on primary phytoplankton production in the Baltic in 1974. - Medd. Havsfiskelab. Lysekil 195 : 1 - 29, 1975. [Chl.]

32566 - ACKER, S., DURANTON, J. : Evidence for a biosynthetic heterogeneity of the chlorophyll b. - In : AKOYUNOGLOU, G., ARGYROUDI-AKOYUNOGLOU, J.H. (ed.) : Chloroplast Development. Pp. 131 - 134. Elsevier/North-Holland Biomedical Press, Amsterdam - New York - Oxford 1978.

32567 - ACKERSON, R.C., CHILCOTE, D.O. : Effects of defoliation and TIBA (triiodobenzoic acid) on tillering, dry matter production, and carbohydrate reserves of two cultivars of Kentucky bluegrass. - Crop Sci. 18 : 705 - 708, 1978. [Growth analysis.]

32568 - ACOCK, B., CHARLES-EDWARDS, D.A., FITTER, D.J., HAND, D.W., LUDWIG, L.J., WARREN WILSON, J., WITHERS, A.C. : The contribution of leaves from different levels within a tomato crop to canopy net photosynthesis: An experimental examination of two canopy models. - J. exp. Bot. 29 : 815 - 827, 1978.

32569 - ACOCK, B., CHARLES-EDWARDS, D.A., FITTER, D.J., HAND, D.W., WARREN WILSON, J. : Modelling canopy net photosynthesis by isolated blocks and rows of chrysanthemum plants. - Ann. appl. Biol. 90 : 255 - 263, 1978.

32570 - ADAMS, M.S., GUILIZZONI, P., ADAMS, S. : Relationship of dissolved inorganic carbon to macrophyte photosynthesis in some Italian lakes. - Limnol. Oceanogr. 23 : 912 - 919, 1978.

*32571 - ADAMS, M.W.W., REEVES, S.G., HALL, D.O., CHRISTOU, G., RIDGE, B., RYDON, H.N.:
Biological activity of synthetic tetranuclear iron-sulphur analogues of the
active sites of ferredoxins. - Biochem. biophys. Res. Commun. *79* : 1184 -
1191, 1977.

32572 - ADAMSON, H. : Evidence for the accumulation of both chlorophyll *a* and *b* in
darkness in an angiosperm. - In : AKOYUNOGLOU, G., ARGYROUDI-AKOYUNOGLOU, J.H.
(ed.) : Chloroplast Development. Pp. 135 - 140. Elsevier/North-Holland Biomedical Press, Amsterdam - New York 1978.

32573 - AĔROV, I.L. : Izmeneniya fotosinteza, soprotivleniĭ diffuzii $CO_2$ i dykhaniya
u svekly v techenie vegetatsii. [Changes in photosynthesis, resistances to
$CO_2$ diffusion and respiration in beet during vegetation.] - Fiziol. Biokhim.
kul't. Rast. *10* : 77 - 82, 1978. [In R, ab : E.]

32574 - AGARWALA, S.C., MEHROTRA, N.K. : Growth and metabolism of rice plants subjected to high alkalinity (SAR) in irrigation waters and soil calcareousness. -
Indian J. Plant Physiol. *21* : 59 - 65, 1978. [Chl.]

32575 - AGHION, J., LEBLANC, R.M. : Interactions of photosynthetic pigments in monolayers at a water-air interface. - J. Membrane Biol. *42* : 189 - 198, 1978.

32576 - AGRAWAL, P.K., FOCK, H. : Carbon dioxide exchange and the fate of recent
photosynthetic fixed carbon in relation to seed shrivelling in two lines of
hexaploid triticale. - In : Inter-Disciplinary Symposium on Photosynthesis
and Productivity. Abstract of Papers. Pp. 59 - 60. Indian nat. Sci. Acad.,
New Delhi 1978.

32577 - AGRAWAL, P.K., FOCK, H. : The specific radioactivity of glycolic acid in relation to the specific activity of carbon dioxide evolved in light in photosynthesizing sunflower leaves. - Planta *138* : 257 - 261, 1978.

32578 - AHARONI, N., RICHMOND, A.E. : Endogenous gibberellin and abscisic acid content as related to senescence of detached lettuce leaves. - Plant Physiol.
*62* : 224 - 228, 1978. [Chl.]

32579 - AHOKAS, H. : Cytoplasmic male sterility in barley II. Physiology and anther
cytology of *msm1*. - Hereditas *89* : 7 - 21, 1978. [Chl, Car.]

*32580 - AHRENS, E.H.Jr., WILLIAMS, D.C., BATTERSBY, A.R. : Biosynthesis of porphyrins
and related macrocycles. Part 11. Studies on biosynthesis of the phytyl
chain of chlorophyll *a* by use of carbon-13. - J. chem. Soc., Perkin Trans.
I *1977* : 2540 - 2545, 1977.

32581 - AĬRAPETYAN, A.L. : [Characteristics of accumulation of chlorophyll *b* in hybrid cucumber plants.] - Izv. sel'skokhoz. Nauk (Erevan) *21* (3) : 42 - 47,
1978. [In Armen., ab : R.]

32582 - AKAO, S., CHATTERTON, N.J., CARLSON, G.E., HUNGERFORD, W.E. : [Studies on
the translocation of photosynthates in alfalfa. Part 2. The effects of phosphorus and adenosine-5'-triphosphate (ATP) on the translocation and distribution of photosynthates.] - Shikoku Nogyo Shikenjo Hokoku [Bul. Shikoku
agr. Exp. Sta.] *31* : 133 - 146, 1978. [In Jap., ab : E.]

32583 - AKAZAWA, T. : Structure and function of ribulose bisphosphate carboxylase. -
In : HALL, D.O., COOMBS, J., GOODWIN, T.W. (ed.) : Proceedings of the Fourth
International Congress on Photosynthesis. Pp. 447 - 456. Biochem. Soc., Lon-
don 1978.

32584 - AKAZAWA, T., NEWCOMB, E.H., OSMOND, C.B. : Pathway and products of $CO_2$-fixa-
tion by green prokaryotic algae in the cloacal cavity of *Diplosoma virens*. -
Mar. Biol. *47* : 325 - 330, 1978.

32585 - AKAZAWA, T., TAKABE, T., ASAMI, S., KOBAYASHI, H. : Ribulose bisphosphate
carboxylases from *Chromatium vinosum* and *Rhodospirillum rubrum* and their role in photosynthetic carbon assimilation. - In : SIEGELMAN, H.W., HIND, G.
(ed.) : Photosynthetic Carbon Assimilation. Pp. 209 - 226. Plenum Press,
New York - London 1978.

32586 - AKERS, C.P., WEYBREW, J.A., LONG, R.C. : Ultrastructure of glandular tricho-
mes of leaves of *Nicotiana tabacum* L., cv. Xanthi. - Amer. J. Bot. *65* : 282
- 292, 1978. [Chloroplast.]

32587 - **AKHMANOV, S.A., BORISOV, A.Yu., DANIELIUS, R.V., KOZLOVSKIJ, V.S., PISKAR-SKAS, A.S., RAZJIVIN, A.P.** : Primary photosynthesis selectively excited by tunable picosecond parametric oscillator. - In : SHANK, C.V., IPPEN, E.P., SHAPIRO, S.L. (ed.) : Picosecond Phenomena. Pp. 134 - 139. Springer-Verlag, Berlin - Heidelberg - New York 1978.

32588 - **AKHMANOV, S.A., BORISOV, A.Yu., DANELYUS, R.V., KOZLOVSKIĬ, V.S., PISKARSKAS, A.S., RAZZHIVIN, A.P.** : Pikosekundnyĭ perenos énergii mezhdu spektral'nymi formami pigmentov reaktsionnogo tsentra iz *Rhodospirillum rubrum*. [Picosecond energy transfer *via* spectral forms of pigments of reaction centres from *Rhodospirillum rubrum*.] - Biofizika *23* : 912 - 913, 1978. [In R, ab : E.]

32589 - **AKIMOTO, Y.** : [Studies on final leaf shape of tobacco in relation to temperature in young stage of leaf. I. Effects of high temperature in young stage of tobacco leaf on its final leaf shape.] - Jap. J. Crop Sci. *47* : 483 - 490, 1978. [In Jap., ab : E.]

*32590 - **AKOYUNOGLOU, G.** : Development of the photosystem II unit in plastids of bean leaves greened in periodic light. - Arch. Biochem. Biophys. *183* : 571 - 580, 1977.

32591 - **AKOYUNOGLOU, G.** : Growth of the PS II unit as monitored by fluorescence measurements, the photo-induced absorbance change at 518 nm, and photochemical activity. - In : AKOYUNOGLOU, G., ARGYROUDI-AKOYUNOGLOU, J.H. (ed.) : Chloroplast Development. Pp. 355 - 366. Elsevier/North-Holland Biomedical Press, Amsterdam - New York - Oxford 1978.

B32592 - **AKOYUNOGLOU, G., ARGYROUDI-AKOYUNOGLOU, J.H.** : Chloroplast Development. - Elsevier/North-Holland Biomedical Press, Amsterdam - New York - Oxford 1978.

32593 - **AKOYUNOGLOU, G., ARGYROUDI-AKOYUNOGLOU, J.H.** : Control of thylakoid growth in *Phaseolus vulgaris*. - Plant Physiol. *61* : 834 - 837, 1978.

32594 - **AKOYUNOGLOU, G., ARGYROUDI-AKOYUNOGLOU, J.H.** : Photosystem II unit: assembly, growth and its interactions during development of higher plants thylakoids. - In : METZNER, H. (ed.) : Photosynthetic Oxygen Evolution. Pp. 453 - 488. Academic Press, London - New York - San Francisco 1978.

32595 - **AKOYUNOGLOU, G., ARGYROUDI-AKOYUNOGLOU, J.H., CHRISTIAS, C., TSAKIRIS, S., TSIMILI-MICHAEL, M.** : Thylakoid growth and differentiation in continuous light as controlled by the duration of preexposure to periodic light. - In : AKOYUNOGLOU, G., ARGYROUDI-AKOYUNOGLOU, J.H. (ed.) : Chloroplast Development. Pp. 843 - 856. Elsevier/North-Holland Biomedical Press, Amsterdam - New York - Oxford 1978.

32596 - **AKULOVA, E.A.** : Élektrontransportnaya tsep' khloroplastov i punkty sopryazheniya. [Electron transport chain and coupling sites of chloroplasts.] - Izv. Akad. Nauk SSSR, Ser. biol. *1978* : 218 - 229, 1978. [In R, ab : E.]

32597 - **AKULOVICH, N.K., PARSHYKAVA, T.A.** : Dzeyanne lipazy na spektral'nyya ŭlastsivastsi protakhlarafilavaga pigmentu étyyaliravanykh listsyaŭ. [Action of lipase on spectral properties of the protochlorophyll pigment of etiolated leaves.] - Vestsi Akad. Navuk belarus. SSR, Ser. biyal. Navuk *1978* (5) : 37 - 39, 138, 1978. [In Belorus., ab : E, R.]

32598 - **ALASAARELA, E.** : Dynamics of the hydrography, nutrients and primary production of the coastal waters of the Bothnian Bay off Oulu. - Aquilo, Ser. bot. *16* : 16 - 38, 1978.

32599 - **ALASAARELA, E.** : Phytoplankton in the coastal waters of the Bothnian Bay off Kemi. - Ann. bot. fenn. *15* : 297 - 308, 1978. [Chl.]

32600 - **ALBERTE, R.S., THORNBER, J.P.** : A rapid procedure for isolating the photosystem I reaction center in a highly enriched form. - FEBS Lett. *91* : 126 - 130, 1978.

32601 - **ALEĬNIKOV, I.M., PRIĬMACHEK, V.R., CHEPELEV, V.V., YASEVICH, Ch.** : Zavisimost' fotosensibilizirovannogo khlorofillom vosstanovleniya metilovogo krasnogo askorbinovoĭ kislotoĭ ot pigmentnogo sostava i sostoyaniya élektrontransportnoĭ tsepi khloroplasta. [Dependence of chlorophyll-photosensibilized reduction of methyl red by ascorbic acid on pigment composition and

state of electron transport chain of chloroplasts. - Fiziol. Biokhim. kul't.
Rast. *10* : 427 - 430, 1978. [In R, ab : E.]

32602 - **ALEKSANDROV, A.Yu., NOVAKOVA, A.A., USPENSKAYA, N.Ya., KURYUSHKIN, A.A.,
KUZ'MIN, R.N., RUBIN, A.B., KONONENKO, A.A.** : Issledovanie sostoyaniya vnu-
trikletochnogo zheleza fotosintezIruyushchikh purpurnykh serobakteriĭ metodom éffekta Messbauera. [Mössbauer effect study of intracellular iron in
photosynthesizing purple sulphur bacteria.] - Mol. Biol. (Moskva) *12* : 55 -
62, 1978. [In R, ab : E.]

*32603 - **ALESHIN, E.P., POROKHNYA, A.D., DANILOVA, T.A., DEMKINA, E.N.** : Otzyvchivost'
risa na mikroélementy. [Response of rice to trace elements.]'- Tr. vses.
nauch.-issled. Inst. Udobreniĭ Agropochvoved. *53* (Vliyanie Mikroélementov
na Urozhaĭ i Obmen Veshchestv Sel'skokhozyaĭstvennykh Kul'tur) : 203 - 208,
1972. [Ps, Chl; In R.]

*32604 - **ALIEV, D.A., AZIZOV, I.V.** : Fotokhimicheskaya aktivnost' khloroplastov ozimoĭ pshenitsy v svyazi s ee optiko-biologicheskoĭ strukturoĭ, produktivnost'·
yu i usloviyami osveshchennosti. [Photochemical activity of chloroplasts
of winter wheat in relation to their optical and biological structure, productivity and irradiance.] - Izv. Akad. Nauk azerb. SSR, Ser. biol. Nauk
*1977* (5) : 69 - 74, 1977. [In R, ab : Azerb.]

32605 - **ALIEV, D.A., SAFAROV, S.A.** : Raspredelenie khlorofilla $a$ i $b$ v list'yakh
razlichnykh sortov pshenitsy po vertikali posevov. [Distribution of chlorophyll $a$ and $b$ in the leaves of different wheat cultivars along crop height.]
- Izv. Akad. Nauk azerb. SSR, Ser. biol. Nauk *1978* (5) : 3 - 7, 1978. [In R,
ab : Azerb.]

32606 - **ALLAN, R.J., KENNEY, B.C.** : Rehabilitation of eutrophic Prairie lakes in
Canada. - Verh. int. Verein. Limnol. *20* : 214 - 224, 1978. [Chl.]

32607 - **ALLEN, J.F.** : Induction of a Mehler reaction in chloroplasts preparations
by flavin mononucleotide: effects on photosynthesis by intact chloroplasts..
- Plant Sci. Lett. *12* : 151 - 159, 1978.

32608 - **ALLEN, J.F.** : Induction of a Mehler reaction in chloroplast preparations
by methyl viologen and by ferredoxin: effects on photosynthesis by intact
chloroplasts. - Plant Sci. Lett. *12* : 161 - 167, 1978.

32609 - **ALLEN, J.F., WHATLEY, F.R.** : Effects of inhibitors of catalase on photosynthesis and on catalase activity in unwashed preparations of intact chloroplasts. - Plant Physiol. *61* : 957 - 960, 1978.

32610 - **ALLEN, M.J.** : Electrochemistry of plant systems. V. The anomalous behaviour
of broad bean leaf epidermis-cuticle preparations. - Bioelectrochem. Bioenerg. *5* : 776 - 779, 1978. [Ps, chloroplast.]

32611 - **ALLEN, R.J.Jr., KIDDER, G., GASCHO, G.J.** : Predicting tons of sugarcane per
acre using solar radiation, temperature and percent plant cane, 1971 through
1976. - Proc. amer. Soc. Sugar Cane Technol. *7* : 18 - 22, 1978.

32612 - **ALLESSIO, M.L., TIESZEN, L.L.** : Translocation and allocation of $^{14}C$-photo-
assimilate by *Dupontia fisheri*. - In : TIESZEN, L.L. (ed.) : Vegetation
and Production Ecology of an Alaskan Arctic Tundra. Pp. 393 - 413. Springer-
-Verlag, New York - Heidelberg - Berlin 1978.

32613 - **ALLOT-DERONNE, M., BLONDON, F., CLABAULT, G.** : Rhizogenèse et activité photosynthétique de la feuille isolée du *Perilla ocymoides* L., cultivée en
photopériodes non inductrices de la floraison. - Compt. rend. Acad. Sci.
Paris, Sér. D *286* : 859 - 862, 1978.

32614 - **ALMASSY, R.J., DICKERSON, R.E.** : *Pseudomonas* cytochrome $c_{551}$ at 2.0 Å resolution: Enlargement of the cytochrome $c$ family. - Proc. nat. Acad. Sci. USA
*75* : 2674 - 2678, 1978. [Ps.]

32615 - **ALMGREN, M.** : Thermodynamic and kinetic limitations on the conversion of
solar energy into storable chemical free-energy. - Photochem. Photobiol.
*27* : 603 - 609, 1978.

32616 - ALSCHER, R., PATTERSON, R., JAGENDORF, A.T. : Activity of thylakoid-bound ribosomes in pea chloroplasts. - Plant Physiol. *62* : 88 - 93, 1978.

32617 - AMANO, H., NODA, H. : [Photosynthetic pigments of five kinds of laver, "nori".] - Bull. jap. Soc. sci. Fish. *44* : 911 - 916, 1978. [In Jap., ab : E.]

32618 - AMES, I.H., TEPPER, H.B. : Seasonal changes in the ultrastructure of aspen bark chloroplasts. - Photosynthetica *12* : 70 - 72, 1978.

32619 - AMESZ, J. : Fluorescence and energy transfer. - In : CLAYTON, R.K., SISTROM, W.R. (ed.) : The Photosynthetic Bacteria. Pp. 333 - 340. Plenum Press, New York - London 1978.

32620 - AMESZ, J., VAN GORKOM, H.J. : Delayed fluorescence in photosynthesis. - Annu. Rev. Plant Physiol. *29* : 47 - 66, 1978.

*32621 - AMEZAGA, E. de, GOLDMAN, C.R., STULL, E.A. : Primary productivity and rate of change of biomass of various species of phytoplankton in Castle Lake, California. - Verh. int. Verein. Limnol. *18* : 1768 - 1775, 1973.

32622 - ANAN'EV, G.M., ZAKRZHEVSKIĬ, D.A. : Vliyanie defitsita margantsa na sootnoshenie faz vspyshechnoĭ i nepreryvnoĭ kinetik vydeleniya kisloroda kletkami khlorelly v protsesse ikh rosta. [Effect of manganese deficiency on the phase relation of flash- and continuous kinetics of oxygen evolution by *Chlorella* cells in their growth process.] - Dokl. Akad. Nauk SSSR *242* : 1429 - 1432, 1978. [In R.]

32623 - ANDERSEN, K., KING, W., VALENTINE, R.C. : Catalytic mutants of ribulose bisphosphate carboxylase/oxygenase. - In : SIEGELMAN, H.W., HIND, G. (ed.) : Photosynthetic Carbon Assimilation. Pp. 379 - 390. Plenum Press, New York - London 1978.

32624 - ANDERSEN, R.A., MEYER, R.L., KIM, K.S. : Ultrastructure observations on the vegetative cell chloroplast of *Hydrurus foetidus (Chrysophyceae).* - J. Phycol. *14* (Suppl.) : 29, 1978.

32625 - ANDERSON, J.E., KREITH, F. : Effects of film-forming and silicone antitranspirants on four herbaceous plant species. - Plant Soil *49* : 161 - 173, 1978. [Ps.]

32626 - ANDERSON, J.M., WALDRON, J.C., THORNE, S.W. : Chlorophyll-protein complexes of spinach and barley thylakoids. Spectral characterization of six complexes resolved by an improved electrophoretic procedure. - FEBS Lett. *92* : 227 - 233, 1978.

32627 - ANDERSON, J.W., DONE, J. : Light-dependent assimilation of nitrite by isolated pea chloroplasts. - Plant Physiol. *61* : 692 - 697, 1978.

32628 - ANDERSON, L.E., NEHRLICH, S.C., CHAMPIGNY, M.-L. : Light modulation of enzyme activity. Activation of the light effect mediators by reduction and modulation of enzyme activity by thiol-disulfide exchange? - Plant Physiol. *61* : 601 - 605, 1978. [Ps.]

32629 - ANDERSON, W.K., SMITH, R.C.G., McWILLIAM, J.R. : A systems approach to the adaptation of sunflower to new environments. II. Effects of temperature and radiation on growth and yield. - Field Crop Res. *1* : 153 - 163, 1978.

B32630 - ANDERSSON, B. : Separation of Spinach Chloroplast Lamellae Fragments by Phase Partition Including the Isolation of Inside-Out Thylakoids. - University of Lund, Lund 1978.

32631 - ANDERSSON, B., ÅKERLUND, H.-E. : Inside-out membrane vesicles isolated from spinach thylakoids. - Biochim. biophys. Acta *503* : 462 - 472, 1978.

32632 - ANDERSSON, B., SIMPSON, D.J., HØYER-HANSEN, G. : Freeze-fracture evidence for the isolation of inside-out spinach thylakoid vesicles. - Carlsberg Res. Commun. *43* : 77 - 89, 1978.

32633 - ANDRÉ, M., MASSIMINO, D., DAGUENET, A. : Daily patterns under the life cycle of a maize crop. I. Photosynthesis, transpiration, respiration. - Physiol. Plant. *43* : 397 - 403, 1978.

32634 - ANDREEV, N.G., KOLOMEĬCHENKO, V.V. : Ispol'zovanie solnechnoĭ ênergii seno-

kosami I pastbishchami nechernozem'ya. [Utilization of solar energy by meadows and pastures on non-chernozem soils.] - Dokl. VASKhNIL *1978* (10) : 5 -
- 7, 1978. [In R.]

32635 - **ANDREEVA, N.E., BARASHKOV, B.I., ZAKHAROVA, G.V., SHUBIN, V.V., CHIBISOV, A.K.**
: Priroda èlektronno-vozbuzhdennogo sostoyaniya v okislitel'no-vosstanovitel'-
nykh reaktsiyakh pigmentov. II. Analiz skhemy pervichnykh protsessov v reaktsii fotookisleniya khlorofillov *a* i *b* i feofitina *a*. [The nature of excited
states in redox reactions of pigments. II. Analysis of primary steps in chlorophylls *a*, *b* and pheophytin *a* photooxidation.] - Biofizika *23* : 565 - 570,
1978. [In R, ab : E.]

32636 - **ANDREO, C.S.** : Inhibition of energy-transducing functions of chloroplasts
by spegazzinine. - Arch. Biochem. Biophys. *186* : 416 - 421, 1978.

32637 - **ANDREU, J.M., WARTH, R., MUÑOZ, E.** : Glycoprotein nature of energy-transducing ATPases. Chemical characterization of glycopeptides isolated from bacterial and chloroplast coupling factors. - FEBS Lett. *86* : 1 - 5, 1978.

32638 - **ANDREWS, T.J., LORIMER, G.H.** : Photorespiration - still unavoidable? - FEBS
Lett. *90* : 1 - 9, 1978.

32639 - **ANGELOV, M., STANEV, V.** : Vliyanie na temperaturata v zonata na kornevata
sistema v"rkhu s"d"rzhanieto na khlorofila i intenznvnostta fotosintezata
v listata na domateni rasteniya. [Effect of temperature in the root zone
on chlorophyll content and photosynthetic rate of tomato leaf leaves.] -
Fiziol. Rast. (Sofia) *4*(4) : 12 - 18, 1978. [In Bulg., ab : E, R.]

32640 - **ANIKIEV, V.V., DONTSOV, V.V.** : Nedostatok vody v pochve i ustoĭchivost' k
nemu kukuruzy v razlichnye periody razvitiya. [Soil water deficit resistance
in the course of maize plant development.] - In : Problemy Zasukhoustoĭchivosti Rasteniĭ. Pp. 100 - 107. Nauka, Moskva 1978. [Growth analysis; in R.]

32641 - **ANITOFF, O.É., LEICKNAM, J.-P.** : Première analyse cinétique d'une variation
de constante diélectrique photo-induite dans les chloroplastes. - Compt.
rend. Acad. Sci. Paris, Sér. B *286* : 203 - 205, 1978.

32642 - **ANTON, J.A., LOACH, P.A., GOVINDJEE** : Transfer of excitation energy between
porphyrin centers of a covalently-linked dimer. - Photochem. Photobiol. *28* :
235 - 242, 1978. [Models.]

*32643 - **ANTONIELLI, M.** : Studio dei pigmenti di *Potamogeton perfoliatus* L. del lago
Trasimeno prelevato a varie profondità. [Pigments of *Potamogeton perfoliatus*
L. of lake Trasimeno, taken from different depths.] - Riv. Idrobiol. *8* : 67 -
75, 1969. [In Ital., ab : E.]

*32644 - **ANTONIELLI, M.** : Contenuto in clorofilla in foglie di *Pinus laricio* POIR e
*Secale cereale* L. ad altitudini diverse. [Chlorophyll content in leaves of
*Pinus laricio* POIR and *Secale cereale* L. at different altitudes.] - Ecol.
agr. *9* : 5 - 11, 1973. [In Ital., ab : E.]

*32645 - **ANTONIELLI, M., CAGIOTTI, M.R.** : Presenza di clorofillide in idrofite ed
alghe del Lago Trasimeno e variabilità del pigmento in *Spirogyra*. [Presence
of chlorophyllide in hydrophytes and algae of Lake Trasimeno and the variability of pigment content in *Spirogyra*.] - Riv. Idrobiol. *14* : 3 - 12, 1975.
[In Ital., ab : E.]

*32646 - **ANTONIELLI, M., GRANDI, A.** : Variazzioni del contenuto in clorofilla *a* e *b*
e loro rapporti durante l'ontogenesi di foglie di *Parthenocissus tricuspidata*
PLANCH. [Variations in chlorophyll *a* and *b* contents and their ratios during
ontogenesis of the leaves of *Parthenocissus tricuspidata* PLANCH.] - Ann. Fac.
agr. Univ. Perugia *26* : 543 - 551, 1971. [In Ital., ab : E.]

32647 - **ANTONIELLI, M., VENANZI, G., POGGIONI, S.** : Contenuto in clorofille di diffe-
renti cultivar di *Zea mays* L. [Chlorophyll content in various maize cultivars.] - Ann. Fac. agr. Univ. Perugia *32-33* : 197 - 206, 1977-1978. [In Ital.,
ab : E.]

32648 - **ANTONIW, L.D., SPRENT, J.I.** : Growth and nitrogen fixation of *Phaseolus vulgaris* L. at two irradiances. I. Growth. - Ann. Bot. *42* : 389 - 397, 1978.
[Ps.]

32649 - APASHEVA, L.M., BUDZHIASHVILI, D.M., NAĬDICH, V.I. : O mekhanizme porazheni-
ya fotosinteticheskogo apparata kletok khlorelly pri deĭstvii alkilsulfatov.
[Mechanism of damage to the photosynthetic apparatus of *Chlorella* cells in
the presence of alkylsulphates.] - Dokl. Akad. Nauk SSSR *239* : 1469 - 1471,
1978. [In R.]

32650 - APEL, K., KLOPPSTECH, K. : Light-induced appearance of mRNA coding for the
apoprotein of the light-harvesting chlorophyll *a/b* protein. - In :
AKOYUNOGLOU, G., ARGYROUDI-AKOYUNOGLOU, J.H. (ed.) : Chloroplast Development.
Pp. 653 - 656. Elsevier/North-Holland Biomedical Press, Amsterdam - New York
- Oxford 1978.

32651 - APEL, K., KLOPPSTECH, K. : The plastid membranes of barley (*Hordeum vulgare*).
Light-induced appearance of mRNA coding for the apoprotein of the light-harvesting chlorophyll *a/b* protein. - Europe. J. Biochem. *85* : 581 - 588, 1978.

32652 - APEL, P. : Untersuchungen über Blattdicke und Ausbildung der Ertragskompo-
nenten bei Sommergerste. - Kulturpflanze *26* : 71 - 79, 1978.

32653 - APEL, P., PEISKER, M. : Einfluß hoher Sauerstoffkonzentrationen auf den $CO_2$-
Kompensationspunkt von $C_4$-Pflanzen. - Kulturpflanze *26* : 99 - 103, 1978.

32654 - APEL, P., TICHÁ, I., PEISKER, M. : $CO_2$-Kompensationspunkt von *Moricandia arvensis* (L.) DC. bei Blättern unterschiedlicher Insertionshöhe und bei verschiedenen $O_2$-Konzentrationen. - Biochem. Physiol. Pflanzen *172* : 547 - 552,
1978.

32655 - APPLEBURY, M.L., PETERS, K.S., RENTZEPIS, P.M. : Primary intermediates in
the photochemical cycle of bacteriorhodopsin. - Biophys. J. *23* : 375 - 382,
1978.

32656 - APTE, S.K., ROWELL, P., STEWART, W.D.P. : Electron donation to ferredoxin
in heterocysts of the $N_2$-fixing alga *Anabaena cylindrica*. - Proc. roy. Soc.
London *B 200* : 1 - 25, 1978.

32657 - ARATA, H., NISHIMURA, M. : Coupling of electron transfer and proton translocation in purple photosynthetic bacteria. - In : DUTTON, P.L., LEIGH, J.S.,
SCARPA, A. (ed.) : Frontiers of Biological Energetics: Electrons to Tissues.
Vol. 1. Pp. 307 - 315. Academic Press, New York - San Francisco - London
1978.

32658 - ARGYRAKIS, P., KOPELMAN, R. : Exciton percolation III. Stochastic and coherent migration in binary and ternary random lattices. - J. theor. Biol. *73* :
205 - 236, 1978. [Ps.]

32659 - ARGYROUDI-AKOYUNOGLOU, J.H. : Isolation of the 25-23 kdalton polypeptide(s)
from SDS-solubilized thylakoids: Their importance in the onset of the cation-
-induced control of spillover and grana formation. - In : AKOYUNOGLOU, G.,
ARGYROUDI-AKOYUNOGLOU, J.H. (ed.) : Chloroplast Development. Pp. 401 - 412.
Elsevier/North-Holland Biomedical Press, Amsterdam - New York - Oxford 1978.

*32660 - ARGYROUDI-AKOYUNOGLOU, J.H., AKOYUNOGLOU, G. : Correlation between cation-
-induced formation of heavy subchloroplast fractions and cation-induced increase in chlorophyll *a* fluorescence yield in Tricine-washed chloroplasts. -
Arch. Biochem. Biophys. *179* : 370 - 377, 1977.

*32661 - ARGYROUDI-AKOYUNOGLOU, J.H., TSAKIRIS, S. : Development of the cation-indu-
ced stacking capacity during the biogenesis of higher plant thylakoids. -
Arch. Biochem. Biophys. *184* : 307 - 315, 1977.

32662 - ARIHARA, J., IWATA, F., WATANABE, K. : [Effect of upright leaves on dry matter production and grain yield of corn (*Zea mays* L.).] - Jap. J. Crop Sci.
*47* : 536 - 540, 1978. [In Jap., ab : E.]

32663 - ARKIN, G.F., RITCHIE, J.T., MAAS, S.J. : A model for calculating light interception by a grain sorghum canopy. - Trans. ASAE *21* : 303 - 308, 1978.

32664 - ARMOND, P.A., BADGER, M.R., BJÖRKMAN, O. : Characteristics of the photosyn-
thetic apparatus developed under different thermal growth regimes. - In :
AKOYUNOGLOU, G., ARGYROUDI-AKOYUNOGLOU, J.H. (ed.) : Chloroplast Development.
Pp. 857 - 862. Elsevier/North-Holland Biomedical Press, Amsterdam - New York
- Oxford 1978.

32665 - ARMOND, P.A., MOONEY, H.A. : Correlation of photosynthetic unit size and density with photosynthetic capacity. - Carnegie Inst.Year Book 77 : 234 - 237, 1978.

32666 - ARMOND, P.A., MOONEY, H.A. : General predictive value of pigment system analysis for determining photosynthetic capacity. - Plant Physiol. 61 (Suppl.) : 82, 1978.

32667 - ARMOND, P.A., SCHREIBER, U., BJÖRKMAN, O. : Photosynthetic acclimation to temperature in the desert shrub, Larrea divaricata. II. Light-harvesting efficiency and electron transport. - Plant Physiol. 61 : 411 - 415, 1978.

32668 - ARMOND, P.A., STAEHELIN, L.A. : Displacement of integral membrane proteins during lipid phase transitions in Anacystis nidulans. - Carnegie Inst. Year Book 77 : 291 - 294, 1978. [Thylakoid.]

32669 - ARMSTRONG, J.E., CALDER, J.A. : Inhibition of light-induced pH increase and $O_2$ evolution of marine microalgae by water-soluble components of crude and refined oil. - Appl. environ. Microbiol. 35 : 858 - 862, 1978.

32670 - ARNHEIM, K., GOLECKI, J.R., OELZE, J. : Adaptation to phototrophic conditions of chemotropic mecillinam sphaeroplasts of Rhodospirillum rubrum. - FEMS Microbiol. Lett. 4 : 279 - 281, 1978.

32671 - ARNOLD, C.-G. : Über Antibiotikaresistenz bei Mitochondrien und Chloroplasten. - Deut. Apoth.-Zeit. 118 : 362 - 364, 1978.

*32672 - ARNOLD, W. : Delayed light in photosynthesis. - Annu. Rev. Biophys. Bioeng. 6 : 1 - 6, 1977.

32673 - ARNON, D.I. : Photosynthetic phosphorylation: Conversion of sunlight to biochemical energy. - In : VAN TAMELEN, E.E. (ed.) : Bioorganic Chemistry. Vol. IV. Pp. 1 - 36. Academic Press, New York - San Francisco - London 1978.

*32674 - ARNON, D.I., TSUJIMOTO, H.Y., HIYAMA, T. : Electron paramagnetic resonance studies of photosynthetic electron transport: Photoreduction of ferredoxin and membrane-bound iron-sulfur centers. - Proc. nat. Acad. Sci. USA 74 : 3826 - 3830, 1977.

32675 - ARNTZEN, C.J. : Dynamic structural features of chloroplast lamellae. - In : SANADI, D.R., VERNON, L.P. (ed.) : Current Topics in Bioenergetics. Vol. 8. Pp. 111 - 160. Academic Press, New York - San Francisco - London 1978.

32676 - ARO, E.-M., VALANNE, N. : Effect of magnesium on chlorophyll-protein complexes. - Physiol. Plant. 43 : 261 - 265, 1978.

32677 - ARONOFF, S. : The association of chlorophyll b and of galactosyl diacyl glyceride with chlorophyll a. - Photosynthetica 12 : 298 - 303, 1978.

32678 - ARTECA, R.N., POOVAIAH, B.W., SMITH, O.E. : Effects of short term $CO_2$ enrichment of the root zone on tuberization and dry matter content in Solanum tuberosum L. - Plant Physiol. 61 (Suppl.) : 8, 1978.

32679 - ASADA, K., KANEMATSU, S. : Distribution of cuprozinc, manganic, and ferric superoxide dismutases in plants and fungi: an evolutionary aspect. - In : MATSUBARA, H., YAMANAKA, T. (ed.) : Evolution of Protein Molecules. Pp. 361 - 372. Jap. sci. Soc. Press, Tokyo 1978.

*32680 - ASADA, K., KANEMATSU, S., TAKAHASHI, M., KONO, Y. : Superoxide dismutases in photosynthetic organisms. - In : YASUNOBU, K.T., MOWER, H.F., HAYAISHI, O. (ed.) : Iron and Copper Proteins. Pp. 551 - 564. Plenum Publ. Co., New York 1976.

*32681 - ASADA, K., KANEMATSU, S., UCHIDA, K. : Superoxide dismutases in photosynthetic organisms: Absence of the cuprozinc enzyme in eukaryotic algae. - Arch. Biochem. Biophys. 179 : 243 - 256, 1977.

32682 - ASADA, K., NAKANO, Y. : Affinity for oxygen in photoreduction of molecular oxygen and scavenging of hydrogen peroxide in spinach chloroplasts. - Photochem. Photobiol. 28 : 917 - 920, 1978.

*32683 - ASAMI, S., AKAZAWA, T. : Enzymic formation of glycolate in Chromatium. Role of superoxide radical in a transketolase-type mechanism. - Biochemistry 16 : 2202 - 2207, 1977. [Ps.]

32684 - ASAMI, S., AKAZAWA, T. : Biosynthetic mechanism of glycolate in *Chromatium*
6. Glycolate formation and metabolism under low $O_2$. - Plant Cell Physiol.
*19* : 1353 - 1362, 1978.

32685 - ASAMI, S., AKAZAWA, T. : Photooxidative damage in photosynthetic activities
of *Chromatium vinosum*. - Plant Physiol. *62* : 981 - 986, 1978.

32686 - ASANA, R.D. : On overview of photosynthesis and productivity. - In : Inter-
-Disciplinary Symposium on Photosynthesis and Productivity. Abstract of Papers. Pp. 1 - 4. Indian nat. Sci. Acad., New Delhi 1978.

32687 - ASANOV, A.N., YAKOBSON, B.I. : Kinetika fotosinteza pri svetovom nasyshchenii.
[Kinetics of photosynthesis during light saturation.] - Dokl. Akad. Nauk SSSR
*243* : 230 - 233, 1978. [In R.]

32688 - ASGHAR, A., SAMI, M., NADEEM, M.T., SATTAR, A. : Effect of some pre-drying
unit operations on the chlorophyll stability and nutritional quality of de-
hydrated peas. - Lebensm.-Wiss. Technol. *11* : 15 - 18, 1978.

32689 - ASLAM, M., HUNT, L.A. : Photosynthesis and transpiration of the flag leaf in
four spring-wheat cultivars. - Planta *141* : 23 - 28, 1978.

32690 - ASTON, M.J. : Differences in the behaviour of adaxial and abaxial stomata
of amphistomatous sunflower leaves: inherent or environmental? - Aust. J.
Plant Physiol. *5* : 211 - 218, 1978. [Stomatal resistance.]

32691 - ATKINS, C.A., FLINN, A.M. : Carbon dioxide fixation in the carbon economy
of developing seeds of *Lupinus albus* (L.). - Plant Physiol. *62* : 486 - 490,
1978.

*32692 - ATKINSON, M.A., COOPER, J.I. : Ultrastructural changes in leaf cells of pop-
lar naturally infected with poplar mosaic virus. - Ann. appl. Biol. *83* :
395 - 398, 1976. [Chloroplast.]

32693 - AUBUCHON, R.R., THOMPSON, D.R., HINCKLEY, T.M. : Environmental influences
on photosynthesis within the crown of a white oak. - Oecologia *35* : 295 -
306, 1978.

*32694 - AUERBACH, S., PRÜFER, P., WEISE, G. : Gasstoffwechselphysiologische Schädi-
gungskriterien bei submersen Makrophyten vom Typ *Fontinalis antipyretica* L.
unter Einwirkung von Schwermetallen oder Phenol. - Int. Rev. ges. Hydrobiol.
*58* : 19 - 32, 1973.

32695 - AUGUSTIN, P., GUTEZEIT, B. : Messung der Gaskonzentration in festen, flüssi-
gen oder gasförmigen Medien durch Ausnutzung der Permeation. - Chem. Techn.
*30* : 102 - 104, 1978. [$CO_2$.]

32696 - AULD, B.A., DENNETT, M.D., ELSTON, J. : The effect of temperature changes on
the expansion of individual leaves of *Vicia faba* L. - Ann. Bot. *42* : 877 -
888, 1978. [Growth analysis.]

32697 - AUSLÄNDER, W., JUNGE, W. : The proton pump in photosynthesis of green plants:
quantitative measurement of the rapid pH-changes in the internal volume of
thylakoids *via* neutral red. - In : AZZONE, G.F., AVRON, M., METCALFE, J.C.,
QUAGLIARIELLO, E., SILIPRANDI, N. (ed.) : The Proton and Calcium Pumps. Pp.
31 - 44. Elsevier/North-Holland Biomedical Press, Amsterdam - Oxford - New
York 1978.

32698 - AUSSENAC, G., GRANIER, A. : Quelques résultats de cinétique journalière du
potentiel de sève chez les arbres forestiers. - Ann. Sci. forest. *35* : 19 -
32, 1978. [Ps.]

32699 - AUSTIN, R.B., KINGSTON, G., LONGDEN, P.C., DONOVAN, P.A. : Gross energy
yields and the support energy requirements for the production of sugar from
beet and cane; a study of four production areas. - J. agr. Sci. *91* : 667 -
675, 1978.

*32700 - AVER'YANOV, A.A. : Kamera dlya odnovremennogo izmereniya khemilyuminestsen-
tsii i kontsentratsii kisloroda. [Chamber for simultaneous measuring of chemiluminescence and oxygen concentration.] - Nauch. Dokl. vyssh. Shkoly,
biol. Nauki *20* (9) : 131 - 134, 1977. [In R.]

*32701 - **AVIRAM, I.** : The role of lysines in *Euglena* cytochrome *c*-552: Chemical modification studies. - Arch. Biochem. Biophys. *181* : 199 - 207, 1977.

32702 - **AVIRAM, I., WEISSMANN, C.** : Spectrophotometric and fluorometric study of the
denaturation of *Euglena* cytochrome *c*-552. - Biochemistry *17* : 2020 - 2025,
1978.

32703 - **AVRON, M.** : Energy transduction in photophosphorylation. - FEBS Lett. *96* :
225 - 232, 1978.

32704 - **AXELSSON, L.** : The photostability of different chlorophyll forms in dark
grown leaves of wheat. IV. The effect of heating, and of freezing and thawing.
- Physiol. Plant. *44* : 57 - 63, 1978.

32705 - **AXLER, R.P., GERSBERG, R.M., PAULSON, L.J.** : Primary productivity in meromictic Big Soda Lake, Nevada. - Great Basin Naturalist *38* : 187 - 192, 1978.

*32706 - **AZIZKHODZHAEV, A., RAKHMANKULOV, S.A., IMAMALIEV, A.I.** : Fotokhimicheskie
reaktsii khloroplasta khlopchatnika v svyazi s geterozisom. [Photochemical
reactions of cotton chloroplast in connection with heterosis.] - Uz. biol.
Zh. *1975* (5) : 19 - 21, 70, 1975. [In R, ab : Uz.]


32707 - **BABAYAN, R.S., GEVORKYAN, A.M.** : Izmenenie soderzhaniya svobodnykh aminokis-
lot i rastvorimykh uglevodov v defektivnykh po sintezu pigmentov prorostkakh
yachmenya. [Changes in the content of free amino acids and soluble carbohydrates in pigment defective barley seedlings.] - Biol. Zh. Arm. *31* : 1207 -
1212, 1978. [In R, ab : Arm., E.]

*32708 - **BABIĬ, V.S., IVANKOV, A.A.** : Vliyanie kaptana na yablonyu. [Effect of captan
on apple trees.] - Sadovod. Vinograd. Vinodel. Mold. *26* (9) : 37 - 39, 1971.
[Chl; in R.]

*32709 - **BABITSKIĬ, A.F.** : Fotofosforilirovanie na izolirovannykh khloroplastakh in-
brednykh liniĭ i gribridov kukuruzy. [Photophosphorylation of isolated chloroplasts from maize inbred lines and hybrids.] - Nauch.-tekh. Byull. vses.
selekts.-genet. Inst. *25* : 33 - 38, 1975. [In R.]

32710 - **BACCARINI-MELANDRI, A., JONES, O.T.G., HAUSKA, G.** : Cytochrome $c_2$ - an electron carrier shared by the respiratory and photosynthetic electron transport
chain of *Rhodopseudomonas capsulata*. - FEBS Lett. *86* : 151 - 154, 1978.

32711 - **BACCARINI-MELANDRI, A., MELANDRI, B.A.** : Coupling factors. - In : CLAYTON,
R.K., SISTROM, W.R. (ed.) : The Photosynthetic Bacteria. Pp. 615 - 628. Plenum Press, New York - London 1978.

32712 - **BACH, S.D., JOSSELYN, M.N.** : Mass blooms of the alga *Cladophora* in Bermuda.
- Mar. Pollut. Bull. *9* (2) : 34 - 37, 1978. [Ps.]

32713 - **BACON, G.J., BACHELARD, E.P.** : The influence of nursery conditioning treatments on some physiological responses of recently transplanted seedlings
of *Pinus caribaea* MOR. var. *hondurensis* B. and G. - Aust. Forest Res. *8* :
171 - 183, 1978. [Ps.]

32714 - **BADER, F.G.** : A structured model of photosynthesis in *Anacystis nidulans*. -
Biotechnol. Bioeng. *20* : 119 - 125, 1978.

32715 - **BADGER, M.R., KAPLAN, A., BERRY, J.A.** : Active inorganic carbon accumulation
in *Anabaena variabilis* and its relationship to photosynthesis. - Plant Physiol. *61* (Suppl.) : 101, 1978.

32716 - **BADGER, M.R., KAPLAN, A., BERRY, J.A.** : A mechanism for concentrating $CO_2$
in *Chlamydomonas reinhardtii* and *Anabaena variabilis* and its role in photosynthetic $CO_2$ fixation. - Carnegie Inst. Year Book *77* : 251 - 261, 1978.

32717 - **BAGAUTDINOVA, R.I.** : Rol' morfofiziologicheskikh korrelyatsiĭ v formirovanii
fotosinteticheskogo apparata. [The role of morpho-physiological correlations
in the formation of photosynthetic apparatus.] - In : Mezostruktura i Funktsional'naya Aktivnost' Fotosinteticheskogo Apparata. Pp. 79 - 92. Ural'skiĭ
gosudarstvennyĭ Universitet, Sverdlovsk 1978. [In R.]

*32718 - BAGNARA, D., TRIOLO, L., CERVIGNI, T., BASSANELLI, C. : Efficienza fotosinte-
tica in *Triticum* : individuazioni di linee di frumenti tetraploidi a bassa
fotorespirazione. [Photosynthetic efficiency in *Triticum* : Identification
of tetraploid lines with low photorespiration.] - Genet. agrar. *29* : 269 -
276, 1975. [In Ital., ab : E.]

32719 - BAHR, J.T. : Activation of RuBP carboxylase by cyanate, $MnCl_2$ or $CaCl_2$. -
Plant Physiol. *61* (Suppl.) : 98, 1978.

32720 - BAHR, J.T., JENSEN, R.G. : Activation of ribulose bisphosphate carboxylase
in intact chloroplasts by $CO_2$ and light. - Arch. Biochem. Biophys. *185* :
39 - 48, 1978.

32721 - BAILLE, A., GUICHERD, R., MERMIER, M. : Le pyranomètre sélectif : étude de
ses possibilités d'utilisation dans le domaine agronomique et dans celui de
la pollution atmosphérique. - Ann. agron. *29* : 59 - 78, 1978.

32722 - BAJAJ, K.L., KAUR, G., BRAR, J.S., SUKHIJA, B.S. : Chemical composition and
keeping-quality of carrot (*Daucus carota* L.) varieties. - Plant Foods Man
*2* : 159 - 165, 1978.

B32723 - BAKER, D.A. :   Transport Phenomena in Plants. - Chapman and Hall Ltd., London 1978. [Ps.]

*32724 - BAKER, E.W., SMITH, G.D. : Fossil porphyrins and chlorins in deep ocean
sediments. - In : YEN, T.F. (ed.) : Chemistry of Marine Sediments. Pp. 73 -
99. Ann Arbor Sci. Publ. Inc., Ann Arbor 1977. [Chl.]

32725 - BAKER, N.R. : Effect of high cation concentrations on Photosystem II activities. - Plant Physiol. *62* : 889 - 893, 1978.

32726 - BAKER, N.R., HEATH, R.L., LEECH, R.M. : Consistently high rates of electron
transport in chloroplasts isolated from hydroponically-grown spinach plants.
- Plant Sci. Lett. *12* : 317 - 322, 1978.

32727 - BAKKER, E.P., CAPLAN, S.R. : Phospholipid substitution of the purple membrane. The stoichiometry of light-induced proton release by phospholipid-sub-
stituted purple membranes. - Biochim. biophys. Acta *503* : 362 - 379, 1978.

*32728 - BAKUMENKO, L.A., STONOV, L.D. : Ispol'zovanie zamedlennoĭ fluorestsentsii
khlorelly dlya opredeleniya gerbitsidnykh svoĭstv soedineniĭ. [Use of delayed fluorescence of *Chlorella* for determining the herbicide properties of
compounds.] - Khim. sel'sk. Khoz. *15* (3) : 67 - 70, 1977. [In R.]

32729 - BALAN, V.V. : Soderzhanie khlorofilla i stepen' aktivnosti katalazy v list'-
yakh yabloni v zavisimosti ot sorta, podvoya i dozy predplantazhnogo udobreniya. [Content of chlorophyll and rate of catalase activity in apple tree
leaves in relation to cultivar, rootstock, and dose of presowing fertilizer.]
- Tr. kishinev. sel'skokhoz. Inst. Im. M.V.Frunze *1978* (Voprosy Intensifikatsii Plodovodstva) : 25 - 29, 1978. [In R.]

*32730 - BÁLINT, E., HEVESI, J., VASS, I. : Degree of polarization in dye-detergent
model systems. - Acta phys. chem. (Szeged) *23* : 123 - 127, 1977. [Chl.]

32731 - BALLESTER, A., ARESES, M.L., PEREZ, F., VIEITEZ, E. : Efecto de la morfacti-
na clorfurenol sobre el contenido de pigmentos en *Phaseolus vulgaris* cv
"Contender". [Effect of morphactin chlorflurenol on the pigment content in
*Phaseolus vulgaris* cv. Contender.] - Turrialba *28* : 113 - 115, 1978.

32732 - BALTSCHEFFSKY, H. : Evolutionary aspects of biological energy conversion. -
In : SCHÄFER, G., KLINGENBERG, M. (ed.) : Energy Conversion in Biological
Membranes. Pp. 3 - 18. Springer-Verlag, Berlin - Heidelberg - New York 1978.
[Ps.]

*32733 - BALTSCHEFFSKY, M. : Energy transduction in the chromatophore membrane. - In :
ABRAHAMSSON, S., PASCHER, I. (ed.) : Structure of Biological Membranes. Pp.
41 - 62. Plenum Press, New York 1977.

32734 - BALTSCHEFFSKY, M. : Photosynthetic phosphorylation. - In : CLAYTON, R.K.,
SISTROM, W.R. (ed.) : The Photosynthetic Bacteria. Pp. 595 - 613. Plenum
Press, New York - London 1978.

*32735 - **BAMBERG, E.** : Rhodopsin and other proteins in artificial lipid membranes. - Biophys. Struct. Mechanism *3* : 39 - 42, 1977. [Bacteriorhodopsin.]

32736 - **BANAI, M., SHAVIT, N., CHIPMAN, D.M.** : Specificity of nucleotide binding sites in isolated chloroplast coupling factor ($CF_1$). - Biochim. biophys. Acta *504* : 100 - 107, 1978.

32737 - **BANBA, H., OHKUBO, T.** : [Ecological studies of individual variation in the various characters of upland-cultured paddy rice. I. Individual variation in the various characters of upland-cultured paddy rice in different varieties and plant density.] - Jap. J. Crop Sci. *47* : 147 - 154, 1978. [Growth analysis; in Jap., ab : E.]

32738 - **BANBA, H., OHKUBO, T.** : [Ecological studies of individual variation in the various characters of upland-cultured paddy rice. II. Individual variation in the various characters of upland-cultured paddy rice in different tillage method, seeding time and plant density.] - Jap. J. Crop Sci. *47* : 155 - 162, 1978. [Growth analysis; in Jap., ab : E.]

32739 - **BANERJI, D., RAUF, A.** : Higher than leaf Hill activity of cotyledonary chloroplasts from developing legume seeds. - Biochem. biophys. Res. Commun. *85* : 921 - 926, 1978.

32740 - **BANSAL, R.P., SEN, D.N.** : A new report on antholysis on *Pedalium murex* L. - Sci. Cult. *44* : 367 - 369, 1978. [Chl, Car.]

32741 - **BAQAR, M.R., LEE, T.H.** : Interaction of CPTA and high temperature on carotenoid synthesis in tomato fruit. - Z. Pflanzenphysiol. *88* : 431 - 435, 1978.

32742 - **BARANKIEWICZ, T.J.** : $CO_2$ exchange rates and $^{14}C$ photosynthetic products of maize leaves as affected by potassium deficiency. - Z. Pflanzenphysiol. *89* : 11 - 20, 1978.

32743 - **BARANKIEWICZ, T.J.** : Potas jako czynnik w fotosyntezie. [Potassium as factor of photosynthesis.] - Wiadom. bot. *22* : 163 - 175, 1978. [In Pol.]

32744 - **BARBER, J.** : Biophysics of photosynthesis. - Rep. Prog. Phys. *41* : 1157 - 1199, 1978.

32745 - **BARBER, J.** : Photosystem-2 and $O_2$ evolution. Report on Symposium 6. - In : HALL, D.O., COOMBS, J., GOODWIN, T.W. (ed.) : Proceedings of the Fourth International Congress on Photosynthesis. Pp. 423 - 433. Biochem. Soc., London 1978.

32746 - **BARBER, J., SEARLE, G.F.W.** : Cation induced increase in chlorophyll fluorescence yield and the effect of electrical charge. - FEBS Lett. *92* : 5 - 8, 1978.

32747 - **BARBER, J., SEARLE, G.F.W., TREDWELL, C.J.** : Picosecond time-resolved study of $MgCl_2$-induced chlorophyll fluorescence yield changes from chloroplasts. - Biochim. biophys. Acta *501*: 174 - 182, 1978.

*32748 - **BARBER, J., TREDWELL, C.J., PORTER, G.** : Application of picosecond spectroscopy to photosynthesis research. - UV Spectrom. Group Bull. *5* : 65 - 76, 1977.

32749 - **BARCIKOWSKI, A.** : Sezonowa zmienność chlorofilu *a* i *b* u *Majanthemum bifolium* w różnych warunkach siedliskowych. [Seasonal changes in chlorophyll *a* and *b* in *Majanthemum bifolium* in different habitat conditions.] - Acta Univ. Nicolai Copernici, Biol. XXI - Nauki mat.-przyrod. *44* : 51 - 59, 1978. [In Pol., ab : E.]

32750 - **BARCKHAUSEN, R.** : Ultrastructural changes in wounded plant storage tissue cells. - In : KAHL, G. (ed.) : Biochemistry of Wounded Plant Tissues. Pp. 1 - 42. Walter de Gruyter & Co., Berlin - New York 1978. [Chloroplast.]

32751 - **BARDEN, J.A.** : Apple leaves, their morphology and photosynthetic potential. - HortScience *13* : 644 - 646, 1978.

32752 - **BARKO, J.W., SMART, R.M.** : The growth and biomass distribution of two emergent freshwater plants, *Cyperus esculentus* and *Scirpus validus*, on different sediments. - Aquat. Bot. *5* : 109 - 117, 1978.

32753 - BARNES, A., HOLE, C.C. : A theoretical basis of growth and maintenance respiration. - Ann. Bot. *42* : 1217 - 1221, 1978. [Model.]

32754 - BARNES, R.B., AMEND, J., SISTROM, W.R., GRIFFITH, O.H. : Quantum yield and
image contrast of bacteriochlorophyll monolayers in photoelectron microscopy. - Biophys. J. *21* : 195 - 202, 1978.

32755 - BARR, R., CRANE, F.L. : Distinction between ferricyanide and class III acceptor sites in photosystem II of spinach chloroplasts. - Plant Physiol. *61*
(Suppl.) : 76, 1978.

32756 - BARRETT, J., HUNTER, C.N., JONES, O.T.G. : Properties of a cytochrome *c*-en-
riched light particulate fraction isolated from the photosynthetic bacterium
*Rhodopseudomonas spheroides*. - Biochem. J. *174* : 267 - 275, 1978.

32757 - BARRETT, J., JONES, O.T.G. : Localization of ferrochelatase and of newly synthesized haem in membrane fractions from *Rhodopseudomonas spheroides*. - Biochem. J. *174* : 277 - 281, 1978. [Chl.]

32758 - BARROS, R.S., COSTA MERCÊS, W., ALVIN, R. : Sink strength and cassava productivity. - HortScience *13* : 474 - 475, 1978. [Growth analysis.]

32759 - BARRY, M., GAUGER, J., TOLBERT, N.E. : Changes in RUBP carboxylase/oxygenase
and peroxisomal enzymes in tomato fruit during development. - Plant Physiol.
*61* (Suppl.) : 50, 1978.

*32760 - BARSKIĬ, E.L., SAMUILOV, V.D. : Élektrokhromnyĭ éffekt bakteriokhlorofilla
kak indikator na énergizovannoe sostoyanie membran fototrofnykh bakteriĭ.
[Electrochromic effect of bacteriochlorophyll as indicator of energized state of membranes of phototrophic bacteria.] - In : Problemy Regulyatsii Obmena Veshchestv u Mikroorganizmov. Pp. 355 - 361. Pushchino-na-Oke 1973. [In
R.]

32761 - BARSUKOV, A.I. : Formirovanie list'ev i zerna rasteniyami yarovoĭ pshenitsy
v zavisimosti ot ploshchadi pitaniya. [Formation of leaves and grain in
wheat plants in dependence on nutrition area.] - Dokl. VASKhNIL *1978* (10) :
7 - 8, 1978. [In R.]

32762 - BARTA, A.L. : Effect of root temperature on dry matter distribution, carbohydrate accumulation, and acetylene reduction activity in alfalfa and birds-
foot trefoil. - Crop Sci. *18* : 637 - 640, 1978.

32763 - BARTELS, P.G., WATSON, C.W. : Inhibition of carotenoid synthesis by fluridone and norflurazon. - Weed Sci. *26* : 198 - 203, 1978.

32764 - BARTSCH, R.G. : Cytochromes. - In : CLAYTON, R.K., SISTROM, W.R. (ed.) : The
Photosynthetic Bacteria. Pp. 249 - 279. Plenum Press, New York - London
1978.

32765 - BASIOUNY, F.M., VAN, T.K., BIGGS, R.H. : Some morphological and biochemical
characteristics of $C_3$ and $C_4$ plants irradiated with UV-B. - Physiol. Plant.
*42* : 29 - 32, 1978.

32766 - BASKIN, J.M., BASKIN, C.C. : A discussion of the growth and competitive ability of $C_3$ and $C_4$ plants. - Castanea *43* (2) : 71 - 76, 1978.

*32767 - BASSHAM, J.A. : Control of photosynthetic carbon metabolism. - In : Rate
Control of Biological Processes. Symposia of the Society for Experimental
Biology. Vol. 27. Pp. 461 - 483. Cambridge Univ. Press, London 1973.

32768 - BASSHAM, J.A., KROHNE, S., LENDZIAN, K. : *In vivo* control mechanism of the
carboxylation reaction. - In : SIEGELMAN, H.W., HIND, G. (ed.) : Photosynthetic Carbon Assimilation. Pp. 77 - 93. Plenum Press, New York - London
1978.

32769 - BASZYŃSKI, T., RUSZKOWSKA, M., KRÓL, M., TUKENDORF, A., WOLIŃSKA, D. : The
effect of copper deficiency on the photosynthetic apparatus of higher plants.
- Z. Pflanzenphysiol. *89* : 207 - 216, 1978.

32770 - BATES, S.S., CRAIGIE, J.S. : Chloroplast pigments of a green phytoplankter
from the Hudson Estuary, U.S.A. - Phycologia *17* : 79 - 84, 1978.

32771 - **BATTERSBY, A.R.** : The discovery of nature's biosynthetic pathways. - Experientia *34* : 1 - 13, 1978. [Chl.]

32772 - **BAUER, H.** : Photosynthesis of ivy leaves (*Hedera helix*) after heat stress. I. $CO_2$-gas exchange and diffusion resistances. - Physiol. Plant. *44* : 400 - 406, 1978.

32773 - **BAUER, H., LARCHER, W.** : $CO_2$-evolution in light and darkness of ivy leaves (*Hedera helix* L.) with depressed photosynthesis after heat stress. - Z. Pflanzenphysiol. *89* : 457 - 460, 1978.

32774 - **BAULINA, O.I., GUSEV, M.V.** : Dinamika ul'trastrukturnykh izmeneniĭ obligatno fototrofnoĭ vodorosli *Anabaena variabilis* v period sokhraneniya eyu zhiznesposobnosti v temnote. [Dynamics of ultrastructure changes in obligatory phototrophic blue-green alga *Anabaena variabilis* during maintenance of viability in the dark.] - Fiziol. Rast. *25* : 1168 - 1171, 1978. [Thylakoids; in R, ab : E.]

32775 - **BAULINA, O.I., SEMENOVA, L.R., MINEEVA, L.A., GUSEV, M.V.** : Osobennosti ul'trastrukturnoĭ organizatsii kletok khemogeterotrofnoĭ sinezelenoĭ vodorosli *Chlorogloea fritschii*. [Characteristics of the cellular ultrastructural organization of the chemoheterotrophic blue-green alga *Chlorogloea fritschii*.] - Mikrobiologiya *47* : 919 - 923, 1978. [In R, ab : E.]

32776 - **BAZZAZ, M.B., REBEIZ, C.A.** : Chloroplast culture : The chlorophyll repair potential of mature chloroplasts incubated in a simple medium. - Biochim. biophys. Acta *504* : 310 - 323, 1978.

32777 - **BEADLE, C.L., TURNER, N.C., JARVIS, P.G.** : Critical water potential for stomatal closure in Sitka spruce. - Physiol. Plant. *43* : 160 - 165, 1978. [Ps.]

32778 - **BEALE, S.I.** : δ-Aminolevulinic acid in plants : Its biosynthesis, regulation, and role in plastid development. - Annu. Rev. Plant Physiol. *29* : 95 - 120, 1978.

32779 - **BEALE, S.I.** : Biosynthesis of photosynthetic pigments - pathways and regulation. - In : HALL, D.O., COOMBS, J., GOODWIN, T.W. (ed.) : Proceedings of the Fourth International Congress on Photosynthesis. Pp. 507 - 516. Biochem. Soc., London 1978.

*32780 - **BEAUMONT, G., BASTIN, R., THERRIEN, H.P.** : Effets physiologiques de l'atrazine à doses sublétales sur *Lemna minor* L. I. Influence sur la croissance, la teneur en chlorophylle, en protéines et en azote soluble et total. - Naturaliste can. *103* : 527 - 533, 1976.

32781 - **BECHER, B., TOKUNAGA, F., EBREY, T.G.** : Ultraviolet and visible absorption spectra of the purple membrane protein and the photocycle intermediate. - Biochemistry *17* : 2293 - 2300, 1978. [Bacteriorhodopsin.]

32782 - **BECKER, J.F., GEACINTOV, N.E., SWENBERG, C.E.** : Photovoltages in suspensions of magnetically oriented chloroplasts. - Biochim. biophys. Acta *503* : 545 - 554, 1978.

32783 - **BECKER, W.M., LEAVER, C.J., WEIR, E.M., RIEZMAN, H.** : Regulation of glyoxysomal enzymes during germination of cucumber. - Plant Physiol. *62* : 542 - 549, 1978. [RuBPC.]

32784 - **BEDDARD, G.S., CARLIN, S., HARRIS, L., PORTER, G., TREDWELL, C.J.** : Quenching of chlorophyll fluorescence by nitrobenzene. - Photochem. Photobiol. *27* : 433 - 438, 1978.

32785 - **BEKASOVA, O.D., KASHENTSEVA, E.M., EVSTIGNEEV, V.B.** : O fotosensibiliziruyushchem deĭstvii fikobiliproteidov v adsorbirovannom sostoyanii. [Photosensitizing effect of phycobiliproteins in absorbed state.] - Izv. Akad. Nauk SSSR, Ser. biol. *1978* : 558 - 564, 1978. [In R, ab : E.]

32786 - **BEKINA, R.M., LYSENKO, G.G.** : Issledovanie aktivnosti nekotorykh protsessov fotodykhatel'nogo tipa u pshenits raznoĭ produktivnosti. [Activity of some respiratory type processes in wheats of different productivity.] - Sel'skokhoz. Biol. *13* : 369 - 374, 1978. [In R, ab : E.]

32787 - **BELAY, A., FOGG, G.E.** : Photoinhibition of photosynthesis in *Asterionella formosa* (*Bacillariophyceae*). - J. Phycol. *14* : 341 - 347, 1978.

32788 - BELETSKIĬ, Yu.D., KESSLER, R.M., KOLOKOLOVA, N.S., YAKUBOVA, N.R. : Aktiv-
         nost' β-galaktozidazy i β-glyukozidazy u vneyadernykh khlorofil'nykh mutan-
         tov podsolnechnika. [β-galactosidase and β-glucosidase activities in extra-
         nuclear chlorophyll mutants of sunflower.] - Genetika *14* : 1387 - 1392, 1978.
         [In R, ab : E.]

32789 - BELKIN, S., PADAN, E. : Sulfide-dependent hydrogen evolution in the cyano-
         bacterium *Oscillatoria limnetica.* - FEBS Lett. *94* : 291 - 294, 1978.

32790 - BELKOV, M.V., LOSEV, A.P. : On the location of the electronic transitions
         of chlorophyll *"a"* and protochlorophyll *"a"* depending on the degree of sol-
         vate state. - Spectroscopy Lett. *11* : 653 - 669, 1978.

32791 - BELL, D.H., HAUG, A., GOOD, N.E. : Stimulation of microsecond-delayed fluo-
         rescence from spinach chloroplasts by uncouplers and by phosphorylation. -
         Biochim. biophys. Acta *504* : 446 - 455, 1978.

*32792 - BELL, L.N., DEMIDOV, È.D. : Kachestvo sveta i pervichnye ènergeticheskie
         protsessy fotosinteza. [Light quality and primary energetic processes of pho-
         tosynthesis.] - In : Itogi Issledovaniya Mekhanizma Fotosinteza. Pp. 197 -
         208. Pushchino 1974. [In R.]

32793 - BELLIVEAU, J.W., LANYI, J.K. : Calcium transport in *Halobacterium halobium*
         envelope vesicles. - Arch. Biochem. Biophys. *186* : 98 - 105, 1978.

*32794 - BELYANIN, V.N., TRISHIN, M.K., KUT'IN, A.M. : Sravnitel'nyĭ analiz rosta i
         fotosinteza termofil'nykh form *Cyanophyta* na osnove parametricheskogo urav-
         neniya. [Comparison of growth analysis and photosynthesis of thermophile
         forms of *Cyanophyta* at the level of parametrical equations.] - In : Intensiv-
         naya Svetokul'tura Rasteniĭ. Pp. 182 - 191. Inst. Fiz. Sib. Otd. Akad. Nauk
         SSSR, Krasnoyarsk 1977. [In R.]

32795 - BEN-AMOTZ, A., AVRON, M. : On the mechanism of osmoregulation in *Dunaliella.* -
         In : CAPLAN, S.R., GINZBURG, M. (ed.) : Energetics and Structure of Halophi-
         lic Microorganisms. Pp. 529 - 541. Elsevier/North-Holland Biomedical Press,
         Amsterdam - New York 1978. [Ps.]

32796 - BEN-ASHER, J., FUCHS, M., GOLDBERG, D. : Radiation and energy balance of
         sprinkler and trickle irrigated fields. - Agron. J. *70* : 415 - 417, 1978.

32797 - BEN-ASHER, J., SAMMIS, T.W. : Radiation and energy balance of a trickle-irri-
         gated lemon grove. - Agron. J. *70* : 568 - 572, 1978.

32798 - BEN-BASSAT, D., MAYER, A.M. : Light-induced Hg volatilization and $O_2$ evolu-
         tion in *Chlorella* and the effect of DCMU and methylamine. - Physiol. Plant.
         *42* : 33 - 38, 1978.

*32799 - BENDALL, D.S. : Development of photosynthetic electron transport in green-
         ing barley. - Biochem. Soc. Trans. *5* : 84 - 88, 1977.

32800 - BENDALL, D.S. : Oxygen evolution in photosynthesis. - Biochem. Soc. Trans.
         *6* : 372 - 375, 1978.

32801 - BENDALL, D.S., WOOD, P.M. : Kinetics of electron transfer through higher-
         -plant plastocyanin. - In : HALL, D.O., COOMBS, J., GOODWIN, T.W. (ed.) :
         Proceedings of the Fourth International Congress on Photosynthesis. Pp. 771
         - 775. Biochem. Soc., London 1978.

*32802 - BENECKE, U., GÖBL, F. : The influence of different mycorrhizae on growth,
         nutrition and gas-exchange of *Pinus mugo* seedlings. - Plant Soil *40* : 21 -
         32, 1974.

32803 - BENEDICT, C.R. : Nature of obligate photoautotrophy. - Annu. Rev. Plant Phy-
         siol. *29* : 67 - 93, 1978.

32804 - BENEDICT, C.R., WONG, W.W. : $\delta^{13}C$ values, leaf structure and photorespiration
         of marine macrophytes. - Plant Physiol. *61* (Suppl.) : 8, 1978.

32805 - BENGTSON, C., KLOCKARE, B., KLOCKARE, R., LARSSON, S., SUNDQVIST, C. : The
         after-effect of water stress on chlorophyll formation during greening and
         the levels of abscisic acid and proline in dark grown wheat seedlings. -
         Physiol. Plant. *43* : 205 - 212, 1978.

32806 - BENGTSON, C., LARSSON, S., SKARBY, L. : Effekter av ozon på fotosyntes-
respirations- och transpirationshastichhet hos unga tallplantor. [Effect of
ozone on photosynthesis, respiration and transpiration rates by pine seedlings.] - Inst. Vatten- Luftvardsforsk. Publ. $B$ $417$ : 1 - 15, 1978. [In
Swed., ab : E.]

32807 - BEN-HAYYIM, G. : $Mg^{2+}$ translocation across the thylakoid membrane : Studies
using the ionophore A23187. - Europe. J. Biochem. $83$ : 99 - 104, 1978.

32808 - BENJAMIN, L.R., WREN, M.J. : Root development and source-sink relations in
carrot, Daucus carota L. - J. exp. Bot. $29$ : 425 - 433, 1978. [Ps, growth
analysis.]

32809 - BENNETT, G.R., WOHLER, J.R., MOORE, J.R. : The use of mid-day $^{14}C$ fixation
studies in the determination of primary productivity in lake systems. - Proc.
Pennsylvania Acad. Sci. $49$ : 73 - 75, 1975.

32810 - BENNETT, K.J., ROOK, D.A. : Stomatal and mesophyll resistances in two clones
of Pinus radiata D. DON known to differ in transpiration and survival rate.
- Aust. J. Plant Physiol. $5$ : 231 - 238, 1978. [Ps.]

32811 - BENNOUN, P., MASSON, A., PICCIONI, R., CHUA, N.H. : Uniparental mutants of
Chlamydomonas reinhardi defective in photosynthesis. - In : AKOYUNOGLOU, G.,
ARGYROUDI-AKOYUNOGLOU, J.H. (ed.) : Chloroplast Development. Pp. 721 - 726.
Elsevier/North-Holland Biomedical Press, Amsterdam - New York - Oxford 1978.

32812 - BENSASSON, R., GOLDSCHMIDT, C.R., LAND, E.J., TRUSCOTT, T.G. : Laser intensity and the comparative method for determination of triplet quantum yields.
- Photochem. Photobiol. $28$ : 277 - 281, 1978.

*32813 - BERDYKULOV, Kh. A. : Fotosintez khlorelly, kul'tiviruemoĭ pod otkrytym nebom.
[Photosynthesis of Chlorella grown under open air.] - In : Kul'tivirovanie
Vodorosleĭ i Vysshikh Vodnykh Rasteniĭ v Uzbekistane. Pp. 76 - 86. Fan, Tash-
kent 1972. [In R.]

32814 - BERDYKULOV, Kh.A., NURIEVA, D. : Fotosinteticheskaya produktivnost' Ankistrodesmus angustus v zavisimosti ot temperatury i prodolzhitel'nosti svetovogo
perioda. [Photosynthetic productivity of Ankistrodesmus angustus in relation
to temperature and length of exposure to light.] - Uz. biol. Zh. $1978$ (3) :
17 - 19, 1978. [In R, ab : Uz.]

*32815 - BERDYKULOV, Kh.A., RAKHIMOV, A.R., SADYKOVA, A.Sh. : Vliyanie impul'snogo
kontsentrirovannogo solnechnogo sveta (IKSS) na urozhaĭnost' i nekotorye
biokhimicheskie svoĭstva stsenedesmusa (Scenedesmus obliquus (TURP.) KÜTZ.)
v kul'ture. [Effect of impulse concentrated solar radiation (ICSR) on yield
and some biochemical properties of Scenedesmus obliquus (TURP.) KÜTZ. in
culture.] - In : Kul'tivirovanie Vodorosleĭ i Vysshikh Vodnykh Rasteniĭ v
Uzbekistane. Pp. 71 - 76. Fan, Tashkent 1972. [In R.]

*32816 - BERDYKULOV, Kh.A., SADYKOV, M.S., ASKAROV, M. : Svetoimpul'snaya stimulyatsi-
ya produktivnosti khlorelly. [Light-induced stimulation of Chlorella productivity.] - In : Ėkologiya i Fiziologiya Mikroorganizmov, Vodorosleĭ i Vodnykh Rasteniĭ. Pp. 228 - 230. Fan, Tashkent 1973. [In R.]

*32817 - BERDYKULOV, Kh.A., SADYKOVA, A.Sh., SADYKOV, M.S. : Vliyanie IKSS na usvo-
enie solnechnoĭ ėnergii stsenedesmusom. [Effect of impulse concentrated solar radiation on solar energy utilization by Scenedesmus.] - In : Vodorosli
i Griby Srednej Azii. Vyp. 1. Pp. 88 - 93. Fan, Tashkent 1974. [In R.]

*32818 - BERDYKULOV, Kh.A., YAKUBOV, Kh.F. : Dnevnoĭ khod fotosinteza i nakoplenie
biomassy mikrovodorosleĭ pri vyrashchivanii ikh na razlichnykh pitatel'nykh
sredakh. [Diurnal course of photosynthesis and microalgae biomass accumulation during cultivation in various nutrient media.] - In : Kul'tivirovanie
Vodorosleĭ i Vysshikh Vodnykh Rasteniĭ v Uzbekistane. Pp. 44 - 46. Fan,
Tashkent 1971. [In R.]

32819 - BEREZNEGOVSKAYA, L.N., IDRISOVA, L.S. : Vliyanie razlichnykh kontsentratsiĭ
sakharozy v pitatel'noĭ srede na zeleneyushchuyu kul'turu tkaneĭ durmana in-
deĭskogo. [Effect of various concentrations of saccharose in nutritive medium on the greening tissue culture of Datura innoxia MILL.] - Nauch. Dokl.
vyssh. Shkoly, biol. Nauki $21$ (5) : 95 - 99, 1978. [Ps, Chl, Car; in R.]

32820 - BERGUM, P.W., NADLER, K.D. :  Effects of isonicotinic acid hydrazide on the synthesis of Δ-aminolevulinic acid and chlorophyll in greening barley shoots. - Plant Physiol. *61* (Suppl.) : 82, 1978.

32821 - BERKOVICH, Yu.A. : Otsenka oshibok pri izmereniyakh vidimogo fotosinteza v zakrytykh ustanovkakh. [Evaluation of errors in measurements of apparent photosynthesis in closed systems.] - Fiziol. Rast. *25* : 196 - 203, 1978. [In R, ab : E.]

32822 - BERLYN, M.B. : A mutational approach to the study of photorespiration. - In : SIEGELMAN, H.W., HIND, G. (ed.) : Photosynthetic Carbon Assimilation. Pp. 153 - 164. Plenum Press, New York - London 1978.

32823 - BERLYN, M.B., ZELITCH, I., BEAUDETTE, P.D. : Photosynthetic characteristics of photoautotrophically grown tobacco callus cells. - Plant Physiol. *61* : 606 - 610, 1978.

32824 - BERNER, T. : Ecophysiological activity of hypolithic desert algae. - J. Phycol. *14* (Suppl.) : 39, 1978. [Ps.]

32825 - BERNER, T., DUBINSKY, Z., SHELEF, G. : Primary productivity in high rate sewage oxidation ponds. - J. Phycol. *14* (Suppl.) : 25, 1978. [Ps.]

32826 - BERRY, J., FARQUHAR, G. : The $CO_2$ concentrating function of $C_4$ photosynthesis. A biochemical model. - In : HALL, D.O., COOMBS, J., GOODWIN, T.W. (ed.) : Proceedings of the Fourth International Congress on Photosynthesis. Pp. 119 - 131. Biochem. Soc., London 1978. [Model.]

32827 - BERRY, J.A., KAPLAN, A., BADGER, M. : Evidence for a $CO_2$ concentrating mechanism in the alga *Chlamydomonas reinhardtii*. - Plant Physiol. *61* (Suppl.) : 38, 1978.

32828 - BERRY, J.A., OSMOND, C.B., LORIMER, G.H. : Fixation of $^{18}O_2$ during photorespiration. Kinetic and steady-state studies of the photorespiratory carbon oxidation cycle with intact leaves and isolated chloroplasts of $C_3$ plants. - Plant Physiol. *62* : 954 - 967, 1978.

32829 - BERZINYA, A.Ya. : Deĭstvie razlichnykh doz medi i zheleza na produktivnost' i khimicheskiĭ sostav rasteniĭ salata. [Effect of various doses of copper and iron on productivity and chemical composition of lettuce plants.] - In : Fiziologo-biokhimicheskie Issledovaniya Rasteniĭ. Pp. 100 - 110, 158, Zinatne, Riga 1978. [Chl, Car; in R.]

32830 - BEST, E.P.H., MEULEMANS, J.T. : Primary productivity in the submerged aquatic macrophyte *Ceratophyllum demersum*. - Verh. int. Verein. Limnol. *20* : 88 - 93, 1978.

32831 - BEST, E.P.H., WITTENBOER, J.P.v.d. : Effects of paraquat on growth and photosynthesis of *Ceratophyllum demersum* and *Elodea canadensis*. - In : Proceedings EWRS 5$^{th}$ Symposium on Aquatic Weeds. Pp. 157 - 162. 1978.

32832 - BETHLENFALVAY, G.J., ABU-SHAKRA, S.S., PHILLIPS, D.A. : Interdependence of nitrogen nutrition and photosynthesis in *Pisum sativum* L. I. Effect of combined nitrogen on symbiotic nitrogen fixation and photosynthesis. - Plant Physiol. *62* : 127 - 130, 1978.

32833 - BETHLENFALVAY, G.J., ABU-SHAKRA, S.S., PHILLIPS, D.A. : Interdependence of nitrogen nutrition and photosynthesis in *Pisum sativum* L. II. Host plant response to nitrogen fixation by *Rhizobium* strains. - Plant Physiol. *62* : 131 - 133, 1978.

32834 - BETHLENFALVAY, G.J., CASTELFRANCO, P.A. : Enhancement of energy conservation by Hill reaction inhibitors in isolated spinach (*Spinacia oleracea*) chloroplast fragments. - Weed Sci. *26* : 84 - 89, 1978.

*32835 - BETHLENFALVAY, G.J., PHILLIPS, D.A. : Photosynthesis and symbiotic nitrogen fixation in *Phaseolus vulgaris* L. - In : HOLLAENDER, A. *et al.* (ed.) : Genetic Engineering for Nitrogen Fixation. Pp. 401 - 409. Plenum Press, New York 1977.

32836 - BETHLENFALVAY, G.J., PHILLIPS, D.A. : Interactions between symbiotic nitrogen fixation, combined-N application, and photosynthesis in *Pisum sativum*. - Physiol. Plant. *42* : 119 - 123, 1978.

32837 - BEWLEY, J.D., HALMER, P., KROCHKO, J.E., WINNER, W.E. : Metabolism of a
         drought-tolerant and a drought-sensitive moss : respiration, ATP synthesis
         and carbohydrate status. - In : CROWE, J.H., CLEGG, J.S. (ed.) : Dry Biological
         Systems. Pp. 185 - 203. Academic Press, New York - San Francisco - London
         1978. [Ps.]

32838 - BEWLEY, J.D., PACEY, J. : Desiccation-induced ultrastructural changes in
         drought-sensitive and drought-tolerant plants. - In : CROWE, J.H., CLEGG, J.S.
         (ed.) : Dry Biological Systems. Pp. 53 - 73. Academic Press, New York -
         San Francisco - London 1978. [Chloroplast.]

32839 - BEYELER, W., BACHOFEN, R. : Initial events of light-dependent ATP synthesis
         in spinach subchloroplast particles. - Europe. J. Biochem. 88 : 61 - 67,
         1978.

32840 - BHAGWAT, A.S., RAMAKRISHNA, J., SANE, P.V. : Comparative studies on ribulo-
         se-1,5-biphosphate carboxylase/oxygenase from C-3, C-4 and Crassulacean
         plants. - In : Interdisciplinary Symposium on Photosynthesis and Producti-
         vity. Abstract of Papers. Pp. 23 - 24. Indian nat.Sci. Acad., New Delhi
         1978.

32841 - BHAGWAT, A.S., RAMAKRISHNA, J., SANE, P.V. : Specific inhibition of oxyge-
         nase activity of ribulose-1,5-diphosphate carboxylase by hydroxylamine. -
         Biochem. biophys. Res. Commun. 83 : 954 - 962, 1978.

32842 - BHAGWAT, A.S., SANE, P.V. : Evidence for the involvement of superoxide ani-
         ons in the oxygenase reaction of ribulose-1,5-diphosphate carboxylase. - Bi-
         ochem. biophys. Res. Commun. 84 : 865 - 873, 1978.

32843 - BHATIA, D.S., MALIK, C.P. : Changing pattern of enzymes in the epidermal
         peelings with opened and closed stomata. - Biochem. Physiol. Pflanzen 172 :
         173 - 176, 1978. [Ps.]

32844 - BHATNAGAR, G.P., PURUSHOTHAMAN, A. : Chlorophyll ratios in Vellar Estuary,
         S. India. - Nat. Acad. Sci. Lett. (India) 1 : 195 - 197, 1978.

32845 - BICKEL, H., PALME, L., SCHULTZ, G. : Incorporation of shikimate and other
         precursors into aromatic acids and prenylquinones of isolated spinach chlo-
         roplasts. - Phytochemistry 17 : 119 - 124, 1978.

32846 - BIDABÉ, B. : Possibilités d'utilisation de l'allométrie pour la différenci-
         ation des organes en croissance chez le Pommier. - Ann. Amélior. Plant. 28 :
         113 - 125, 1978. [Leaf-area measurement.]

32847 - BIDDINGTON, N.L., THOMAS, T.H. : Influence of different cytokinins on the
         transpiration and senescence of excised oat leaves. - Physiol. Plant. 42 :
         369 - 374, 1978. [Chl.]

32848 - BIGGINS, J. : Functional homogeneity of P-700 in cyclic and non-cyclic elec-
         tron transport reactions in thylakoids. - Biochim. biophys. Acta 504 : 288 -
         297, 1978.

32849 - BIGGINS, J., SVEJKOVSKÝ, J. : Reorientation of a long-wavelength chlorophyll-
         -a-protein by divalent cations as revealed by the linear dichroism of mag-
         neto-oriented thylakoids. - FEBS Lett. 89 : 201 - 204, 1978.

32850 - BILLETT, E.E., BURNETT, J.H. : The host-parasite physiology of the maize
         smut fungus, Ustilago maydis. 2. Translocation of $^{14}C$-labelled assimilates
         in smutted maize plants. - Physiol. Plant Pathol. 12 : 103 - 112, 1978.

32851 - BINDER, A., JAGENDORF, A., NGO, E. : Isolation and composition of the sub-
         units of spinach chloroplast coupling factor protein. - J. biol. Chem. 253 :
         3094 - 3100, 1978.

B32852 - BINDER, J., ORTNER, K.M. : Die Abhängigkeit der Erträge vom Witterungsver-
         lauf. Eine Quantifizierung für Prognosezwecke. - Agrarwirtschaftliches Insti-
         tut des Bundesministeriums für Land- und Forstwirtschaft, Wien 1978.

32853 - BINDER, R.G., SELMAN, B.R. : Ferredoxin catalyzed cyclic photophosphoryla-
         tion : Reversal of dibromothymoquinone inhibition by N,N,N',N',-tetramethyl-
         -p-phenylenediamine. - Z. Naturforsch. 33C : 261 - 265, 1978.

32854 - BINGHAM, S., SCHIFF, J.A. : Labeling of thylakoid polypeptides in *Euglena gracilis* var. B. - Plant Physiol. *61* (Suppl.) : 83, 1978.

32855 - BIRD, I.F., CORNELIUS, M.J., KEYS, A.J., WHITTINGHAM, C.P. : Intramolecular labelling of sucrose made by leaves from [$^{14}$C]carbon dioxide or [3-$^{14}$C]serine. - Biochem. J. *172* : 23 - 27, 1978. [Ps.]

32856 - BIRKY, C.W. Jr. : Transmission genetics of mitochondria and chloroplasts. - Annu. Rev. Genet. *12* : 471 - 512, 1978.

32857 - BIRRELL, G.B., LEE, T.D., GRIFFITH, O.H., KEANA, J.F.W. : Synthesis and properties of chlorophyll-derived nitroxide spin labels. - Bioorg. Chem. *7* : 409 - 420, 1978.

32858 - BIRRELL, G.B., SISTROM, W.R., GRIFFITH, O.H. : Lipid-protein associations in chromatophores from the photosynthetic bacterium *Rhodopseudomonas sphaeroides*. - Biochemistry *17* : 3768 - 3773, 1978.

32859 - BISHOP, D.G., NOLAN, W.G., JOHNS, S.R., WILLING, R.I. : Evolution of the lipid components of chloroplast membranes : the role of lipid fluidity in membrane-associated reactions. - In : DEAMER, D.W. (ed.) : Light Transducing Membranes. Pp. 269 - 288. Acad. Press, New York - San Francisco - London 1978.

32860 - BISHOP, N.I., JONES, L.W. : Alternate fates of the photochemical reducing power generated in photosynthesis : Hydrogen production and nitrogen fixation. - In : SANADI, D.R., VERNON, L.P. (ed.) : Current Topics in Bioenergetics. Vol. 8. Pp. 3 - 31. Academic Press, New York - San Francisco - London 1978.

32861 - BISHOP, N.I., SICHER, R.C., METZ, J.G. : Perturbation of the electron transport system of photosystem II as revealed by temperature- and light-sensitive mutations of *Scenedesmus*. - In : HALL, D.O., COOMBS, J., GOODWIN, T.W. (ed.) : Proceedings of the Fourth International Congress on Photosynthesis. Pp. 373 - 386. Biochem. Soc., London 1978.

32862 - BISWAL, U.C., MOHANTY, P. : Changes in the ability of photophosphorylation and activities of surface-bound adenosine triphosphatase and ribulose diphosphate carboxylase of chloroplasts isolated from the barley leaves senescing in darkness. - Physiol. Plant. *44* : 127 - 133, 1978.

32863 - BISWAS, A.K., CHOUDHURI, M.A. : Differential behaviour of flag leaf of intact rice plant during ageing. - Biochem. Physiol. Pflanzen *173* : 220 - 228, 1978.

32864 - BISWAS, A.K., MUKHERJI, S. : Penicillin induced regulation of chlorophyll formation and Hill activity of isolated chloroplasts in rice (*Oryza sativa* L.). - Curr. Sci. *47* : 555 - 556, 1978.

32865 - BIYASHEV, G.Z., ABDURAKHMANOV, A.A., ZHUMABEKOV, E.Zh., BEKMUKHAMBETOVA, B.A.: Pigmenty fotosinteza i khlorofillaznaya aktivnost' u sakharnoĭ svekly s razlichnym urovnem ploidnosti. [Pigments of photosynthesis and the chlorophyllase activity in sugar beet with the different ploidy level.] - Sel'skokhoz. Biol. *13* : 865 - 867, 954, 1978. [In R, ab : E.]

32866 - BJÖRKMAN, O., BADGER, M., ARMOND, P.A. : Thermal acclimation of photosynthesis : Effect of growth temperature on photosynthetic characteristics and components of the photosynthetic apparatus in *Nerium oleander*. - Carnegie Inst. Year Book *77* : 262 - 276, 1978.

32867 - BLACK, C.C., BROWN, R.H., MOORE, R.C. : Plant photosynthesis. - In : DÖBEREINER, J., BURRIS, R.H., HOLAENDER, A., FRANCO, A.A., NEYRA, C.A., SCOTT, D.B. (ed.) : Limitations and Potentials for Biological Nitrogen Fixation in the Tropics. Pp. 95 - 110. Plenum Press, New York 1978.

*32868 - BLACKMAN, G.E., COOPER, J.P. : Photosynthesis and solar energy conversion. - Phil. Trans. roy. Soc. London *B 274* : 321 - 329, 1976.

32869 - BLAD, B.L., STEADMAN, J.R., WEISS, A. : Canopy structure and irrigation influence white mold disease and microclimate of dry edible beans. - Phytopathology *68* : 1431 - 1437, 1978. [Growth analysis.]

32870 - **BLAHA, J.** : Variabilita tvarů listu v klonové selekci révy vinné. [Variability of the shape of leaves in clone selection of grapevine.] - Sborník ÚVTIZ Genet. Šlechtění (Praha) *14* : 267 - 274, 1978. [In Czech, ab : E, G, R. ]

32871 - **BLANCHET, R., GELFI, N.** : Relation entre développement foliaire, transpiration et production chez le soja (cv Amsoy 71 et Hodgson). - Ann. agron. *29* : 223 - 242, 1978.[ Ps.]

32872 - **BLANK, R., GROBE, B., ARNOLD, C.-G.** : Time-sequence of nuclear and chloroplast fusions in the zygote of *Chlamydomonas reinhardii*. - Planta *138* : 63 - 64, 1978.

32873 - **BLATT, M.R., BRIGGS, W.R.** : Blue-light-induced chloroplast aggregation and cortical fiber reticulation in the alga *Vaucheria*. - Carnegie Inst. Year Book *77* : 333 - 336, 1978.

32874 - **BLAXTER, K.L.** : Energy flow in agriculture. - In : HALL, D.O., COOMBS, J., GOODWIN, T.W. (ed.) : Proceedings of the Fourth International Congress on Photosynthesis. Pp. 685 - 694. Biochem. Soc., London 1978.

32875 - **BLEDSOE, C.S., ROSS, C.W.** : Metabolism of mevalonic acid in vegetative and induced plants of *Xanthium strumarium*. - Plant Physiol. *62* : 683 - 686, 1978. [Chl, Car.]

32876 - **BLEE, E., SCHANTZ, R.** : Biosynthesis of galactolipids in *Euglena gracilis*. II, Changes in fatty acid composition of galactolipids during chloroplast development. - Plant Sci. Lett. *13* : 257 - 267, 1978. [Ps, Chl.]

32877 - **BLOCK, F.R., STERZELMEIER, K.** : Kontinuierliche Messung der Blatttranspiration mit Hilfe eines neuen, elektronischen Gasfeuchtemeßgerätes. - Gartenbauwissenschaft *43* : 142 - 144, 1978. [Ps.]

32878 - **BLOK, M.C., HELLINGWERF, K.J., KAPTEIN, R., DE KRUIJFF, B.** : Light-induced pH-changes inside bacteriorhodopsin vesicles as measured by $^{31}P$ NMR. - Biochim. biophys. Acta *514* : 178 - 184, 1978.

32879 - **BLOK, M.C., VAN DAM, K.** : Association of bacteriorhodopsin-containing phospholipid vesicles with phospholipid-impregnated Millipore filters. - Biochim. biophys. Acta *507* : 48 - 61, 1978.

32880 - **BLOK, V.R.** : Rol' sostoyaniya s perenosom zaryada dimera khlorofilla v pervichnom akte fotosinteza. [Role of the charge-transfer state of a chlorophyll dimer in the primary act of photosynthesis.] - Zh. fiz. Khim. *52* : 245, 1978. [In R.]

32881 - **BLUM, A., SCHERTZ, K.F., TOLER, R.W., WELCH, R.I., ROSENOW, D.T., JOHNSON, J.W., CLARK, L.E.** : Selection for drought avoidance in sorghum using aerial infrared photography. - Agron. J. *70* : 472 - 477, 1978. [Chl.]

32882 - **BLUM, U., SENECA, E.D., STROUD, L.M.** : Photosynthesis and respiration of *Spartina* and *Juncus* salt marshes in North Carolina: Some models. - Estuaries *1* : 228 - 238, 1978.

32883 - **BLUME, D.E., McCLURE, J.W.** : Photocontrol of phenylalanine ammonia-lyase in barley seedlings treated with pyridazinone inhibitors of chloroplast development. - Phytochemistry *17* : 1545 - 1547, 1978. [Chl.]

*32884 - **BLUMENFELD, L.A., CHERNAVSKII, D.S.** : Tunnelling of electrons in biological processes. - J. theor. Biol. *39* : 1 - 7, 1973. [Ps.]

32885 - **BLYUMENFEL'D, L.A., BURBAEV, D.Sh., KHANGULOV, S.V., TSAPIN, A.I.** : Struktury tipa spinovykh stekol v khloroplastakh. [Structures of the type of spin glasses in chloroplasts.] - Biofizika *23* : 614 - 619, 1978. [In R, ab : E.]

32886 - **BLYUMENFEL'D, L.A., GOL'DFEL'D, M.G., DMITROVSKIĬ, L.G.** : Fosforilirovanie v khloroplastakh v usloviyakh kislotnogo udara. [Phosphorylation in chloroplasts under acid pulse.] - Biofizika *23* : 549 - 551, 1978. [In R, ab : E.]

32887 - **BOALCH, G.T., HARBOUR, D.S., BUTLER, E.I.** : Seasonal phytoplankton production in the western English Channel 1964 - 1974. - J. mar. biol. Assoc. UK *58* : 943 - 953, 1978. [Ps.]

32888 - BOARDMAN, N.K. : Solar energy conversion in photosynthesis and its potential
         contribution to world demand for liquid and gaseous fuels. - In : HALL, D.O.,
         COOMBS, J., GOODWIN, T.W. (ed.) : Proceedings of the Fourth International
         Congress on Photosynthesis. Pp. 635 - 644. Biochem. Soc., London 1978.

32889 - BOARDMAN, N.K., ANDERSON, J.M. : Composition, structure and photochemical
         activity of developing and mature chloroplasts. - In : AKOYUNOGLOU, G.,
         ARGYROUDI-AKOYUNOGLOU, J.H. (ed.) : Chloroplast Development. Pp. 1 - 14. El-
         sevier/North-Holland Biomedical Press, Amsterdam - New York - Oxford 1978.

32890 - BOARDMAN, N.K., ANDERSON, J.M., GOODCHILD, D.J. : Chlorophyll-protein comple-
         xes and structure of mature and developing chloroplasts. - In : SANADI, D.R.,
         VERNON, L.P. (ed.) : Current Topics in Bioenergetics. Vol. 8. Pp. 36 - 109.
         Academic Press, New York - San Francisco - London 1978.

32891 - BOBEK, B., BERGSTROM, R. : A rapid method of browse biomass estimation in a
         forest habitat. - J. Range Manage. $31$ : 456 - 458, 1978.

32892 - BOBODZHANOV, V.A., SIDOROVA, K.K., NASYROV, Yu.S. : Fiziologicheskie osoben-
         nosti i produktivnost' mutantov gorokha v razlichnykh ékologicheskikh uslovi-
         yakh. [Physiological characteristics and productivity of pea mutants under
         different ecological conditions.] - Dokl. Akad. Nauk tadzh. SSR $21$ (3) :
         54 - 56, 1978. [Chl; in R, ab : Tajik.]

32893 - BOCHAROV, E.A., DZHANUMOV, D.A. : Sintez polyarnykh lipidov v khloroplastakh
         ozimykh pshenits v svyazi s ikh morozoustoĭchivost'yu. [Synthesis of polar
         lipids in chloroplasts of winter wheat plants with respect to their frost-
         -resistance.] - Fiziol. Rast. $25$ : 756 - 760, 1978. [In R, ab : E.]

32894 - BÖCHER, M., KLUGE, M. : Der $C_4$-Weg der C-Fixierung bei *Spinacia oleracea*.
         II. Pulse-chase Experimente mit suspendierten Blattstreifen. - Z. Pflanzen-
         physiol. $86$ : 405 - 421, 1978.

32895 - BÖCHER, T.W., OLESEN, P. : Structural and ecophysiological pattern in the
         xero-halophytic $C_4$ grass, *Sporobolus rigens* (TR.) DESV. - Kongel. danske
         Vidensk. Selsk., biol. Skr. $22$ (3) : 1 - 48, 1978. [Ps.]

32896 - BOGENRIEDER, A., KLEIN, R. : Die Abhängigkeit der UV-Empfindlichkeit von
         von der Lichtqualität bei der Aufzucht (*Lactuca sativa* L.).- Angew. Bot. $52$ :
         283 - 293, 1978. [Ps.]

32897 - BÖGER, P. : Photobiologische Umwandlung der Sonnenenergie. - Naturwissen-
         schaften $65$ : 407 - 412, 1978. [Ps.]

32898 - BÖGER, P. : Some properties of plastocyanin and its function in algal chlo-
         roplasts. - In : HALL, D.O., COOMBS, J., GOODWIN, T.W. (ed.) : Proceedings
         of the Fourth International Congress on Photosynthesis. Pp. 755 - 764. Bio-
         chem. Soc., London 1978.

32899 - BÖGER, P., KUNERT, K.J. : Phytotoxic action of paraquat on the photosynthe-
         tic apparatus. - Z. Naturforsch. $33$ C : 688 - 694, 1978.

32900 - BÖGER, P., VETTER, H. : Herbicides in modern crop farming. - Plant Res. De-
         velopment $8$ : 79 - 101, 1978. [Ps.]

32901 - BOGOMOLNI, R.A., STUBBS, L., LANYI, J.K. : Illumination-dependent changes
         in the intrinsic fluorescence of bacteriorhodopsin. - Biochemistry $17$ : 1037
         - 1041, 1978.

32902 - BOGORAD, L., BEDBROOK, J.R., COEN, D.M., KOLODNER, R., LINK, G. : Genes for
         chloroplast proteins and rRNAs. - In : AKOYUNOGLOU, G., ARGYROUDI-AKOYUNOGLOU,
         J.H. (ed.) : Chloroplast Development. Pp. 541 - 551. Elsevier/North-Holland
         Biomedical Press, Amsterdam - New York - Oxford 1978.

*32903 - BOGUSLAVSKIĬ, L.I., BOITSOV, V.G., VOLKOV, A.G., KOZLOV, I.A., METEL'SKIĬ,
         S.T. : Fotoindutsirovannyĭ perenos protonov iz vody v oktan, kataliziruemyĭ
         bakteriorodopsinom. [Light-dependent translocation of proton from water to
         octane catalyzed by bacteriorhodopsin.] - Bioorg. Khim. $2$ : 1125 - 1131,
         1976. [In R, ab : E.]

32904 - BOGUSLAVSKIĬ, L.I., ZHURAVLEV, L.T., KANDELAKI, M.D., SHENGELIA, K.Ya. :
         Issledovanie fotookisleniya vody na granitse razdela faz oktan/voda v pri-
         sutstvii khlorofilla mass-spektrometricheskim metodom. [Study of water pho-

tooxidation at the octane/water interface in the presence of chlorophyll by a mass-spectrometric method.] - Dokl. Akad. Nauk SSSR *240* : 1453 - 1456, 1978. [In R.]

32905 - BÖHME, H. : Reactions of antibodies against ferredoxin, ferredoxin-NADP$^+$ reductase and plastocyanin with spinach chloroplasts. - Europe. J. Biochem. *84*: 87 - 93, 1978.

32906 - BÖHME, H. : Quantitative determination of ferredoxin, ferredoxin-NADP$^+$ reductase and plastocyanin in spinach chloroplasts. - Europe. J. Biochem. *83* : 137 - 141, 1978.

32907 - BÖHME, H., BÖGER, P. : Dual-wavelength spectroscopy in photosynthesis. - Z. Naturforsch. *33 C* : 161 - 168, 1978.

32908 - BÖHME, H., KUNERT, K.J., BÖGER, P. : The role of plastidic cytochrome *c* in algal electron transport and photophosphorylation. - Biochim. biophys. Acta *501* : 275 - 285, 1978.

32909 - BOHNER, H., BÖGER, P. : Reciprocal formation of cytochrome *c*-553 and plastocyanin in *Scenedesmus*. - FEBS Lett. *85* : 337 - 339, 1978.

32910 - BOHRA, O.P., ALI, K.H., DWIVEDI, S.N. : Diurnal distribution of photosynthetic pigments and planktons in relation to environmental parameters in Malad Creek, Bombay. - Compar. Physiol. Ecol. *3* : 207 - 211, 1978.

32911 - BOĬCHENKO, E.A., UDEL'NOVA, T.M., ZARIN', V.Ė. : Metally v ėvolyutsii assimilyatsii uglekisloty v biosfere. [Metals in the evolution of carbon dioxide assimilation in the biosphere.] - Izv. Akad. Nauk SSSR, Ser. biol. *1978* (1) : 12 - 18, 1978. [In R, ab : E.]

32912 - BOLAND, R.L., GARNER, G.B., NELSON, C.J., ASAY, K.H. : Organic acid metabolism in leaves of tall fescue genotypes. - Phyton *36* : 61 - 68, 1978. [Ps.]

32913 - BOLTON, J., BROWN, R.H. : Effect of nitrogen nutrition on photosynthesis and associated characteristics in $C_3$, $C_4$ and intermediate grass species. - Plant Physiol. *61* (Suppl.) : 38, 1978.

32914 - BOLTON, J.R. : Photochemical storage of solar energy. - Solar Energy *20* : 181 - 183, 1978.

32915 - BOLTON, J.R. : Primary electron acceptors. - In : CLAYTON, R.K., SISTROM, W.R. (ed.) : The Photosynthetic Bacteria. Pp. 419 - 429. Plenum Press, New York - London 1978.

32916 - BOLTON, J.R. : Solar energy conversion efficiency in photosynthesis - or - why two photosystems? - In : HALL, D.O., COOMBS, J., GOODWIN, T.W. (ed.) : Proceedings of the Fourth International Congress on Photosynthesis. Pp. 621 - 634. Biochem. Soc., London 1978.

32917 - BOLTON, P., WHARFE, J., HARWOOD, J.L. : The lipid composition of a barley mutant lacking chlorophyll *b*. - Biochem. J. *174* : 67 - 72, 1978.

32918 - BONDARENKO, V.P. : Vodopotreblenie i produktivnost' sakharnogo sorgo pri raznykh polivnykh rezhimakh. [Water consumption and productivity of sweet sorghum under different irrigation regimes.] - Byull. vsesoyuz. nauch.-issl. Inst. Kukuruzy *1978* (2-3) : 54 - 58, 1978. [In R.]

32919 - BONHOMME, R., VARLET-GRANCHER, C. : Estimation of the gramineous crop geometry by plant profiles including leaf width variations. - Photosynthetica *12* : 193 - 196, 1978.

32920 - BONOTTO, S., NUYTS, G., BAUGNET-MATHIEU, L., GILLES, J., DUJARDIN, E., SIRONVAL, C. : Evidence for heterogeneity of chloroplasts in *Acetabularia* : Fluorescence emission and protein-synthesis. - In : AKOYUNOGLOU, G., ARGYROUDI-AKOYUNOGLOU, J.H. (ed.) : Chloroplast Development. Pp. 333 - 344. Elsevier/North-Holland Biomedical Press, Amsterdam - New York - Oxford 1978.

32921 - BOOKJANS, G., BÖGER, P. : Modification of ferredoxin-NADP$^+$ reductase from the alga *Bumilleriopsis* with butanedione and dansyl chloride. - Arch. Biochim. Biophys. *190* : 459 - 465, 1978.

32922 - **BOOKJANS, G., SAN PIETRO, A., BÖGER, P.** : Resolution and reconstitution of spinach ferredoxin-NADP⁺ reductase. - Biochem. biophys. Res. Commun. *80* : 759 - 765, 1978.

32923 - **BOOTE, K.J., GALLAHER, R.N., ROBERTSON, W.K., HINSON, K., HAMMOND, L.C.** : Effect of foliar fertilization on photosynthesis, leaf nutrition, and yield of soybeans. - Agron. J. *70* : 787 - 791, 1978.

32924 - **BORISEVICH, G.P., KNOX, P.P., KONONENKO, A.A., RUBIN, A.B., VOZARY, E.** : Electric field-induced polarization of photosynthetic membranes and reaction centers of *Rhodopseudomonas sphaeroides*, strain 1760-1. - Acta biochim. biophys. Acad. Sci. hung. *13* : 67 - 72, 1978.

32925 - **BORISOV, A.Yu.** : Energy-migration mechanisms in antenna chlorophylls. - In : **CLAYTON, R.K., SISTROM, W.R.** (ed.) : The Photosynthetic Bacteria. Pp. 323 - 331. Plenum Press, New York - London 1978.

32926 - **BORISOV, A.Yu.** : Fotosintez kak proobraz solnechnoĭ ênergetiki novogo tipa. I. Khlorofil'nyĭ apparat fotosinteza. [Photosynthesis as a prototype of new solar energetics. I. Chlorophyll apparatus of photosynthesis.] - Mol. Biol. (Moskva) *12* : 267 - 275, 1978. [In R, ab : E.]

32927 - **BOROWITZKA, M.A.** : Plastid development and floridean starch grain formation during carposporogenesis in the coralline red alga *Lithothrix aspergillum* GRAY. - Protoplasma *95* : 217 - 228, 1978.

32928 - **BOROWITZKA, M.A., VESK, M.** : Ultrastructure of the *Corallinaceae* I. The vegetative cells of *Corallina officinalis* and *C. cuvierii*. - Mar. Biol. *46* : 295 - 304, 1978.

32929 - **BORZENKOVA, R.A.** : Fitokhromnaya i gormonal'naya regulyatsiya khloroplastogeneza pri differentsirovke tkaneĭ. [Phytochrome and hormonal regulation of chloroplastogenesis during the differentiation of tissues.] - In : Mezostruktura i Funktsional'naya Aktivnost' Fotosinteticheskogo Apparata. Pp. 46 - 60. Ural'skiĭ gos. Univ., Sverdlovsk 1978. [In R.]

32930 - **BORZENKOVA, R.A., BORTNIKOVA, I.F.** : Svetozavisimost' deĭstviya kinetina v protsesse khloroplastogeneza. [The effect of illumination on the action of kinetin in the process of chloroplastogenesis.] - Fiziol. Rast. *25* : 254 - 261, 1978. [In R, ab : E.]

*32931 - **BORZENKOVA, R.A., MOKRONOSOV, A.T.** : Rol' fitogormonov v biogeneze khloroplastov. [Role of phytohormones in chloroplast biogenesis.] - Fiziol. Rast. *23* : 490 - 496, 1976. [In R, ab : E.]

32932 - **BOSCHETTI, A., DIEZI, R., EICHENBERGER, W., SCHAFFNER, J.C.** : Characterization of thylakoid proteins by two-dimensional separation. - In : **AKOYUNOGLOU G., ARGYROUDI-AKOYUNOGLOU, J.H.** (ed.) : Chloroplast Development. Pp. 195 - 200. Elsevier/North-Holland Biomedical Press, Amsterdam - New York - Oxford 1978.

32933 - **BOSCHETTI, A., SAUTON-HEINIGER, E., SCHAFFNER, J.-C., EICHENBERGER, W.** : A two-dimensional separation of proteins from chloroplast thylakoids and other membranes. - Physiol. Plant. *44* : 134 - 140, 1978.

32934 - **BOSE, S., ARNTZEN, C.J.** : Reversible inactivation of Photosystem II reaction centers in cation-depleted chloroplast membranes. - Arch. Biochem. Biophys. *185* : 567 - 575, 1978.

32935 - **BOSE, S., HOCH, G.E.** : $Mg^{2+}$-$H^+$ exchange in chloroplast membranes in dark. - Z. Naturforsch. *33 C* : 105 - 107, 1978.

32936 - **BOSE, S., HOCH, G.E.** : 9-aminoacridine binding to chloroplast membranes in dark. Reversal by $Mg^{2+}$. - Z. Naturforsch. *33 C* : 108 - 112, 1978.

32937 - **BOTHE, H., DISTLER, E., EISBRENNER, G.** : Hydrogen metabolism in blue-green algae. - Biochimie *60* : 277 - 289, 1978.

32938 - **BOTHE, H., FLOENER, L.** : Physiological characterization of *Cyanophora paradoxa*, a flagellate containing cyanelles in endosymbiosis. - Z. Naturforsch. *33 C* : 981 - 987, 1978. [Ps.]

32939 - BOTMAN, K.S., GIRSHEVICH, E.I. : Summarnoe isparenie vlagi orekhovym lesom
v Zapadnom Tyan'-Shane. [Total evaporation by a walnut forest in the Western
Tien-Shan.] - Lesovedenie *1978* (3) : 19 - 27, 1978. [Growth analysis; in R,
ab : E.]

32940 - BOTO, K.G., BUNT, J.S. : Selective excitation fluorometry for the determination of chlorophylls and pheophytins. - Anal. Chem. *50* : 392 - 395, 1978.

32941 - BOTT, T.L., BROCK, J.T., CUSHING, C.E., GREGORY, S.V., KING, D., PETERSEN,
R.C. : A comparison of methods for measuring primary productivi.ty and community respiration in streams. - Hydrobiologia *60* : 3 - 12, 1978.

32942 - BÖTTCHER,R. : Demonstration der Chlorophyllfluorescenz im Schulversuch. -
Math. naturwiss. Unterr. *31* : 271 - 274, 1978.

32943 - BOTTOMLEY, P.J., VAN BAALEN, C. : Characteristics of heterotrophic growth
in the blue-green alga *Nostoc* sp. strain Mac. - J. gen. Microbiol. *107* : 309
- 318, 1978. [Ps.]

32944 - BOTTOMLEY, W., WHITFELD, P.R. : The products of cell free-transcription and
translation of total spinach chloroplast DNA. - In : AKOYUNOGLOU, G., ARGY-
ROUDI-AKOYUNOGLOU, J.H. (ed.) : Chloroplast Development. Pp. 657 - 662. El-
sevier/North-Holland Biomedical Press, Amsterdam - New York - Oxford 1978.

32945 - BOUGES-BOCQUET, B. : Absorption changes from 437 nm to 530 nm in *Chlorella
pyrenoidosa* under flash excitation. Probable detection of ferredoxin-NADP-
-reductase. - FEBS Lett. *85* : 340 - 344, 1978.

32946 - BOUGES-BOCQUET, B. : Reaction leading to the photo-induced reduction of fer-
redoxin-NADP-reductase (FNR) in *Chlorella* cells. - FEBS Lett. *94* : 95 - 99,
1978.

32947 - BOUGES-BOCQUET, B., DELOSME, R. : Evidence for a new electron donor to $P-700$
in *Chlorella pyrenoidosa*. - FEBS Lett. *94* : 100 - 104, 1978.

*32948 - BOULTER, D., HASLETT, B.G., PEACOCK, D., RAMSHAW, J.A.M., SCAWEN, M.D. :
Chemistry, function, and evolution of plastocyanin. - Int. Rev. Biochem.
(Plant Biochem. II) *13* : 1 - 40, 1977.

32949 - BOUTHYETTE, P.-Y., JAGENDORF, A.T. : The site of synthesis of pea chloroplast coupling factor 1. - Plant Cell Physiol. *19* : 1169 - 1174, 1978.

32950 - BOUTTON, T.W., CAMERON, G.N., SMITH, B.N. : Insect herbivory on $C_3$ and $C_4$
grasses. - Oecologia *36* : 21 - 32, 1978.

32951 - BOUTTON, T.W., SMITH, B.N. : The use of stable carbon isotopes in studying
plant-animal relationships. - Plant Physiol. *61* (Suppl.) : 6, 1978. [$C_3$-$C_4$
plants.]

32952 - BOWDEN, L., LORD, J.M. : Purification and comparative properties of microsomal and glyoxysomal malate synthase from castor bean endosperm. - Plant
Physiol. *61* : 259 - 265, 1978.

32953 - BOWES, G., HOLADAY, A.S., VAN, T.K., HALLER, W.T. : Photosynthetic and photorespiratory carbon metabolism in aquatic plants. - In : HALL, D.O., COOMBS,
J., GOODWIN, T.W. (ed.) : Proceedings of the Fourth International Congress
on Photosynthesis. Pp. 289 - 298. Biochem. Soc., London 1978.

32954 - BOWES, J.M., CROFTS, A.R. : Interactions of protons with transitions of the
watersplitting enzyme of photosystem II as measured by delayed fluorescence.
- Z. Naturforsch. *33 C* : 271 - 275, 1978.

32955 - BOWYER, J.R., BACCARINI-MELANDRI, A., MELANDRI, B.A., CROFTS, A.R. : The role of ubiquinone-10 in cyclic electron transport in *Rhodopseudomonas capsulata* Ala pho$^+$: Effects of lyophilization and extraction. - Z. Naturforsch.
*33 C* : 704 - 711, 1978.

*32956 - BRACH, E.J., TINKER, R.W., AMOUR, G.T.S. : Improvements in spectral reflec-
tance measurements of field crops. - Can. agr. Eng. *19* (2) : 78 - 83, 1977.

32957 - BRADBEER, J.W. : Report on the development of photosynthetic systems. - In :
HALL, D.O., COOMBS, J., GOODWIN, T.W. (ed.) : Proceedings of the Fourth International Congress on Photosynthesis. Pp. 549 - 555. Biochem. Soc., London
1978.

32958 - BRADBEER, J.W., BÖRNER, T. : Activities of glyceraldehyde-phosphate dehydrogenase (NADP⁺) and phosphoribulokinase in two barley mutants deficient in
chloroplast ribosomes. - In : AKOYUNOGLOU, G., ARGYROUDI-AKOYUNOGLOU, J.H.
(ed.) : Chloroplast Development. Pp. 727 - 732. Elsevier/North-Holland Biomedical Press, Amsterdam - New York - Oxford 1978.

32959 - BRADBEER, J.W., HARGRAVE, D.F., LANGMAN, L. : The photomorphogenetic control
of the development of plastid glyceraldehyde-phosphate dehydrogenase and
phosphoribulokinase activities in cotyledons of mustard seedlings. - Photochem. Photobiol. *27* : 183 - 187, 1978.

32960 - BRADBURY, I.K., MALCOLM, D.C. : Dry matter accumulation by *Picea sitchensis*
seedlings during winter. - Can. J. Forest Res. *8* : 207 - 213, 1978.

32961 - BRADFORD, J.M., ROBERTS, P.E. : Distribution of reactive phosphorus and
plankton in relation to upwelling and surface circulation around New Zealand.
- New Zeal. J. mar. Freshw. Res. *12* : 1 - 15, 1978. [Chl.]

32962 - BRÄNDÉN, R. : Ribulose-1,5-diphosphate carboxylase and oxygenase from green
plants are two different enzymes . - Biochem. biophys. Res. Commun. *81* :
539 - 546, 1978.

32963 - BRÄNDÉN, R., BRÄNDÉN, C.I. : Separation of ribulose 1,5-bisphosphate carbo-
xylase and oxygenase activities. - In : SIEGELMAN, H.W., HIND, G. (ed.) :
Photosynthetic Carbon Assimilation. Pp. 391 - 397. Plenum Press, New York -
London 1978.

*32964 - BRANGEON, J. : Structural modifications in the lamellar system of isolated
*Zea mays* chloroplasts under different ionic conditions, - J. Microscopie *21* :
75 - 84, 1974.

32965 - BRAR, G.S., LENZ, F., THIES, W. : Photosynthesis and respiration of two genotypes of rapeseed (*Brassica napus* L.) - Z. Acker- Pflanzenbau *147* : 147 -
153, 1978.

32966 - BRAUMANN, T., MAHRO, B., GRIMME, L.H. : Hochdruck-Flüssigkeits-chromatographi-
sche (HPLC) Analyse eines Gesamtpigment-Extrakts des Photosynthese-Apparates.
- Ber. deut. bot. Ges. *91* : 563 - 567, 1978.

32967 - BRAVDO, B., PALLAS, J.E. Jr. : Photosynthetic efficiency of several peanut
cultivars. Analysis of leaf factors which determine the net photosynthetic
rate. - Plant Physiol. *61* (Suppl.) : 7, 1978.

32968 - BREAZEALE, V.D., BUCHANAN, B.B., WOLOSIUK, R.A. : Chloroplast sedoheptulose
1,7-bisphosphatase: Evidence for regulation by the ferredoxin/thioredoxin
system. - Z. Naturforsch. *33 C* : 521 - 528, 1978.

32969 - BREEZE, V., ELSTON, J. : Some effects of temperature and substrate content
upon respiration and the carbon balance of field beans (*Vicia faba* L.). -
Ann. Bot. *42* : 863 - 876, 1978. [Ps.]

32970 - BREIDERT, D., KRAUSSE, H.J., NARAHARI, P., SCHÖN, W.J. : Analytical and biochemical problems in breeding programmes for improved seed protein. - In :
Seed Protein Improvement by Nuclear Techniques (Proc. Meeting Baden/Vienna
1977). Pp. 355 - 363. IAEA, Vienna 1978. [Ps.]

32971 - BREMNER, P.M., RAWSON, H.M. : The weights of individual grains of the wheat
ear in relation to their growth potential, the supply of assimilate and interaction between grains. - Aust. J. Plant Physiol. *5* : 61 - 72, 1978.

32972 - BRILL, A.S. : Activation of electron transfer reactions of the blue proteins.
- Biophys. J. *22* : 139 - 142, 1978. [Plastocyanin.]

32973 - BRINKMAN, M.A., FREY, K.J. : Flag leaf physiological analysis of oat isolines
that differ in grain yield from their recurrent parents. - Crop Sci. *18* :
69 - 73, 1978. [Ps.]

32974 - BRINKMANN, G., SENGER, H. : Light-dependent formation of thylacoid membranes
during the development of the photosynthetic apparatus in pigment mutant
C-2A' of *Scenedesmus obliquus*. - In : AKOYUNOGLOU, G., ARGYROUDI-AKOYUNOGLOU,
J.H. (ed.) : Chloroplast Development. Pp. 201 - 206. Elsevier/North-Holland
Biomedical Press, Amsterdam - New York - Oxford 1978.

32975 - BRINKMANN, G., SENGER, H. : The development of structure and function in chloroplasts of greening mutants of *Scenedesmus* IV. Blue light-dependent carbohydrate and protein metabolism. - Plant Cell Physiol. *19* : 1427 - 1437, 1978.

32976 - BRITTON, C.M., DODD, J.D., WEICHERT, A.T. : Net aerial primary production of an *Andropogon-Paspalum* grassland ecosystem. - J. Range Manage. *31* : 381 - 386, 1978.

32977 - BROCH-DUE, M., ORMEROD, J.G. : Isolation of a BChl *c* mutant from *Chlorobium* with BChl *d* by cultivation at low light intensity. - FEMS Microbiol. Lett. *3* : 305 - 308, 1978.

32978 - BROCH-DUE, M., ORMEROD, J.G., FJERDINGEN, B.S. : Effect of light intensity on vesicle formation in *Chlorobium*. - Arch. Microbiol. *116* : 269 - 274, 1978.

32979 - BROCK, T.D. : Use of fluorescence microscopy for quantifying phytoplankton, especially filamentous blue-green algae. - Limnol. Oceanogr. *23* : 158 - 160, 1978.

32980 - BROCKINGTON, N.R. : Simulation models in crop production research. - Acta Agr. scand. *28* : 33 - 40, 1978.

32981 - BROCKMANN, H. Jr. : Stereochemistry and absolute configuration of chlorophylls and linear tetrapyrroles. - In : DOLPHIN, D. (ed.) : Porphyrins. Vol. II. Part B. Pp. 287 - 326. Academic Press, New York - San Francisco - London 1978.

*32982 - BRODA, E. : Die Entwicklung der bioenergetischen Prozesse. - Phys. Blätter *30* : 444 - 456, 1974. [Ps.]

32983 - BRODA, E. : Erfindungen der lebenden Zelle. - Naturwiss. Rundsch. *31* : 356 - 363, 1978. [Ps.]

32984 - BROECKER, H.-C., PETERMANN, J., SIEMS, W. : The influence of wind on $CO_2$-exchange in a wind-wave tunnel, including the effects of monolayers. - J. mar. Res. *36* : 595 - 610, 1978.

32985 - BROUERS, M., SIRONVAL, C. : The reduction of protochlorophyllide into chlorophyllide. VII. Relations between energy transfer, 690 nm fluorescence emission, and reduction; a theory. - Photosynthetica *12* : 399 - 405, 1978.

*32986 - BROUWER, R. : A comparison of the effect of drought and low root temperatures on leaf elongation and photosynthesis in maize. - Acta Horticult. *39* : 141 - 152, 1974.

32987 - BROVCHENKO, M.I., ZAVYALOVA, T.F. : K voprosu o putyakh transporta veshchestv v list'yakh rastenii s $C_4$-tipom fotosinteza. [Pathways of substance transport in leaves of plants with $C_4$-type of photosynthesis.] - Fiziol. Rast. *25* : 1144 - 1150, 1978. [in R, ab : E.]

32988 - BROWN, C.A. Jr., FARMER, F.H., JARRETT, O. Jr., STATON, W.L. : Laboratory studies of *in vivo* fluorescence of phytoplankton. - In : Proceedings of the Fourth Joint Conference on Sensing of Environmental Pollutants. Pp. 482 - 488. Amer. chem. Soc., Washington 1978.

32989 - BROWN, D.H., HOUSE, K.L. : Evidence of a copper-tolerant ecotypes of the hepatic *Solenostoma crenulatum*. - Ann. Bot. *42* : 1383 - 1392, 1978. [Ps.]

32990 - BROWN, D.H., SMIRNOFF, N. : Observations on the effect of ozone on *Cladonia rangiformis*. - Lichenologist *10* : 91 - 94, 1978. [Ps.]

32991 - BROWN, H.M., KINGZEIT, P.C., GRIFFITH, O.H. : Photoelectron quantum yield spectrum and photoelectron microscopy of β-carotene. - Photochem. Photobiol. *27* : 445 - 449, 1978.

32992 - BROWN, J.S. : Coupled P700 photooxidation and cytochrome $b_6$ reduction in a chlorophyll-protein complex. - Photosynthetica *12* : 310 - 315, 1978.

32993 - BROWN, J.S. : Photoreactive chlorophyll-protein complexes isolated from non-green algae. - Plant Physiol. *61* (Suppl.) : 102, 1978.

32994 - BROWN, J.S., VAN GINKEL, G. : Composition of isolated P700-chlorophyll-protein complexes and their photoactivity when combined with lipid vesicles or Triton micelles. - Carnegie Inst. Year Book *77* : 298 - 302, 1978.

32995 - BROWN, L.M., HELLEBUST, J.A. : Sorbitol and proline as intracellular osmotic
solutes in green alga *Stichococcus bacillaris*. - Can. J. Bot. *56* : 676 - 679,
1978. [Ps.]

32996 - BROWN, R.H. : A difference in N use efficiency in $C_3$ and $C_4$ plants and its
implications in adaptation and evolution. - Crop Sci. *18* : 93 - 98, 1978.

32997 - BROWN, R.H., SIMMONS, R.E. : Water use efficiency in grass species differing
in $CO_2$ fixation cycle. - Plant Physiol. *61* (Suppl.) : 38, 1978.[Ps, resistances.]

*32998 - BROWN, S.R. : Paleolimnological evidence from fossil pigments. - Mitt. int.
Verein. Limnol. *17* : 95 - 103, 1969. [Chl, Car.]

*32999 - BROWN, S.R., DALEY, R.J., McNEELY, R.N. : Composition and stratigraphy of
the fossil phorbin derivatives of Little Round Lake, Ontario. - Limnol. Oceanogr. *22* : 336 - 348, 1977. [Chl.]

33000 - BROWNE, C.L., FANG, S.C. : Estimation of whole-plant resistance to gaseous
exchange independent of leaf temperature measurement. - Plant Physiol. *61* :
231 - 235, 1978. [Resistances.]

33001 - BRÜGGEMANN, M., WEIGER, C., GIMMLER, H. : Synchronized culture of the halo-
tolerant unicellular green alga *Dunaliella parva*. - Biochem. Physiol. Pflanzen *172* : 487 - 506, 1978. [Ps, Chl.]

33002 - BRUN, W.A. : Assimilation. - In : NORMAN, A.G. (ed.) : Soybean Physiology,
Agronomy, and Utilization. Pp. 45 - 76. Academic Press, New York - San Francisco - London 1978. [Ps.]

33003 - BRYAN, A.M., OLAFSSON, P.G. : The effect of polychlorobiphenyls (Aroclor
1242) on bicarbonate-$C^{14}$ uptake by *Euglena gracilis*. - Bull. environm. Contam. Toxicol. *19* : 374 - 381, 1978.

33004 - BUCHANAN, B.B., CRAWFORD, N.A., WOLOSIUK, R.A. : Ferredoxin/thioredoxin system functions with effectors in activation of NADP-glyceraldehyde 3-phospha-
te dehydrogenase of barley seedlings. - Plant Sci. Lett. *12* : 257 - 264,
1978.

33005 - BUCHANAN, B.B., WOLOSIUK, R.A., CRAWFORD, N.A., YEE, B.C. : Evidence for
three thioredoxins in leaves. - Plant Physiol. *61* (Suppl.) : 38, 1978.

*33006 - BUCHECKER, R., LIAAEN-JENSEN, S. : Absolute configuration of heteroxanthin
and diadinoxanthin. - Phytochemistry *16* : 726 - 733, 1977.

33007 - BUCHOLTZ, D.L., LAVY, T.L. : Pesticide interactions in oats (*Avena sativa*
L. 'Neal'). - Agr. Food Chem. *26* : 520 - 524, 1978. [Ps.]

*33008 - BUDIMIR, M., PLESNIČAR, M., KLJAJIĆ, R. : Effects of herbicide 2,4-DB and
fungicide captan on reactions of mitochondria and chloroplasts. - Arch. environm. Contam. Toxicol. *4* : 166 - 174, 1976.

30009 - BUDZIKIEWICZ, H., JOHANNES, B., WEIGEL, H., HEIMANN, M. : Zur Problematik
der $^{18}O$-Analyse bei Carotinoiden. - Fresenius Z. anal. Chem. *290* : 382 - 388,
1978.

30010 - BUESCHER, R.W., DOHERTY, J.H. : Color development and carotenoid levels in
*rin* and *nor* tomatoes as influenced by ethephon, light and oxygen. - J. Food
Sci. *43* : 1816 - 1818, 1825, 1978.

33011 - BUGGELN, R.G. : Physiological investigations on *Alaria esculenta (Laminaria-
les, Phaeophyceae)*. III. Exudation by the blade. - J. Phycol. *14* : 54 - 56,
1978. [Ps.]

33012 - BÜHLER, B., DRUMM, H., MOHR, H. : Investigations on the role of ethylene
in phytochrome-mediated photomorphogenesis. II. Enzyme levels and chlorophyll
synthesis. - Planta *142*   : 119 - 122, 1978.

33013 - BUICULESCU, I., POPESCU, D., IORDAN, M., PEICEA, I.M., ŞERBĂNESCU, G. :
The influence of air-polluting gases on some plant metabolism. - Rev. roum.
Biol., Ser. Biol. vég. *23* : 187 - 193, 1978. [Ps, Chl.]

*33014 - BUINOVA, M.G. : Khlorofill i karotin v rasteniyakh estestvennoĭ i kul'tur-
noĭ flory Zabaĭkal'ya. [Chlorophyll and carotene in plants of natural and

cultured flora of the Transbaikal region.] - In : Fiziologiya I Produktivnost' Rasteniĭ v Zabaĭkal'e. Pp. 51 - 89. Ulan-Udê 1977. [In R.]

*33015 - BUKHBINDER, A.A. : Vliyanie vozrasta lista na nakoplenie nekotorykh plasti-
cheskikh veshchestv, pigmentov I êfirnykh masel v gerani, vozdelyvaemoĭ na
allyuvial'nykh pochvakh Kolkhidskoĭ nizmennosti. [Effect of leaf age on the
accumulation of certain plastid substances, pigments, and essential oils in
geraniums grown on alluvial soils of the Colchis Lowlands.] - Subtrop. Kul't.
*1975* (1) : 106 - 110, 1975. [In R.]

33016 - BUKHOV, N.G., KARAPETYAN, N.V. : Issledovanie aktseptornoĭ chasti fotosistemy I po temperaturnoĭ zavisimosti fotoprevrashcheniĭ *P*700. [Investigation
of the acceptor part of photosystem I by temperature dependence of *P*700 photoreactions.] - Mol. Biol. (Moskva) *12* : 868 - 878, 1978. [In R, ab : E.]

33017 - BULLEID, N.C. : An improved method for the extraction of adenosine triphosphate from marine sediment and seawater. - Limnol. Oceanogr. *23* : 174 - 178,
1978.

33018 - BUL'ON, V.V. : Soderzhanie feopigmentov v planktone. [Content of pheopigments
in plankton.] - Gidrobiol. Zh. *14* (3) : 62 - 70, 1978. [In R, ab : E.]

33019 - BUNCE, J.A. : Effects of shoot environment on apparent root resistance to
water flow in whole soybean and cotton plants. - J. exp. Bot. *29* : 595 - 601,
1978. [Growth analysis.]

33020 - BUNCE, J.A. : Effects of water stress on leaf expansion, net photosynthesis,
and vegetative growth of soybeans and cotton. - Can. J. Bot. *56* : 1492 -
1498, 1978.

33021 - BUNCE, J.A. : Interrelationships of diurnal expansion rates and carbohydrate
accumulation and movement in soya beans. - Ann. Bot. *42* : 1463 - 1466, 1978.

33022 - BURCH, G.J., JOHNS, G.G. : Root absorption of water and physiological responses to water deficits by *Festuca arundinacea* SCHREB. and *Trifolium repens*
L. - Aust. J. Plant Physiol. *5* : 859 - 871, 1978. [Growth analysis.]

33023 - BURINGH, P. : Limits to the productive capacity of the biosphere. - In :
ST-PIERRE, L.E., BROWN, G.R. (ed.) : Future Sources of Organic Raw Materials.
Chemrawn I. Pp. 325 - 332. Pergamon Press, Oxford 1978. [Ps.]

33024 - BURKE, J.J., DITTO, C.L., ARNTZEN, C.J. : Involvement of the light-harvesting
complex in cation regulation of excitation energy distribution in chloroplasts. - Arch. Biochem. Biophys. *187*: 252 - 263, 1978.

33025 - BURKE, J.J., STEINBACK, K.E., ARNTZEN, C.J. : Cation regulation of grana
stacking and excitation energy distribution in a chlorophyll *b*-less barley
mutant. - Plant Physiol. *61* (Suppl.) : 102, 1978.

33026 - BURKE, J.J., STEINBACK, K.E., OHAD, I., ARNTZEN, C.J. : Control of photosynthetic competence in the Y-1 mutant of *Chlamydomonas reinhardi*. - In :
AKOYUNOGLOU, G., ARGYROUDI-AKOYUNOGLOU, J.H. (ed.) : Chloroplast Development. Pp. 413 - 418. Elsevier/North-Holland Biomedical Press, Amsterdam -
New York - Oxford 1978.

*33027 - BURRIS, R.H. : Energetics of biological $N_2$ fixation. - In : MITSUI, A.,
MIYACHI, S., SAN PIETRO, A., TAMURA, S. (ed.) : Biological Solar Energy
Conversion. Pp. 275 - 289. Academic Press, New York - San Francisco - London
1977. [Ps.]

33028 - BUSBY, J.R., BLISS, L.C., HAMILTON, C.D. : Microclimate control of growth
rates and habitats of the boreal forest mosses, *Tomenthypnum nitens* and *Hylocomium splendens*. - Ecol. Monogr. *48* : 95 - 110, 1978.

33029 - BUSBY, J.R., WHITFIELD, D.W.A. : Water potential, water content, and net
assimilation of some boreal forest mosses. - Can. J. Bot. *56* : 1551 - 1558,
1978.

33030 - BUSCHMANN, C., MEIER, D., KLEUDGEN, H.K., LICHTENTHALER, H.K. : Regulation
of chloroplast development by red and blue light. - Photochem. Photobiol.
*27* : 195 - 198, 1978.

33031 - BUSCHMANN, C., SIRONVAL, C. : Influence of kinetin on protochlorophyll(ide)
accumulation and on the Shibata shift in *Raphanus* seedlings. - Planta *139* :
127 - 132, 1978.

33032 - BUSE, G. : Sequence homology of cytochrome oxidase subunits to electron car-
riers of photophosphorylation. - In : SCHÄFER, G., KLINGENBERG, M. (ed.) :
Energy Conservation in Biological Membranes. Pp. 53 - 55. Springer-Verlag,
Berlin - Heidelberg - New York 1978.

33033 - BUSH, T.F., ULABY, F.T. : An evaluation of radar as a crop classifier. - Remote Sensing Environm. *7* : 15 - 36, 1978.

33034 - BUTCHER, H. : Total glycoalkaloids and chlorophyll in potato cultivars bred
in New Zealand. - N. Zeal. J. exp. Agr. *6* : 127 - 130, 1978.

33035 - BUTLER, W.L. : Energy distribution in the photochemical apparatus of photosynthesis. - Annu. Rev. Plant Physiol. *29* : 345 - 378, 1978.

33036 - BUTLER, W.L. : On the role of cytochrome $b_{559}$ in oxygen evolution in photosynthesis. - FEBS Lett. *95* : 19 - 25, 1978.

33037 - BUTLER, W.L., STRASSER, R.J. : Effect of divalent cations on energy coupling
between the light-harvesting chlorophyll *a/b* complex and photosystem II. -
In : HALL, D.O., COOMBS, J., GOODWIN, T.W. (ed.) : Proceedings of the Fourth
International Congress on Photosynthesis. Pp. 11 - 20. Biochem. Soc., London
1978.

33038 - BYKOV, O.D. : Prostoĭ metod opredeleniya parametrov $^{12}CO_2$ i $^{14}CO_2$ fotosinteticheskogo gazoobmena pri nizkoĭ kontsentratsii uglekislogo gaza v usloviyakh zakrytoĭ sistemy. [A simple method for the determination of $^{12}CO_2$ and
$^{14}CO_2$ parameters in photosynthetic gas exchange at low $CO_2$ concentration
under a closed system conditions.] - Tr. priklad. Bot., Genet. Selektsii *61*
(3) : 3 - 12, 1978. [In R, ab : E.]

33039 - BYKOV, O.D., LEVIN, E.S. : Primenenie analogovykh vychislitel'nykh mashin
dlya modelirovaniya protsessov uglekislotnogo gazoobmena pri fotosinteze i
dykhanii. [Application of analogue computers for simulation of the processes
of $CO_2$ gas exchange in the course of photosynthesis and respiration.] - Tr.
priklad. Bot., Genet. Selektsii *61* (3) : 17 - 44, 1978. [In R, ab : E.]

33040 - BYKOV, O.D., LIMAR', R.S. : Ispol'zovanie ugleroda $^{14}C$ dlya izucheniya transporta assimilyatov v rasteniyakh pshenitsy. [Using $^{14}C$ for studies of photosynthate transport in wheat plants.] - Tr. priklad. Bot., Genet. Selektsii *61* (3) : 45 - 53, 1978. [In R, ab : E.]

33041 - BYSTRZEJEWSKA,G. : Wpływ deficytu fosforu na fotosyntetyczną funkcję roślin.
[Effect of phosphorus deficiency on plant photosynthetic functions.] - Postępy Nauk roln. *25* (5) : 43 - 54, 1978. [In Pol.]

33042 - BYTNIEWSKA, K. : Karboksylaza/oksygenaza rybulozodwufosforanu. [Ribulose-
bisphosphate carboxylase/oxygenase.] - Postępy Biochem. *24* : 333 - 346, 1978.
[In Pol.]


33043 - CABANETTES, A., RAPP, M. : Biomasse, minéralomasse et productivité d'un éco-
système à Pin pignons (*Pinus pinea* L_) du littoral méditerranéen. - Oecol.
Plant. *13* : 271 - 286, 1978.

33044 - CADÉE, G.C. : On the origin of organic matter accumulating on tidal flats
of Balgzand, Dutch Wadden Sea. - Hydrobiol. Bull. *12* : 145 - 150, 1978.
[Chl.]

33045 - CADÉE, G.C. : Primary production and chlorophyll in the Zaire river, estuary and plume. - Neth. J. Sea Res. *12* : 368 - 381, 1978.

33046 - CAHEN, D., GARTY, H., CAPLAN, S.R. : Spectroscopy and energetics of the purple membrane of *Halobacterium halobium*. A photoacoustic study. - FEBS Lett.
*91* : 131 - 134, 1978.

33047 - CAHEN, D., MALKIN, S., GUREVITZ, M., OHAD, I. : Development and repair of
photosystem II activity in normal and chloramphenicol-treated *Euglena gracilis* cells. - Plant Physiol. *62* : 1 - 5, 1978.

33048 - CAHEN, D., MALKIN, S., LERNER, E.I. : Photoacoustic spectroscopy of chloroplast membranes; listening to photosynthesis. - FEBS Lett. *91* : 339 - 342,
1978.

*33049 - CAIRNS, J. Jr. : Aquatic ecosystem assimilative capacity. - Fisheries 2 (2) :
5 - 7, 1977.

33050 - CALDWELL, M.M., JOHNSON, D.A., FAREED, M. : Constraints on tundra productivity : Photosynthetic capacity in relation to solar radiation utilization
and water stress in arctic and alpine tundras. - In : TIESZEN, L.L. (ed.) :
Vegetation and Production Ecology of an Alaskan Arctic Tundra. Pp. 323 - 342.
Springer-Verlag, New York - Heidelberg - Berlin 1978.

*33051 - CALDWELL, M.M., MOORE, R.T. : A portable small-stage photoelectric planimeter
for leaf area measurements. - J. Range Manage. *24* : 394 - 395, 1971.

33052 - CALLAGHAN, T.V., COLLINS, N.J., CALLAGHAN, C.H. : Photosynthesis, growth and
reproduction of *Hylocomium splendens* and *Polytrichum commune* in Swedish Lapland. Strategies of growth and population dynamics of tundra plants 4. -
Oikos *31* : 73 - 88, 1978.

*33053 - CALLIS, J., WALBOT, V. : Determination of the number of ribosomal cistrons
in chloroplasts of $C_3$ and $C_4$ plants. - In : BOGORAD, L., WEIL, J.H. (ed.) :
Acides Nucléiques et Synthèse des Protéines chez les Végétaux. Coll. CNRS
no. 261. Pp. 137 - 141. Édit. CNRS, Paris 1977.

33054 - CALVIN, G.J., CALVIN, M. : The only source of energy. - Lawrence Berkeley
Lab. Univ. California Rep. *LBL-7548* : 1 - 13, 1978. [Ps.]

33055 - CALVIN, M. : Petroleum plantations. - Lawrence Berkeley Lab. Univ. California
Rep. *LBL-8236* : 1 - 37, 1978. [Ps.]

33056 - CALVIN, M. : Simulating photosynthetic quantum conversion. - Accounts chem.
Res. *11* : 369 - 374, 1978.

33057 - CAMARA, B., MONÉGER, R. : Free and esterified carotenoids in green and red
fruits of *Capsicum annuum*. - Phytochemistry *17*: 91 - 93, 1978. [Chl.]

33058 - CAMM, E.L., GREEN, B.R. : Can isolated chloroplasts make the proteins of the
light-harvesting complex? - Plant Physiol. *61* (Suppl.) : 103, 1978.

33059 - CAMPBELL, D.C., KONDRA, Z.P. : Relationships among growth patterns, yield
components and yield of rapeseed. - Can. J. Plant Sci. *58* : 87 - 93, 1978.

33060 - CAMPBELL, D.E., WATTERS, G., CORSE, J.W. : Effect of relative humidity and
mineral nutrition on $CO_2$ exchange and senescence of lettuce and soybean. -
Plant Physiol. *61* (Suppl.) : 5, 1978.

33061 - CAMPBELL, P.J. : Primary productivity of a hypersaline Antarctic lake. -
Aust. J. mar. Freshwater Res. *29* : 717 - 724, 1978.

33062 - CAMPBELL, W.H., BLACK, C.C. : The relationship of $CO_2$ assimilation pathways
and photorespiration to the physiological quantum requirement of green plant
photosynthesis. - BioSystems *10* : 253-264, 1978.

33063 - CAMPILLO, A.J., SHAPIRO, S.L. : Picosecond fluorescence studies of exciton
migration and annihilation in photosynthetic systems. A review. - Photochem.
Photobiol. *28* : 975 - 989, 1978.

33064 - CANAANI, O.D., SAUER, K. : Absorption and circular dichroism spectra of chloroplast membrane fragments from spinach, barley and a barley mutant at room
temperature and liquid nitrogen temperature. - Biochim. biophys. Acta *501* :
545 - 551, 1978.

*B33065 - CANALE, R.P. (ed.) : Modeling Biochemical Processes in Aquatic Ecosystems. -
Ann Arbor Sci. Publ., Ann Arbor, Mich. 1976. [Ps, production.]

33066 - CANNELL,M.G.R., BRIDGWATER, F.E., GREENWOOD, M.S. : Seedling growth rates,
water stress responses and root-shoot relationships related to eight-year
volumes among families of *Pinus taeda* L. - Silvae Genet. *27* : 237 - 248,
1978.

33067 - CANNISTRARO, S., VAN DE VORST, A., JORI, G. : EPR studies on singlet oxygen
production by porphyrins. - Photochem. Photobiol. *28* : 257 - 259, 1978. [Chl
precursors.]

33068 - CANVIN, D.T. : Photorespiration and the effect of oxygen on photosynthesis.
- In : SIEGELMAN, H.W., HIND, G. (ed.) : Photosynthetic Carbon Assimilation.
Pp. 61 - 76. Plenum Press, New York - London 1978.

33069 - CANVIN, D.T., FOCK, H., LLOYD, N.D.H. : Environmental treatments and carbon
flow in the glycolate pathway. - In : HALL, D.O., COOMBS, J., GOODWIN, T.W.
(ed.) : Proceedings of the Fourth International Congress on Photosynthesis.
Pp. 323 - 334. Biochem. Soc., London 1978.

33070 - CAPBLANCQ, J., DAUTA, A. : Phytoplancton et production primaire de la rivi-
ère Lot. - Ann. Limnol. 14 : 85 - 112, 1978.

33071 - CAPBLANCQ, J., DÉCAMPS, H. : Dynamics of the phytoplankton in the river Lot.
- Verh. int. Verein. Limnol. 20 : 1479 - 1484, 1978. [Chl.]

33072 - CAPERON, J., SMITH, D.F. : Photosynthetic rates of marine algae as a function
of inorganic carbon concentration. - Limnol. Oceanogr. 23 : 704 - 708, 1978.

33073 - CAPIEL, M. : Effect of various meteorological indices on the yield and nutrient composition of napiergrass (Pennisetum purpureum L.). - J. Agr. Univ.
Puerto Rico 62 : 76 - 89, 1978.

33074 - CAPIEL, M., BRENES, E., LUGO-LÓPEZ, M.A., SCHOCH, P.G., GUZMAN, V.L. : An
evaluation of the growth and water consumption rate of grain sorghum ( Sorghum bicolor ) at four climatic sites in the tropics and subtropics. - J.
Agr. Univ. Puerto Rico 62 : 10 - 28, 1978.

33075 - CAPLAN, S.R., EISENBACH, M., GARTY, H. : Processes involved in light-induced
pH changes in bacteriorhodopsin-containing particles. - In : CAPLAN, S.R.,
GINZBURG, M. (ed.) : Energetics and Structure of Halophilic Microorganisms.
Pp. 49 - 66. Elsevier/North-Holland Biomedical Press, Amsterdam - New York
1978.

33076 - CAPPY, J.J., NOBLE, R.D. : Photosynthetic response of soybeans with genetically altered chlorophyll. - Ohio J. Sci. 78 : 267 - 271, 1978.

33077 - CARDE, J.-P. : Ultrastructural studies of Pinus pinaster needles : the endo-
dermis. - Amer. J. Bot. 65 : 1041 - 1054, 1978. [Chloroplast.]

33078 - CARMELI, C., LIFSHITZ, Y., GUTMAN, M. : Control of kinetic changes in ATPase
activity of soluble coupling factor I from chloroplasts. - FEBS Lett. 89 :
211 - 214, 1978.

33079 - CARMI, A., KOLLER, D. : Effects of the roots on the rate of photosynthesis
in primary leaves of bean (Phaseolus vulgaris L.). - Photosynthetica 12 :
178 - 184, 1978.

33080 - CAROLIN, R.C., JACOBS, S.W.L., VESK, M. : Kranz cells and mesophyll in the
Chenopodiales. - Aust. J. Bot. 26 : 683 - 698, 1978.

33081 - CARPENTER, E.J., McCARTHY, J.J. : Benthic nutrient regeneration and high
rate of primary production in continental shelf waters. - Nature 274 : 188 -
189, 1978.

*33082 - CARPENTER, W.J. : Photosynthetic supplementary lighting of spray pompon,
Chrysanthemum morifolium RAMAT. - J. amer. Soc. 101 : 155 - 158, 1976.

33083 - CARSON, J.L., BROWN, R.M. Jr. : Studies of Hawaiian freshwater and soil algae II. Algal colonization and succession on a dated volcanic substrate. -
J. Phycol. 14 : 171 - 178, 1978. [Chl.]

33084 - CARSTAIRS, A.G., OECHEL, W.C. Effects of several microclimatic factors and
nutrients on net carbon dioxide exchange in Cladonia alpestris (L.) RABH.
in the Subarctic. - Arctic alp. Res. 10 : 81 - 94, 1978.

33085 - CARVALHO, F.I.F. de, QUALSET, C.O. : Genetic variation for canopy architecture and its use in wheat breeding. - Crop Sci. 18 : 561 - 567, 1978.

33086 - CASADIO, R., BACCARINI MELANDRI, A., MELANDRI, B.A. : Limited cooperativity
in the coupling between electron flow and photosynthetic ATP synthesis. A
comparative study in chromatophores phosphorylating at very different rates.
- FEBS Lett. 87 : 323 - 328, 1978.

33087 - CASADORO, G., RASCIO, N. : Chloroplast ontogenesis in Helianthus annuus L. -
Protoplasma 97 : 165 - 172, 1978.

33088 - CASADORO, G., RASCIO, N. : Chloroplast ontogenesis in *Lippia citriodora* L. - In : Ninth International Congress on Electron Microscopy. Vol. 2. Pp. 416 - 417. Toronto 1978.

33089 - CASADORO, G., RASCIO, N. : Thylakoid membranes in sunflower and in other plants. - J. Ultrastruct. Res. *65* : 30 - 35, 1978.

33090 - CASHMORE, A.R., BROADHURST, M.K., GRAY, R.E. : Cell-free synthesis of leaf protein: Identification of an apparent precursor of the small subunit of ribulose-1,5-bisphosphate carboxylase. - Proc. nat. Acad. Sci. USA *75* : 655 - 659, 1978.

*33091 - CASPER, S.J., SCHÖNBORN, W. : Ansätze für eine Bilanzierung und Berechnung des Sauerstoffhaushaltes der mittleren Saale. - Limnologica (Berlin) *10* : 191 - 195, 1976. [Ps.]

33092 - CASSEL, D.K., BAUER, A., WHITED, D.A. : Management of irrigated soybeans on a moderately coarse-textured soil in the Upper Midwest. - Agron. J. *70* : 100 - 104, 1978. [Growth analysis.]

33093 - CASSELLS, A.C., BARLASS, M. : A method for the isolation of stable mesophyll protoplasts from tomato leaves throughout the year under standard conditions. - Physiol. Plant. *42* : 236 - 242, 1978.

33094 - CASTELFRANCO, P.A., SCHWARCZ, S. : Glutamate requirement for the synthesis of Mg protoporphyrin IX by isolated greening chloroplasts that cannot be replaced by $\delta$-aminolevulinic acid (ALA) (1). - Plant Physiol. *61* (Suppl.) : 82, 1978.

33095 - CASTELFRANCO, P.A., SCHWARCZ, S. : Mg-protoporphyrin-IX and $\delta$-aminolevulinic acid synthesis from glutamate in isolated greening chloroplasts. Mg-protoporphyrin-IX synthesis. - Arch. Biochem. Biophys. *186* : 365 - 375, 1978.

33096 - CASTRO, P.R.C. : Effect of salinity on cotton plants treated with chemicals. - Plant Physiol. *61* (Suppl.) : 95, 1978. [Ps.]

33097 - CĂTĂNESCU, V. : Relațiile dintre procesul de fotosinteză și recolta de porumb, în condițiile solului podzolic-exogleic de la Albota-Argeș. [Relationship between the photosynthesis process and corn yield under the conditions of the exogley podzolic soil at Albota-Argeș.] - An.Inst. Cercetări Cereale Plante tehnice Fundulea *43* : 449 - 461, 1978. [In Roum., ab : E, R.]

33098 - CATTOLICO, R.A. : Variation in plastid number. Effect on chloroplast and nuclear deoxyribonucleic acid complement in unicellular alga *Olisthodiscus luteus*. - Plant Physiol. *62* : 558 - 562, 1978.

33099 - CAZZULO, J.J. : Regulatory properties of enzymes from marine and extremely halophilic bacteria - malic enzyme and citrate synthase. - In : CAPLAN, S.R., GINZBURG, M. (ed.) : Energetics and Structure of Halophilic Microorganisms. Pp. 371 - 378. Elsevier/North-Holland Biomedical Press, Amsterdam - New York 1978. [Ps.]

33100 - CEDERBLAD, A.V., VASCONCELOS, A.C. : PSI reaction center protein-site of synthesis of constituent polypeptides. - Plant Physiol. *61* (Suppl.): 83, 1978.

33101 - CENCI, C.A. : Produttività, pabularità e produzione di seme in *Brachypodium pinnatum*. (L.) BEAUV. [Productivity, palatability and seed yield in *Brachypodium pinnatum* (L.) BEAUV.] - Genet. agr. *82* : 1 - 12, 1978. [In Ital., ab : E.]

33102 - CERFF, R. : Glyceraldehyde-3-phosphate dehydrogenase(NADP) from *Sinapis alba* L. NAD(P)-induced conformation changes of the enzyme. - Europe. J. Biochem. *82* : 45 - 53, 1978.

33103 - CERUTI, A. : L'agricoltura e la selvicoltura fissatrici dell'energia solare per l'uomo. [Agriculture and forestry: fixers of solar energy for humans.] - Ann. Accad. Agr. Torino *120* : 1 - 19, 1977/78. [In Ital., ab : E, F.]

33104 - ČERVENKA, K. : Orientierung der Blätter in der Apfelkrone und Absorption der Sonnenstrahlung. - Arch. Gartenbau *26* : 373 - 385, 1978.

33105 - CEULEMANS, R., IMPENS, I., LEMEUR, R., MOERMANS, R., SAMSUDDIN, Z. : Water movement in the soil-poplar-atmosphere system. 1. Comparative study of stomatal morphology and anatomy, and the influence of stomatal density and dimensions on the leaf diffusion characteristics in different poplar clones. - Oecol. Plant. *13* : 1 - 12, 1978.

33106 - CEULEMANS, R., IMPENS, I., LEMEUR, R., MOERMANS, R., SAMSUDDIN, Z. : Water movement in the soil-poplar-atmosphere system. II. Comparative study of the transpiration regulation during water stress situations in four different poplar clones. - Oecol. Plant. *13* : 139 - 146, 1978. [Resistances.]

33107 - CHABOT, B.F. : Environmental influences on photosynthesis and growth in *Fragaria vesca*. - New Phytol. *80* : 87 - 98, 1978.

33108 - CHAĬKA, M.Ts. : Biyasintêz khlarafilu ǔ suvyazi z biyagenezam plastydnykh membran. [Chlorophyll biosynthesis in relation to biogenesis of plastid membranes.] - Vestsi Akad. Navuk belarus. SSR, Ser. biyal. Navuk *1978* (6) : 44 - 51, 141, 1978. [In Belorus., ab : E, R.]

*33109 - CHAKRABARTI, S.N., SEN, S. : Segregation of induced chlorophyll mutants in rice. - Indian J. Genet.Plant Breed. *36* : 162 - 165, 1976.

33110 - CHAKRABORTY, S., SAHA, S. : Studies on the photosynthesis of some rice cultivars. - In : Inter-Disciplinary Symposium on Photosynthesis and Productivity. Abstract of Papers. P. 45. Indian nat. Sci. Acad., New Delhi 1978.

33111 - CHALMERS, D.J., WILSON, I.B. : Productivity of peach trees : tree growth and water stress in relation to fruit growth and assimilate demand. - Ann. Bot. *42* : 285 - 292, 1978.

*33112 - CHAMBROY, Y., FERRY, P., FLANZY, C. : Absorption de gaz carbonique par les baies de raisin. - Compt. rend. Acad. Agr. Fr. *56* : 1470 - 1474, 1970.

33113 - CHAMMAI, A., SCHANTZ, R. : Biosynthesis of phosphatidylglycerol in *Euglena gracilis*. - In : AKOYUNOGLOU, G., ARGYROUDI-AKOYUNOGLOU, J.H. (ed.) : Chloroplast Development. Pp. 297 - 302. Elsevier/North-Holland Biomedical Press, Amsterdam - New York - Oxford 1978. [Chloroplast.]

*33114 - CHAMOROVSKIĬ, S.K., ANDREENKO, T.I., PYT'EVA, N.F., RUBIN, A.B. : Vliyanie razobshchitelya fosforilirovaniya karboniltsianid-M-khlorfenilgidrazona na fotosinteticheskie êlektrontransportnye reaktsii u bakteriĭ *Ectothiorhodospira shaposhnikovii*. [Effect of uncoupler of photophosphorylation, carbonyl cyanide M-chlorophenyl hydrazone, on photosynthetic electron transport reactions in the bacteria *Ectothiorhodospira shaposhnikovii*.] - Nauch. Dokl. vyssh. Shkoly, biol. Nauki *19* (2) : 73 - 79, 1976. [In R.]

33115 - CHAMP, P. : Estimation de la production secondaire planctonique d'après les données sur la production primaire. - Cahiers ORSTOM, Sér. Hydrobiol. *12* : 173 - 175, 1978.

33116 - CHAMPIGNY, M.-L. : Adenine nucleotides and the control of photosynthetic activities. - In : HALL, D.O., COOMBS, J., GOODWIN, T.W. (ed.) : Proceedings of the Fourth International Congress on Photosynthesis. Pp. 479 - 488. Biochem. Soc., London 1978.

33117 - CHAMPIGNY, M.-L., JOYARD, J. : Analyse du mécanisme de stimulation, par l'ATP, de la pénétration du carbone inorganique dans les chloroplastes intacts, isolés de l'Épinard. - Compt. rend. Acad. Sci. Paris, Sér. D *286* : 1791 - 1794, 1978.

33118 - CHAN, A.T. : Comparative physiological study of marine diatoms and dinoflagellates in relations to irradiance and cell size. I. Growth under continuous light. - J. Phycol. *14* : 396 - 402, 1978. [Ps, Chl.]

33119 - CHANCE, B., WARING, A., SARONIO, C. : Low temperature electron transport in cytochrome *c* in the cytochrome *c*-cytochrome oxidase reaction : evidence for electron tunneling. - In : SCHÄFER, G., KLINGENBERG, M. (ed.) : Energy Conservation in Biological Membranes. Pp. 56 - 73. Springer-Verlag, Berlin - Heidelberg - New York 1978. [Ps.]

33120 - CHANG, J.C., DAS, T.P. : Theory of proton hyperfine interactions in bacterial photosynthetic systems. Bacteriopheophytin cation and anion. - Biochim. biophys. Acta *502* : 61 - 79, 1978.

33121 - CHAPMAN, A.R.O., MARKHAM, J.W., LÜNING, K. : Effects of nitrate concentration on the growth and physiology of *Laminaria saccharina (Phaeophyta)* in
culture. - J. Phycol. *14* : 195 - 198, 1978. [Ps, Chl.]

33122 - CHAPMAN, D.J., SIMPSON, E.E., WILLIAMS, J.P. : Extra-thylakoid synthesis of
galactolipids. - Plant Physiol. *61* (Suppl.) : 105, 1978.

33123 - CHARLES-EDWARDS, D.A. : An analysis of the photosynthesis and productivity
of vegetative crops in the United Kingdom. - Ann. Bot. *42* : 717 - 731, 1978.

33124 - CHARLES-EDWARDS, D.A. : Leaf carbon dioxide compensation points at high light
flux densities. - Ann. Bot. *42* : 733 - 739, 1978.

33125 - CHAROLAIS, N. : Influence du sel sur la photorespiration chez le cotonnier.
- Bull. Soc. bot. Fr. *125*(Actual. bot. 3/4): 199 - 214, 1978.

33126 - CHARPY-ROUBAUD, C.J., CHARPY, L.J., MAESTRINI, S.Y. : Étude de la production
primaire des eaux des golfes nord-patagoniques (Argentine). Détermination
des facteurs nutritionnels limitant la fertilité du phytoplancton du golfe
"San José". - Compt. rend. Acad. Sci. Paris, Sér. D *287* : 539 - 542, 1978.

33127 - CHARPY-ROUBAUD, C.J., CHARPY, L.J., MAESTRINI, S.Y., PIZARRO, M.J. : Étude
de la production primaire des eaux des golfes nord-patagoniques (Argentine).
Estimation de la fertilité potentionelle au moyen de tests biologiques. -
Compt. rend. Acad. Sci. Paris, Sér. D *287* : 1031 - 1034, 1978.

33128 - CHARTIER, P. : Cereals residues as biomass source to produce clean fuel and
related products. - In : HALL, D.O., COOMBS, J., GOODWIN, T.W. (ed.) : Proceedings of the Fourth Internatuonal Congress on Photosynthesis. Pp. 645 -
656. Biochem. Soc., London 1978.

33129 - CHATTOPADHYAY, N.C., NANDI, B. : Changes in total contents of saccharides,
proteins and chlorophyll in malformed mango inflorescence induced by *Fusarium moniliforme* var. *subglutinans*. - Biol. Plant. *20* : 468 - 471, 1978.

33130 - CHAZDON, R.L. : Ecological aspects of the distribution of $C_4$ grasses in selected habitats of Costa Rica. - Biotropica *10* : 265 - 269, 1978.

*33131 - CHEMERIS, Yu.K., UGOLKOVA, N.G., PAKHORUKOVA, L.V., VENEDIKTOV, P.S., SHEN-
DEROVA, L.V. : Ustanovka dlya issledovaniya zamedlennoǐ fluorestsentsii khlorofilla v kul'ture sinkhronno delyashchikhsya kletok vodorosleǐ. [A set up
for the analysis of delayed chlorophyll fluorescence in a culture of syn-
chronously dividing algae cells.] - Nauch. Dokl. vyssh. Shkoly, biol. Nauki
*20* (3) : 138 - 142, 1977. [In R.]

*33132 - CHEMIKOSOVA, S.B., PIGULEVSKAYA, T.K. : Vliyanie khlora na intensivnost'
fotosinteza pshenitsy. [Effect of chlorine on the rate of photosynthesis in
wheat.] - In : Rasteniya i Promyshlennaya Sreda. Pp. 138 - 139. Naukova Dumka, Kiev 1976. [In R.]

33133 - CHEN, C.-H., BERNS, D.S. : Comparison of the stability of phycocyanins from
thermophilic, mesophilic, psychrophilic and halophilic algae. - Biophys.
Chem. *8* : 203 - 213, 1978.

*33134 - CHEN, C.-H., KAO, O.H.W., BERNS, D.S. : Denaturation of phycocyanin by urea
and determination of the enthalpy of denaturation by microcalorimetry. -
Biophys. Chem. *7* : 81 - 86, 1977.

33135 - CHEN, K., UCHIMIYA, H., WILDMAN, S.G. : Segregation analysis of genes coding
for fraction 1 protein small subunit polypeptides. - Plant Physiol. *61*
(Suppl.) : 108, 1978.

*33136 - CHEN, K., WILDMAN, S.G., SMITH, H.H. : Chloroplast DNA distribution in para-
sexual hybrids as shown by polypeptide composition of fraction I protein. -
Proc. nat. Acad. Sci. USA *74* : 5109 - 5112, 1977.

33137 - CHENIAE, G.M., MARTIN, I.F. : Studies on the mechanism of Tris-induced inactivation of oxygen evolution. - Biochim. biophys. Acta *502*: 321 - 344,
1978.

33138 - CHEREZOV, S.N., GUSEV, N.A., STUPISHINA, E.A. : Spektroskopiya organell rastitel'noǐ kletki v infrakrasnoǐ oblasti. [Infra-red spectroscopy of plant
cell organelles.] - In : Vodnyǐ Rezhim Rastenǐǐ v Svyazi s Raznymi Ékologi-

cheskimi Usloviyami. Pp. 363 - 369. Izdatel'stvo Kazanskogo Universiteta, Kazan' 1978. [In R.]

*33139 - **CHERNAVSKAYA, N.M., CHERNAVSKII, D.S.** : On the role of membranes for electron tunelling. - Studia biophys. *35* : 149 - 153, 1973. [Chl, Car.]

*B33140 - **CHERNAVSKAYA, N.M., CHERNAVSKIĬ, D.S.** : Tunel'nyĭ Transport Ėlektronov v Fotosinteze. [Tunnel Transport of Electrons in Photosynthesis.] - Izd. mosk. Univ., Moskva 1977. [In R.]

*33141 - **CHERNAVSKAYA, N.M., CHERNAVSKY, D.S., GÜNTHER, K., HACHE, A., KILYACHKOV, A. A.** : On the mechanism of energy transformation in photosynthesis. - Stud. biophys. *62* : 109 - 126, 1977.

*33142 - **CHERNAVSKY, D.S., CHERNAVSKAYA, N.M., GÜNTHER, K., HACHE, A.** : A model of the active transport of protons in photosynthesis. - Stud. biophys. *49* : 91 - 109, 1975.

*33143 - **CHERNISHEVA, S., MUSTÁRDY, L.A., GARAB, G.I., FALUDI-DÁNIEL, Á.** : The effect of sodium ions on the low temperature fluorescence spectra and ultrastructure of the granum membrane. - Acta phys. chem. *23* : 129 - 133, 1977.

33144 - **CHERNOKOLEV, A.T., KUKUSHKIN, A.K., SOLNTSEV, M.K.** : Vliyanie okislitel'no--vosstanovitel'nykh svoĭstv sredy na termolyuminestsentsiyu khloroplastov vysshikh rasteniĭ. [Effect of the redox properties of the medium on the thermoluminescence of chloroplasts from higher plants.] - Biofizika *23* : 554 - 556, 1978. [In R, ab : E.]

33145 - **CHERNOMORSKIĬ, S.A., FRAGINA, A.I.** : Vliyanie proizvodnykh khlorofilla na aktivnost' nekotorykh pishchevaritel'nykh fermentov *in vitro*. [Effect of chlorophyll derivatives on the activity of some digestive enzymes *in vitro*.] - Nauch. Dokl. vyssh. Shkoly, biol. Nauki *21* (3) : 29 - 31, 1978. [In R.]

*33146 - **CHERNOV, I.A., KRAĬNOVA, N.N., PEKH, S.M., YUDANOVA, G.A.** : Obespechenie dvuokis'yu ugleroda fotosinteticheskogo apparata kletok khlorelly pri fotosinteze. [Carbon dioxide requirement of the photosynthetic apparatus of *Chlorella* cells during photosynthesis.] - Nauch. Dokl. vyssh. Shkoly, biol. Nauki *20* (9) : 109 - 114, 1977. [In R.]

33147 - **CHERNOVA, S.I., ZVALINSKIĬ, V.I.** : Deĭstvie sveta vysokoĭ intensivnosti na sostoyanie khlorofilla v morskikh vodoroslyakh. [Effect of high irradiance on chlorophyll state in marine algae.] - Sb. Rabot Akad. Nauk SSSR, dal'nevost. nauch. Tsentr, Inst. Biol. Morya (Vladivostok) *11* (Ėkologicheskie Aspekty Fotosinteza Morskikh Makrovodorosleĭ) : 167 - 174, 187, 1978. [In R, ab : E.]

33148 - **CHERNYAD'EV, I.I., TEREKHOVA, I.V., AL'BITSKAYA, O.N., GORONKOVA, O.I., DOMAN, N.G.** : Osveshchennost' kak faktor dinamicheskoĭ regulyatsii fotosinteticheskogo metabolizma ugleroda u spiruliny. [Illumination as a factor of the dynamic regulation of photosynthetic carbon metabolism in *Spirulina*.] - Fiziol. Rast. *25* : 815 - 820, 1978. [In R, ab : E.]

33149 - **CHERNYAD'EV, I.I., ZIMINA, T.A., DOMAN, N.G.** : Intensifikatsiya rosta rasteniĭ i fotosintez list'ev gorokha. [Intensification of plant growth and photosynthesis of pea leaves.] - Dokl. Akad. Nauk SSSR *238* : 748 - 750, 1978. [In R.]

*33150 - **CHESNOKOV, V.A., MIROSLAVOVA, S.A., MOSHKANOVA, V.V.** : Vliyanie predvaritel'noĭ teplovoĭ obrabotki na gazoobmen list'ev rasteniĭ s C-3 i C-4 putem fotosinteza. [Effect of preliminary heating on gas exchange of plant leaves with $C_3$ and $C_4$ pathway of photosynthesis.] - Vest. leningrad. Univ., Ser. biol. *1974* (9) : 108 - 115, 1974. [In R, ab : E.]

33151 - **CHICHEV, P.** : Vliyanie na ekstremni polozhitelni temperaturi v"rkhu fotosintezata. II. Skorost na $CO_2$ fiksatsiyata i razpredelenie na $^{14}C$ v zavisimost ot silata i prod"lzhitelnostta na toplinnoto v"zdeĭstvie. [Effect of extreme positive temperatures on photosynthesis. II. Rate of $CO_2$ fixation and $^{14}C$ distribution in relation to the intensity and duration of heat treatment.] - Fiziol. Rast. (Sofia) *4* (2) : 36 - 43, 1978. [In Bulg., ab : E, R.]

*33152 - **CHKHOMELIDZE, O.I., TSISKARISHVILI, L.P.** : O transformatsii solnechnoĭ ėnergii v nekotorykh vodoemakh Gruzii. [Solar energy transformation in some water reservoirs in Georgia.] - Izv. Akad. Nauk gruz. SSR, Ser. biol. *1* (1) : 57 - 62, 1975. [Ps; In R, ab : E, Georg.]

33153 - CHOLLET, R. : Evaluation of the light/dark $^{14}$C assay of photorespiration. Tobacco leaf disk studies with glycidate and glyoxylate. - Plant Physiol. *61* :
929 - 932, 1978.

33154 - CHOLLET, R. : Cyanate modification of essential lysyl residues in the catalytic subunit of tobacco RuBP carboxylase. - Plant Physiol. *61* (Suppl.) :
108, 1978.

33155 - CHOLLET, R. : Inactivation of tobacco ribulosebisphosphate carboxylase by
2,3-butanedione. - Biochem. biophys. Res. Commun. *83* : 1267 - 1274, 1978.

*33156 - CHOLLET, R., ANDERSON, L.L. : Regulation of ribulose 1,5-bisphosphate car-
boxylase-oxygenase activities by temperature pretreatment and chloroplast
metabolites. - Arch. Biochem. Biophys. *176* : 344 - 351, 1976.

33157 - CHOLLET, R., ANDERSON, L.L. : Cyanate modification of essential lysyl residues in the catalytic subunit of tobacco ribulosebisphosphate carboxylase. -
Biochim. biophys. Acta *525* : 455 - 467, 1978.

33158 - CHONAN, N. : A comparative anatomy of mesophyll among the leaves of gramineous crops. - Jap. agr. Res. quart. *12* : 128 - 131, 1978.

33159 - CHOW, P.N.P., LaBERGE, D.E. : Wild oat herbicide studies. 2. Physiological
and chemical changes in barley and wild oats treated with diclofop-methyl
herbicide in relation to plant tolerance. - Agr. Food Chem. *26* : 1134 - 1137,
1978. [Ps, Chl.]

33160 - CHOW, W.S., THORNE, S.W., BOARDMAN, N.K. : Formation of the proton gradient
across the chloroplast thylakoid membrane in relation to ATP synthesis. -
In : DUTTON, P.L., LEIGH, J.S., SCARPA, A. (ed.) : Frontiers of Biological
Energetics : Electrons to Tissues. Vol.1. Pp. 287 - 296. Academic Press,
New York - San Francisco - London 1978.

33161 - CHOW, W.S., THORNE, S.W., BOARDMAN, N.K. : The movement of protons during
energy transduction in chloroplast thylakoid membrane. - In : DEAMER, D.W.
(ed.) : Light Transducing Membranes. Structure, Function, Evolution. Pp.
253 - 268. Academic Press, New York - San Francisco - London 1978.

33162 - CHOWDHURY, S.I., WARDLAW, I.F. : The effect of temperature on kernel development in cereals.-Aust. J. agr. Res. *29* : 205 - 223, 1978. [Ps.]

33163 - CHRÉTIENNOT-DINET, M.J., VACELET, E. : Séparation fractionnée du phytoplancton et estimation de l'assimilation autotrophe et hétérotrophe de $^{14}$C. -
Oceanol. Acta *1* : 407 - 413, 1978.

33164 - CHRIST, R.A. : The elongation rate of wheat leaves I. Elongation rates during day and night. - J. exp. Bot. *29* : 603 - 610, 1978.

33165 - CHRIST, R.A. : The elongation rate of wheat leaves II. Effect of sudden
light change on the elongation rate. - J. exp. Bot. *29* : 611 - 618, 1978.

33166 - CHRISTENSEN, B. : Biomass and primary production of *Rhizophora apiculata* BL.
in a mangrove in southern Thailand. - Aquat. Bot. *4* : 43 - 52, 1978.

33167 - CHRISTIE, E.K. : Ecosystem processes in semiarid grasslands. I Primary production and water use of two communities possessing different photosynthetic pathways. - Aust. J. agr. Res. *29* : 773 - 787, 1978.

33168 - CHRISTY, A.L., FISHER, D.B. : Kinetics of $^{14}$C-photosynthate translocation
in morning glory vines. - Plant Physiol. *61* : 283 - 290, 1978.

33169 - CHRŌST, R.J. : The estimation of extracellular release by phytoplankton and
heterotropic activity of aquatic bacteria. - Acta microbiol. pol. *27* : 139 -
146, 1978. [Photosynthates.]

*33170 - CHUA, N.-H., GILLHAM, N.W. : The sites of synthesis of the principal thyla-
koid membrane polypeptides in *Chlamydomonas reinhardtii*. - J. Cell Biol.
*74* : 441 - 452, 1977.

33171 - CHUA, N. H., SCHMIDT, G.W. : *In vitro* synthesis, transport, and assembly of
ribulose 1,5-bisphosphate carboxylase subunits. - In : SIEGELMAN, H.W.,
HIND, G. (ed.) : Photosynthetic Carbon Assimilation. Pp. 325 - 347. Plenum
Press, New York - London 1978.

33172 - CHUA, N.-H., SCHMIDT, G.W. : Post-translational transport into intact chlo-
        roplasts of a precursor to the small subunit of ribulose-1,5-bisphosphate
        carboxylase. - Proc. nat. Acad. Sci. USA 75 : 6110 - 6114, 1978.

*33173 - CHUCHALIN, A.I. : O radiatsionnom rezhime tsenoza v usloviyakh svetokul'tury.
        [The radiation regime of a coenosis under conditions of light cultivation.] -
        In : Intensivnaya Svetokul'tura Rasteniĭ. Pp. 118 - 133. Inst. Fiz. Sib. Otd.
        Akad. Nauk SSSR, Krasnoyarsk 1977. [Chl, Car; in R.]

*33174 - CHUCHALIN, A.I., EROSHIN, N.S., TIKHOMIROV, A.A., ANISTRATOVA, N.A., SHUR,
        L.A., SHILENKO, M.P. : Soderzhanie pigmentov i opticheskie svoĭstva list'ev
        pshenitsy v usloviyakh intensivnoĭ svetokul'tury rasteniĭ. [Pigment concen-
        tration and optical properties of leaves of wheat grown in intense light.] -
        Izv. sib. Otd. Akad. Nauk SSSR, Ser. biol. Nauk 1977 (2) : 38 - 43, 1977.
        [In R, ab : E.]

*33175 - CHUCHALIN, A.I., SID'KO, F.Ya. : O ratsional'nom ispol'zovanii FAR tsenozom
        pshenitsy v usloviyakh svetokul'tury. [The economic use of PhAR by a wheat
        coenose under conditions of light cultivations.] - In : Intensivnaya Sveto-
        kul'tura Rasteniĭ. Pp. 100 - 117. Inst. Fiz. Sib. Otd. Akad. Nauk SSSR, Kras-
        noyarsk 1977. [In R.]

33176 - CHUNG, H.-H., TRLICA, M.J. : Combustion and nonaqueous titration determinat-
        ion of carbon in plant tissues. - Anal. Biochem. 88 : 123 - 127, 1978.

33177 - CHUPAKHINA, G.N., OKUNTSOV, M.M., ARKHIPOVA, N.D. : Biosintez askorbinovoĭ,
        degidroaskorbinovoĭ i diketogulonovoĭ kislot list'yami yachmenya pri ingibi-
        rovanii biosinteza pigmentov streptomitsinom. [Biosynthesis of ascorbic,
        dehydroascorbic, and diketoglutaric acids in barley leaves under inhibition
        of pigment biosynthesis with streptomycin.] - Fiziol. Rast. 25 : 1179 -
        1184, 1978. [In R, ab : E.]

*33178 - CHUPAKHINA, K.G. : Karotin v lugovykh travakh. [Carotene in meadows grasses.] -
        In : Materialy 18 Nauchnoĭ Konferentsii Blagoveshchenskogo Sel'skokhozyaĭstven-
        nogo Instituta, Sekts. Agron. Pp. 88 - 90. Blagoveshchensk 1970. [In R.]

*33179 - CHUPAKHINA, K.G., STUKOVA, L.I., KORCHAGINA, L.D. : Karotin v lugovykh tra-
        vakh. [Carotene in meadow grasses.] - Tr. blagoveshchensk. sel'skokhoz. Inst.
        6 (2) : 148 - 150, 1971. [In R.]

33180 - CIFERRI, O. : The chloroplast DNA mystery. - Trends biochem. Sci. 3 : 256 -
        258, 1978.

33181 - CIHA, A.J., BRUN, W.A. : Effect of pod removal on nonstructural carbohydra-
        te concentration in soybean tissue. - Crop Sci. 18 : 773 - 776, 1978. [Pho-
        tosynthate and dry-matter accumulation.]

*33182 - CLAAS, M. : Dünnschichtchromatographische Auftrennung von Blattfarbstoffen. -
        Apothekerpraktikant pharmazeutisch-tech. Assistent 17 : 81 - 86, 1971.

33183 - CLARK, J.R., MESSENGER, D.I., DICKSON, K.L., CAIRNS, J. Jr. : Extraction of
        ATP from Aufwuchs communities. - Limnol. Oceanogr. 23 : 1055 - 1059, 1978.

33184 - CLARKE, J.M., SIMPSON, G.M. : Growth analysis of Brassica napus cv. Tower. -
        Can. J. Plant Sci. 58 : 587 - 595, 1978.

33185 - CLARKE, R.H., CONNORS, R.E., FRANK, H.A. : Investigation of the structure
        of the reaction center in photosynthetic bacteria by optical detection of
        triplet state magnetic resonance. - In : FIALA, J. (ed.) : Third Internatio-
        nal Seminar on Energy Transfer in Condensed Matter. Pp. 67 - 71. Univ. Karlo-
        va, Praha 1978.

33186 - CLAYTON, B.J., CLAYTON, R.K. : Properties of photochemical reaction centers
        purified from Rhodopseudomonas gelatinosa. - Biochim. biophys. Acta 501 :
        470 - 477, 1978.

33187 - CLAYTON, R.K. : Effects of dehydration on reaction centers from Rhodopseudo-
        monas sphaeroides. - Biochim. biophys. Acta 504 : 255 - 264, 1978.

33188 - CLAYTON, R.K. : Physicochemical mechanisms in reaction centers of photosyn-
        thetic bacteria. - In : CLAYTON, R.K., SISTROM, W.R. (ed.) : The Photosyn-
        thetic Bacteria. Pp. 387 - 396. Plenum, New York - London 1978.

33189 - CLAYTON, R.K., CLAYTON, B.J. : Molar extinction coefficients and other properties of an improved reaction center preparation from *Rhodopseudomonas viridis*. - Biochim. biophys. Acta *501* : 478 - 487, 1978.

33190 - CLAYTON, R.K., RAFFERTY, C.N., CLAYTON, B.J., BAROUCH, Y. : New data on composition, structure and electron transfer in photosynthetic membranes and
reaction centers. - In : HALL, D.O., COOMBS, J., GOODWIN, T.W. (ed.) : Proceedings of the Fourth International Congress on Photosynthesis. Pp. 45 - 54.
Biochem. Soc., London 1978.

B33191 - CLAYTON, R.K., SISTROM, W.R. (ed.) : The Photosynthetic Bacteria. - Plenum
Press, New York - London 1978.

33192 - CLEGG, M.D., SULLIVAN, C.Y., EASTIN, J.D. : A sensitive technique for the
rapid measurement of carbon dioxide concentrations. - Plant Physiol. *62* :
924 - 926, 1978.

33193 - CLEMENT, C.R., HOPPER, M.J., JONES, L.H.P., LEAFE, E.L. : The uptake of nitrate by *Lolium perenne* from flowing nutrient solution. II. Effect of light,
defoliation, and relationship to $CO_2$ flux. - J. exp. Bot. *29* : 1173 - 1183,
1978. [Ps.]

33194 - CLEZY, P.S., FOOKES, C.J.R. : Chemistry of pyrollic compounds. XLI. The synthesis of the phaeophorbides $c_1$ and $c_2$ methyl esters. - Aust. J. Chem. *31* :
2491 - 2504, 1978.

33195 - CLOUGH, J.M., ALBERTE, R.S., TEERI, J.A. : Adaptive response of *Solanum dul-
camara* L. to a range of environmental conditions. - Plant Physiol. *61* (Suppl.)
: 5, 1978. [Ps.]

33196 - CLUTTERBUCK, B.J., SIMPSON, K. : The interactions of water and fertilizer
nitrogen in effects on growth pattern and yield of potatoes. - J. agr. Sci.
*91* : 161 - 172, 1978.

33197 - COBB, A.H. : Inorganic polyphosphate involved in the symbiosis between chloroplasts of alga *Codium fragile* and mollusc *Elysia viridis*. - Nature *272* :
554 - 555, 1978.

33198 - COBB, A.H., ROTT, J. : The carbon fixation characteristics of isolated *Codium fragile* chloroplasts. Chloroplast intactness, the effect of photosynthetic carbon reduction cycle intermediates and the regulation of $RuBP^+$ carboxylase *in vitro*. - New Phytol. *81* : 527 - 541, 1978.

33199 - COCHRANE, L.A., FORD, E.D. : Growth of a Sitka spruce plantation : analysis
and stochastic description of the development of the branching structure. -
J. appl. Ecol. *15* : 227 - 244, 1978. [Canopy architecture, simulation.]

33200 - CODD, G.A., COSSAR, J.D. : The site of inhibition of photosystem II by
3-(3,4-dichlorophenyl)-N-N'-dimethylurea in thylakoids of the cyanobacterium
*Anabaena cylindrica*. - Biochem. biophys. Res. Commun. *83* : 342 - 346, 1978.

33201 - CODD, G.A., SALLAL, A.-K.J. : Glycollate oxidation by thylakoids of the cyanobacteria *Anabaena cylindrica, Nostoc muscorum* and *Chlorogloea fritschii*.
- Planta *139* : 177 - 181, 1978.

33202 - COEN, D.M., BEDBROOK, J.R., LINK, G., GREBANIER, A., STEINBACK, K., BEATON,
A., RICH, A., BOGORAD, L. : Genes and mRNA's for maize chloroplast proteins:
Changes. - In : AKOYUNOGLOU, G., ARGYROUDI-AKOYUNOGLOU, J.H. (ed.) : Chloro-
plast Development. Pp. 553 - 558. Elsevier/North-Holland Biomedical Press,
Amsterdam - New York - Oxford 1978.

33203 - COGDELL, R.J. : Carotenoids in photosynthesis. - Phil. Trans. roy. Soc. London B *284* : 569 - 579, 1978.

33204 - COGDELL, R.J., CROFTS, A.R. : Analysis of the pigment content of an antenna
pigment-protein complex from three strains of *Rhodopseudomonas sphaeroides*.
- Biochim. biophys. Acta *502* : 409 - 416, 1978.

33205 - COHEN, C.E., REBEIZ, C.A. : Chloroplast biogenesis. XXII. Contribution of
short wavelength and long wavelength protochlorophyll species to the greening of higher plants. - Plant Physiol. *61* : 824 - 829, 1978.

33206 - **COHEN, W.S.** : Bicarbonate and the conformational state of the chloroplast coupling factor. - Plant Physiol. *61* (Suppl.) : 102, 1978.

33207 - **COHEN, W.S.** : The coupling of electron flow to ATP synthesis in pea chloroplasts stored in the presence of glycerol at -70° C. - Plant Sci. Lett. *11* : 191 - 197, 1978.

*33208 - **COJENEANU, N., PISICĂ-DONOSE, A.** : Variaţia diurnă şi sezonieră a fotosintezei la frunzele soiului de viţă de vie Aligoté altoit pe berlandieri × riparia Kober 5 BB. [Diurnal and seasonal variation in photosynthesis of leaves of the vine cultivar Aligoté, grafted on berlandieri × riparia Kober 5 BB.] - Lucrări şti. Inst. Agron. "Nicolae Balcescu", Horticult. *1972* : 229 - 236, 1972. [In Roum., ab : E, F, R.]

33209 - **COLDEA, G., PLAMADĂ, E.** : Ecosystem processes in a stand of *Pinus mugo* TURRA. I. Standing crop, dry matter production and litter fall. - Flora *167* : 249 - 255, 1978.

33210 - **COLE, J., FISHER, S.G.** : Annual metabolism of a temporary pond ecosystem. - Amer. Midland Naturalist *100* : 15 - 22, 1978. [Seasonal course of primary production.]

33211 - **COLE, K., SHEATH, R.G.** : Ultrastructural observation of spermatial differentiation in *Bangia (Rhodophyta)*. - J. Phycol. *14* (Suppl.) : 29, 1978. [Chloroplast.]

33212 - **COLLATZ, G.J.** : The interaction between steady state photosynthesis and [RuP$_2$] as a function of irradiance, $CO_2$ and $O_2$. - Plant Physiol. *61* (Suppl.) : 109, 1978.

33213 - **COLLATZ, G.J.** : The interaction between photosynthesis and ribulose-P$_2$ concentration - effect of light, $CO_2$, and $O_2$. - Carnegie Inst. Year Book *77* : 248 - 251, 1978.

33214 - **COLLATZ, G.J., BADGER, M.R.** : The use of isolated whole leaf cells for photosynthesis studies. - Carnegie Inst. Year Book *77* : 245 - 247, 1978.

33215 - **COLLINS, C.R., FARRAR, J.F.** : Structural resistances to mass transfer in the lichen *Xanthoria parietina*. - New Phytol. *81* : 71 - 83, 1978. [Ps.]

33216 - **COLLINS, D., WEAVER, T.** : Effects of summer weather modification (irrigation) in *Festuca idahoensis* - *Agropyron spicatum* grasslands. - J. Range Manage. *31* : 264 - 269, 1978. [Dry-matter production.]

33217 - **COLMAN, B.** : Exretion of glycolate by a species of *Chlorella (Chlorophyceae)*. - J. Phycol. *14* : 434 - 437, 1978.

33218 - **COLMAN, B., COLEMAN, J.R.** : Inhibition of photosynthetic $CO_2$ fixation in blue-green algae by malonate. - Plant Sci. Lett. *12* : 101 - 105, 1978.

33219 - **COLMAN, B., MAWSON, B.T.** : The role of plasmolysis in the isolation of photosynthetically active leaf mesophyll cells. - Z. Pflanzenphysiol. *86* : 331 - 338, 1978.

33220 - **COLMAN, P.M., FREEMAN, H.C., GUSS, J.M., MURATA, M., NORRIS, V.A., RAMSHAW, J.A.M., VENKATAPPA, M.P.** : The structure of plastocyanin determined by X-ray diffraction at 2.7 Å resolution. - In : HALL, D.O., COOMBS, J., GOODWIN, T.W. (ed.) : Proceedings of the Fourth International Congress on Photosynthesis. Pp. 810 - 813. Biochem. Soc., London 1978.

33221 - **CONCIN, R., BINDER, H., BRUNNER, P., BOBLETER, O.** : Wachstumskammer für die Anzucht von verholzenden Pflanzen unter radioaktiver Kohlendioxidatmosphäre. Growth chamber for the cultivation of ligneous plants in a radioactive carbon dioxide atmosphere. - Kerntechnik *20* : 32 - 38, 1978.

33222 - **CONSTABLE, G.A., HEARN, A.B.** : Agronomic and physiological responses of soybean and sorghum crops to water deficits. I. Growth, development and yield. - Aust. J. Plant Physiol. *5* : 159 - 167, 1978.

33223 - **CONSTANTINI, A., IANNONE, R.** : Contributo allo studio del bilancio energetico delle piante nell'ambiente climatico di Roma. Nota III. Uso del modello resistivo per il calcolo del calore sensibile di scambio in un blocco di vegetazione. [Energetic balance of plants in the climatic environment of Rome.

Note III. A resistance model to calculate the sensible heat exchange in a vegetation block.] - Ann. Ist. sperim. Nutr. Piante *8*(3) : 1 - 8, 1978. [In Ital., ab : E, F, G.]

33224 - CONWAY, H.L. : Sorption of arsenic and cadmium and their effects on growth, micronutrient utilization, and photosynthetic pigment composition of *Asterionella formosa*. - J. Fish. Res. Board Can. *35* : 286 - 294, 1978.

*33225 - COOK, J.R., HAGGARD, S.S., HARRIS, P. : Cyclic changes in chloroplast structure in synchronized *Euglena gracilis*. - J. Protozool. *23* : 368 - 373, 1976.

33226 - COOK, M.G., EVANS, L.T. : Effect of relative size and distance of competing sinks on the distribution of photosynthetic assimilates in wheat. - Aust. J. Plant Physiol. *5* : 495 - 509, 1978.

*33227 - COOKE, R.J. : Mevalonate-activating enzymes in wheat etioplasts. - New Phytol. *78* : 91 - 94, 1977.

33228 - COOKINGHAM, R.E., LEWIS, A., LEMLEY, A.T. : A vibrational analysis of rhodopsin and bacteriorhodopsin chromophore analogues : Resonance Raman and infrared spectroscopy of chemically modified retinals and Schiff bases. - Biochemistry *17* : 4699 - 4711, 1978.

*33229 - CORLEY, R.H.V., GRAY, B.S., NG, S.K. : Productivity of the oil palm (*Elaeis guineensis* JACQ.) in Malaysia. - Exp. Agr. *7* : 129 - 136, 1971.

*33230 - CORNELIA, D., ŞTIRBAN, M. : Annual dynamics of assimilatory pigments in *Viscum album* L. and in its host plant, *Populus tremula* L. - Contrib. bot. Gradina bot. Univ. Babeş-Bolyai Cluj *1976* : 243 - 249, 1976.

33231 - CORNIC, G. : La photorespiration se déroulant dans un air sans $CO_2$ a-t-elle une fonction? - Can. J. Bot. *56* : 2128 - 2137, 1978.

33232 - CORRALL, A.J., FENLON, J.S. : A comparative method for describing the seasonal distribution of production from grasses. - J. agr. Sci. *91* : 61 - 67, 1978.

33233 - COSNER, J.C. : Phycobilisomes in spheroplasts of *Anacystis nidulans*. - J. Bacteriol. *135* : 1137 - 1140, 1978.

33234 - COSNER, J.C., TROXLER, R.F. : Phycobiliprotein synthesis in protoplasts of the unicellular cyanophyte, *Anacystis nidulans*. - Biochim. biophys. Acta *519* : 474 - 488, 1978.

*33235 - COSSINS, E.A., LOR, K.L. : Interrelationships between C-1 metabolism and photosynthesis in *Euglena gracilis*. - In : PFLEIDERER, W. (ed.) : Chemistry and Biology of Pteridines. Pp. 321 - 328. W. de Gruyter, Berlin - New York 1975.

B33236 - COSTES, C. (ed.) : Photosynthèse et Production Végétale. 2nd Ed. - Gauthier--Villars, Paris 1978.

33237 - COSTES, C., BAZIER, R., BALTSCHEFFSKY, H., HALLBERG, C. : Mild extraction of lipids and pigments from *Rhodospirillum rubrum* chromatophores. - Plant Sci. Lett. *12* : 241 - 249, 1978.

33238 - CÔTÉ, R., LACROIX, G. : Variabilité à court terme des propriétés physiques, chimiques et biologiques du Saguenay, fjord subarctique du Québec (Canada). - Int. Rev. ges. Hydrobiol. *63* : 25 - 39, 1978. [Chi.]

33239 - CÔTÉ, R., LACROIX, G. : Capacité photosynthétique du phytoplancton de la couche aphotique dans le fjord du Saguenay. - Int. Rev. ges. Hydrobiol. *63* : 233 - 246, 1978.

33240 - COTTON, T.M., LOACH, P.A., KATZ, J.J., BALLSCHMITER, K. : Studies of chlorophyll-chlorophyll and chlorophyll-ligand interactions by visible absorption and infrared spectroscopy at low temperatures. - Photochem. Photobiol. *27* : 735 - 749, 1978.

33241 - COTTON, T.M., VAN DUYNE, R.P. : Resonance Raman spectroelectrochemistry of bacteriochlorophyll and bacteriochlorophyll cation radical. - Biochem. biophys. Res. Commun. *82* : 424 - 433, 1978.

33242 - COWLING, D.W., KOZIOL, M.J. : Growth of ryegrass (*Lolium perenne* L.) exposed to $SO_2$. I. Effects on photosynthesis and respiration. - J. exp. Bot. *29* : 1029 - 1036, 1978.

33243 - COWPER, S.W. : The drift algae community of seagrass beds in Redfish Bay, Texas. - Contrib. mar. Sci. *21* : 125 - 132, 1978. [Primary productivity.]

33244 - COYNE, P.I., BINGHAM, G.E. : Photosynthesis and stomatal light responses in snap beans exposed to hydrogen sulfide and ozone. - J. Air Pollut. Cont. Assoc. *28* : 1119 - 1123, 1978.

33245 - COYNE, P.I., KELLEY, J.J. : Meteorological assessment of $CO_2$ exchange over an Alaskan arctic tundra. - In : TIESZEN, L.L. (ed.) : Vegetation and Production Ecology of an Alaskan Arctic Tundra. Pp. 299 - 321. Springer Verlag, New York - Heidelberg - Berlin 1978.

33246 - CRĂCIUN, C., MARTON, A., PĚTERFI, Ş. : The culture of some filamentous green algae in different conditions of light and nutritive medium. III. Ultrastructural peculiarities of the algae *Ulothrix variabilis* and *Stigeoclonium subsecundum*. - Rev. roum. Biol., Sér. Biol. vég. *23* : 17 - 22, 1978.

33247 - CRESTI, M., CIAMPOLINI, F., PACINI, E., SARFATTI, G. : Phytoferritin in plastids of the style of *Olea europaea* L. - Acta bot. neerl. *27* : 417 - 423, 1978.

33248 - CRITCHLEY, C., SMILLIE, R.M., PATTERSON, B.D. : Effect of temperature on photoreductive activity of chloroplasts from passionfruit species of different chilling sensitivity. - Aust. J. Plant Physiol. *5* : 443 - 448, 1978.

33249 - CRITTENDEN, P.D., KERSHAW, K.A.: A procedure for the simultaneous measurement of net $CO_2$-exchange and nitrogenase activity in lichens. - New Phytol. *80* : 393 - 401, 1978.

33250 - CROFTS, A.R. : Organization of electron transport - a report. - In : HALL, D.O., COOMBS, J., GOODWIN, T.W. (ed.) : Proceedings of the Fourth International Congress on Photosynthesis. Pp. 223 - 232. Biochem. Soc., London 1978.

33251 - CROFTS, A.R., BOWYER, J. : Electron and hydrogen transfer through the antimycin sensitive site in *Rhodopseudomonas capsulata*. - In : AZZONE, G.F., AVRON, M., METCALFE, J.C., QUAGLIARIELLO, E., SILIPRANDI, N. (ed.) : The Proton and Calcium Pumps. Pp. 55 - 64. Elsevier/North-Holland Biomedical Press, Amsterdam - Oxford - New York 1978.

*33252 - CROFTS, A.R., JACKSON, J.B., COGDELL, R.J., EVANS, E.H. : The mechanism of $H^+$ - uptake in photosynthetic bacteria. - Pestic. Sci. *4* : 615 - 616, 1973.

33253 - CROFTS, A.R., WOOD, P.M. : Photosynthetic electron-transport chains of plants and bacteria and their role as proton pumps. - In : SANADI, D.R., VERNON, L.P. (ed.) : Current Topics in Bioenergetics. Vol. 7. Pp. 175 - 244. Academic Press, New York - San Francisco - London 1978.

33254 - CROSBIE, T.M., MOCK, J.J., PEARCE, R.B. : Inheritance of photosynthesis in a diallel among eight maize inbred lines from Iowa Stiff Stalk Synthetic. - Euphytica *27* : 657 - 664, 1978.

33255 - CROSBIE, T.M., PEARCE, R.B., MOCK, J.J. : Relationship among $CO_2$-exchange rate and plant traits in Iowa Stiff Stalk Synthetic maize population. - Crop Sci. *18* : 87 - 90, 1978.

33256 - CROUSE, E.J., SCHMITT, J.M., BOHNERT, H.-J., GORDON, K., DRIESEL, A.J., HERRMANN, R.G. : Intramolecular compositional heterogeneity of *Spinacia* and *Euglena* chloroplast DNAs. - In : AKOYUNOGLOU, G., ARGYROUDI-AKOYUNOGLOU, J.H. (ed.) : Chloroplast Development. Pp. 565 - 572. Elsevier/North-Holland Biomedical Press, Amsterdam - New York - Oxford 1978.

33257 - CSATORDAY, K. : Fluorescence from sensitizing phycobilin chromophores in the blue-green alga *Anacystis nidulans*. - Biochim. biophys. Acta *504* : 341 - 343, 1978.

33258 - CSATORDAY, K., KLEINEN HAMMANS, J.W., GOEDHEER, J.C. : Excitation energy transfer in *Anacystis nidulans*. - Biochem. biophys. Res. Commun. *81* : 571 - 575, 1978.

33259 - CUENDET, P.A., ZÜRRER, H., SNOZZI, M., ZUBER, H. : On the localization of a
bacteriochlorophyll-associated polypeptide in the chromatophore membrane of
*Rhodospirillum rubrum* G-9. - FEBS Lett. *88* : 309 - 312, 1978.

33260 - CUNNINGHAM, G.L., REYNOLDS, J.F.: A simulation model of primary production
and carbon allocation in the creosotebush (*Larrea tridentata* [DC] COV.). -
Ecology *59* : 37 - 52, 1978.

33261 - CUTLER, J.M., RAINS, D.W. : Effects of water stress and hardening on the internal water relations and osmotic constituents of cotton leaves. - Physiol.
Plant. *42* : 261 - 268, 1978. [Photosynthates.]

33262 - CZARNOWSKI, M. : Produkcja fotosyntetyczna roślin w ekosystemie lasu grądo-
wego. [Plant photosynthetic production in oak-hornbeam forest ecosystem.] -
Stud. Nat. Ser. A *14* : 165 - 190, 1978. [In Pol., ab : E.]

33263 - CZECZUGA, B. : Lutein - a carotenoid dominating in *Desmococcus vulgaris*
(*Chaetophoraceae*). - Bull. Acad. pol. Sci., Sér. Sci. biol., Cl. II *26* :
453 - 455, 1978.

33264 - CZECZUGA, B. : The carotenoid content in certain plants from Abisco National
Park (Swedish Lapland). - Acta Soc. Bot. Pol. *47* : 205 - 209, 1978.

*33265 - CZECZUGA, B., BOBIATYŃSKA, E. : The effect of DDT on the carotenoids in uni-
cellular algae. - Rocz. Akad. Med. Juliana Marchlewskiego Białymstoku *22* :
57 - 63, 1977.

33266 - CZERPAK, R., CZECZUGA, B. : Występowanie, biosynteza i rola biologiczna karotenoidów u glonów. [Presence, biosynthesis and biological role of carotenoids in algae.] - Wiadom. bot. *22* : 47 - 59, 1978. [In Pol.]

33267 - CZOCHRALSKA, B., SZWEYKOWSKA, M., DENCHER, N.A., SHUGAR, D. : Electrochemical studies on *trans*- and *cis*-retinal and bacteriorhodopsin. - Bioelectrochem.
Bioenerg. *5* : 713 - 722, 1978.

33268 - CZUCHAJOWSKA, Z., NIEMTUR, S. : Seasonal variability of chlorophyll content
and the changes of the acidity and buffer capacity in the needles of *Larix
decidua* and *Larix leptolepis* grown in different air pollution conditions. -
Acta Soc. Bot. Pol. *47* : 41 - 49, 1978.

33269 - CZUCHAJOWSKA, Z., PRZYBYLSKI, T. : The seasonal changes of chlorophylls and
carotenoids in unpolluted and polluted needles of *Pinus silvestris*. - Bull.
Acad. pol. Sci., Sér. Sci. biol. *26* : 369 - 376, 1978.


33270 - DA CRUZ, G.S., AUDUS, L.J. : Studies of hormone-directed transport in deca-
pitated stolons of *Saxifraga Sarmentosa*. - Ann. Bot. *42* : 1009 - 1027, 1978.
[Photosynthates.]

33271 - DALEY, L., DAILEY, F., CRIDDLE, R. : Light activation of ribulose bisphosphate carboxylase. - Plant Physiol. *61* (Suppl.) : 108, 1978.

33272 - DALEY, L.S., BIDWELL, R.G.S. : Separation of phosphohydroxypyruvate, 3-phos-
phoglyceric acid and O-phosphoserine by paper chromatography and chemical
derivatization. - J. Chromatogr. *147* : 233 - 241, 1978.

33273 - DALEY, L.S., DAILEY, F., CRIDDLE, R.S. : Light activation of ribulose bisphosphate carboxylase. Purification and properties of the enzyme in tobacco.
- Plant Physiol. *62* : 718 - 722, 1978.

*33274 - DALEY, R.J., BROWN, S.R., McNEELY, R.N. : Chromatographic and SCDP measure-
ments of fossil phorbins and the postglacial history of Little Round Lake,
Ontario. - Limnol. Oceanogr. *22* : 349 - 360, 1977. [Chl.]

33275 - DANCSHÁZY, Zs., DRACHEV, L.A., ORMOS, P., NAGY, K., SKULACHEV, V.P. : Kinetics of the blue light-induced inhibition of photoelectric activity of bacteriorhodopsin. - FEBS Lett. *96* : 59 - 63, 1978.

33276 - DANCSHÁZY, Zs., ORMOS, P., DRACHEV, L.A., SKULACHEV, V.P. : Investigation
by focused laser beam scanning of the photoelectric activity of bacteriorhodopsin-containing lipid bilayers. - Biophys. J. *24* : 423 - 428, 1978.

33277 - DANON, A., DEGANI, H., CAPLAN, S.R. : The effect of the purple membrane of
*H. halobium* on the osmotic fragility and water permeability of the cells. -
In : CAPLAN, S.R., GINZBURG, M. (ed.) : Energetics and Structure of Halophi-
lic Microorganisms. Pp. 217 - 224. Elsevier/North-Holland Biomedical Press,
Amsterdam - New York 1978.

33278 - D'AOUST, A.L. : La physiologie des semis d'épinette noire *Picea mariana*
(MILL.) B.S.P. en contenants. - Can. Centre Rech. Forest. Laurentides Rapp.
Inf. *LAU-X-35* : 1 - 26, 1978. [Ps.]

33279 - DARWINKEL, A. : Patterns of tillering and grain production of winter wheat
at a wide range of plant densities. - Neth. J. agr. Sci. *26* : 383 - 398,
1978. [Photosynthates.]

33280 - DAS, V.S.R., SANTAKUMARI, M. : The incomplete evolution of $C_4$-photosynthesis
within the pantropical taxon, *Boerhaavia (Nyctaginaceae)*. - Photosynthetica
*12* : 418 - 422, 1978.

B33281 - DAVENPORT, D.C., HAGAN, R.M., GAY, L.W., KYNARD, B.E., BONDE, E.K., KREITH,
F., ANDERSON, J.E. : Factors Influencing Usefulness of Antitranspirants
Applied on Phreatophytes to Increase Water Supplies. - California Water Re-
sources Center, Davis 1978. [Ps.]

33282 - DAVIDSON, J.N., HANSON, M.R., BOGORAD, L. : Erythromycin resistance and the
chloroplast ribosome in *Chlamydomonas reinhardi*. - Genetics *89* : 281 - 297,
1978.

*33283 - DAVIES, B.H. : $C_{30}$ carotenoids. - Int. Rev. Biochem. *14* : 51 - 100, 1977.

*33284 - DAVIES, B.H. : Lipid-soluble leaf pigments. - Biochem. Soc. Trans. *5* : 1256
- 1259, 1977.

33285 - DAVIES, F.S., LAKSO, A.N. : Effect of prestress drying cycles on subsequent
water stress responses of apple trees. - HortScience *13* : 276, 1978. [Ps.]

33286 - DAVIES, F.S., LAKSO, A.N. : Water relations in apple seedlings : Changes in
water potential components, abscisic acid levels and stomatal conductances
under irrigated and non-irrigated conditions. - J. amer. Soc. hort. Sci.
*103* : 310 - 313, 1978.

33287 - DAVIES, W.J., GILL, K., HALLIDAY, G. : The influence of wind on the behaviour
of stomata of photosynthetic stems of *Cytisus scoparius* (L.) LINK. - Ann.Bot.
*42* : 1149 - 1154, 1978.

33288 - DAVIES, W.J., MANSFIELD, T.A., ORTON, P.J. : Strategies employed by plants
to conserve water : can we improve on them? - In : Proceedings Joint BCPC
and BPGRG Symposium - "Oportunities for Chemical Plant Growth Regulation".
Pp. 45 - 54. British Crop Protection Council, Croydon, Surrey 1978. [Ps.]

*33289 - DAVIS, D.J., SAN PIETRO, A. : Evidence for the role of sulfhydryl groups in
a pH-dependent transition of ferredoxin:NADP oxidoreductase. - Arch. Biochem.
Biophys. *184* : 572 - 577, 1977.

33290 - DAVIS, S.D., McCREE, K.J. : Photosynthetic rate and diffusion conductance
as a function of age in leaves of bean plants. - Crop Sci. *18* : 280 - 282,
1978.

*33291 - DAVTYAN, V.A., CHILINGARYAN, A.A. : [Effect of diminuition of leaf biomass
on growth and photosynthesis of plants.] - Izv. sel'skokhoz. Nauki *1975* (1) :
89 - 93, 1975. [In Arm., ab : R.]

33292 - DAWES, C.J., MOON, R.E., DAVIS, M.A. : The photosynthetic and respiratory
rates and tolerances of benthic algae from a mangrove and salt marsh estuary:
a comparative study. - Estuar. coast. mar. Sci. *6* : 175 - 185, 1978.

33293 - DAY, W., LEGG, B.J., FRENCH, B.K., JOHNSTON, A.E., LAWLOR, D.W., JEFFERS,
W. De C. : A drought experiment using mobile shelters : the effect of drought
on barley yield, water use and nutrient uptake. - J. agr. Sci. *91* : 599 -
623, 1978. [Growth analysis.]

33294 - DAYHOFF, M.O., SCHWARTZ, R.M. : Evolution of early life inferred from protein and ribonucleic acid sequences. - In : NODA, H. (ed.) : Origin of Life.
Pp. 547 - 560. Bus. Cent. Acad. Soc. Japan 1978. [Ps.]

B33295 - DEAMER, D.W. (ed.) : Light Transducing Membranes : Structure, Function, and Evolution. - Acad. Press, New York - London 1978.

33296 - DECHANT, S. : Zum gegenwärtigen Stand der Kenntnisse über Enzyme und Enzym-aktivitäten während der Keimung von Spermatophyta unter besonderer Berücksichtigung der Coniferophytina. - Wiss. Z. tech. Univ. Dresden *27* : 819 - 822, 1978.

33297 - DeCLERCQ, D.R., SHEARER, J.A. : Phytoplankton primary production, chlorophyll and suspended carbon in the Experimental Lakes Area - 1977 data. - Fish. mar. Serv. Data Rep. *74* : 1 - 62, 1978.

33298 - DE GREEF, J.A. : Studies on the Shibata-shift in relation to light quality and seedling age. - In : AKOYUNOGLOU, G., ARGYROUDI-AKOYUNOGLOU, J.H. (ed.) : Chloroplast Development. Pp. 817 - 822. Elsevier/North-Holland Biomedical Press, Amsterdam - New York - Oxford 1978.

33299 - DE GREEF, J.A., MOEREELS, E., SPRUYT, E., NEELS, L. : Eigenschappen van enzymen uit het energie-metabolisme tijdens de morfogenese van geetioleerde bonekiemplanten (*Phaseolus vulgaris* L. cv. Limburg). [Properties of enzymes in energy metabolism during morphogenesis of etiolated bean plants (*Phaseolus vulgaris* L. cv. Limburg).] - Bul. Soc. roy. Bot. Belg. *111* : 99 - 118, 1978. [Ps; in Flem., ab : E.]

33300 - DE GROOTH, B.G., VAN GRONDELLE, R., ROMIJN, J.C., PULLES, M.P.J. : The mechanism of reduction of the ubiquinone pool in photosynthetic bacteria at different redox potentials. - Biochim. biophys. Acta *503*: 480 - 490, 1978.

33301 - DEI, M. : Inter-organ control of greening in etiolated cucumber cotyledons. - Physiol. Plant. *43* : 94 - 98, 1978.

33302 - DEI, M., TSUJI, H. : The influence of various plant hormones on the stimulatory action of red light on greening in excised etiolated cotyledons of cucumber. - Plant Cell Physiol. *19* : 1407 - 1414, 1978.

33303 - DE JONG, D.W., WOODLIEF, W.G. : High-speed, low-pressure liquid chromatography of chloroplast pigments from tobacco mutants. - J. agr. Food Chem. *26* : 1281 - 1288, 1978.

33304 - DE JONG, T. : Comparative effects of salinity on gas exchange characteristics of $C_3$ and $C_4$ coastal strand species. - Plant Physiol. *61* (Suppl.) : 86, 1978.

33305 - DE JONG, T.M. : Comparative gas exchange of four California beach taxa. - Oecologia *34* : 343 - 351, 1978.

33306 - DE JONG, T.M. : Comparative gas exchange and growth responses of $C_3$ and $C_4$ beach species grown at different salinities. - Oecologia *36* : 59 - 68, 1978.

*33307 - DE KOK, J., BUTLER, J., BRAAMS, R., VAN GELDER, B.F. : The reduction of porphyrin cytochrome *c* by hydrated electrons and the subsequent electron transfer reaction from reduced porphyrin cytochrome *c* to ferricytochrome *c*. - Biochim. biophys. Acta *460* : 290 - 298, 1977.

33308 - DE KOK, L.J., VAN HASSELT, P.R., KUIPER, P.J.C. : Photo-oxidative degradation of chlorophyll-*a* and unsaturated lipids in liposomal dispersions at low-temperature. - Physiol. Plant. *43* : 7 - 12, 1978.

33309 - DELANEY, M.E., JONES, M., ROGERS, L.J. : Effect of 1,1,1-trichloro-2,2-bis-(*p*-chlorophenyl)ethane on cytochromes of the photosynthetic electron transport system. - J. exp. Bot. *28* : 25 - 30, 1978.

33310 - DELANEY, M.E., WALKER, D.A. : Comparison of the kinetic properties of ribulose bisphosphate carboxylase in chloroplast extracts of spinach, sunflower and four other reductive pentose phosphate-pathway species. - Biochem. J. *171* : 477 - 482, 1978.

33311 - DELEPELAIRE, P., BENNOUN, P. : Energy transfer and site of energy trapping in Photosystem I. - Biochim. biophys. Acta *502* : 183 - 187, 1978.

*33312 - DELÉPINE, R., ASENSI, A., GUGLIELMI, G. : Nouveaux types d'ultrastructure plastidiale chez les Phéophycées. - Phycologia *15* : 425 - 434, 1976.

33313 - DELLWEG, H.-G., SUMPER, M. : Selective formation of bacterio-opsin trimers
by chemical cross-linking of purple membrane. - FEBS Lett. *90* : 123 - 126,
1978.

33314 - DELOSME, R., ZICKLER, A., JOLIOT, P. : Turnover kinetics of Photosystem I
measured by the electrochromic effect in *Chlorella*. - Biochim. biophys. Acta
*504* : 165 - 174, 1978.

33315 - DELRIEU, M.J. : Oscillation après 1 ou 2 éclairs de la quantité de centres
photochimiques du Système II dans les états $S_2$ et $S_3$, dans des condition
variées. - Compt. rend. Acad. Sci. Paris, Sér. D *287* : 563 - 566, 1978.

33316 - DELRIEU, M.-J. : Oscillatory kinetics of the number of photosynthetic system II centers in $S_2$ and $S_3$ states after flashes under various conditions. -
Plant Cell Physiol. *19* : 1447 - 1456, 1978.

*33317 - DEL RÍO, L.A., GÓMEZ, M., LÓPEZ-GORGÉ, J. : Catalase and peroxidase activi-
ties, chlorophyll and proteins during storage of pea plants at chilling temperatures. - Rev. españ. Fisiol. *33* : 143 - 148, 1977.

33318 - DELROT, S., BONNEMAIN, J.-L. : Étude du mécanisme de l'accumulation des pro-
duits de la photosynthèse dans les nervures. - Compt. rend. Acad. Sci. Sér.
D *287* : 125 - 130, 1978.

33319 - DE LUCA, V., DENNIS, D.T. : Isoenzyme of pyruvate kinase in proplastids from
developing castor bean endosperm. - Plant Physiol. *61* : 1037 - 1039, 1978.

33320 - DE MARCH, L. : Permanent sedimentation of nitrogen, phosphorus, and organic
carbon in a high arctic lake. - J. Fish. Res. Board Can. *35* : 1089 - 1094,
1978. [Primary production.]

33321 - DeMICHELE, D.W., SHARPE, P.J.H., GOESCHL, J.D. : Toward the engineering of
photosynthetic productivity. - CRC crit. Rev. Bioeng. *3* : 23 - 91, 1978.

*33322 - DEMIDENKO, R.N., LYSENKO, M.K. : Izmenenie opticheskikh svoĭstv list'ev ya-
bloni sorta Aport Aleksandriĭskiĭ v zavisimosti ot usloviĭ mineral'nogo pitaniya. [Changes in leaf optical properties of apple tree cultivar Aport
Aleksandriĭskiĭ in dependence on mineral nutrition.] - Biol. Nauki (Alma-Ata)
*8* : 156 - 161, 1975. [In R.]

33323 - DEMIDOV, E.D., BELL, L.N. : Light quality and the primary energy storing
processes of photosynthesis. - Photosynthetica *12* : 158 - 165, 1978.

33324 - DEMIDOV, É.D., KERIMOV, S.Kh., BELL, L.N. : O svyazi mezhdu vydeleniem gli-
kolevoĭ kisloty i énergeticheskoĭ obespechennost'yu kletok khlorelly na svetu. [The relation between glycolate excretion and energy storage in illuminated *Chlorella* cells.] - Fiziol. Rast. *25* : 1089 - 1095, 1978. [In R, ab :
E.]

*33325 - DEMINA, N.S., IVANOV, I.D. : Rol' protsessa vydeleniya molekulyarnogo vodo-
roda v énergetike fotosinteziruyushchikh bakteriĭ. [Role of production of
molecular hydrogen in energetic of photosynthesizing bacteria.] - In : Novoe
v Izuchenii Biologicheskoĭ Fiksatsii Azota. Pp. 29 - 32. Nauka, Moskva 1971.
[In R.]

33326 - DENCHER, N.A. : Light-induced behavioural reactions of *Halobacterium halobium* : Evidence for two rhodopsins acting as photopigments. - In : CAPLAN,
S.R., GINZBURG, M. (ed.) : Energetics and Structure of Halophilic Microorganisms. Pp. 67 - 88. Elsevier/North-Holland Biomedical Press, Amsterdam -
New York 1978.

33327 - DENCHER, N.A., HEYN, M.P. : Solubilization of purple membrane by the non-
-ionic detergents Triton X-100 and octyl-β-D-glucoside. - In : CAPLAN, S.R.,
GINZBURG, M. (ed.) : Energetics and Structure of Halophilic Microorganisms.
Pp. 233 - 238. Elsevier/North-Holland Biomedical Press, Amsterdam - New York
1978.

33328 - DENCHER, N.A., HEYN, M.P. : Formation and properties of bacteriorhodopsin
monomers in the non-ionic detergens octyl-β-D-glucoside and Triton X-100. -
FEBS Lett. *96* : 322 - 326, 1978.

33329 - DENESH, M., ANDRIANOV, V.K., BULYCHEV, A.A., KURELLA, G.A., URAZMANOV, R.I. : Vliyanie ditsiklogeksilkarbodiimida na fotoindutsirovannyĭ transport $H^+$ v kletkakh i izolirovannykh khloroplastakh *Nitellopsis obtusa*. [Effect of dicyclohexylcarbodiimide on light-induced $H^+$-transport in cells and isolated chloroplasts of *Nitellopsis obtusa*.] - Fiziol. Rast. *25* : 1163 - 1167, 1978. [In R, ab : E.]

33330 - DENNETT,M.D., MILFORD, J.R., ELSTON, J. : The effect of temperature on the relative leaf growth rate of crops of *Vicia faba* L. - Agr. Meteorol. *19* : 505 - 514, 1978.

33331 - DENNIS, J.G., TIESZEN, L.L., VETTER, M.A. : Seasonal dynamics of above- and belowground production of vascular plants at Barrow, Alaska. - In : TIESZEN, L.L. (ed.) : Vegetation and Production Ecology of an Alaskan Arctic Tundra. Pp. 113 - 140. Springer-Verlag, New York - Heidelberg - Berlin 1978. [Chl.]

33332 - DENNISS, I.S., SANDERS, J.K.M. : Synthesis of a chlorophyll which does not aggregate. - Tetrahedron Lett. *1978* (3) : 295 - 298, 1978.

33333 - DENNY, P., HARMAN, J., ABRAHAMSSON, J., BRYCESON, I. : Limnochemical and phytoplankton studies on Nyumba ya Mungu reservoir, Tanzania. - Biol. J. Linn. Soc. *10* : 29 - 48, 1978. [Chl.]

*33334 - DERA, J., BOJANOWSKI, R. : Wstępne badania warunków fotosyntezy w wodach Zatoki Gdańskiej. [Initial investigation of conditions affecting photosynthesis in the Bay of Gdańsk.] - Acta geophys. polon. *14* : 23 - 31, 1966. [In Pol., ab : R.]

33335 - DETLING, J.K., PARTON, W.J., HUNT, H.W. : An empirical model for estimating $CO_2$ exchange of *Bouteloua gracilis* (H.B.K.) LAG. in the shortgrass prairie. - Oecologia *33* : 137 - 147, 1978.

33336 - DEUTCH, B. : Light-regulated body ion balance in marine slug *Elysia viridis* (Montagu). - Nature *274* : 159 - 160, 1978. [Chl, Car.]

33337 - DeVAULT, D., KUNG, M.C. : Interactions among photosynthetic antenna excited states. - Photochem. Photobiol. *28* : 1029 - 1038, 1978.

33338 - DE VILLIERS, O.T., FOURIE, M.P., WIID, I.J.F. : The effect of methabenzthiazuron on biochemical changes in lucerne and clover. - Agroplantae *10* : 87 - 90, 1978. [Chl, Car.]

33339 - DEVLIN, R.M., SARAS, C.N., KISIEL, M.J., KOSTUSIAK, A.S. : Influence of fluridone on chlorophyll content of wheat (*Triticum aestivum*) and corn (*Zea mays*). - Weed Sci. *26* : 432 - 433, 1978.

33340 - DEVOL, A.H., PACKARD, T.T. : Seasonal changes in respiratory enzyme activity and productivity in Lake Washington microplankton. - Limnol. Oceanogr. *23* : 104 - 111, 1978. [Chl.]

*33341 - DE WAAL, F.E.B., GOEDHEER, J.C., THOMAS, J.B. : Slow transients of light-induced pH changes at the outer surface of photosynthetic membranes. - Acta bot. neer. *26* : 89 - 93, 1977.

*33342 - DE WIT, C.T. : Modelle der Ertragsbildung als Brücke zwischen Prozeß und System. - In : UNGER, K. (ed.) : Biophysikalische Analyse pflanzlicher Systeme. Pp. 19 - 30. VEB Gustav Fischer Verlag, Jena 1977.

B33343 - DE WIT, C.T., GOUDRIAAN, J. : Simulation of Ecological Processes. - Pudoc, Wageningen 1978. [Primary production.]

33344 - DHINDSA, R.S. : Hormonal regulation of enzymes of nonautotrophic $CO_2$ fixation in unfertilized cotton ovules. - Z. Pflanzenphysiol. *89* : 355 - 362, 1978.

33345 - DIAMANTOGLOU, S., MELETIOU-CHRISTOU, M.S. : Kohlenhydratgehalte und osmotische Verhältnisse bei Blättern und Rinden von *Ceratonia siliqua* und *Quercus coccifera* im Jahresgang. - Flora *167* : 472 - 479, 1978.

*33346 - DICKMANN, D.I., GJERSTAD, D.H. : Application to woody plants of a rapid method for determining leaf $CO_2$ compensation concentrations. - Can. J. Forest. Res. *3* : 237 - 242, 1973.

33347 - DICKS, J.W. : Inhibition of the Hill reaction of isolated chloroplasts by herbicidal phenylureas. - Pestic. Sci. *9* : 59 - 62, 1978.

33348 - DIEPENBROCK, W., GEISLER, G. : Untersuchungen zur Bedeutung der Fruchtwand der Rapsschote als Organ der Assimilatbildung und als Stickstoffreservoir für die Samen. - Z. Acker- Pflanzenbau *146* : 54 - 67, 1978. [Chl.]

33349 - DIHORU, A. : Cercetări chemotaxonomice la unele specii de *Arum* şi *Iris* din flora României. [Chemotaxonomic studies on some *Arum* and *Iris* species of Roumanian flora.] - Stud. Cercet. Biol., Ser. Biol. veg. *30* (1) : 15 - 18, 1978. [Chl; in Roum., ab : E.]

33350 - DILEY, R.A., PROCHASKA, L.J. : Two separate and non-interacting intramembrane domains for protons released in electron transport in chloroplast thylakoid membranes. - In : AZZONE, G.F., AVRON, M., METCALFE, J.C., QUAGLIARIELLO, E., SILIPRANDI, N. (ed.) : The Proton and Calcium Pumps. Pp. 45 - 54. Elsevier/North-Holland Biomedical Press, Amsterdam - Oxford - New York 1978.

*33351 - DILOV, Kh., ĬORDANOVA, S. : Issledovanie pigmentov mestnogo termofil'nogo shtamma *Chlorella vulgaris* S.L. 8/1. [Pigments of the local thermophilic strain *Chlorella vulgaris* S.L. 8/1.] - Dokl. Akad. sel'skokhoz. Nauk Bolg. *4* (2) : 163 - 167, 1971. [In R.]

*33352 - DILOV, Kh., R"OSLER, M., BOZHKOVA, M. : Posledeĭstvie na tretiraneto s razlichni temperaturi na t"mno v"rkhu rastezha i khimicheskiya s"stav na ednoklet"chni vodorasli. [Aftereffect of darkness and various temperatures on growth and chemical composition of unicellular algae.] - Prilozh. Mikrobiol. *3* : 91 - 100, 1974. [Chl; in Bulg., ab : E, R.]

33353 - DILOVA, S., ZEINALOV, Yu., PETKOVA, R. : State of the pigment-protein complex in higher plants II. Action of phospholipase D. - Photosynthetica *12* : 304 - 309, 1978.

33354 - DINER, B.A.: Double hitting in photosystem II viewed from the donor side. - In: HALL, D.O., COOMBS, J., GOODWIN, T.W. (ed.) : Proceedings of the Fourth International Congress on Photosynthesis. Pp. 359 - 372. Biochem. Soc., London 1978.

33355 - DISMUKES, G.C., McGUIRE, A., BLANKENSHIP, R., SAUER, K. : Electron spin polarization in photosynthesis and the mechanism of electron transfer in photosystem I. Experimental observations. - Biophys. J. *21* : 239 - 254, 1978.

33356 - DISMUKES, G.C., SAUER, K. : The orientation of membrane bound radicals. An EPR investigation of magnetically ordered spinach chloroplasts. - Biochim. biophys. Acta *504* : 431 - 445, 1978.

*33357 - DI TORO, D.M. : Combining chemical equilibrium and phytoplankton models - a general methodology. - In : CANALE, R.P. (ed.) : Modeling Biochemical Processes in Aquatic Ecosystems. Pp. 233 - 255. Ann Arbor Science, Ann Arbor 1976.

33358 - DIXON, K.R., LUXMOORE, R.J., BEGOVICH, C.L. : CERES - a model of forest stand biomass dynamics for predicting trace contaminant, nutrient, and water effects. I. Model description. - Ecol. Modelling *5* : 17 - 38, 1978.

33359 - DIXON, K.R., LUXMOORE, R.J., BEGOVICH, C.L. : CERES - a model of forest stand biomass dynamics for predicting trace contaminant, nutrient, and water effects. II. Model application. - Ecol. Modelling *5* : 93 - 114, 1978.

33360 - DMITRIEV, A.P., GUCSHA, N.I., GRODZINSKY, D.M. : Intracellular oxygen and radiosensitization of the blue-green algae. - Plant Sci. Lett. *12* : 361 - 364, 1978. [Ps.]

33361 - DOBRINSKIĬ, L.N., KRYAZHIMSKIĬ, F.V., MALAFEEV, Yu.M. : Izpol'zovanie optiko--akusticheskikh gazoanalizatorov v biogeotsenologicheskikh issledovaniyakh. [Use of IRGAs in biogeocenological studies.] - In : Ispol'zovanie Optiko--Akusticheskikh Gazoanalizatorov v Ėkologo-Fiziologicheskikh i Biogeotsenologicheskikh Issledovaniyakh. Pp. 46 - 62. Inst. Ėkol. Rast. Zhivot. Akad. Nauk SSSR, Sverdlovsk 1978. [In R.]

33362 – **DODA, D.D., GREEN, A.E.S.** : Spectral sunphotometry using a compact spectrometer. – Remote Sens. Environ. *7* : 97 – 104, 1978.

33363 – **DOEHLERT, D.C., KU, M.S.B., EDWARDS, G.E.** : Effects of $CO_2$, $O_2$, and temperature on post-illumination burst in *Triticum aestivum* L. – Plant Physiol. *61* : (Suppl.) : 7, 1978.

*33364 – **DOKULIL, M.** : Der Neusiedler See (Österreich). – Ber. naturhist. Ges. (Hannover) *118* : 205 – 211, 1974. [Chl.]

*33365 – **DOKULIL, M.** : Phytoplankton, Primärproduktion und Bakterien im Wörthersee. – Carinthia II *164/84* :199 – 203, 1974.

*33366 – **DOKULIL, M.** : Horizontal- und Vertikalgradienten in einem Flachsee (Neusiedlersee, Österreich). – In : Verhandlungen der Gesellschaft für Ökologie. Pp. 177 – 187. Wien 1975. [Chl.]

33367 – **DOKULIL, M., HAMMER, L., JEWSON, D.H.** : Vergleichende Untersuchungen zur Primärproduktion des Phytoplanktons im Neusiedlersee. $O_2$, $^{14}C$ und Experimente mit künstlicher Zirkulation. – Biol. Forsch. Inst. Burgenland *29* : 60 – 73, 1978.

33368 – **DOLEY, D.** : Effects of shade on gas exchange and growth in seedlings of *Eucalyptus grandis* HILL *ex* MAIDEN. – Aust. J. Plant Physiol. *5* : 723 – 738, 1978.

33369 – **DOMINGUEZ, C., HUME, D.J.** : Flowering, abortion and yield of early-maturing soybeans at three densities. – Agron. J. *70* : 801 – 805, 1978. [Photosynthates.]

33370 – **DOMINY, P.J., BAKER, N.R.** : Effect of high salinity on primary photochemistry and energy transfer in pea thylakoids. – Plant Physiol. *61* (Suppl.) : 88, 1978.

33371 – **DONAGHAY, P.L., DeMANCHE, J.M., SMALL, L.F.** : On predicting phytoplankton growth rates from carbon : nitrogen ratios. – Limnol. Oceanogr. *23* : 359 – 362, 1978.

*33372 – **DOODSON, J.K.** : The plants reaction to disease. – J. nat. Inst. agr. Bot. *14* : 204 – 206, 1976. [Ps, Chl.]

33373 – **DOUCE, R., JOYARD, J.** : Importance of the envelope in chloroplast biogenesis. – In : AKOYUNOGLOU, G., ARGYROUDI-AKOYUNOGLOU, J.H. (ed.) : Chloroplast Development. Pp. 283 – 296. Elsevier/North-Holland Biomedical Press, Amsterdam – New York – Oxford 1978.

33374 – **DOUILLARD, R., BERGERON, É.** : Activité lipoxygénasique de chloroplastes de plantules de Blé. – Compt. rend. Acad. Sci. Paris, Sér. D *286* : 753 – 755, 1978.

33375 – **DOVNAR, V.S.** : Metodika polucheniya svetovykh krivykh chistoĭ produktivnosti fotosinteza v vegetatsionnykh opytakh. [Methods for obtaining light curves of net productivity of photosynthesis in vegetative experiments.] – Tr. priklad. Bot. Genet. Selektsii *61* (3) : 72 – 81, 1978. [In R, ab : E.]

33376 – **DRACHEV, L.A., KAULEN, A.D., SKULACHEV, V.P.** : Time resolution of the intermediate steps in the bacteriorhodopsin-linked electrogenesis. – FEBS Lett. *87* : 161 – 167, 1978.

33377 – **DREW, A.P., BAZZAZ, F.A.** : Variations in distribution of assimilate among plant parts in three populations of *Populus deltoides*. – Silvae Genet. *27* : 189 – 193, 1978.

33378 – **DREW, E.A.** : Carbohydrate and inositol metabolism in the seagrass, *Cymodocea nodosa*. – New Phytol. *81* : 249 – 264, 1978. [Ps.]

33379 – **DREW, E.A.** : Factors affecting photosynthesis and its seasonal variation in the seagrasses *Cymodocea nodosa* [UCRIA] ASCHERS, and *Posidonia oceanica* [L.] DELILE in the Mediterranean. – J. exp. mar. Biol. Ecol. *31* : 173 – 194, 1978.

33380 – **DREWS, G.** : Structure and development of the membrane system of photosynthetic bacteria. – In : SANADI, D.R., VERNON, L.P. (ed.) : Current Topics in Bioenergetics. Vol. 8. Pp. 161-207. Academic Press, New York – San Francisco – London 1978.

33381 - DREWS, G., FEICK, R., SCHUMACHER, A., FIRSOW, N. : Organization and formation of the photosynthetic apparatus of *Rhodopseudomonas capsulata* and *Rhodopseudomonas palustris*. - In : HALL, D.O., COOMBS, J., GOODWIN, T.W. (ed.) : Proceedings of the Fourth International Congress on Photosynthesis. Pp. 83 - 93. Biochem. Soc., London 1978.

*33382 - DROBA, M. : Modelowe układy jednostek fotosyntetycznych. [Model systems of photosynthetic units.] - Zesz. nauk. Univ. Jagiell. *464* (Prace Biol. mol. 4) : 63 - 68, 1977. [In Pol., ab : E.]

33383 - DROBA, M., WIĘCKOWSKI, S. : Effects of sodium octyl, decyl, and tetradecyl sulphates on the release of different protein fractions from spinach thylakoid membranes. - Biochem. Physiol. Pflanz. *172* : 343 - 349, 1978.

33384 - DROMGOOLE, F.I. : The effects of pH and inorganic carbon on photosynthesis and dark respiration of *Carpophyllum (Fucales, Phaeophyceae)*. - Aquat. Bot. *4* : 11 - 22, 1978.

33385 - DROMGOOLE, F.I. : The effect of oxygen on dark respiration and apparent photosynthesis of marine macro-algae. - Aquat. Bot. *4* : 281 - 297, 1978.

*33386 - DROZDOVA, N.N., KUZNETSOV, B.A., MESTECHKINA, N.M., SHUMAKOVICH, G.P., PUSHKINA, E.M., KRASNOVSKIĬ, A.A. : Élektrokhimicheskoe okislenie i vosstanovlenie bakteriokhlorofilla *b* i ego feofitina.[Electrochemical oxidation and reduction of bacteriochlorophyll *b* and its pheophytin.] - Dokl. Akad. Nauk SSSR *235* : 1437 - 1440, 1977. [In R.]

33387 - DUBACQ, J.-P., KADER, J.-C. : Free flow electrophoresis of chloroplasts. - Plant Physiol. *61* : 465 - 468, 1978.

33388 - DUBBE, D.R., FARQUHAR, G.D., RASCHKE, K. : Effect of abscisic acid on the gain of the feedback loop involving carbon dioxide and stomata. - Plant Physiol. *62* : 413 - 417, 1978. [Ps.]

33389 - DUBERTRET, G., LEFORT-TRAN, M. : Structural and functional organization of chlorophyll in the developing thylakoids of *Euglena gracilis*. - In : AKOYUNOGLOU, G., ARGYROUDI-AKOYUNOGLOU, J.H. (ed.) : Chloroplast Development. Pp. 419 - 425. Elsevier/North-Holland Biomedical Press, Amsterdam - New York - Oxford 1978.

33390 - DUBERTRET, G., LEFORT-TRAN, M. : Functional and structural organization of chlorophyll in the developing photosynthetic membranes of *Euglena gracilis* Z. I. Formation of System II photosynthetic units during greening under optimal light intensity. - Biochim. biophys. Acta *503* : 316 - 332, 1978.

33391 - DUCRUET, J.M., GASQUEZ, J. : Observation de la fluorescence sur feuille entière et mise en évidence de la résistance chloroplastique à l'atrazine chez *Chenopodium album* L. et *Poa annua* L. - Chemosphère *7* : 691 - 696, 1978.

33392 - DUJARDIN, E. : Emerson-like effect in protochlorophyll(ide) photoreduction in bean leaves. - In : METZNER, H. (ed.) : Photosynthetic Oxygen Evolution. Pp. 511 - 520. Academic Press, London - New York - San Francisco 1978.

33393 - DUJARDIN, E., SIRONVAL, C. : The mechanisms of photoreduction of protochlorophyll(ide). - In : AKOYUNOGLOU, G., ARGYROUDI-AKOYUNOGLOU, J.H. (ed.) : Chloroplast Development. Pp. 83 - 98. Elsevier/North-Holland Biomedical Press, Amsterdam - New York - Oxford 1978.

33394 - DUNCAN, W.G., McCLOUD, D.E., McGRAW, R.L., BOOTE, K.J. : Physiological aspects of peanut yield improvement. - Crop Sci. *18* : 1015 - 1020, 1978. [Growth analysis.]

33395 - DUNIN, F.X., ASTON, A.R., REYENGA, W. : Evaporation from a *Themeda* grassland. II. Resistance model of plant evaporation. - J. appl. Ecol. *15* : 847 - 858, 1978. [Growth analysis.]

33396 - DÜRING, H. : Untersuchungen zur Umweltabhängigkeit der stomatären Transpiration bei Reben. II. Ringelungs- und Temperatureffekte. - Vitis *17* : 1 - 9, 1978.

33397 - DURMISHIDZE, S.V., BERIASHVILI, T.V. : Peredvizhenie assimilyatov iz grozdi
v pobegi i list'ya vinogradnoĭ lozy. [Transport of assimilates from clusters
to shoots and leaves of grapevine.] - Fiziol. Rast. 25 : 49 - 54, 1978.
[In R, ab : E.]

33398 - DURRANT, M.J., DRAYCOTT, A.P., MILFORD, G.F.J. : Effect of sodium fertilizer
on water status and yield of sugar beet. - Ann. appl. Biol. 88 : 321 - 328,
1978. [Growth analysis.]

33399 - DUTTON, J.E., ROGERS, L.J. : Isoelectric focusing of ferredoxins, flavodo-
xins and a rubredoxin. - Biochim. biophys. Acta 537 : 501 - 506, 1978.

33400 - DUTTON, P.L., BASHFORD, C.L., van den BERG, W.H., BONNER, H.S., CHANCE, B.,
JACKSON, J.B., PETTY, K.M., PRINCE, R.C., SORGE, J.R., TAKAMIYA, K. :
Electron and proton translocation in the reaction center-ubiquinone-cytochro-
mes $b/c_2$ oxidoreductase of Rhodopseudomonas sphaeroides. - In : HALL, D.O.,
COOMBS, J., GOODWIN, T.W. (ed.) : Proceedings of the Fourth International
Congress on Photosynthesis. Pp. 159 - 171. Biochem. Soc., London 1978.

33401 - DUTTON, P.L., EVANS, W.C. : Metabolism of aromatic compounds by Rhodospiril-
laceae. In : CLAYTON, R.K., SISTROM, W.R. (ed.) : The Photosynthetic Bacteria.
Pp. 719 - 726. Plenum Press, New York - London 1978. [Ps.]

33402 - DUTTON, P.L., PRINCE, R.C. : Reaction-center-driven cytochrome interactions
in electron and proton translocation and energy coupling. - In : CLAYTON,
R.K., SISTROM, W.R. (ed.) : The Photosynthetic Bacteria. Pp. 525 - 570. Plenum Press, New York - London 1978.

33403 - DUTTON, P.L., PRINCE, R.C., TIEDE, D.M. : The reaction center of photosynthetic bacteria.-Photochem. Photobiol. 28 : 939 - 949, 1978.

*33404 - DUVAL, W.S., BROCKINGTON, P.J., MELVILLE, M.S. von, GEEN, G.H. : Spectropho-
tometric determination of dissolved oxygen concentration in water. - J. Fish.
Res. Board Can. 31 : 1529 - 1530, 1974.

33405 - DUYSEN, M.E., FREEMAN, T.P. : Thylakoid development and chlorophyll distribution in water-stressed and hormone-treated wheat leaves. - Can. J. Bot.
56 : 1941 - 1945, 1978.

33406 - DUYSENS, L.N.M., VAN GRONDELLE, R., VALLE-TASCON, S. del : Electron transport and photophosphorylation associated with primary reactions in purple
bacteria. - In : HALL, D.O., COOMBS, J., GOODWIN, T.W. (ed.) : Proceedings
of the Fourth International Congress on Photosynthesis. Pp. 173 - 183. Biochem. Soc., London 1978.

33407 - DVORNIKOV, S.S., KNYUKSHTO, V.N., SEVCHENKO, A.N., SOLOV'EV, K.N., TSVIRKO,
M.P. : Polyarizatsionnye spektry fosforestsentsii i fluorestsentsii khlorofillov a i b i ikh feofitinov. [Polarization spectra of the phosphorescence
and fluorescence of chlorophylls a and b and their pheophytins.] - Dokl.
Akad. Nauk SSSR 240 : 1457 - 1460, 1978. [In R.]

33408 - DVORNIKOV, S.S., KNYUKSHTO, V.N., SEVCHENKO, A.N., SOLOV'EV, K.N., TSVIRKO,
M.Ts. : Dual'naya fosforestsentsiya tsinkovykh kompleksov feofitinov a i b .
[Double phosphorescence of Zn-complexes of pheophytins a and b .] - Dokl.
Akad. Nauk SSSR 242 : 1060 - 1063, 1978. [In R.]

33409 - DWIVEDI, R.S., MATHEW, O., MICHAEL, K.J., RAY, P.K., SUMATHY KUTTY AMMA, B.,
NINAN, S. : Carbonic anhydrase, carbon assimilation rate and canopy structure
in relation to nut yield of coconut (Cocos nucifera L.). - In : Inter-Disciplinary Symposium on Photosynthesis and Productivity. Abstract of Papers.
Pp. 54 - 56. Indian nat. Sci. Acad., New Delhi 1978.

33410 - D'YACHENKO, A.P. : Sravnitel'nyĭ analiz strukturnykh i funktsional'nykh oso-
bennosteĭ fotosinteticheskogo apparata razlichnykh ěkologicheskikh grupp
vysshikh rasteniĭ. [Comparative analysis of structural and functional pecul-
arities of photosynthetic apparatus of various ecological groups of higher
plants.] - In : Mezostruktura i Funktsional'naya Aktivnost' Fotosinteticheс-
kogo Apparata. Pp. 93 -  102. Ural'skiĭ gosudarstvennyĭ Universitet, Sverdlovsk 1978. [In R.]

33411 - DYKYJOVÁ, D. : Plant growth and estimates of production. - In : DYKYJOVÁ, D., KVĚT, J. (ed.) : Pond Littoral Ecosystems. Pp. 159 - 163. Springer-Verlag, Berlin - Heidelberg - New York 1978.

33412 - DYKYJOVÁ, D. : Determination of energy content and net efficiency of solar energy conversion by fishpond helophytes. - In : DYKYJOVÁ, D., KVĚT, J. (ed.) : Pond Littoral Ecosystems. Pp. 216 - 220. Springer-Verlag, Berlin - Heidelberg - New York 1978.

B33413 - DYKYJOVÁ, D., KVĚT, J. (ed.) : Pond Littoral Ecosystems. Structure and Func-. tioning. Methods and Results of Quantitative Ecosystem Research in the Cze- choslovakian IBP (International Biological Program) Wetland Project. Ecolo- gical Studies, Analysis and Synthesis. Vol. 28. Springer-Verlag, Berlin - Heidelberg - New York 1978. [Ps.]

33414 - DZHAGAROV, B.M., SALOKHIDDINOV, K.I., BONDAREV, S.L. : Tushenie molekulyar- nym kislorodom singletnogo i tripletnogo sostoyaniĭ porfirinov. [Quenching of singlet and triplet states of porphyrins by molecular oxygen.] - Biofizi- ka 23 : 762 - 767, 1978. [In R, ab : E.]

*33415 - DZIĘCIOŁ, U. : Studies on the distribution of $^{14}C$-assimilates in strawberry plants. - In : Annu. Rep. Isotope Lab. Fruit Biochem. Part II. Pp. 1 - 11. Res. Inst. Pomology, Skierniewice 1969.

33416 - EASTON, H.S. : A comparative study of genetic effects in isogenic diploid and tetraploid plants of *Festuca pratensis* HUDS. II. Leaf elongation and seedling growth studies. - Ann. Amélior. Plant. 28 : 609 - 622, 1978. [Growth analysis.]

33417 - EBRINGER, L. : Effects of drugs on chloroplasts. - Progress mol. subcell. Biol. 6 : 271 - 350, 1978.

33418 - EBRINGER, L., TRUBAČÍK, S., TUYET, T. : The photocatalytic breakage of nit- rofurans and its consequences on *Euglena gracilis*. - Acta Fac. Rerum nat. Univ. Comenianae, Microbiol. 6 : 1 - 16, 1978. [Chl.]

33419 - EDELMAN, M., REISFELD, A. : Characterization, translation and control of the 32000 dalton chloroplast membrane protein in *Spirodela*. - In : AKOYUNOGLOU, G., ARGYROUDI-AKOYUNOGLOU, J.H. (ed.) : Chloroplast Development. Pp. 641 - - 652. Elsevier/North-Holland Biomedical Press, Amsterdam - New York - Ox- ford 1978.

33420 - EDER, J., WAGENMANN, R., RÜDIGER, W. : Immunological relationship between phycoerythrins from various blue-green algae. - Immunochemistry 15 : 315 - - 321, 1978.

33421 - EDGERTON, M.E., MOORE, T.A., GREENWOOD, C. : Salt reversal of the acid-indu- ced changes in purple membrane from *Halobacterium halobium*. - FEBS Lett. 95 : 35 - 39, 1978.

*33422 - EDMONSON, W.T. : Lake Washington. - In : SEYB, L., RANDOLPH, K. (ed.) : North American Project - A study of U.S. Water Bodies.(Environm. Res. Lab. tech. Rep. *EPA-600/3-77-086*.) Pp. 288 - 300. Corvallis, Oregon 1977. [Ps, Chl.]

33423 - EDMUNDS, L.N., Jr., LAVAL-MARTIN, D.L., SHUCH, D.J. : Cell cycle-related and endogenous circadian rhythms of photosynthesis in *Euglena*. - Plant Phy- siol. 61 (Suppl.) : 14, 1978.

33424 - EDWARDS, G., ROBINSON, S., WALKER, D. : Requirements for isolating functio- nal chloroplasts of wheat and sunflower. - Plant Physiol. 61 (Suppl.) : 38, 1978.

33425 - EDWARDS, G.E., HUBER, S.C. : Usefulness of isolated cells and protoplasts for photosynthetic studies. - In : HALL, D.O., COOMBS, J., GOODWIN, T.W. (ed.) : Proceedings of the Fourth International Congress on Photosynthesis. Pp. 95 - 106. Biochem. Soc., London 1978.

33426 - EDWARDS, G.E., ROBINSON, S.P., TYLER, N.J.C., WALKER, D.A. : A requirement for chelation in obtaining functional chloroplasts of sunflower and wheat. - Arch. Biochem. Biophys. 190: 421 - 433, 1978.

33427 - EDWARDS, G.E., ROBINSON, S.P., TYLER, N.J.C., WALKER, D.A. : Photosynthesis
by isolated protoplasts, protoplast extracts, and chloroplasts of wheat. Influence of orthophosphate, pyrophosphate, and adenylates. - Plant Physiol.
*62* : 313 - 319, 1978.

33428 - EDWARDS, R.R.C. : Ecology of a coastal lagoon complex in Mexico. - Estuarine
coastal mar. Sci. *6* : 75 - 92, 1978. [Primary production.]

*33429 - EFIMTSEV, E.I., ARKHIPOV, V.N., LELËTKIN, V.A., LITVIN, F.F. : Issledovanie
biosinteza khlorofilla v ètiolirovannykh list'yakh metodom skorostnoĭ spek-
trofotometrii. [High speed spectrophotometric study of chlorophyll biosynthesis in etiolated leaves.] - Nauch. Dokl. vyssh. Shkoly, biol. Nauki *20*
(5) : 45 - 49, 1977. [In R.]

33430 - EFIMTSEV, E.I., BOĬCHENKO, V.A., EFIMTSEVA, È.P., LITVIN, F.F. : Polyarogra-
ficheskiĭ metod issledovaniya deĭstviya i kinetiki gazoobmena $O_2$ pri fotosinteticheskikh protsessakh. [Polarographic method for investigating the
action spectra and kinetics of $O_2$ exchange in photosynthesis and other biological processes.] - Fiziol. Rast. *25* : 860 - 868, 1978. [In R, ab : E.]

33431 - EGOROVA, L.I., SEMIKHATOVA, O.A., YUDINA, O.S. : Vliyanie temperatury na
reaktivatsiyu fotosinteza posle teplovogo povrezhdeniya. [Influence of temperature on the reactivation of photosynthesis after heat injury.] - Bot.
Zh. *63* : 356 - 362, 1978. [In R, ab : E.]

33432 - EGUCHI, H., MATSUI, T. : Computer control of plant growth by image processing. III. Image processing for evaluation of plant growth in practical cultivation. - Environ. Control Biol. *16* : 47 - 55, 1978.

33433 - EHLERINGER, J.R. : Implications of quantum yield differences on the distributions of $C_3$ and $C_4$ grasses. - Oecologia *31* : 255 - 267, 1978.

33434 - EHLERINGER, J.R., BJÖRKMAN, O. : Pubescence and leaf spectral characteristics in a desert shrub, *Encelia farinosa*. - Oecologia *36* : 151 - 162, 1978.

33435 - EHLERINGER, J.R., BJÖRKMAN, O. : A comparison of photosynthetic characteristics of *Encelia* species possessing glabrous and pubescent leaves. - Plant
Physiol. *62* : 185 - 190, 1978.

33436 - EHLERINGER, J.R., MOONEY, H.A. : Leaf hairs : effects on physiological activity and adaptive value to a desert shrub. - Oecologia *37* : 183 - 200, 1978.
[Ps.]

33437 - EHRENBERG, B., LEWIS, A. : The pK of Schiff base deprotonation in bacteriorhodopsin. - Biochem. biophys. Res. Commun. *82* : 1154 - 1159, 1978.

33438 - EHRLER, W.L., IDSO, S.B., JACKSON, R.D., REGINATO, R.J. : Diurnal changes
in plant water potential and canopy temperature of wheat as affected by
drought. - Agron. J. *70* : 999 - 1004, 1978. [Leaf temperature.]

33439 - EICHHORN, M. : Die Wirkung von Blau- und Rotlicht auf die Stoffproduktion
im Chemostat kultivierter *Scenedesmus*-Populationen. - Biochem. Physiol.
Pflanzen *172* : 417 - 420, 1978.

33440 - EICKMEIER, W.G. : Photosynthetic pathway distributions along an aridity
gradient in Big Bend National Park, and implications for enhanced resource
partitioning. - Photosynthetica *12* : 290 - 297, 1978.

33441 - EICKMEIER, W.G., ADAMS, M.S. : Gas exchange in *Agave lecheguilla* TORR.
(*Agavaceae*) and its ecological implications. - Southwestern Natur. *23* :
473 - 486, 1978.

33442 - EISBRENNER, G., DISTLER, E., FLOENER, L., BOTHE, H. : The occurrence of the
hydrogenase in some blue-green algae. - Arch. Microbiol. *118* : 177 - 184,
1978.

33443 - EISENBACH, M., GARTY, H., BAKKER, E.P., KLEMPERER, G., ROTTENBERG, H.,
CAPLAN, S.R. : Kinetic analysis of light-induced pH changes in bacteriorhodopsin-containing particles from *Halobacterium halobium*. - Biochemistry *17* :
4691 - 4698, 1978.

33444 - EISENBACH, M., KUPPERMANN, B., FLEISCHER, N., ROBINSON, T., CAPLAN, S.R.,
TANNY, G. : Purple membrane in solvents : solubility parameter mapping. -
In : CAPLAN, S.R., GINZBURG, M. (ed.) : Energetics and Structure of Halophi-
lic Microorganisms. Pp. 239 - 252. Elsevier/North-Holland Biomedical Press,
Amsterdam - New York 1978.

33445 - EISENBERG, D., BAKER, T.S., SUH, S.W., SMITH, W.W. : Structural studies of
ribulose 1,5-bisphosphate carboxylase/oxygenase. - In : SIEGELMAN, H.W.,
HIND, G. (ed.) : Photosynthetic Carbon Assimilation. Pp. 271 - 281. Plenum
Press, New York - London 1978.

*33446 - ELDER, J.F. : Iron uptake by freshwater algae and its diel variation. - In :
DRUCKER, H., WILDUNG, R.E. (ed.) : Biological Implications of Metals in the
Environment. Pp. 346 - 357. Technical Information Center, Energy Research
and Development Administration, Springfield 1977. [Ps, Chl.]

B33447 - Élektronnaya Mikroskopiya v Botanicheskikh Issledovaniyakh. [Electron Microscopy in Botanical Research.] - Zinatne, Riga 1978. [Chl, chloroplast; in R.]

33448 - ELEMA, R.P., MICHELS, P.A.M., KONINGS, W.N. : Response of 9-aminoacridine
fluorescence to transmembrane pH-gradients in chromatophores from *Rhodopseudomonas sphaeroides*. - Europe. J. Biochem. *92* : 381 - 387, 1978.

*33449 - EL-FOULY, M.M., ABO-EL LEL, G., EL-HINDI, M.H. : Effect of cycocel on growth,
flowering and yield of *Vicia Faba* I. - Agrochimica *19* : 374 - 379, 1975.
[Growth analysis.]

33450 - ELIAS, B.A., GIVAN, C.V. : Density gradient and differential centrifugation
methods for chloroplast purification and enzyme localization in leaf tissue.
The case of citrate synthase in *Pisum sativum* L. - Planta *142* : 317 - 320,
1978.

33451 - ELIÁŠ, P. : Transpiračné odpory listov a ich meranie. [Transpiration resistances of leaves and their measuring.] - Acta ecol. *7* : 53 - 115, 1978. [In
Slovak, ab : E, R.]

33452 - ELLENSON, J.L., PHEASANT, D.J., LEVINE, R.P. : Light/dark labeling differences in chloroplast membrane polypeptides associated with chloroplast coupling factor O. - Biochim. biophys. Acta *504* : 123 - 135, 1978.

33453 - ELLIS, R.J., BARACLOUGH, R. : Synthesis and transport of chloroplast prote-
ins inside and outside the cell. - In : AKOYUNOGLOU, G., ARGYROUDI-AKOYUNO-
GLOU, J.H. (ed.) : Chloroplast Development. Pp. 185 - 194. Elsevier/North-
-Holland Biomedical Press, Amsterdam - New York - Oxford 1978.

33454 - ELLIS, R.J., HIGHFIELD, P.E., SILVERTHORNE, J. : The synthesis of chloroplast proteins by subcellular systems. - In : HALL, D.O., COOMBS, J., GOOD-
WIN, T.W. (ed.) : Proceedings of the Fourth International Congress on Photosynthesis. Pp. 497 - 506. Biochem. Soc., London 1978.

33455 - ELLSWORTH, R.K., MURPHY, S.J. : Enzymatic preparation of Mg-protoporphyrin
IX monomethyl ester. - Photosynthetica *12* : 81 - 82, 1978.

33456 - ELLSWORTH, R.K., ST. PIERRE, M.E. : Enzymatic preparation of [$^3$H]NADH. -
Anal. Biochem. *88* : 338 - 339, 1978.

33457 - EL-SAYYAD, S., WAGNER, H. : A phytochemical study of *Calligonum comosum* L.
HENRY. - Planta med. *33* : 262 - 264, 1978.[Car.]

33458 - ELSTNER, E.F., FROMMEYER, D. : Analysis of different mechanisms of photosynthetic oxygen reduction. - Z. Naturforsch. *33 C* : 276 - 279, 1978.

33459 - ELSTNER, E.F., FROMMEYER, D. : Production of hydrogen peroxide by photosystem II of spinach chloroplast lamellae. - FEBS Lett. *86* : 143 - 146, 1978.

33460 - ELSTNER, E.F., FROMMEYER, D. : Three mechanisms of photosynthetic oxygen
reduction. - Plant Physiol. *61* (Suppl.) : 88, 1978.

33461 - ELSTNER, E.F., SARAN, M., BORS, W., LENGFELDER, E. : Oxygen activation
in isolated chloroplasts. Mechanism of ferredoxin-dependent ethylene formation from methionine. - Europe. J. Biochem. *89* : 61 - 66, 1978.

33462 - ELSTNER, E.F., ZELLER, H. : Bleaching of p-nitrosodimethylaniline by photosystem I of chloroplast lamellae. - Plant Sci. Lett. *13* : 15 - 20, 1978.

33463 - EL-ZEFTAWI, B.M., GARRETT, R.G. : Effects of ethephon, GA and light exclu-
sion on rind pigments, plastid ultrastucture and juice quality of Valencia
oranges. - J. hort. Sci. *53* : 215 - 223, 1978.

33464 - ENAMI, I., MATSUBAYASI, T., FUKUDA, I. : A new sensitive polarographic method for determination of hydroquinone in the Hill reaction; with special
reference to the case of thermal alga *Cyanidium*. - Photosynthetica *12* : 83 -
- 86, 1978.

33465 - ENOCH, H.Z., SACKS, J.M. : An empirical model of $CO_2$ exchange of a $C_3$ plant
in relation to light, $CO_2$ concentration and temperature. - Photosynthetica
*12* : 150 - 157, 1978.

33466 - EREZ, J. : Vital effect on stable-isotope composition seen in poraminifera
and coral skeletons. - Nature *273* : 199 - 202, 1978. [Ps.]

33467 - ERICSSON, A. : Seasonal changes in translocation of $^{14}C$ from different age-
-classes of needles on 20-year-old Scots pine trees (*Pinus silvestris*). -
Physiol. Plant. *43* : 351 - 358, 1978.

33468 - EROKHIN, Yu. E., CHUGUNOV, V.A., MAKHNEVA, Z.K., AGRIKOVA, I.M., VASIL'EV,
B.G. : Sravnitel'noe issledovanie pigment-lipoproteinovykh kompleksov B890
iz sernykh (*Chromatium minutissimum*) i nesernykh (*Rhodopseudomonas palustris*)
purpurnykh fotosintezI ruyushchikh bakteriĭ. [Comparison of pigment-lipoprotein B890 complexes from sulphur (*Chromatium minutissimum*) and non-sulphur
(*Rhodopseudomonas palustris*) purple photosynthesizing bacteria.] - Biokhimiya *43* : 669 - 677, 1978. [In R, ab : E.]

33469 - EROKHIN, Yu.E., VASIL'EV, B.G. : Molekulyarnaya organizatsiya dlinnovolno-
vykh kompleksov purpurnykh fotosintezIruyushchikh bakteriĭ. Issledovanie
deĭstviya pronazy na kompleksy B890 *Chromatium minutissimum* i *Rhodopseudomonas palustris*. [Molecular organization of long-wave complexes of photosynthesizing purple bacteria. Investigation of the pronase action on B890 complexes of *Chromatium minutissimum* and *Rhodopseudomonas palustris*.] - Mol.
Biol. (Moskva) *12* : 759 - 765, 1978. [In R, ab : E.]

*33470 - ESAU, K., CRONSHAW, J. : Plastids and mitochondria in the phloem of *Cucurbi-
ta*. - Can. J. Bot. *46* : 877 - 880, 1968.

33471 - ESCOBAR, D.E., GAUSMAN, H.W. : Effect of lead on reflectance of Mexican
squash plant leaves. - J. Rio Grande Valley hort. Soc. *32* : 81 - 88, 1978.

33472 - ESKINS, K., THOMAS, F.L., DUTTON, H.J. : Interaction of chloroplast pigment
ratios. - Plant Physiol. *61* (Suppl.) : 104, 1978.

33473 - ESTEP, M.F., TABITA, F.R., PARKER, P.L., VAN BAALEN, C. : Carbon isotope
fractionation by ribulose-1,5-bisphosphate carboxylase from various organisms. - Plant Physiol. *61* : 680 - 687, 1978.

33474 - ESTEP, M.F., TABITA, F.R., VAN BAALEN, C. : Purification of ribulose 1,5-bis-
phosphate carboxylase and carbon isotope fractionation by whole cells and
carboxylase from *Cylindrotheca* sp. (*Bacillariophyceae*). - J. Phycol. *14* :
183 - 188, 1978.

33475 - ESTÈVE, J., NICOLAS, P., NIGON, V. : Fitting survival curves with theoretical models; goodness-of-fit tests and parameter estimation. Application to
the clonal survival of *Chlorella* and the chloroplastic survival of *Euglena
gracilis* after irradiation. - Math. Biosci. *42* : 279 - 298, 1978.

*33476 - ESTEVEZ, M.P., MARTINEZ, M., VICENTE, C. : Photodestruction of β-carotene
and protection of the photo-oxydation by chloroatranorine. - An. Inst. bot.
Cavanilles *34* : 309 - 316, 1977.

33477 - ESTEVEZ LOPEZ, M.P., VINCENTE, C. : Ferredoxina y ferredoxin-NADP⁺-reductasa
de *Evernia prunastri* (L.) ACH. [Ferredoxin and ferredoxin-NADP⁺-reductase

from *Evernia prunastri* (L.) ACH.] - Rev. Bryol. Lichénol. *44* : 111 - 121, 1978. [In Span, ab : F.]

33478 - **ESYUNINA, A.I.** : Opredelenie aktivnosti oksidazy glikolevoǐ kisloty. [Determination of glycolic acid oxidase activity.] - Tr. priklad. Bot. Genet. Selektsii *61* (3) : 82 - 85, 1978. [In R, ab : E.]

B33479 - **ETHERINGTON, J.R.** : Plant Physiological Ecology. (The Institute of Biology's Studies in Biology, No. 98.) - Edward Arnold, London 1978. [Ps.]

33480 - **ETTL, H.** : Teilungsverhalten der Chromatophoren in bezug auf die Mitose während des Lebenszyklus von *Diatoma hiemale* var. *mesodon*, 'I. - Plant Syst. Evol. *129* : 315 - 322, 1978.

33481 - **EVANGELATOS, G.P., VAKIRTZI-LEMONIAS, C., AKOYUNOGLOU, G.** : Galactolipid changes in the developing thylakoids during etioplast-protochloroplast-chloroplast differentiation. - In : AKOYUNOGLOU, G., ARGYROUDI-AKOYUNOGLOU, J.H. (ed.) : Chloroplast Development. Pp. 303 - 308. Elsevier/North-Holland Biomedical Press, Amsterdam - New York - Oxford 1978.

33482 - **EVANS, E.H., CARR, N.G., EVANS, M.C.W.** : Changes in photosynthetic activity in the cyanobacterium *Chlorogloea fritschii* following transition from dark to light growth. - Biochim. biophys. Acta *501* : 165 - 173, 1978.

*33483 - **EVANS, E.H., CARR, N.G., RUSH, J.D., JOHNSON, C.E.** : Identification of a non-magnetic iron centre and an iron-storage or -transport material in blue-green algal membranes by Mössbauer spectroscopy. - Biochem. J. *166* : 547 - 551, 1977. [Ps.]

*33484 - **EVANS, E.H., CROFTS, A.R.** : Delayed fluorescence, the carotenoid shift and the light-induced membrane potential. - Biochem. Soc. Trans. *2* : 159 - 162, 1974.

33485 - **EVANS, M.C.W.** : Electron-paramagnetic-resonance studies of photosystem II. - Biochem. Soc. Trans. *6* : 906 - 908, 1978.

33486 - **EVANS, M.C.W.** : Light harvesting and reaction centers. - In : HALL, D.O., COOMBS, J., GOODWIN, T.W. (ed.) : Proceedings of the Fourth International Congress on Photosynthesis. Pp. 71 - 79. Biochem. Soc., London 1978.

33487 - **EVANS, M.C.W., HEATHCOTE, P., TELFER, A., BARBER, J.** : Redox potential dependence of electron transport and variable fluorescence in photosystem I. - In : DUTTON, P.L., LEIGH, J.S., SCARPA, A. (ed.) : Frontiers of Biological Energetics : Electrons to Tissues. Vol. 1. Pp. 241 - 248. Academic Press, New York - San Francisco - London 1978.

33488 - **EVANS, P.S.** : Plant root distribution and water use patterns of some pasture and crop species. - New Zeal. J. agr. Res. *21* : 261 - 265, 1978. [Dry-matter accumulation.]

33489 - **EVEN-CHEN, Z., ATSMON, D., ITAI, C.** : Hormonal aspects of senescence in detached tobacco leaves. - Physiol. Plant. *44* : 377 - 382, 1978. [Ps, Chl.]

33490 - **EVSTIGNEEV, V.B., PROSKURYAKOV, I.I., OLOVYANISHNIKOVA, G.D., VOZNYAK, V.M.** : Sravnitel'noe izuchenie fotopotentsiala i signala EPR, voznikayushchikh pri fotookislenii metallzameshchennykh feofitinov. [Comparative study of photopotential and EPR signal arising under photooxidation of metal-substituted pheophytins.] - Biofizika *23* : 379 - 380, 1978. [In R, ab : E.]

33491 - **FABIAN-GALAN, G., ATANASIU, L.** : Products of photosynthetic $^{14}CO_2$ incorporation in the lichens *Peltigera horizontalis* BAUMG. and *Usnea florida* WIGG. - Rev. roum. Biol., Sér. Biol. vég. *23* : 23 - 29, 1978.

*33492 - **FADRUS, H., MALÝ, J.** : Zur Problematik der Wirkung und Beseitigung der Nitrite bei der Sauerstoffbestimmung in Wässern. - Acta hydrochim. hydrobiol. *3* : 333 - 345, 1975.

33493 - **FAIR, P.** : An investigation into the effect of varying nitrogen regimes on the abundance of peroxisomes and activities on nitrate reductase and catalase in barley (*Hordeum vulgare* L.) and maize (*Zea mays* L.). - Ann. Bot. *42* : 101 - 107, 1978.

33494 - FAJER, J., DAVIS, M.S., BRUNE, D.C., FORMAN, A., THORNBER, J.P. : Optical and paramagnetic identification of a primary electron acceptor in bacterial photosynthesis. - J. amer. chem. Soc. 100 : 1918 - 1920, 1978.

33495 - FAKOREDE, M.A.B., MOCK, J.J. : Changes in morphological and physiological traits associated with recurrent selection for grain yield in maize. - Euphytica 27 : 397 - 409, 1978. [Ps.]

33496 - FALKOWSKI, P.G., OWENS, T.G. : Effects of light intensity on photosynthesis and dark respiration in six species of marine phytoplankton. - Mar. Biol. 45 : 289 - 295, 1978.

*33497 - FAL'TSMAN, A.V. : O vozmozhnosti povysheniya absolyutnoĭ tochnosti izmereniya vremeni zhizni fluorestsentsii fazovym metodom. [The possibility of increasing absolute accuracy of the measurement of life time of fluorescence by the phase method.] - Nauch. Dokl. vyssh. Shkoly, biol. Nauki 20 (7) : 123 - 126, 1977. [In R.]

33498 - FALUDI-DÁNIEL, Á., BIALEK, G.E., HORVÁTH, G., RÓZSA, Z.Sz., GREGORY, R.P.F. : Differential light-scattering of granal and agranal chloroplasts and their fragments. - Biochem. J. 174 : 647 - 651, 1978.

33499 - FALUDI-DÁNIEL, Á., HORVÁTH, G., DROPPA, M. : Energization of thylakoids in greening chloroplasts in maize. - In : AKOYUNOGLOU, G., ARGYROUDI-AKOYUNO-GLOU, J.H. (ed.) : Chloroplast Development. Pp. 427 - 432. Elsevier/North--Holland Biomedical Press, Amsterdam - New York - Oxford 1978.

33500 - FARES, Y., DeMICHELE, D.W., GOESCHL, J.D., BALTUSKONIS, D.A. : Continuously produced, high specific activity $^{11}C$ for studies of photosynthesis, transport and metabolism. - Int. J. appl. Radiat. Isotop. 29 : 431 - 441, 1978.

33501 - FARINEAU, J. : Field indicating absorbance changes in etiochloroplasts of maize leaves (mesophyll cells). - In : AKOYUNOGLOU, G., ARGYROUDI-AKOYUNO-GLOU, J.H. (ed.) : Chloroplast Development. Pp. 433 - 438. Elsevier/North--Holland Biomedical Press, Amsterdam - New York - Oxford 1978.

33502 - FARINEAU, J., GUILLOT-SALOMON, T., POPOVIC, R. : Etude de la structure et des activités photochimiques des deux types d'etiochloroplastes formés chez la feuille de Zea Mays L. au cours d'un verdissement en régime d'éclairs brefs. - Biol. celT. 32 : 307 - 316, 1978.

33503 - FARINEAU, N., HOFFELT, M., ROUSSAUX, J. : Interactions entre le chloramphénicol et la 6-benzylaminopurine au cours du verdissement de cotylédons de concombre. - Can. J. Bot. 56 : 1186 - 1197, 1978.

33504 - FARQUHAR, G.D. : Feedforward responses of stomata to humidity. - Aust. J. Plant Physiol. 5 : 787 - 800, 1978. [Model.]

33505 - FARQUHAR, G.D., DUBBE, D.R., RASCHKE, K. : Gain of the feedback loop involving carbon dioxide and stomata. Theory and measurement. - Plant Physiol. 62 : 406 - 412, 1978.

33506 - FARQUHAR, G.D., RASCHKE, K. : On the resistance to transpiration of the sites of evapotranspiration within the leaf. - Plant Physiol. 61 : 1000 - 1005, 1978. [Stomatal resistance.]

33507 - FARRAR, J.F. : Ecological physiology of the lichen Hypogymnia physodes. IV. Carbon allocation at low temperatures. - New Phytol. 81 : 65 - 69, 1978.

*33508 - FARRAR, J.F., RELTON, J., RUTTER, A.J. : Sulphur dioxide and the growth of Pinus sylvestris. - J. appl. Ecol. 14 : 861 - 875, 1977. [Ps.]

33509 - FAUST, M.A., GOFF, N.M., MIKLAS, J.J. : Carbon and phosphorus assimilation by phytoplankton and bacteria in the Rhode river estuary. - J. Phycol. 14 (Suppl.) : 32, 1978.

B33510 - FEDDES, R.A., KOWALIK, P.J., ZARADNY, H. : Simulation of Field Water Use and Crop Yield. - Pudoc, Wageningen 1978. [Ps.]

*33511 - FEDOROV, V.V., TRENKENSHU, R.P., BERESNEV, G.F. : Produktivnost' i éffektivnost' fotosinteza mikrovodorosleĭ pri obluchenii bezrtutnymi lyuminestsentnymi lampami krasnogo sveta. [Productivity and effectivity of photosynthesis of microalgae on irradiation with red light luminescent lamps.] - In : Intensivnaya Svetokul'tura Rasteniĭ. Pp. 211 - 214. Inst. Fiz. Sib. Otd. Akad. Nauk SSSR, Krasnoyarsk 1977. [In R.]

*33512 - FEDOROVA, E.I. : Nekotorye osobennosti tsikla prevrashcheniya organicheskogo veshchestva v vodoemakh-okhladitelyakh. [Some peculiarities of the cycle of organic matter transformation in cooling water reservoirs.] - In : Tipologiya Ozernogo Nakopleniya Organicheskogo Veshchestva. Pp. 108 - 129. Nauka, Moskva 1976. [Primary production ; In R.]

33513 - FEDOSEEVA, G.P. : Fenotipicheskaya izmenchivost' mezostruktury i funktsional'noĭ aktivnosti fotosinteticheskogo apparata. [Phenotypical variability of mesostructure and functional activity of photosynthetic tissues.] - In : Mezostruktura i Funktsional'naya Aktivnost' Fotosinteticheskogo Apparata. Pp. 112 - 131. Ural'skiĭ Gosudarstvennyĭ Universitet, Sverdlovsk 1978. [In R.]

33514 - FEE, E.J. : Studies of hypolimnion chlorophyll peaks in the Experimental Lakes Area, northwestern Ontario. - Fish. mar. Serv. Tech. Rep. *754* : 1 - 21, 1978.

33515 - FEHER, G., OKAMURA, M.Y. : Chemical composition and properties of reaction centers. - In : CLAYTON, R.K., SISTROM, W.R. (ed.) : The Photosynthetic Bacteria. Pp. 349 - 386. Plenum Press, New York - London 1978.

33516 - FEICK, R., DREWS, G. : Isolation and characterization of light harvesting bacteriochlorophyll·protein complexes from *Rhodopseudomonas capsulata*. - Biochim. biophys. Acta *501* : 499 - 513, 1978.

33517 - FEIERABEND, J. : Cooperation of cytoplasmic and plastidic protein synthesis in rye leaves. - In : AKOYUNOGLOU, G., ARGYROUDI-AKOYUNOGLOU, J.H. (ed.) : Chloroplast Development. Pp. 207 - 213. Elsevier/North-Holland Biomedical Press, Amsterdam - New York - Oxford 1978.

33518 - FEIERABEND, J., DE BOER, J. : Comparative analysis of the action of cytokinin and light on the formation of ribulosebisphosphate carboxylase and plastid biogenesis. - Planta *142* : 75 - 82, 1978.

33519 - FEIERABEND, J., SCHUBERT, B. : Comparative investigation of the action of several chlorosis-inducing herbicides on the biogenesis of chloroplasts and leaf microbodies. - Plant Physiol. *61* : 1017 - 1022, 1978.

33520 - FEIERABEND, J., WILDNER, G. : Formation of the small subunit in the absence of the large subunit of ribulose 1,5-bisphosphate carboxylase in 70 S ribosome-deficient rye leaves. - Arch. Biochem. Biophys. *186* : 283 - 291, 1978.

33421 - FEIGE, G.B. : Photosynthetische [14]C-Markierung heterocystenspezifischer Lipide in den Blaualgennestern der Haloragaceae *Gunnera manicata*. - Z. Pflanzenphysiol. *87* : 181 - 186, 1978.

*33522 - FEJER, S.O., PHILPOTTS, L.E., SPANGELO, L.P.S. : Precision of aerial photography in apple tree measurements. - Can. J. Plant Sci. *52* : 1083 - 1084, 1972.

33423 - FELLER, U., ERISMANN, K.H. : Veränderungen des Gaswechsels und der Aktivitäten proteolytischer Enzyme während der Seneszenz von Weizenblättern (*Triticum aestivum* L.). - Z. Pflanzenphysiol. *90* : 235 - 244, 1978.

33524 - FELLOWS, R.J., BOYER, J.S. : Altered ultrastructure of cells of sunflower leaves having low water potentials. - Protoplasma *93* : 381 - 395, 1978. [Chloroplast.]

33525 - FELLOWS, R.J., EGLI, D.B., LEGGETT, J.E. : A pod leakage technique for phloem translocation studies in soybean (*Glycine max* [L.] MERR.). - Plant Physiol. *62* : 812 - 814, 1978. [Photosynthates.]

33526 - FENERCIOGLU, H., CREAN, D.E. : Carotene content of green snap beans. - Ohio agr. Res. Dev. Cent. Res. Circ. *240* : 44 - 45, 1978.

33527 - FERERES, E., ACEVEDO, E., HENDERSON, D.W., HSIAO, T.C. : Seasonal changes in water potential and turgor maintenance in sorghum and maize under water stress. - Physiol. Plant. *44* : 261 - 267, 1978. [Stomatal resistance.]

33528 - FERGUSON, I.S., LEECH, J.II. : Generalized least squares estimation of yield functions. - Forest Sci. *24* : 27 - 42, 1978.

33529 - FERNANDEZ, J. : A simple system to determine photosynthesis in field conditions by means of $^{14}CO_2$. - Photosynthetica *12* : 145 - 149, 1978.

*33530 - FERNÁNDEZ, J., SANCHO, C. : Estudio comparativo de la inhibición que produce el CMU sobre la fotofosforilación, fotocarboxilación y reacción de hill en cebada. (*Hordeum vulgare* L.). [Comparative study of the inhibition produced by CMU on the photophosphorylation, photocarboxylation and Hill reaction in barley (*Hordeum vulgare* L.).] - Junta Energia nuclear, Rep..J.E.N. *382* : 1 - - 98, 1977. [In Span., ab : E.]

33531 - FERRARI, I., VILLANI, M. : Ricerche su fitoplancton e fitobentos in un lago di montagna, il Lago Santo Parmense. [Phytoplankton and phytobenthos investigations in mountain lake, Lago Santo Parmense.] - G. bot. ital. *112* : 229 - - 237, 1978. [Ps, Chl; in Ital., ab : E.]

33532 - FERREE, D.C. : Cultural factors influencing net photosynthesis of apple trees. - HortScience *13* : 650 - 652, 1978.

33533 - FERRI, G., COMERIO, G., IADAROLA, P., ZAPPONI, M.C., SPERANZA, M.L. : Subunit structure and activity of glyceraldehyde-3-phosphate dehydrogenase from spinach chloroplasts. - Biochim. biophys. Acta *522* : 19 - 31, 1978.

33534 - FERRIER, J.M. : Further theoretical analysis of concentration-pressure-flux waves in phloem transport systems. - Can. J. Bot. *56* : 1086 - 1090, 1978.

33535 - FEUILLADE, J., FEUILLADE, M., DRUART, J.-C., MENTHON, M. : Dosage de pigments, mesures possible du biovolume de Cyanophycées en Limnologie. - Bull. Soc. Phycol. France *23* : 62 - 87, 1978. [Car, biliproteins.]

33536 - FIALA, K. : Seasonal development of helophyte polycormones and relationship between underground and aboveground organs. - In : DYKYJOVÁ, D., KVĚT, J. (ed.) : Pond Littoral Ecosystems. Pp. 174 - 181. Springer-Verlag, Berlin - Heidelberg - New York 1978. [Dry-matter production.]

33537 - FICK, W.H., MOSER, L.E. : Carbon-14 translocation in three warm-season grasses as affected by stage of development. - J. Range Manage. *31* : 305 - 308, 1978.

33538 - FIKSDAHL, A., MORTENSEN, J.T., LIAAEN-JENSEN, S. : High-pressure liquid chromatography of carotenoids. - J. Chromatogr. *157* : 111 - 117, 1978.

33539 - FILATOV, G.V., SUPONINA, S.L., KOTOVA, G.P. : Fotosinteticheskaya aktivnost' gibridov kukuruzy v zavisimosti ot materinskoĭ formy. [Photosynthetic activity of maize hybrids as affected by the maternal form.]- Sel'skokhoz. Biol. *13* : 201 - 204, 1978. [In R, ab : E.]

*33540 - FILIMONOV, V.S., SOKOLOV, V.I., SID'KO, A.F. : Izuchenie spektral'noĭ yarkosti posevov pshenitsy. [Spectral properties of wheat canopies.] - In : Intensivnaya Svetokul'tura Rasteniĭ. Pp. 133 - 148. Inst. Fiz. sib. Otd. Akad. Nauk SSSR, Krasnoyarsk 1977. [In R.]

33541 - FILIPPOV, G.L. : Puti dal'neĭshego povysheniya k.p.d. fotosinteza posevov kukuruzy pri dostatochnom uvlazhnenii. [Ways for further increase of photosynthetic efficiency of maize stands under sufficient irrigation.] - Fiziol. Biokhim. kul't. Rast. *10* : 492 - 498, 1978. [In R, ab : E.]

*33542 - FINCH-SAVAGE, W.E., ELSTON, J. : The death of leaves in crops of field beans. - Ann. appl. Biol. *85* : 463 - 465, 1977. [Leaf-area formation.]

33543 - FINDENEGG, G.R., FISCHER, K. : Apparent photorespiration of *Scenedesmus obliquus* : Decrease during adaptation to low $CO_2$ level. - Z. Pflanzenphysiol. *89* : 363 - 371, 1978.

*33544 - FINENKO, Z.Z., LANSKAYA, L.A. : Rost i skorost' deleniya vodorosleĭ v limitirovannykh ob"emakh vody. [Growth and division rate of algae in limited water volumes.] - In : KHAĬLOVA, K.M. (ed.) : Ėkologicheskaya Fiziologiya Morskikh Planktonnykh Vodorosleĭ (v Usloviyakh Kul'tur). Pp. 22 - 50. Naukova Dumka, Kiev 1971. [Ps; in R.]

*33545 - FINENKO, Z.Z., TEN, V.S., AKININA, D.K., SERGEEVA, L.M., BERSENEVA, G.M. :
          Pigmenty v morskikh odnokletochnykh vodoroslyakh i intensivnost' fotosinte-
          za. [Pigments in unicellular marine algae and photosynthetic rate.] - In :
          KHAĬLOV, K.M.  (ed.) : Ėkologicheskaya Fiziologiya Morskikh Planktonnykh Vo-
          dorosleĭ (v Usloviyakh Kul'tur). Pp. 51 - 92. Naukova Dumka, Kiev 1971. [In
          R.]

33546 - FINNIGAN, J.J., MULHEARN, P.J. : A simple mathematical model of airflow in
          waving plant canopies. - Boundary-Layer Meteorol. 14 : 415 - 431, 1978.

33547 - FISCHER, R.A., TURNER, N.C. : Plant productivity in the arid and semiarid
          zones. - Annu. Rev. Plant Physiol. 29 : 277 - 317, 1978. [Ps.]

33548 - FISHER, D.B. : An evaluation of the Münch hypothesis for phloem transport in
          soybean. - Planta 139 : 25 - 28, 1978. [Photosynthates.]

33549 - FISHER, D.B., HOUSLEY, T.L., CHRISTY, A.L. : Source pool kinetics for $^{14}C$-
          -photosynthate translocation in morning glory and soybean. - Plant Physiol.
          61 : 291 - 295, 1978.

33550 - FITZGERALD, M.P., HUSAIN, A., ROGERS, L.J. : A constitutive flavodoxin from
          a eucaryotic alga. - Biochem. biophys. Res. Commun. 81 : 630 - 635, 1978.

33551 - FLEISCHHACKER, P., SENGER, H. : Adaptation of the photosynthetic apparatus
          of Scenedesmus obliquus to strong and weak light conditions. II. Differences
          in photochemical reactions, the photosynthetic electron transport and photo-
          synthetic units. - Physiol. Plant. 43 : 43 - 51, 1978.

33552 - FLEISCHMANN, D. : Delayed fluorescence and chemiluminescence. - In : CLAY-
          TON, R.K., SISTROM, W.R. (ed.) : The Photosynthetic Bacteria. Pp. 513 - 523.
          Plenum Press, New York - London 1978.

33553 - FLIEGE, R., FLÜGGE, U.-I., WERDAN, K., HELDT, H.W. : Specific transport of
          inorganic phosphate, 3-phosphoglycerate and triosephosphates across the inner
          membrane of the envelope in spinach chloroplasts. - Biochim. biophys. Acta
          502 : 232 - 247, 1978.

33554 - FLOWERS, T.J., HALL, J.L. : Salt tolerance in the halophyte, Suaeda mariti-
          ma (L.) DUM. : The influence of the salinity of the culture solution on the
          content of various organic compounds. - Ann. Bot. 42 : 1057 - 1063, 1978.
          [Ps.]

33555 - FLOWERS, T.J., HALL, J.L., WARD, M.E. : Salt tolerance in the halophyte,
          Suaeda maritima (L.) DUM. : Properties of malic enzyme and PEP carboxylase.
          - Ann. Bot. 42 : 1065 - 1074, 1978.

33556 - FLÜCKIGER, W., FLÜCKIGER-KELLER, H. : Veränderungen im Gehalt an β-Carotin
          und Vitamin C in der Petersilie im Nahbereich einer Autobahn. - Qual. Plant.
          - Plant Foods hum. Nutr. 28 : 1 - 9, 1978.

33557 - FLÜCKIGER, W., FLÜCKIGER-KELLER, H., OERTLI, J.J. : Der Einfluß verkehrsbe-
          dingter Luftverunreinigungen auf die Peroxydaseaktivität, das ATP-Bildungs-
          vermögen isolierter Chloroplasten und das Längenwachstum von Mais. - Z.
          Pflanzenkr. Pflanzenschutz 85 : 41 - 47, 1978.

33558 - FLÜCKIGER, W., FLÜCKIGER-KELLER, H., OERTLI, J.J. : Inhibition of regulatory
          ability of stomata caused by exhaust gases. - Experientia 34 : 1274, 1978.
          [Stomatal resistance.]

33559 - FLÜCKIGER, W., OERTLI, J.J., FLÜCKIGER-KELLER, H. : The effect of wind gusts
          on leaf growth and foliar water relations of aspen. - Oecologia 34 : 101 -
          - 106, 1978. [Growth analysis.]

33560 - FLÜGGE, U.I., HELDT, H.W. : Specific labelling of the active site of the
          phosphate translocator in spinach chloroplasts by 2,4,6-trinitrobenzene sul-
          fonate. - Biochem. biophys. Res. Commun. 84 : 37 - 44, 1978.

33561 - FOCK, H., KLUG, K., KRAMPITZ, M.-J. : Die spezifische Radioaktivität der
          Glykolsäure und der photorespiratorischen $CO_2$-Entwicklung von Helianthus-
          und Bohnenblättern nach $^{14}CO_2$-Aufnahme bei verschiedenen $CO_2$-Konzentrationen.
          - Ber. deut. bot. Ges. 91 : 539 - 549, 1978.

33562 - FOMISHYNA, R.M. : Vplyv zasolennya na pigmentnyĭ aparat tsukrovogo buryaku. [Effect of salinization on pigment apparatus of sugar beet.] - Ukr. bot. Zh. *35* : 652 - 656, 672, 1978. [In Ukr., ab : E,R.]

33563 - FONG, F.K., GALLOWAY, L. : The primary water splitting light reaction. Mass spectrometric determination of gaseous hydrogen and oxygen evolution from water photolysis by platinized chlorophyll *a* dihydrate polycrystals. - J. amer. chem. Soc. *100* : 3594 - 3596, 1978.

33564 - FONG, F.K., HOFF, A.J., BRINKMAN, F.A. : Electron spin resonance observation of the photooxidation of hydrated chlorophyll *a* dimers by water. *In vitro* photochemical characterization of reaction centers in photosynthesis. - J. amer. chem. Soc. *100* : 619 - 621, 1978.

33565 - FONTENO, W.C., McWILLIAMS, E.L. : Light compensation points and acclimatization of four tropical foliage plants. - J. amer. Soc. hort. Sci. *103* : 52 - - 56, 1978. [Ps.]

33566 - FOOTE, K.C., SCHAEDLE, M. : The contribution of aspen bark photosynthesis to the energy balance of the stem. - Forest Sci. *24* : 569 - 573, 1978.

*33567 - FORD, E.D., NEWBOULD, P.J. : The biomass and production of ground vegetation and its relation to tree cover through a deciduous woodland cycle. - J. Ecol. *65* : 201 - 212, 1977.

33568 - FORD, G.A., CATANZARO, B., FORK, D.C. : Computer analysis of kinetic data. - Carnegie Inst. Year Book *77* : 307 - 310, 1978. [Cytochrome absorbance changes.]

33569 - FORK, D.C., MURATA, N., SATO, N. : The lipid and fatty acid composition, electron transport, and light-energy redistribution in the thermophilic blue- -green alga *Synechococcus lividus* grown at 55 °C and 38 °C. - Carnegie Inst. Year Book *77* : 283 - 289, 1978.

33570 - FORNO, I.W., BOURNE, A.S. : An approach to estimating the standing crop of waterhyacinth. - J. aquat. Plant Manage. *16* : 50 - 52, 1978.

*33571 - FORTINI, S., MARIANI, B.M. : Ribulosio 1,5-difosfato e fosfoenolpiruvato carbossilasi e anidrasi carbonica in foglie di mais normale e *opaque*-2. [Ribulose 1,5-bisphosphate and phosphoenolpyruvate carboxylases and carbonic anhydrase in leaves of normal and *opaque*-2 maize.] - Maydica *16* : 83 - 94, 1971. [In Ital., ab : E.]

33572 - FOTEENKOVA, Z.D. : Rost i razvitie morkovi v zavisimosti ot gustoty stoyani- ya i doz udobreniĭ. [Growth and development of carrot as affected by stand density and fertilizer supply.] - In : Agrotekhnicheskie Priemy Promyshlennoĭ Tekhnologii v Ovoshchevodstve. Pp. 33 - 41. Min. sel'sk. Khoz. SSSR, Kishinev 1978. [Chl; in R.]

33573 - FOURNIER, D., PRALAVORIO, R., ARAMBOURG, Y. : Essai de mise au point de mé- thodes d'estimation simultanées des populations de feuilles, fleurs, fruits de l'Olivier et de leur ravageur. - Rev. Zool. agr. Pathol. vég. *77* : 25 - - 36, 1978.

33574 - FRACKOWIAK, D. : Local electric field effect on the chlorophyll fluorescen- ce. - Photochem. Photobiol. *28* : 377 - 382, 1978.

33575 - FRACKOWIAK, D. : Polarization of bacteriochlorophyll absorption in liquid crystal cell. - Acta phys. pol. *A 54* : 757 - 760, 1978.

33576 - FRACKOWIAK, D., SKOWRON, A. : Photoelectrochemical properties of biliproteins and biliprotein chromophoric groups. - Photosynthetica *12* : 76 - 80, 1978.

33577 - FRADKIN, L.I., KALININA, L.M., SAĬ, P.K. : Ab nakaplenni pigmentaŭ i ab ikh daŭgakhvalevaĭ fluarèstsèntsyi ŭ ètyyaliravanykh listsyakh u prysutnastsi èkzagennaĭ dèl'ta-aminalevulinavaĭ kislaty. [Pigment accumulation and long- -wave fluorescence in etiolated leaves in the presence of delta-levulinic acid.] - Vestsi Akad. Navuk belarus. SSR, Ser. biyal. Navuk *1978* (1) : 36 - - 41, 139, 1978. [In Belorus., ab : E, R.]

33578 - FRADKIN, L.I., KOLYAGO, V.M., NISENBAUM, G.D., DOMANSKAYA, I.N. : Dezinte- gratsiya i fraktsionirovanie khloroplastnykh membran yachmenya pri razlich-

nom soderzhanii digitonina i khloroplastov. [Disintegration and fractionation
of barley chloroplast membranes at different concentrations of digitonin and
chloroplasts.] - Biokhimiya *43* : 723 - 733, 1978. [In R, ab : E.]

33579 - FRADKIN, L.I., SHLYK, A.A. : Spektral'noe issledovanie gruppovoĭ lokalizatsii
molekul protokhlorofillida i khlorofillov v tsentrakh biosinteza. [Spectral
study of group localization of molecules of protochlorophyllide and chlorophylls in centres of biosynthesis.] - Zh. prikl. Spektroskopii *29* : 1029 -
- 1039, 1978. [In R, ab : E.]

33580 - FRAGATA, M. : Interaction of chlorophyll *a* with β-carotene in phosphatidyl
choline bilayers. - J. Colloid Interface Sci. *66* : 470 - 474, 1978.

33581 - FRALEY, R.T., JAMESON, D.M., KAPLAN, S. : The use of the fluorescent probe
α-parinaric acid to determine the physical state of the intracytoplasmic
membranes of the photosynthetic bacterium, *Rhodopseudomonas sphaeroides*. -
Biochim. biophys. Acta *511* : 52 - 69, 1978.

33582 - FRANCIS, K. : Photochemical activities of *Parthenium* chloroplasts. - Curr.
Sci. *47* : 899 - 900, 1978.

B33583 - FRANCOIS, L.E., MAAS, E.V. : Plant Responses to Salinity: An Indexed Biblio-
graphy. - Office of the Regional Administrator for Agricultural Research
(Western Region), Science and Education Administration, US Department of
Agriculture, Berkeley 1978. [Chl, Car.]

33584 - FRANK, G., SIDLER, W., WIDMER, H., ZUBER, H. : The complete amino acid sequence of both subunits of C-phycocyanin from the cyanobacterium *Mastigocladus laminosus*. - Hoppe-Seyler's Z. physiol. Chem. *359* : 1491 - 1507, 1978.

33585 - FRASCH, W., CHENIAE, G. : Flash induced inactivation of $O_2$ evolution by
Tris. - Plant Physiol. *61* (Suppl.) : 75, 1978.

*33586 - FRASER, J.E. : Water quality and phytoplankton productivity of Summersville
reservoir. - Proc. West Virginia Acad. Sci. *46* : 8 - 16, 1974.

33587 - FRASER, P.J.B., FRANCEY, R.J., PEARMAN, G.I. : Stable carbon isotopes in
tree rings as climatic indicators. - In : ROBINSON, B.W. (ed.) : Stable Isotopes in the Earth Sciences. DSIR Bull. Vol. 220. Pp. 67 - 73. New Zeal.
Dep. Sci. Ind. Res., Wellington 1978.

33588 - FRÉCHETTE, M., LEGENDRE, L. : Photosynthèse phytoplanctonique: réponse à un
stimulus simple, imitant les variations rapides de la lumière engendrées
par les vagues. - J. exp. mar. Biol. Ecol. *32* : 15 - 25, 1978.

33589 - FREEMAN, B.A., PLATT-ALOIA, K, MUDD, J.B., THOMSON, W.W. : Ultrastructural
and lipid changes associated with the aging of citrus leaves. - Protoplasma
*94* : 221 - 233, 1978.

33590 - FREEMAN, H.C., NORRIS, V.A., RAMSHAW, J.A.M., WRIGHT, P.E. : High resolution proton magnetic resonance studies of plastocyanin. - FEBS Lett. *86* :
131 - 135, 1978.

33591 - FREEMAN, H.C., NORRIS, V.A., RAMSHAW, J.A.M., WRIGHT, P.E. : High resolution
proton magnetic resonance studies of plastocyanin. - In : HALL, D.O.,
COOMBS, J., GOODWIN, T.W. (ed.) : Proceedings of the Fourth International
Congress on Photosynthesis. Pp. 805 - 809. Biochem. Soc., London 1978.

33592 - FREEMAN, T.P., DUYSEN, M.E. : The significance of protein synthesis early
in the etioplast to chloroplast conversion of wheat leaves as demonstrated
by cycloheximide treatment. - Protoplasma *97* : 111 - 124, 1978.

33593 - FREVERT, J., KINDL, H. : Plant microbody proteins. Purification and glycoprotein nature of glyoxysomal isocitrate lyase from cucumber cotyledons. -
Europe. J. Biochem. *92* : 35 - 43, 1978.

*33594 - FREY, D.G. : Paleolimnology. - Mitt. int. Ver. Limnol. *20* : 95 - 123, 1974.
[Chl.]

33595 - FREYER, H.D. : Preliminary evaluation of past $CO_2$ increase as derived from
$^{13}C$ measurements in tree rings. - In : WILLIAMS, J. (ed.) : Carbon Dioxide,
Climate and Society. Pp. 69 - 77. Pergamon Press, Oxford - New York - Toronto - Sydney - Paris - Frankfurt 1978.

33596 - FREYSSINET, G. : Le conditionnement général du verdissement d'*Euglena gracilis*. - Année biol. *17* : 245 - 280, 1978.

33597 - FREYSSINET, G., HARRIS, G., BINGHAM, S., STILLER, J., SCHIFF, J.A. : Cytochrome 552 formation in *Euglena gracilis* var. *bacillaris*. - Plant Physiol. *61* (Suppl.) : 82, 1978.

33598 - FREYSSINET, G., HARRIS, G., SCHIFF, J.A. : Synthesis of cytochrome *c*-552 in *Euglena gracilis*. - In : AKOYUNOGLOU, G., ARGYROUDI-AKOYUNOGLOU, J.H. (ed.) : Chloroplast Development. Pp. 267 - 270. Elsevier/North-Holland Biomedical Press, Amsterdam - New York - Oxford 1978.

33599 - FRIČ, F. : Transport fotoasimilátov označených $^{14}$C a zlúčenin označených $^{32}$P v jačmeni po infekcii hubou *Erysiphe graminis*, f.sp. *hordei* MARCHAL. [Transport of $^{14}$C-photosynthates and $^{32}$P-compounds in barley infected with *Erysiphe graminis*, f.sp. *hordei* MARCHAL.] - Acta bot. slov. Acad. Sci. slov. Ser. B 2 : 71 - 86, 1978. [In Slovak, ab : E, R.]

33600 - FRICK, H. : Compartmentation of pyrimidine metabolism in *Lemna minor*. - Plant Physiol. *61* (Suppl.) : 82, 1978. [Chl.]

33601 - FRICK, H. : Pyrimidine metabolism in *Lemna minor*. I. Functional compartmentation of chloroplast pyrimidine metabolism in a higher plant. - Plant Physiol. *61* : 989 - 992, 1978. [Chl.]

33602 - FROLOV, A.K. : Assimilyatsionnyľ apparat kustarnikov lesostepnoľ dubravy v raznykh usloviyakh osveshchennosti. [Assimilating apparatus of shrubs in forest-steppe oak-forests under different illumination.] - Bot. Zh. *63* : 1202 - 1206, 1978. [In R.]

33603 - FROSCH, S., BERGFELD, R., MOHR, H. : Control by phytochrome of ribulosebisphosphate carboxylase levels in mustard seedling cotyledons in the presence of norflurazon (SAN 9789). - In : AKOYUNOGLOU, G., ARGYROUDI-AKOYUNOGLOU, J.H. (ed.) : Chloroplast Development. Pp. 781 - 786. Elsevier/North-Holland Biomedical Press, Amsterdam - New York - Oxford    1978.

33604 - FRYDRYCH, J. : The effect of increased $CO_2$ concentration on photosynthetic rate of leaves of kohlrabi clones (*Brassica oleracea* var. *gongylodes* L.). - Z. Pflanzenzüchtung *80* : 158 - 161, 1978.

33605 - FUHRHOP, J.-H. : Molecular oxygen, light and metalloporphyrins. - In : METZNER, H. (ed.) : Photosynthetic Oxygen Evolution. Pp. 381 - 391. Academic Press, London - New York - San Francisco 1978. [Ps, Chl.]

33606 - FUJII, S., SHIMMEN, T., TAZAWA, M. : Light-induced changes in membrane potential in *Spirogyra*. - Plant Cell Physiol. *19* : 573 - 590, 1978.

33607 - FUJII, Y., KUROKAWA, T., YAMAGUCHI, I., MISATO, T. : Selective herbicidal activity of 3,3'-dimethyl-4-methoxybenzophenone (methoxyphenone, NK-049); absorption, translocation and metabolism. - Nippon Noyaku Gakkaishi [J. Pesticide Sci.] *3* : 291 - 298, 1978. [Chl, Car.]

*33608 - FUJIMOTO, K. : [On growth of regeneration trees and environmental factors in selection forests (III) Hydrophysiological conditions and chlorophyll content in leaves of Sugi seedlings in the model of group-selection stand.] - Bull. Ehime Univ. Forest. *14* : 25 - 34, 1977. [In Jap., ab : E.]

33609 - FUJITA, I., DAVIS, M.S., FAJER, J. : Anion radicals of pheophytin and chlorophyll *a* : their role in the primary charge separations of plant photosynthesis. - J. amer. chem. Soc. *100* : 6280 - 6282, 1978.

33610 - FUJITA, Y. : Carotenoid photobleaching induced by photosystem II action in the membrane fragments of the blue-green alga *Anabaena variabilis*: Action of $O_2$. - Plant Cell Physiol. *19* : 1129 - 1136, 1978.

33611 - FUKAI, S., SILSBURY, J.H. : A growth model for *Trifolium subterraneum* L. swards. - Aust. J. agr. Res. *29* : 51 - 65, 1978.

33612 - FULLER, R.C. : Photosynthetic carbon metabolism in the green and purple bacteria. - In : CLAYTON, R.K., SISTROM, W.R. (ed.) : The Photosynthetic Bacteria. Pp. 691 - 705. Plenum Press, New York - London 1978.

*33613 - FURYAEV, E.A., TERSKOV, I.A., BELYANIN, V.N. : Spektrofotometricheskie kha-
rakteristiki otdel'nykh kletok *Porphyridium cruentum* v periodicheskoĭ kul'-
ture. [Spectrophotometric characteristics of individual *Porphyridium cruentum* cells in periodical culture.] - In : Intensivnaya Svetokul'tura Rasteniĭ.
Pp. 201 - 211. Inst. Fiz. sib. Otd. Akad. Nauk SSSR, Krasnoyarsk 1977. [Chl,
biliproteins; In R.]

33614 - FUSSELL, L.K., PEARSON, C.J. : Effect of thermal history on photosynthate
translocation and photosynthesis. - Aust. J. Plant Physiol. *5* : 547 - 551,
1978.

33615 - GABRYŚ, H., WALCZAK, T. : The effect of temperature and $CO_2$ concentration on
light-induced chloroplast movements in *Tradescantia* leaves. - Acta protozool.
*18* : 131 - 132, 1978.

*33616 - GACHKOVSKII, V.F. : Changes in the fluorescent spectra of Mg-phthalocyanine
complexes with their transformation to the adsorbed state. - Adv. mol. Relax.
Processes *5* : 107 - 113, 1973. [Chl.]

33617 - GACHKOVSKIĬ, V.F. : O sostoyanii khlorofilla v zhivykh list'yakh rasteniĭ.
[State of chlorophyll in the living leaves of plants.] - Dokl. Akad. Nauk
SSSR *243* : 684 - 687, 1978. [In R.]

33618 - GACHKOVSKY, V.F. : On "the red shift" in the spectra of chlorophyll in plant
leaves. - J. mol. Struct. *47* : 443 - 461, 1978.

33619 - GÄCHTER, R., DAVIS, J.S., MARÈS, A. : Regulation of copper availability to
phytoplankton by macromolecules in lake water. - Environ. Sci. Technol. *12* :
1416 - 1421, 1978. [Ps.]

33620 - GALKIN, V.I. : Metodika otsenki potentsiala fotosinteticheskoĭ i ėkonomiches-
koĭ produktivnosti sortov zernovykh kul'tur. [Methods for the estimation of
photosynthetic and economic potential productivity of grain crop varieties.]
- Tr. priklad. Bot. Genet. Selektsii *61* (3) : 54 - 61, 1978. [In R, ab : E.]

33621 - GALKIN, V.I., KOSHKIN, V.A., POLYAKOV, M.I., GAMMERMAN, A.Ya., LANIN, M.I. :
Regresionnyĭ analiz sortovykh razlichiĭ v skorosti foto- i biokhimicheskikh
reaktsiĭ fotosinteza rasteniĭ. [Regression analysis of varietal differences
as regards the rate of photosynthetic photo- and biochemical reactions in
plants.] - Tr. priklad. Bot. Genet. Selektsii  *61*(3) : 62 - 67, 1978. [In R,
ab : E.]

*33622 - GALKINA, V.N. : Vozdeĭstvie rastvorimykh organicheskikh soedineniĭ ėkskre-
mentov morskikh kolonial'nykh ptits na fotosintez fitoplanktona. [Action
of soluble organic substances in excrements of marine colonial birds on phytoplankton photosynthesis.] - Ėkologiya *1977*(5) : 77 - 82, 1977. [In R.]

33623 - GALLACHER, A.E., SPRENT, J.I. : The effect of different water regimes on
growth and nodule development of greenhouse-grown *Vicia faba*. - J. exp. Bot.
*29* : 413 - 423, 1978.

33624 - GALLAGHER, J.L. : Estuarine angiosperms : Productivity and initial photosynthate dispersion in the ecosystem. - In : WILEY, M.L. (ed.) : Estuarine Interactions. Pp. 131 - 143. Academic Press, New York - San Francisco - London
1978.

33625 - GALLING, G. : Development of chloroplast structure and function in a tempe-
rature-sensitive mutant of *Chlorella*. - In : AKOYUNOGLOU, G., ARGYROUDI-AKO-
YUNOGLOU, J.H. (ed.) : Chloroplast Development. Pp. 439 - 444. Elsevier/
North-Holland Biomedical Press, Amsterdam - New York - Oxford 1978.

33626 - GALLOWAY, L., ROETTGER, J., FRUGE, D.R., FONG, F.K. : Photochemical upconver-
sion in the chlorophyll *a* water splitting light reaction: Causative factors
underlying the two-quanta/electron requirement in plant photosynthesis. -
J. amer. chem. Soc. *100* : 4635 - 4638, 1978.

33627 - GALSTON, A.W. : Leaf movements and the analysis of plant behavior. - Bot.
Mag. (Tokyo) *1978*(Special Issue 1) : 243 - 253, 1978.

B33628 - GAMALEĬ, Yu.V., KULIKOV, G.V. : Razvitie Khlorenkhimy Lista. [Development of Leaf Chlorenchyma.] - Nauka, Leningrad 1978. [In R.]

33629 - GAMBURG, K.Z. : The influence of 1-naphthaleneacetic acid and (2-chloro-ethyl)-trimethylammonium chloride on the carotenoid content of tobacco tissue in suspension culture. - Biol. Plant.$20$ : 93 - 97, 1978.

33630 - GAPONENKA, V.I., BALEVA, E.F., SHAŬCHUK, S.M. : Abnaŭlenne khlarafilu i in-ténsiŭnasts' fotasintezu listsyaŭ kukuruzy roznaga ŭzrostu. [Chlorophyll recovery and photosynthetic rate of maize leaves of different age.] - Vestsi Akad. Navuk belarus. SSR, Ser. biyal. Navuk $1978$ (1) : 21 - 25, 138, 1978. [In Belorus., ab : E, R.]

*33631 - GARCIA-MARTINEZ, J.L. : Fotorrespiracion y productividad vegetal. [Photorespiration and plant productivity.] - Rev. Agroquím. Tecnol. alim. $14$ : 519 - 528, 1974. [In Span.]

33632 - GARRETT, H.E., COX, G.S., ROBERTS, J.E. : Spatial and temporal variations in carbon dioxide concentrations in an oak-hickory forest ravine. - Forest Sci. $24$ : 180 - 190, 1978.

33633 - GARRETT, M.K. : Control of photorespiration at RuBP carboxylase/oxygenase level in ryegrass cultivars. - Nature $274$ : 913 - 915, 1978.

33634 - GARSIDE, C., HULL, G., YENTSCH, C.S. : Coastal source waters and their role as a nitrogen source for primary production in an estuary in Maine. - In : WILEY, M.L. (ed.) : Estuarine Interactions. Pp. 565 - 575. Academic Press, New York - San Francisco - London 1978.

33635 - GARSIDE, C., MALONE, T.C. : Monthly oxygen and carbon budget of the New York Bight Apex. - Est. coast. mar. Sci. $6$ : 93 - 104, 1978.

33636 - GARTY, H., CAHEN, D., CAPLAN, S.R. : Use of photoacoustic spectroscopy in the study of the bioenergetics of purple membrane. - In : CAPLAN, S.R., GINZBURG, M. (ed.) : Energetics and Structure of Halophilic Microorganisms. Pp. 253 - 259. Elsevier/North-Holland Biomedical Press, Amsterdam - New York 1978.

33637 - GARTY, H., EISENBACH, M., SHULDMAN, R., CAPLAN, S.R. : Effects of ionophores on light-induced pH changes in sub-bacterial particles of *Halobacterium halobium*. - In : CAPLAN, S.R., GINZBURG, M. (ed.) : Energetics and Structure of Halophilic Microorganisms. Pp. 261 - 267. Elsevier/North-Holland Biomedical Press, Amsterdam - New York 1978.

33638 - GARTY, H., FLUHR, R., EISENBACH, M., CAPLAN, S.R. : Effects of Triton-X detergents on purple-membrane fragments. - In : CAPLAN, S.R., GINZBURG, M. (ed.) : Energetics and Structure of Halophilic Microorganisms. Pp. 269 - 276. Elsevier/North-Holland Biomedical Press, Amsterdam - New York 1978.

33639 - GASSMAN,M.L., DUGGAN, J.X., STILLMAN, L.C., VLCEK, L.M., CASTELFRANCO, P.A., WEZELMAN, B. : Oxidation of chlorophyll precursors and its relation to the control of greening. - In : AKOYUNOGLOU, G., ARGYROUDI-AKOYUNOGLOU, J.H.(ed.): Chloroplast Development. Pp. 167 - 181. Elsevier/North-Holland Biomedical Press, Amsterdam - New York - Oxford 1978.

33640 - GAST, P., HOFF, A.J. : Determination of the decay rates of the triplet state of *Rhodopseudomonas sphaeroides* by fast laser-flash ESR spectroscopy. - FEBS Lett. $85$ : 183 - 188, 1978.

*33641 - GASYMOV, S.G. : Izuchenie deĭstviya atrazina v posevakh kukuruzy na fone mineral'nykh udobreniĭ. [Atrazin action in maize crops related to mineral fertilization.] - Temat. Sbornik Trudov azerb. nauch.—issled. Inst. Zemled. (Baku) $16$ : 33 - 36, 1976. [Chl, growth analysis; in R, ab : Azerb.]

33642 - GATENBY, A.A. : A comparison of the polypeptide isoelectric points and antigenic determinant sites of the large subunit of Fraction 1 protein from *Lycopersicon esculentum, Nicotiana tabacum* and *Petunia hybrida*. - Biochim. biophys. Acta $534$ : 169 - 172, 1978.

33643 - GATENBY, A.A., COCKING, E.C. : The evolution of Fraction 1 protein and the distribution of the small subunit polypeptide coding sequence in the genus *Brassica*. - Plant Sci. Lett. $12$ : 299 - 303, 1978.

33644 - **GATENBY, A.A., COCKING, E.C.** : The polypeptide composition of the sub-units of Fraction 1 protein in the genus *Lycopersicon*. - Plant Sci. Lett. *13* : 171 - 176, 1978.

33645 - **GATENBY, A.A., COCKING, E.C.** : Fraction 1 protein and the origin of the European potato. - Plant Sci. Lett. *12* : 177 - 181, 1978.

33646 - **GAUTHIER, D., BELOT, Y.** : Mesure de la résistance stomatique des aiguilles de Conifères. - Oecol. Plant. *13* : 13 - 25, 1978.

*33647 - **GAVRILENKO, V.F., ZHIGALOVA, T.V., AGAFODOROVA, M.N.** : Izuchenie svoĭstv so-pryagayushchikh belkov khloroplastov pshenitsy razlichnoĭ produktivnosti. [Properties of coupling chloroplast proteins of wheat of different productivity.] - Nauch. Dokl. vyssh. Shkoly, biol. Nauki *20* (5) : 99 - 108, 1977. [In R.]

33648 - **GAYNOR, J.J., PRICE, C.A.** : Synthesis of proteins by chloroplasts from iron--deficient *Euglena gracilis*. - Plant Physiol. *61* (Suppl.) : 83, 1978.

33649 - **GEACINTOV, N.E., SHENBERG, C.E., CAMPILLO, A.J., HYER, R.C., SHAPIRO, S.L., WINN, K.R.** : A picosecond pulse train study of exciton dynamics in photosynthetic membranes. - Biophys. J. *24* : 347 - 359, 1978.

33650 - **GEIGER, B., MEVARECH, M., WERBER, M.M.** : Immunochemical characterization of ferredoxin from *Halobacterium* of the Dead Sea. - Europe. J. Biochem. *84* : 449 - 455, 1978.

33651 - **GEIKE, F.** : Effect of hexachlorobenzene on the activity of some enzymes from *Tetrahymena pyriformis*. - Bull. environ. Contam. Toxicol. *20* : 640 - 646, 1978. [Chl, Ps.]

33652 - **GEIKE, F., PARASHER, C.D.** : Effect of hexachlorobenzene (HCB) on photosynthetic oxygen evolution and respiration of *Chlorella pyrenoidosa*. - Bull. environm. Contam. Toxicol. *20* : 647 - 651, 1978.

33653 - **GEIKE, F., PARASHER, C.D., KLOKE, A.** : Schäden an Poinsettien (*Euphorbia pulcherrima*) nach praxisüblicher Anwendung von Tamaron und deren mögliche biochemische Ursachen. - Z. Pflanzenkr. Pflanzenschutz *85* : 321 - 327, 1978. [Ps.]

33654 - **GEISLER, G., STAMP, P.** : Chlorophyllgehalt und PEP-Carboxylase-Aktivität von Blättern junger Maispflanzen bei unterschiedlicher Kaliumversorgung. - Z. Acker- Pflanzenbau *147* : 181 - 189, 1978.

33655 - **GEJ, B., WŁODKOWSKI, M.** : Wpływ CCC na asymilację $^{14}CO_2$ i wzrost niektórych roślin uprawnych. [CCC effect on the $^{14}CO_2$ assimilation and the growth of some crops.] - Rocz. Nauk rol. *103* : 29 - 46, 1978. [In Pol., ab : E, R.]

33656 - **GELIN, C., RIPL, W.** : Nutrient decrease and response of various phytoplankton size fractions following the restoration of Lake Trummen, Sweden. - Arch. Hydrobiol. *81* : 339 - 367, 1978.

33657 - **GENDEL, S., OHAD, I., BOGORAD, L.** : Control of phycoerythrin synthesis during chromatic adaptation in the cyanophyte *Fremyella diplosiphon*. - Plant Physiol. *61* (Suppl.) : 15, 1978.

33658 - **GENDRAUD, M., LAFLEURIEL, J.** : Étude comparée des métabolismes glucidiques et nucléotidiques des axes de très jeunes plantes de Betteraves fourragères et sucrières. - Physiol. vég. *16* : 679 - 691, 1978. [Photosynthates.]

33659 - **GENEROZOVA, I.P.** : Strukturno-funktsional'naya kharakteristika khloroplastov v usloviyakh zasukhi. [Structural and functional peculiarities of chloroplasts under drought conditions.] - In : Problemy Zasukhoustoĭchivosti Rasteniĭ. Pp. 183 - 205. Nauka, Moskva 1978. [In R.]

33660 - **GENKEL', P.A.** : Adaptatsiya rasteniĭ k ĕkstremal'nym usloviyam okruzhayush-cheĭ sredy. [Plant adaptation to unfavourable environmental conditions.] - Fiziol. Rast. *25* : 889 - 902, 1978. [Ps; in R, ab : E.]

33661 - **GEPSHTEIN, A., CARMELI, C., NELSON, N.** : Purification and properties of adenosine triphosphatase from *Chromatium vinosum* chromatophores. - FEBS Lett. *85* : 219 - 223, 1978.

B33662 - GERHARDT, B. : Microbodies / Peroxisomen pflanzlicher Zellen. - Springer-Verlag, Wien - New York 1978.

33663 - GERWICK, B.C., SPALDING, M.H., WILLIAMS, G.J. III., EDWARDS, G.E. : Temperature response of $CO_2$ fixation in isolated *Opuntia* cells. - Plant Physiol.
*61* (Suppl.) : 37, 1978.

33664 - GERWICK, B.C., WILLIAMS, G.J. III. : Temperature and water regulation of gas
exchange of *Opuntia polyacantha*. - Oecologia *35* : 149 - 159, 1978. [Ps.]

33665 - GERWICK, B.C., WILLIAMS, G.J., SPALDING, M.H., EDWARDS, G.E. : Temperature
response of $CO_2$ fixation in isolated *Opuntia* cells. - Plant Sci. Lett. *13* :
389 - 396, 1978.

33666 - GEUTLER, G., KROCHMANN, J. : Die Messung der für die Photosynthese wirksamen
Bestrahlungsstärke. - Gartenbauwissenschaft *43* : 271 - 275, 1978.

33667 - GHEORGHE, V., ȚUGULEA, L., BĂLĂNESCU, C., SIMPLĂCEANU, V. : NMR evidence of
the chlorophyll *a*-water interaction. - Rev. roum. Physiol. *23* : 1179 - 1194,
1978.

33668 - GHILDIYAL, M.C., TOMAR, O.P.S., SIROHI, G.S. : Response of cowpea genotypes
to zinc in relation to photosynthesis, nodulation and dry matter distribution. - Plant Soil *49* : 505 - 516, 1978.

33669 - GHOLZ, H.L. : Assessing stress in *Rhododendron macrophyllum* through an analysis of leaf physical and chemical characteristics. - Can. J. Bot. *56* : 546
- 556, 1978. [Growth analysis.]

33670 - GIANNOPOLITIS, C.N., AYERS, G.S. : Enhancement of chloroplast photooxidations with photosynthesis-inhibiting herbicides and protection with NADH or
NADPH. - Weed Sci. *26* : 440 - 443, 1978.

33671 - GIAQUINTA, R. : Source and sink leaf metabolism in relation to phloem translocation. Carbon partitioning and enzymology. - Plant Physiol. *61* : 380 -
385, 1978.

33672 - GIBBONS, G.C. : On the mechanical instability of ribulose-1,5-bisphosphate
carboxylase. - Carlsberg Res. Commun. *43* : 195 - 202, 1978.

33673 - GIBBONS, G.C. : The mechanical instability of ribulose 1,5-bisphosphate carboxylase. - In : SIEGELMAN, H.W., HIND, G. (ed.) : Photosynthetic Carbon
Assimilation. P. 419. Plenum Press, New York - London 1978.

33674 - GIBBS, S.P. : The chloroplasts of *Euglena* may have evolved from symbiotic
green algae. - Can. J. Bot. *56* : 2883 - 2889, 1978.

33675 - GIDDINGS, J., EDDLEMON, G.K. : Photosynthesis/respiration ratios in aquatic
microcosmos under arsenic stress. - Water Air Soil Pollut. *9* : 207 - 212,
1978.

33676 - GIESKES, W.W.C., KRAAY, G.W., TIJSSEN, S.B. : Chlorophylls and their degradation products in the deep pigment maximum layer of the tropical North
Atlantic. - Neth. J. Sea Res. *12* : 195 - 204, 1978.

*33677 - GIGAURI, D.G. : Fotosinteticheskiĭ apparat I intensivnost' fotosinteza pod-
rosta eli vostochnoĭ (*Picea orientalis*) v svyazi s osvetitel'nymi rubkami
D.Kravchinskogo. [Photosynthetic apparatus and photosynthetic rate of *Picea
orientalis* regrowth in relation with the secondary fellings of D.Kravchin-
skiĭ.] - In : Voprosy Gornogo Lesovedeniya I Lesovodstva v Gruzii. Vol. 25.
Pp. 112 - 116. Sabchota Adzhara, Batumi 1976. [In R, ab : E, Georg.]

33678 - GILBERT, C.W., BUETOW, D.E. : One- and two-dimensional gel electrophoretic
analysis of polypeptides synthesized *in vivo* by *Euglena* chloroplasts. -
In : AKOYUNOGLOU, G., ARGYROUDI-AKOYUNOGLOU, J.H. (ed.) : Chloroplast Development. Pp. 271 - 276. Elsevier/North-Holland Biomedical Press, Amsterdam -
New York - Oxford 1978.

33679 - GILES, K.L. : The uptake of organelles and microorganisms by plant protoplasts : old ideas but new horizons. - In : THORPE, T.A. (ed.) : Frontiers
of Plant Tissue Culture 1978. Pp. 67 - 74. Int. Assoc. Plant Tissue Culture,
Calgary 1978. [Chloroplast.]

33680 - GILL, K.S., SINGH, O.S. : Physiological response of dwarf wheat to chloro-
cholinechloride under soil moisture stress. - Biol. Plant. 20 : 421 - 424,
1978. [Chl.]

33681 - GILLBRO, T. : Flash kinetic study of the last steps in the photoinduced reaction cycle of bacteriorhodopsin. - Biochim. biophys. Acta 504 : 175 - 186,
1978.

33682 - GILLBRO, T. : Kinetics of late intermediates in the photocycle of bacteriorhodopsin. - In : CAPLAN, S.R., GINZBURG, M. (ed.) : Energetics and Structure
of Halophilic Microorganisms. Pp. 277 - 282. Elsevier/North-Holland Biomedical Press, Amsterdam - New York, 1978.

33683 - GILLER, Yu.E. : Model' pigment-belkovoĭ globuly khloroplastnoĭ membrany.
[Model of pigment-protein globule of chloroplast membrane.] - Stud. biophys.
71 : 99 - 114, 1978. [In R, ab : E.]

33684 - GILLER, Yu.E., VAKHIDOVA, L.R. : O deĭstvii ingibitorov transkriptsii i
translyatsii na soderzhanie plastidnykh pigmentov v prorostkakh khlopchatnika i gorokha. [Effect of transcription and translation inhibitors on the
plastid pigment content in seedlings of cotton and pea plants.] - Fiziol.
Biokhim. kul't. Rast. 10 : 547 - 551, 1978. [In R, ab : E.]

B33685 - GILLHAM, N.W. : Organelle Heredity. - Raven Press, New York 1978. [Chloro-
plast.]

33686 - GILLHAM, N.W., BOYNTON, J.E., CHUA, N.-H. : Genetic control of chloroplast
proteins. - In : SANADI, D.R., VERNON, L.P. (ed.) : Current Topics in Bioenergetics. Vol. 8. Pp. 211 - 260. Academic Press, New York - San Francisco
- London 1978.

33687 - GIMMLER, H., KÜHNL, E.M., CARL, G. : Salinity dependent resistance of *Dunaliella parva* against extreme temperatures. I. Salinity and thermoresistance.
- Z. Pflanzenphysiol. 90 : 133 - 153, 1978. [Ps.]

33688 - GINGRAS, G. : A comparative review of photochemical reaction center preparations from photosynthetic bacteria. - In : CLAYTON, R.K., SISTROM, W.R. (ed.) :
The Photosynthetic Bacteria. Pp. 119 - 131. Plenum Press, New York - London
1978.

33689 - GINZBURG, B.Z. : Regulation of cell volume and osmotic pressure in *Dunaliella.*. - CAPLAN, S.R., GINZBURG, M. (ed.) : Energetics and Structure of Halophilic Microorganisms. Pp. 543 - 560. Elsevier/North-Holland Biomedical Press,
Amsterdam - New York 1978. [Carbonic anhydrase.]

33690 - GIRAULT, G., GALMICHE, J.M. : Chlortetracycline as a fluorescent probe of the
first nucleotide binding site of the coupling factor $CF_1$ of spinach chloroplasts. - FEBS Lett. 95 : 135 - 139, 1978.

33691 - GIRAULT, G., GALMICHE, J.M. : Effect of nucleotides on potential and pH changes across the thylakoid membrane of spinach chloroplasts. - Biochim. biophys. Acta 502: 430 - 444, 1978.

33692 - GIRNTH, C., BERGFELD, R., KASEMIR, H. : Phytochrome-mediated control of grana and stroma thylakoid formation in plastids of mustard cotyledons. - Planta
141 : 191 - 198, 1978.

33693 - GIRS, G.I., ZUBAREVA, O.N. : Izmenenie aktivnosti khlorofillazy v khvoe sosny obyknovennoĭ pod deĭstviem vysokikh temperatur. [Change in chlorophyllase
activity in pine needles induced by high temperature.] - Izv. sib. Otd. Akad.
Nauk SSSR, Ser. biol. Nauk 1978 (2) : 116 - 118, 1978. [In R, ab : E.]

33694 - GIURGEVICH, J.R., DUNN, E.L. : Seasonal patterns of $CO_2$ and water vapor exchange of *Juncus roemerianus* SCHEELE in a Georgia salt marsh. - Amer. J. Bot.
65 : 502 - 510, 1978.

33695 - GIVNISH, T.J. : Ecological aspects of plant morphology : leaf form in relation to environment. - In : SATTLER, R. (ed.) : Theoretical Plant Morphology.
Pp. 83 - 142. Leiden University Press, Leiden 1978. [Ps.]

33696 - GLACOLEVA, T.A., ZALENSKY, O.V., MOKRONOSOV, A.T. : Oxygen effects on photosynthesis and $^{14}C$ metabolism in desert plants. - Plant Physiol. 62 : 204 -
- 209, 1978.

*33697 - GLADYSHEV, N.P., KON'SHIN, A.T. : Produktivnost' fotosinteza list'ev v raznykh uchastkakh krony yabloni. [Photosynthesis productivity of leaves in different parts of apple-tree crown.] - Nauch. Tr. voronezhsk. sel'skokhoz. Inst. *73* (Biologiya, Agrotekhnika i Selektsiya Plodovykh Rastenii) : 131 - 138, 1975. [In R.]

*33698 - GLADYSHEV, N.P., LEMESHKO, N.E. : Produktivnost' fotosinteza list'ev yabloni v raznykh usloviyakh osveshcheniya. [Photosynthesis productivity of apple leaves under different irradiance.] - Nauch. Tr. voronezhsk. sel'skokhoz. Inst. *73* (Biologiya, Agrotekhnika i Selektsiya Plodovykh Rastenii) : 122 - 130, 1975. [In R.]

33699 - GLAGOLEVA, T.A., MOKRONOSOV, A.T., ZALENSKII, O.V. : Vliyanie kisloroda na fotosinteticheskii metabolizm $C^{14}$ u pustynnykh rastenii. [The effect of oxygen on photosynthetic $^{14}C$ metabolism in desert plants.] - Bot. Zh. *63* : 170 - 182, 1978. [In R, ab : E.]

33700 - GLASE, J.C., GRANET, K. : Bark chlorophyll in the American beech (*Fagus grandifolia*). - Amer. Midland Naturalist *100* : 510 - 512, 1978.

33701 - GLIEM, G., NIEHAUS, F. : Plastidenentwicklung in Tabakhybriden nach Infektion mit dem Tabak-Tumorvirus und nach *Agrobacterium tumefaciens*-Induktion. - Z. Pflanzenkrankheiten Pflanzenschutz *85* : 247 - 253, 1978.

33702 - GLOE, A., RISCH, N. : Bacteriochlorophyll $c_B$, a new bacteriochlorophyll from *Chloroflexus aurantiacus*. - Arch. Microbiol. *118* : 153 - 156, 1978.

33703 - GLOSER, J. : Net photosynthesis and dark respiration of reed estimated by gas-exchange measurements. - In : DYKYJOVÁ, D., KVĚT, J. (ed.) : Pond Littoral Ecosystems. Pp. 227 - 234. Springer-Verlag, Berlin - Heidelberg - New York 1978.

33704 - GNAIGER, E., GLUTH, G., WIESER, W. : pH fluctuation in an intertidal beach in Bermuda. - Limnol. Oceanogr. *23* : 851 - 857, 1978. [Ps.]

33705 - GNANAM, A. : Photorespiration : Is it a necessary evil? - In : Inter-Disciplinary Symposium on Photosynthesis and Productivity. Abstract of Papers. Pp. 17 - 21. Indian nat. Sci. Acad., New Delhi 1978.

33706 - GÖBEL, F. : Direct measurement of pure absorbance spectra of living phototrophic microorganisms. - Biochim. biophys. Acta *538* : 593 - 602, 1978.

33707 - GÖBEL, F. : Quantum efficiencies of growth. - In : CLAYTON, R.K., SISTROM, W.R. (ed.) : The Photosynthetic Bacteria. Pp. 907 - 925. Plenum Press, New York - London 1978.

33708 - GOCHEV, A.D. : Tunneling in adiabatic electron transfer in biological systems. - Dokl. bolg. Akad. Nauk *31* : 695 - 698, 1978. [Chi.]

33709 - GODIK, V.I., SAMUILOV, V.D., BORISOV, A.Yu. : Promezhutochnye sostoyaniya, obrazuyushchiesya pri razdelenii zaryadov v reaktsionnykh tsentrakh *Rhodospirillum rubrum* v usloviyakh nizkogo okislitel'no-vosstanovitel'nogo potentsiala. [The intermediate states arising in reaction centres of *Rhodospirillum rubrum* at low redox potentials.] - Mol. Biol. (Moskva) *12* : 290 - 296, 1978. [In R, ab : E.]

33710 - GOGOTOV, I.N. : Relationships in hydrogen metabolism between hydrogenase and nitrogenase in phototrophic bacteria. - Biochimie *60* : 267 - 275, 1978.

33711 - GOGOTOV, I.N., ZORIN, N.A., SEREBRIAKOVA, L.T., KONDRATIEVA, E.N. : The properties of hydrogenase from *Thiocapsa roseopersicina*. - Biochim. biophys. Acta *523* : 335 - 343, 1978.

33712 - GOLBECK, J.H., KOK, B. : Further studies of the membrane bound iron-sulfur proteins and *P700* in a photosystem I subchloroplast particle. - Arch. Biochem. Biophys. *188* : 233 - 242, 1978.

33713 - GOLBECK, J.H., VELTHUYS, B.R., KOK, B. : Evidence that the intermediate electron acceptor, $A_2$, in Photosystem I is a bound iron-sulfur protein. - Biochim. biophys. Acta *504* : 226 - 230, 1978.

33714 - GOL'DFEL'D, M.G., DMITROVSKIĬ, L.G., BLYUMENFEL'D, L.A. : Éffektivnost' fotofosforilirovaniya v khloroplastakh v statsionarnom i impul'snom rezhimakh osveshcheniya. [Effectiveness of photophosphorylation and the schemes of energy coupling in chloroplasts.] - Mol. Biol. (Moskva) 12 : 179 - 190, 1978. [In R, ab : E.]

33715 - GOL'DFEL'D, M.G., DMITROVSKIĬ, L.G., BLYUMENFEL'D, L.A. : Temperaturnaya zavisimost' fotofosforilirovaniya v khloroplastakh pri impul'snom vozbuzhdenii i mekhanizm sopryazheniya. [Temperature dependence of photophosphorylation in chloroplasts under pulse excitation and the coupling mechanism.] - Mol. Biol. (Moskva) 12 : 857 - 862, 1978. [In R, ab : E.]

33716 - GOL'DFEL'D, M.G., KHANGULOV, S.V. : Vliyanie ferritsianida, temnovoĭ adaptatsii i stareniya na svoĭstva signala ÉPR I chloroplastov. [Effect of ferricyanide, dark adaptation and aging on the properties of ESR signal I in chloroplasts.] - Biofizika 23 : 467 - 473, 1978. [In R, ab : E.]

33717 - GOL'DFEL'D, M.G., KHANGULOV, S.V., BLYUMENFEL'D, L.A. : Flash-induced P700 electron spin resonance signal changes and the two-electron switch in electron transfer in chloroplasts. - Photosynthetica 12 : 21 - 34, 1978.

33718 - GOL'DFEL'D, M.G., MIKOYAN, V.D., GUSEV, M.V., NIKITINA, K.A. : pH-zavisimost' signala É.P.R. tsentrov P700+ kak indikatora uchastiya plastokhinona v fotosinteticheskom transporte élektronov. [pH-dependence of the EPR signal of P700+ centres as an indicator of plastoquinone participation in photosynthetic electron transport.] - Dokl. Akad. Nauk SSSR 239 : 721 - 724, 1978.[In R.]

33719 - GOL'DFEL'D, M.G., TSAPIN, A.I., VOZVYSHAEVA, L.V., KHALILOV, R.I. : Paramagnitnye tsentry i fotokhimicheskie reaktsii subkhloroplastnykh fragmentov fotosistemy 2. [Paramagnetic centres and photochemical reactions of subchloroplast fragments of the photosystem 2.] - Biofizika 23 : 266 - 271, 1978. [In R, ab : E.]

33720 - GOLDFIELD, M.G., HALILOV, R.I., HANGULOV, S.V., KONONENKO, A.A., KNOX, P.P. : Correlation of the light-induced change of absorbance with ESR signal of Photosystem II in presence of silicomolybdate. - Biochem. biophys. Res. Commun. 85 : 1199 - 1203, 1978.

*33721 - GOLDMAN, C.R.: The use of absolute activity for eliminating serious errors in the measurement of primary productivity with C14. - J. Cons. Int. Explor. Mer 32 : 172 - 179, 1968.

*33722 - GOLDMAN, C.R., AMEZAGA, E. DE : Spatial and temporal changes in the primary productivity of Lake Tahoe, California-Nevada between 1959 and 1971. - Verh. Int. Verein. Limnol. 19 : 812 - 825, 1975.

*33723 - GOLDMAN, J.C., RYTHER, J.H. : Mass production of algae : bioengineering aspects. - In : MITSUI, A., MIYACHI, S., SAN PIETRO, A., TAMURA, S. (ed.) : Biological Solar Energy Conversion. Pp. 367 - 378. Academic Press, New York - - San Francisco - London 1977.

33724 - GOLDSWORTHY, A. : An instrument for measuring crop density by light absorbance. - Ann. Bot. 42 : 1315 - 1325, 1978.

33725 - GOLOMAZOVA, G.M., MININA, E.G., SHEMBERG, M.A. : Intensivnost' fotosinteza uzkokronnykh i shirokokronnykh form Pinus silvestris L. [Photosynthetic rate of narrow-crown and broad-crown forms of pine (Pinus silvestris L.).] - Fiziol. Rast. 25 : 85 - 90, 1978. [In R, ab : E.]

*33726 - GOLUBKOVA, B.M., KISLYAKOVA, T.E., BOGACHEVA, I.I. : Strukturno-funktsional'nye osobennosti fotosinteticheskogo apparata u nekotorykh predstaviteleĭ filogeneticheski drevnikh grupp rasteniĭ. [Structural and functional properties of photosynthetic apparatus in some species belonging to phylogenetically ancient plant groups.] - In : Élektronnaya Mikroskopiya v Botanicheskikh Issledovaniyakh. Pp. 141 - 143. Petrozavodsk 1974. [In R.]

33727 - GOMEZ-NAVARRETE, G., MOORE, T.C. : Effects of protein synthesis inhibitors on ent-kaurene biosynthesis during photomorphogenesis of etiolated pea seedlings. - Plant Physiol. 61 : 889 - 892, 1978. [Chl.]

33728 - GOMM, F.B. : Growth and development of meadow plants as affected by environmental variables. - Agron. J. 70 : 1061 - 1065, 1978. [Growth analysis.]

33729 - GONCHARIK, M.N., MARSHAKOVA, M.I., LAGUN, L.P. : Vliyanie form kaliĭnykh udo-
brenlĭ na fotosinteticheskuyu produktivnost' kartofelya. [Effect of forms of
potassium fertilizers on photosynthetic productivity of potato.] - In : Fiziologo-biokhimicheskie Usloviya Povysheniya Produktivnosti Sel'skokhozyaĭstvennykh Rasteniĭ. Pp. 86 - 91. Nauka i Tekhnika, Minsk 1978. [In R.]

33730 - GORDEEVA, T.K. : Nakoplenie i struktura fitomassy soobshchestv pustynnykh
stepeĭ MNR. [Biomass accumulation and structure in desert-steppe communities
in Mongolia.] - In : Geografiya i Dinamika Rastitel'nogo i Zhivotnogo Mira
MNR. Pp. 35 - 39. Nauka, Moskva 1978. [In R.]

33731 - GORDON, K.H.J., PEOPLES, M.B., MURRAY, D.R. : Ageing-linked changes in photosynthetic capacity and in Fraction I protein content of the first leaf of
pea *Pisum sativum* L. - New Phytol. *81* : 35 - 42, 1978.

33732 - GORE, M.G., JORDAN, P.M., CHAUDHRY, A.G. : Automated assay for δ-aminolevulinic acid dehydratase. - Anal. Biochem. *87* : 141 - 147, 1978.

*33733 - GORIN, N., ZONNEVELD, H. : Fixation of epidermis of Golden Delicious apples
for chlorophyll analysis. - Lebensm.-Wiss. Technol. *10* : 50, 1977.

33734 - GORRELL, T.E., UFFEN, R.L. : Light-dependent and light-independent production
of hydrogen gas by photosynthesizing *Rhodospirillum rubrum* mutant C. - Photochem. Photobiol. *27* : 351 - 358, 1978.

33735 - GORTER DE VRIES, H., HOFF, A.J. : Magnetic field effect on the fluorescence
intensity of *Rhodopseudomonas sphaeroides* at 1.4 K. - Chem. Phys. Lett. *55* :
395 - 398, 1978.

33736 - GORYSHINA, T.K. : Sezonnye izmeneniya fotosinteza u travyanistykh rasteniĭ
lesostepnoĭ dubravy v svyazi s dinamikoĭ plastidnogo apparata lista. [Seaso-
nal changes in photosynthesis in herbs of oak forest-steppe in relation to
the dynamics of leaf plastid apparatus.] - Ėkologiya *1978* (6) : 14 - 19,
1978. [In R.]

33737 - GORYSHINA, T.K., PRUZHINA, E.G. : Sezonnaya dinamika plastidnogo apparata u
travyanistykh rasteniĭ lesostepnoĭ dubravy. [Seasonal dynamics of plastid
apparatus in herbs of oak forest-steppe.] - Tr. petergof. biol. Inst. *27* (Voprosy Ėkologicheskoĭ Anatomii i Fiziologii Rasteniĭ) : 60 - 83, 1978. [In R,
ab : E.]

33738 - GOSZ, J.R., HOLMES, R.T., LIKENS, G.E., BORMAN, F.H. : The flow of energy in
a forest ecosystem. - Sci. Amer. *238* (3) : 92 - 102, 1978. [Ps.]

33739 - GOTO, K. : Mutually inverse rhythmic and sigmoidal changes in activity of
cytoplasmic and chloroplast glyceraldehyde 3-phosphate dehydrogenases in
*Lemna gibba* G3. - Plant Cell Physiol. *19* : 749 - 758, 1978.

33740 - GOTTLIEB, L.D. : Allocation, growth rates and gas exchange in seedlings of
*Stephanomeria exigua* ssp. *coronaria* and its recent derivative *S. malheuren-
sis*. - Amer. J. Bot. *65* : 970 - 977, 1978.

33741 - GOUDRIAAN, J., VAN LAAR, H.H. : Calculations of daily totals of the gross
$CO_2$ assimilation of leaf canopies. - Neth. J. agr. Sci. *26* : 373 - 382, 1978.

33742 - GOUDRIAAN, J., VAN LAAR, H.H. : Relations between leaf resistance, $CO_2$-con-
centration and $CO_2$-assimilation in maize, beans, lalang grass and sunflower.
- Photosynthetica *12* : 241 - 249, 1978.

33743 - GOULD, J.M. : Dithiol-specific reversal of triphenyltin inhibition of $CF_0$-
-catalyzed transmembrane proton transfer in chloroplasts. - FEBS Lett. *94* :
90 - 94, 1978.

33744 - GOULD, J.M. : $Hg^{2+}$-induced turnover of the chloroplast ATP synthetase complex in the absence of ADP and phosphate. - FEBS Lett. *95* : 197 - 201, 1978.

33745 - GOVINDJEE : Introduction : Ultrafast reactions in photosynthesis. - Photochem. Photobiol. *28* : 935 - 938, 1978.

*33746 - GOVINDJEE, GOVINDJEE, R. : Light energy conversion by photosynthesis. -
Biochem. Rev. (India) *48* : 25 - 34, 1977.

*33747 - GOVINDJEE, GOVINDJEE, R. : Light energy conversion by photosynthesis. - J. sci. ind. Res. *36* : 663 - 671, 1977.

33748 - GOVINDJEE, KHANNA, R. : Bicarbonate : its role in photosystem II. - In : METZNER, H. (ed.) : Photosynthetic Oxygen Evolution. Pp. 269 - 282. Academic Press, London - New York - San Francisco 1978.

33749 - GOVINDJEE, VAN RENSEN, J.J.S. : Bicarbonate effects on the electron flow in isolated broken chloroplasts. - Biochim. biophys. Acta *505* : 183 - 213, 1978.

33750 - GOVINDJEE, WONG, D. : Regulation of excitation energy transfer among the two pigment systems in photosynthesis. - In : FIALA, J. (ed.) : Third International Seminar on Energy Transfer in Condensed Matter. Pp. 19 - 28. Univerzita Karlova, Praha 1978.

33751 - GOVINDJEE, WYDRZYNSKI, T., MARKS, S.B. : Manganese and chloride : their roles in photosynthesis. - In : METZNER, H. (ed.) : Photosynthetic Oxygen Evolution. Pp. 321 - 344. Academic Press, London - New York - San Francisco 1978.

33752 - GOVINDJEE, R., BECHER, B., EBREY, T.G. : The fluorescence from the chromophore of the purple membrane protein. - Biophys. J. *22* : 67 - 77, 1978.

33753 - GRÄBER, P., SAPHON, S. : Conformational changes of the membrane-bound ATPase of bacterial chromatophores revealed by fluorescence changes of fluorescamine-labelled coupling factors. - Z. Naturforsch. *33 C* : 421 - 427, 1978.

33754 - GRÄBER, P., SCHLODDER, E., WITT, H.T. : Control of the rate of ATP synthesis by conformational changes in the chloroplast ATPase induced by the transmembrane electric field. - In : HALL, D.O., COOMBS, J., GOODWIN, T.W. (ed.) : Proceedings of the Fourth International Congress on Photosynthesis. Pp. 197 - 210. Biochem. Soc., London 1978.

33755 - GRÄBER, P., ZICKLER, A., ÅKERLUND, H.-E. : Electric evidence for the isolation of inside-out vesicles from spinach chloroplasts. - FEBS Lett. *96* : 233 - 237, 1978.

33756 - GRABHERR, G., MÄHR, E., REISIGL, H. : Nettoprimärproduktion und Reproduktion in einem Krumseggenrasen (*Caricetum curvulae*) der Ötztaler Alpen, Tirol. - Oecol. Plant. *13* : 227 - 251, 1978.

33757 - GRABOWSKI, J., GANTT, E. : Photophysical properties of phycobiliproteins from phycobilisomes : fluorescence lifetimes, quantum yields, and polarization spectra. - Photochem. Photobiol. *28* : 39 - 45, 1978.

33758 - GRABOWSKI, J., GANTT, E. : Excitation energy migration in phycobilisomes : comparison of experimental results and theoretical predictions. - Photochem. Photobiol. *28* : 47 - 54, 1978.

33759 - GRACE, J.B., WETZEL, R.G. : The production biology of Eurasian watermilfoil (*Myriophyllum spicatum* L.) : A review. - J. aquat. Plant Manage. *16* : 1 - 11, 1978. [Ps.]

33760 - GRADCHANINOVA, O.D. : Metodika anatomicheskogo issledovaniya lista pshenitsy v svyazi s fotosintezom. [Methods for anatomical studies of a wheat leaf associated with its photosynthetic function.] - Tr. priklad. Bot., Genet. Selektsii *61* (3) : 68 - 71, 1978. [In R, ab : E.]

33761 - GRADINARSKI, L., GEORGIEV, Kh., SP"TARU, A. : Fotosintetichna aktivnost na genotipi domati pri razlichni usloviya na osvetlenie. [Photosynthetic activity of tomato genotypes under various conditions of illumination.] - Fiziol. Rast. (Sofia) *4* (2) : 44 - 51, 1978. [In Bulg., ab : E, R.]

33762 - GRADMANN, D. : Green light (550 nm) inhibits electrogenic Cl-pump in the *Acetabularia* membrane by permeability increase for the carrier ion. - J. Membr. Biol. *44* : 1 - 24, 1978.

33763 - GRADYUSHKO, A.T., SOLOV'EV, K.N., TSVIRKO, M.P. : Vliyanie molekulyarnoĭ struktury na veroyatnosti izluchatel'nykh i bezyzluchatel'nykh perekhodov v molekulakh proizvodnykh khlorofilla. [Influence of molecular structure on the probabilities of radiative and radiationless transitions in the molecules of chlorophyll derivatives.] - Biofizika *23* : 757 - 761, 1978. [In R, ab : E.]

33764 - GRAHAM, D. : Interactions of chloroplasts and mitochondria : effect of light on respiration. - Proc. aust. biochem. Soc. *11* : Q6 - Q7, 1978.

33765 - GRAHAM, L.E., GRAHAM, J.M. : Ultrastructure of endosymbiotic *Chlorella* in a *Vorticella*. - J. Protozool. *25* : 207 - 210, 1978. [Chloroplast.]

33766 - GRAY, E.D. : Ribosomes and RNA metabolism. - In : CLAYTON, R.K., SISTROM, W.R. (ed.) : The Photosynthetic Bacteria. Pp. 885 - 897. Plenum Press, New York - London 1978. [Chromatophores.]

33767 - GRAY, J.C. : Serological reactions of Fraction I proteins from interspecific hybrids in the genus *Nicotiana*. - Plant Systematics Evolution *129* : 177 - 183, 1978.

33768 - GRAY, J.C., KUNG, S.D., WILDMAN, S.G. : Polypeptide chains of the large and small subunits of fraction I protein from tobacco. - Arch. Biochem. Biophys. *185* : 272 - 281, 1978.

33769 - GREBANIER, A., BOGORAD, L. : Membrane proteins synthesized, but not processed, by isolated *Zea mays* chloroplasts. - Plant Physiol. *61* (Suppl.) : 104, 1978.

33770 - GREBANIER, A., CHAMPAGNE, D., ROY, H. : Cross-linking of ribulose 1,5-bisphosphate carboxylase from *Pisum sativum*. - In : SIEGELMAN, H.W., HIND, G. (ed.) : Photosynthetic Carbon Assimilation. Pp. 419 - 420. Plenum Press, New York - London 1978.

33771 - GREBANIER, A., CHAMPAGNE, D., ROY, H. : Cross-linking of ribulose bisphosphate carboxylase from *Pisum sativum* with tetranitromethane. - Plant Physiol. *61* (Suppl.) : 99, 1978.

33772 - GREBANIER, A.E., CHAMPAGNE, D., ROY, H. : Effects of $Mg^{2+}$ and substrates on the conformation of ribulose-1,5-bisphosphate carboxylase. - Biochemistry *17* : 5150 - 5155, 1978.

33773 - GREBANIER, A.E., COEN, D.M., RICH, A., BOGORAD, L. : Membrane proteins synthesized but not processed by isolated maize chloroplasts. - J. Cell Biol. *78* : 734 - 746, 1978.

33774 - GREEN, B.R. : Synthesis of membrane polypeptides by isolated chloroplasts of *Acetabularia*. - Plant Physiol. *61* (Suppl.) : 104, 1978.

33775 - GREEN, J.F., MUIR, R.M. : The effect of potassium on cotyledon expansion induced by cytokinins. - Physiol. Plant. *43* : 213 - 218, 1978. [Chl.]

33776 - GREENWALD, L.S., ZILINSKAS, B.A. : Allophycocyanin-B from *Nostoc* sp. phycobilisomes. - Plant Physiol. *61* (Suppl.) : 88, 1978.

33777 - GREENWAY, H., WINTER, K., LÜTTGE, U. : Phosphoenolpyruvate carboxylase during development of Crassulacean acid metabolism and during a diurnal cycle in *Mesembryanthemum crystallinum*. - J. exp. Bot. *29* : 547 - 559, 1978.

33778 - GRIBOVA, Z.P., ZAKHAROVA, N.I., MURZA, L.I. : Vliyanie okislitelei i vosstanovitelei na skorost' relaksatsii protonov vody v khloroplastakh bobov i chastitsakh fotosistemy 2. [Effect of oxidizing and reducing agents on the rate of water proton relaxation in bean chloroplasts and particles of the Photosystem 2.] - Mol. Biol. (Moskva) *12* : 157 - 164, 1978. [In R, ab : E.]

33779 - GRIFFITH, O.H., BROWN, H.M., LESCH, H.H. : Photoelectron microscopy of photosynthetic membranes. - In : DEAMER, D.W. (ed.) : Light Transducing Membranes. Pp. 313 - 334. Acad. Press, New York - San Francisco - London 1978.

33780 - GRIFFITHS, M. : Specific blue-green algal carotenoids in sediments of Esthwaite Water. - Limnol. Oceanogr. *23* : 777 - 784, 1978.

*33781 - GRIFFITHS, W.T. : Studies *in vitro* on plastid development. - Biochem. Soc. Trans. *5* : 88 - 90, 1977.

33782 - GRIFFITHS, W.T. : Reconstitution of chlorophyllide formation by isolated etioplast membranes. - Biochem. J. *174* : 681 - 692, 1978.

33783 - GRIFFITHS, W.T., MAPLESTON, R.E. : NADPH - protochlorophyllide oxidoreductase. - In : AKOYUNOGLOU, G., ARGYROUDI-AKOYUNOGLOU, J.H. (ed.) : Chloroplast Development. Pp. 99 - 104. Elsevier/North-Holland Biomedical Press, Amsterdam - New York - Oxford 1978.

33784 - GRIMME, L.H. : The regreening and the development of photosynthetic activity
        in orange *Chlorella fusca* cells are separable physiological processes. - In :
        AKOYUNOGLOU, G., ARGYROUDI-AKOYUNOGLOU, J.H. (ed.) : Chloroplast Development.
        Pp. 445 - 448. Elsevier/North-Holland Biomedical Press, Amsterdam - New York
        - Oxford 1978.

33785 - GRODZINSKI, B. : Glyoxylate decarboxylation during photorespiration. - Plan-
        ta *144* : 31 - 37, 1978.

33786 - GROMET-ELHANAN, Z., GEST, H. : A comparison of electron transport and photo-
        phosphorylation systems of *Rhodopseudomonas capsulata* and *Rhodospirillum
        rubrum*. Effects of antimycin A and dibromothymoquinone. - Arch. Microbiol.
        *116* : 29 - 34, 1978.

33787 - GROMET-ELHANAN, Z., LEISER, M. : Energy conservation by electrochemical gra-
        dients in chromatophore membranes of photosynthetic bacteria. - In : AZZONE,
        G.F., AVRON, M., METCALFE, J.C., QUAGLIARIELLO, E., SILIPRANDI, N. (ed.) :
        The Proton and Calcium Pumps. Pp. 81 - 91. Elsevier/North-Holland Biomedical
        Press, Amsterdam - Oxford - New York 1978.

33788 - GROSS, E.L., GRENIER, J. : Regulation of excitation energy distribution in
        subchloroplast particles : Photosystem I. - Arch. Biochem. Biophys. *187* :
        387 - 398, 1978.

33789 - GROSS, E.L., YOUNGMAN, D.R., WINEMILLER, S.L. : An FMN-photosystem I photo-
        voltaic cell. - Photochem. Photobiol. *28* : 249 - 256, 1978.

33790 - GROUT, B.W.W., ASTON, M.J. : Transplanting of cauliflower plants regenerated
        from meristem culture. II. Carbon dioxide fixation and the development of
        photosynthetic ability. - Hort. Res. *17* : 65 - 71, 1978.

33791 - GROUZIS, J.-P. : Étude comparée de la fixation du calcium par les chloro-
        plastes isolés de Lupin jaune (calcifuge) et de Féverole (calcicole). -
        Physiol. vég. *16* : 81 - 92, 1978.

33792 - GROUZIS, J.-P. : Fixation du calcium par les thylakoïdes isolés de Lupin
        jaune (calcifuge) et de Féverole (calcicole) : nature des groupements impli-
        qués et corrélation entre la quantité de calcium fixée et la teneur en grou-
        pements carboxyles des membranes. - Physiol. vég. *16* : 593 - 604, 1978.

33793 - GROVES, M. : Physics in photosynthesis at Flinders University. - Aust. Phy-
        sicist *15* (3) : 41 - 42, 1978.

33794 - GROVES, M.R. : A pulsed laser system for the study of micro-second delayed
        fluorescence from plants. - Photochem. Photobiol. *27* : 491 - 496, 1978.

33795 - GRUMBACH, K.H., LICHTENTHALER, H.K. : Incorporation of $^{14}CO_2$ in prenylquino-
        nes of *Chlorella pyrenoidosa*. - Planta *141* : 253 - 258, 1978.

33796 - GRUMBACH, K.H., LICHTENTHALER, K.H., ERISMANN, K.H. : Incorporation of
        $^{14}CO_2$ in photosynthetic pigments of *Chlorella pyrenoidosa*. - Planta *140* :
        37 - 43, 1978.

33797 - GUBAR', G.D., KRISTKALNE, S.Kh., VITOLA, A.K. : Skorost' izmeneniya prirosta
        biomassy i morfologii rasteniya ogurtsa pri usilenii intensivnosti sveta.
        [Rate of changes in the growth of biomass and morphology of cucumber plants
        at increasing irradiance.] - In : Fiziologo-biokhimicheskie Issledovaniya
        Rasteniĭ. Pp. 7 - 16, 155. Zinatne, Riga 1978. [In R.]

*33798 - GUBAR, G.D., VOĬTSEKHOVICH, Z.V. : Parametry svetovykh krivykh fotosinteza
        v pervye sroki posle vozdeĭstviya usloviĭ zateneniya. [Parameters of light
        curves of photosynthesis in the initial period of shading effect.] - In :
        Adaptatsiya Fiziologo-Biokhimicheskikh Sistem Rasteniya k Peremene Osveshche-
        niya. Pp. 68 - 87. Zinatne, Riga 1977.

33799 - GUDKOV, N.D. : Balans êntropii pri fotosinteze. [Balance of entropy in pho-
        tosynthesis.] - Biofizika *23* : 859 - 863, 1978. [In R, ab : E.]

33800 - GUDKOV, N.D., STOLOVITSKIĬ, Yu.M., EVSTIGNEEV, V.B. : Impul'snaya fotopro-
        vodimost' rastvorov khlorofilla i ego analogov. IV. Ob absolyutnom kvanto-
        vom vykhode obrazovaniya ion-radikalov pri fotookislenii khlorofilla *a*
        *p*-benzokhinonom. [Impulse photoconductance of chlorophyll solutions and its

analogs. IV. Absolute quantum yield of ion-radical formation during photooxidation of chlorophyll $a$ with $p$-benzoquinone.] - Biofizika $23$ : 571 - 575,
1978. [In R, ab : E.]

33801 - GUIGNERY, G., DURANTON, J. : Origin of mRNAs directing chloroplast protein
synthesis $in$ $vitro$. - In : AKOYUNOGLOU, G., ARGYROUDI-AKOYUNOGLOU, J.H.
(ed.) : Chloroplast Development. Pp. 663 - 668. Elsevier/North-Holland Biomedical Press, Amsterdam - New York - Oxford 1978.

33802 - GUIKEMA, J.A., YOCUM, C.F. : Evidence for two sites of inhibition of photosynthetic electron transport by dibromothymoquinone. - Arch. Biochem. Biophys.
$189$ : 508 - 515, 1978.

33803 - GUIKEMA, J.A., YOCUM, C.F. : Rate saturation kinetics and interrelationships
of acceptors to photosystem II. - Plant Physiol. $61$ (Suppl.) : 103, 1978.

*33804 - GUILIZZONI, P., SARACENI, C. : Popolamento a macrofite. [Macrophytic popula-
tion.] - In : BARBANTI, L., BONACINA, C., CALDERONI, A., CAROLLO, A., DE BER-
NARDI, R., GUILIZZONI, P., NOCENTINI, A.M., RUGGIU, D., SARACENI, C., TONO-
LLI, L. : Indagini Ecologiche sul Lago d'Endine. Pp. 183 - 224. Ed. Ist.
Ital. Idrobiologia, Verbania Pallanza 1974. [In Ital.]

33805 - GUILLOT-SALOMON, T., TUQUET, C., LUBAC, M. DE, HALLAIS, M.-F., SIGNOL, M. :
Analyse comparative de l'ultrastructure et de la composition lipidique des
chloroplastes de plantes d'ombre et de soleil. - Cytobiologie $17$ : 442 -
452, 1978.

33806 - GULIEV, F.A., KOROBKOV, M.E., KOCHUBEĬ, S.M. : Primenenie metoda proizvod-
noĭ spektroskopii dlya issledovaniya tonkoĭ struktury spektrov fluorestsentsii fragmentov khloroplastov. [Use of the derivative spectroscopy method
for studying the fine structure of fluorescence spectra of chloroplast fragments.] - Zh. prikl. Spektroskop. $29$ : 646 - 651, 1978. [In R, ab : E.]

33807 - GULIEV, N.M., FEDENKO, E.P., KOMAROVA, T.I., DOMAN, N.G. : Vliyanie usloviĭ
vyrashchivaniya fototrofnykh bakteriĭ na aktivnost' fermentov sinteza i ras-.
pada tsiklicheskogo 3': 5'-AMP. [Effect of growing conditions on the activity of the enzymes of cyclic 3': 5'-AMP synthesis and decay in phototrophic
bacteria.] - Biokhimiya $43$ : 928 - 934, 1978. [In R, ab : E.]

33808 - GULYAEV, B.I. : Otsenka roli êndogennykh faktorov v formirovanii listovogo
apparata rasteniĭ. [Estimation of the role of endogenous factors in the formation of plant leaf apparatus.] - Fiziol. Biokhim. kul't. Rast. $10$ : 366 -
374, 1978. [In R, ab : E.]

33809 - GUPTA, V.K., ANDERSON, L.E. : Light modulation of the activity of carbon metabolism enzymes in the Crassulacean acid metabolism plant $Kalanchoë$. -
Plant Physiol. $61$ :469 - 471, 1978.

33810 - GUREVITZ, M., CAHEN, D. : The fall and rise of $Euglena$ $gracilis;$ study of
PSII and membranal polypeptide development in chloroplast membranes. - In :
AKOYUNOGLOU, G., ARGYROUDI-AKOYUNOGLOU, J.H. (ed.) : Chloroplast Development.
Pp. 449 - 454. Elsevier/North-Holland Biomedical Press, Amsterdam - New York
- Oxford 1978.

33811 - GUSEV, M.V., NIKITINA, K.A. : Fiziologiya i biokhimiya tsianobakteriĭ. [Phy-
siology and biochemistry of cyanobacteria.] - Uspekhi Mikrobiol. $13$ : 30 -
49, 1978. [Ps, Chl, Car; In R.]

33812 - GUSEV, N.A. : Nekotorye itogi i perspektivy izucheniya vodoobmena i sostoyaniya vody rasteniĭ. [Some results and perspectives in studying water exchange and water status in plants.] - Nauch. Dokl. vyssh. Shkoly, biol. Nauki $21$ (2) : 7 - 19, 1978. [Ps; In R.]

33813 - GUTERSTAM, B., WALLENTINUS, I., ITURRIAGA, R. : $In$ $situ$ primary production
of $Fucus$ $vesiculosus$ and $Cladophora$ $glomerata$. - Kieler Meeresforsch. Son-
derheft $4$ : 257 - 266, 1978.

33814 - GUTSCHICK, V.P. : Concentration quenching in chlorophyll-$a$ and relation to
functional charge transfer $in$ $vivo$. - J. Bioenerg. Biomembranes $10$ : 153 -
170, 1978.

33815 - **GWYNN, G.R.** : Chlorophyll disappearance in yellow and green tobaccos. - Tobacco Sci. *22* : 141 - 143, 1978. Tobacco International *180* (24) : 69 - 71,
1978.

*33816 - **GYURJÁN, I., KERESZTES, Á., RAKOVÁN, J.N., SCHRÓTH, Á.** : Kloroplasztisz mu-
tációk strukturális és funkcionális vizsgálata *Chlamydomonas reinhardii*-ban.
I. Kloroplasztisz finomszerkezet, pigment-bartalomés fotoszintetikus aktivi-
tás. [Structural and functional investigations of chloroplast mutants in
*Chlamydomonas reinhardii*. I. Fine structure of chloroplasts pigment content
and photosynthetic activity.] - Biológia (Budapest) *25* : 50 - 71, 1977.
[In Hung., ab : E.]

33817 - **GYURJÁN, I., NAGY, A.H., RAKOVÁN, J.N.** : Ribosome-deficient mutants of *Zea
mays*. - Biochem. Physiol. Pflanz. *173* : 429 - 439, 1978. [Chloroplast, Ps.]

33818 - **HABERMANN, H.M., SHOEMAKER, E.W.** : Water utilization, metabolism and stomatal function in copper deficient *Helianthus annuus*. - Plant Physiol. *61*
(Suppl.) : 88, 1978. [Stomatal resistance.]

33819 - **HACHE, A., SEITZ, H.-P.** : Stability and oscillations in the steady state of
a model system of active proton transport in photosynthesis. - Stud. biophys. *69* : 109 - 118, 1978.

33820 - **HACHTEL, W.** : Sites of synthesis of thylakoidal membrane proteins. - Ber.
deut. bot. Ges. *91* : 509 - 512, 1978.

33821 - **HAEHNEL, W.** : Reactions of plastocyanin in chloroplasts. - In : HALL, D.O.,
COOMBS, J., GOODWIN, T.W. (ed.) : Proceedings of the Fourth International
Congress on Photosynthesis. Pp. 777 - 786. Biochem. Soc., London 1978.

33822 - **HAEHNEL, W., HEUPEL, A., HENGSTERMANN, D.** : Investigations on a galvanic
cell driven by photosynthetic electron transport. - Z. Naturforsch. *33 C* :
392 - 401, 1978.

33823 - **HAGAR, W.G., FREEBERG, J.A.** : A technique for the measurement of photosynthetic rates of sporophytes and gametophytes of the fern *Todea barbara*. -
Plant Physiol. *61* (Suppl.) : 101, 1978.

33824 - **HÄGELE, W., DRISSLER, F., SCHMID, D., WOLF, H.C.** : ESR studies of the photoexcited triplet states of chlorophyll *a* and chlorophyll *b* in PMMA and MTHF
at 4.2 K. - In : FIALA, J. (ed.) : Third International Seminar on Energy
Transfer in Condensed Matter. Pp. 83 - 90. Univerzita Karlova, Praha 1978.

33825 - **HÄGELE, W., SCHMID, D., WOLF, H.C.** : Triplet-state electron spin resonance
of chlorophyll *a* and *b* molecules and complexes in PMMA and MTHF. I : Experimental determination of fine-structure and rate constants. - Z. Naturforsch.
*33 A* : 83 - 93, 1978.

33826 - **HÄGELE, W., SCHMID, D., WOLF, H.C.** : Triplet-state electron spin resonance
of chlorophyll *a* and *b* molecules and complexes in PMMA and MTHF. II : Interpretation of the experimental results. - Z. Naturforsch. *33 A* : 94 - 97,
1978.

33827 - **HAGEMANN, R., BÖRNER, T.** : Plastid ribosome-deficient mutants of higher
plants as a tool in studying chloroplast biogenesis. - In : AKOYUNOGLOU, G.,
ARGYROUDI-AKOYUNOGLOU, J.H. (ed.) : Chloroplast Development. Pp. 709 - 720.
Elsevier/North-Holland Biomedical Press, Amsterdam - New York - Oxford 1978.

33828 - **HAGIHARA, B., ISHIBASHI, F., SASAKI, K., KAMIGAWARA, Y.** : Cellulose acetate
coatings for the polarographic oxygen electrode. - Anal. Biochem. *86* : 417 -
431, 1978.

33829 - **HAJIBRAHIM, S.K., TIBBETTS, P.J.C., WATTS, C.D., MAXWELL, J.R., EGLINTON,G.,
COLIN, H., GUICHON, G.** : Analysis of carotenoid and porphyrin pigments of
geochemical interest by high-performance liquid chromatography. - Anal.
Chem. *50* : 549 - 553, 1978.

33830 - **HÁLA, J., DAMBYN, M., VAVŘINEC, E.** : Low temperature properties of chlorophyll *a* . - In : FIALA, J. (ed.) : Third International Seminar on Energy
Transfer in Condensed Matter. Pp. 136 - 140. Univerzita Karlova, Praha 1978.

*33831 - HALEVY, A.H. : Light energy flux and distribution of assimilates as factors
          controlling the flowering of flower crops. - In : Proceedings of the XIX
          International Horticultural Congress. Vol. 4. Pp. 125 - 134. Warszawa 1974.

33832 - HALL, A.J., MILTHORPE, F.L. : Assimilate source-sink relationships in *Capsi-
          cum annuum* L. III. The effects of fruit excision on photosynthesis and leaf
          and stem carbohydrates. - Aust. J. Plant Physiol. *5* : 1 - 13, 1978.

33833 - HALL, D.O. : Solar energy conversion through biology - could it be a practical
          energy source ? - Fuel *57* : 322 - 333, 1978. [Ps.]

33834 - HALL, E.A.H., MOSS, G.P., UTLEY, J.H.P., WEEDON, B.C.L. : Electrochemical
          reductive acylation of astacene ; a route to the carotenoid astaxanthin. -
          J. chem. Soc. chem. Commun. *1978* : 387 - 388, 1978.

33835 - HALL, N.P., KEYS, A.J., MERRETT, M.J. : Ribulose-1,5-diphosphate carboxy-
          lase protein during flag leaf senescence. - J. exp. Bot. *29* : 31 - 37, 1978.

33836 - HALL, N.P., KOIVUNIEMI, P., TOLBERT, N.E. : The ratio of RuBP carboxylase
          to RuBP oxygenase in crude and purified preparations from tobacco leaves. -
          Plant Physiol. *61* (Suppl.) : 99, 1978.

33837 - HALL, N.P., TOLBERT, N.E. : A rapid procedure for the isolation of ribulose
          bisphosphate carboxylase/oxygenase from spinach leaves. - FEBS Lett. *96* :
          167 - 169, 1978.

33838 - HALL, R.L., DOORLEY, P.F., NIEDERMAN, R.A. : *Trans*-membrane localization of
          reaction center proteins in *Rhodopseudomonas sphaeroides* chromatophores. -
          Photochem. Photobiol. *28* : 273 - 276, 1978.

33839 - HALLAM, N.D., GAFF, D.F. : Reorganization of fine structure during rehydra-
          tion of desiccated leaves of *Xerophyta villosa*. - New Phytol. *81* : 349 - 355,
          1978.

33840 - HALLAM, N.D., GAFF, D.F. : Regeneration of chloroplast structure in *Talbotia
          elegans* : a desiccation-tolerant plant. - New Phytol. *81* : 657 - 662, 1978.

33841 - HALLICK, R.B., GRAY, P.W., CHELM, B.K., RUSHLOW, K.E., OROZCO, E.M. Jr. :
          *Euglena gracilis* chloroplast DNA structure, gene mapping, and RNA transcrip-
          tion. - In : AKOYUNOGLOU, G., ARGYROUDI-AKOYUNOGLOU, J.H. (ed.) : Chloroplast
          Development. Pp. 619 - 622. Elsevier/North-Holland Biomedical Press, Amster-
          dam - New York - Oxford 1978.

33842 - HALLIER, U.W., SCHMITT, J.M., HEBER, U., CHAIANOVA, S.S., VOLODARSKY, A.D. :
          Ribulose-1,5-bisphosphate carboxylase-deficient plastome mutants of *Oenothe-
          ra*. - Biochim. biophys. Acta *504* : 67 - 83, 1978.

33843 - HALLIWELL, B. : A personal assessment of the current state of knowledge of
          photorespiration and $C_4$ metabolism. - In : HALL, D.O., COOMBS, J., GOODWIN,
          T.W. (ed.) : Proceedings of the Fourth International Congress on Photosyn-
          thesis. Pp. 347 - 355. Biochem. Soc., London 1978.

33844 - HALLIWELL, B. : The chloroplast at work. A review of modern developments in
          our understanding of chloroplast metabolism. - Prog. Biophys. mol. Biol.
          *33* : 1 - 54, 1978.

33845 - HALLMAN, E., HARI, P., RÄSÄNEN, P.K., SMOLANDER, H. : Effect of planting
          shock on the transpiration, photosynthesis, and height increment of Scots
          pine seedlings. - Acta forest. fenn. *161* : 4 - 26, 1978.

*33846 - HAM, G.E., LAWN, R.J., BRUN, W.A. : Influence of inoculation, nitrogen fer-
          tilizers and photosynthetic source-sink manipulations on field grown soy-
          beans. - In : NUTMAN, P.S. (ed.) : Symbiotic Nitrogen Fixation in Plants.
          IBP Vol. 7. Pp. 239 - 253. Cambridge Univ. Press, Cambridge - London - New
          York - Melbourne 1975.

33847 - HAMEEDI, M.J. : Aspects of water column primary productivity in Chukchi sea
          during summer. - Mar. Biol. *48* : 37 - 46, 1978.

33848 - HAMMER, P.A., LANGHANS, R.W. : Modeling of plant growth in horticulture. -
          HortScience *13* : 456 - 458, 1978.

33849 - HAMPP, R. : Kinetics of membrane transport during chloroplast development. - Plant Physiol. *62* : 735 - 740, 1978.

33850 - HAMPP, R., WELLBURN, A.R. : Development of photochemical activities in preparations of unresolved internal membranes, enriched prolamellar bodies, and protylakoid vesicles during etioplast chloroplast transformation. - Ber. deut. bot. Ges. *91* : 551 - 561, 1978.

*33851 - HANDA, N., SAIJO, Y. : Composition of organic matter in subsurface chlorophyll maxima. - In : SUGAWARA, K. (ed.) : The Kuroshio II. Proceedings of the Second Symposium on the Results of the Cooperative Study of the Kuroshio and Adjacent Regions. Pp. 177 - 183. Saikon Publ., Tokyo 1972.

33852 - HANIFFA, M.A., PANDIAN, T.J. : Morphometry, primary productivity and energy flow in a tropical pond. - Hydrobiologia *59* : 23 - 48, 1978.

33853 - HANKS, J., MILES, C.D. : Photophosphorylation mutants of *Zea mays*. - Plant Physiol. *61* (Suppl.) : 102, 1978.

33854 - HANKS, R.J., ASHCROFT, G.L., RASMUSSEN, V.P., WILSON, G.D. : Corn production as influenced by irrigation and salinity - Utah studies. - Irrig. Sci. *1* : 47 - 59, 1978.

33855 - HANNA, W.W., BURTON, G.W., POWELL, J.B. : Genetics of mutagen induced non--lethal chlorophyll mutants in pearl millet. - J. Hered. *69* : 273 - 274, 1978.

*33856 - HANNAN, P.J., PATOUILLET, C. : Effect of mercury on algal growth rates. - Biotechnol. Bioeng. *14* : 93 - 101, 1972. [Chl.]

33857 - HANSCOM, Z. III, TING, I.P. : Irrigation magnifies CAM-photosynthesis in *Opuntia basilaris* (*Cactaceae*). - Oecologia *33* : 1 - 15, 1978.

33858 - HANSCOM, Z. III, TING, I.P. : Responses of succulents to plant water stress. - Plant Physiol. *61* : 327 - 330, 1978.

33859 - HANSEN, G.K. : Utilization of photosynthates for growth, respiration, and storage in tops and roots of *Lolium multiflorum*. - Physiol. Plant. *42* : 5 - 13, 1978.

33860 - HANSEN, P., GRAUSLUND, J. : Levels of sorbitol in bleeding sap and in xylem sap in relation to leaf mass and assimilate demand in apple trees. - Physiol. Plant. *42* : 129 - 133, 1978.

33861 - HANSON, M.R., BOGORAD, L. : The *ery*-M2 group of *Chlamydomonas reinhardii* : Cold-sensitive, erythromycin-resistant mutants deficient in chloroplast ribosomes. - J. gen. Microbiol. *105* : 253 - 262, 1978.

*33862 - HARASHIMA, K. : [Recent progress in carotenoid chemistry and biochemistry.] - Kagaku no Ryoiki *31* : 163 - 178, 1977. [In Jap.]

33863 - HARASHIMA, K., SHIBA, T., TOTSUKA, T., SIMIDU, U., TAGA, N. : Occurrence of bacteriochlorophyll *a* in a strain of an aerobic heterotrophic bacterium. - Agr. biol. Chem. *42* : 1627 - 1628, 1978.

33864 - HARAUX, F., DE KOUCHKOVSKY, Y. : Evaluation of the total free proton ratio $\gamma$ in the internal compartment of the thylakoids. - Plant Physiol. *61* (Suppl.) : 88, 1978.

33865 - HARBOUR, J.R., BOLTON, J.R. : The involvement of the hydroxyl radical in the destructive photooxidation of chlorophylls *in vivo* and *in vitro*. - Photochem. Photobiol. *28* : 231 - 234, 1978.

33866 - HARDING, L.W. Jr., COX, J.L., REQUEGNAT, J.E. : Spring-summer phytoplankton production in Humboldt Bay, California. - California Fish Game *64* : 53 - 59, 1978.

33867 - HARDING, L.W. Jr., PHILLIPS, J.H. Jr. : Polychlorinated biphenyls : transfer from microparticulates to marine phytoplankton and the effects on photosynthesis. - Science *202* : 1189 - 1192, 1978.

*33868 - HARDON, J.J., CORLEY, R.H.V., OOI, S.C. : Analysis of growth in oil palm. II. Estimation of genetic variances of growth parameters and yield of fruit bunches. - Euphytica *21* : 257 - 264, 1972.

33869 - HARDT, H., KOK, B. : Comparison of photosynthetic activities of spinach chloroplasts with those of corn mesophyll and corn bundle sheath tissue. - Plant Physiol. *62* : 59 - 63, 1978.

33870 - HARDWICK, R.C., HARDAKER, J.M., INNES, N.L. : Yields and components of yield of dry beans (*Phaseolus vulgaris* L.) in the United Kingdom. - J. agr. Sci. *90* : 291 - 297, 1978.

*33871 - HARDY, R.W.F., HAVELKA, U.D. : Possible routes to increase the conversion of solar energy to food and feed by grain legumes and cereal grains (crop production) : $CO_2$ and $N_2$ fixation, foliar fertilization, and assimilate partitioning. - In : MITSUI, A., MIYACHI, S., SAN PIETRO, A., TAMURA, S. (ed.) : Biological Solar Energy Conversion. Pp. 299 - 322. Academic Press, New York - San Francisco - London 1977.

33872 - HARDY, R.W.F., HAVELKA, U.D., QUEBEDEAUX, B. : Increasing crop productivity : The problem, strategies, approach, and selected rate-limitations related to photosynthesis. - In : HALL, D.O., COOMBS, J., GOODWIN, T.W. (ed.) : Proceedings of the Fourth International Congress on Photosynthesis. Pp. 695 - 719. Biochem. Soc., London 1978.

33873 - HARDY, R.W.F., HAVELKA, U.D., QUEBEDEAUX, B. : The opportunity for and significance of alteration of ribulose 1,5-bisphosphate carboxylase activities in crop production. - In : SIEGELMAN, H.W., HIND, G. (ed.) : Photosynthetic Carbon Assimilation. Pp. 165 - 178. Plenum Press, New York - London 1978.

33874 - HAREL, E. : Chlorophyll biosynthesis and its control. - In : REINHOLD, R., HARBORNE, J.B., SWAIN, T. (ed.) : Progress in Phytochemistry. Vol. 5. Pp. 127 - 180. Pergamon Press, Oxford - New York - Toronto - Sydney - Paris - Frankfurt 1978.

33875 - HAREL, E. : Initial steps in chlorophyll synthesis - problems and open questions. - In : AKOYUNOGLOU, G., ARGYROUDI-AKOYUNOGLOU, J.H. (ed.) : Chloroplast Development. Pp. 33 - 44. Elsevier/North-Holland Biomedical Press, Amsterdam - New York - Oxford 1978.

33876 - HAREL, E., MELLER, E., ROSENBERG, M. : Synthesis of 5-aminolevulinic acid--[$^{14}$C] by cell-free preparations from greening maize leaves. - Phytochemistry *17* : 1277 - 1280, 1978.

33877 - HARGRAVE, B.T., TAGUCHI, S. : Origin of deposited material sedimented in a marine bay. - J. Fish. Res. Board Can. *35* : 1604 - 1613, 1978. [Chl.]

33878 - HARI, P., SALMINEN, R., PELKONEN, P., HUHTAMAA, M., POHJONEN, V. : A new approach for measuring light inside the canopy in photosynthesis studies. - Silva fenn. *10* : 94 - 102, 1976.

33879 - HARIRI, M., PRIOUL, J.L. : Light-induced adaptive responses under greenhouse and controlled conditions in the fern *Pteris cretica* var. *ouvrardii*. - Physiol. Plant. *42* : 97 - 102, 1978.

*33880 - HARNISCHFEGER, G. : The use of fluorescence emission at 77 °K in the analysis of the photosynthetic apparatus of higher plants and algae. - In : PRESTON, R.D., WOOLHOUSE, H.W. (ed.) : Advances in Botanical Research. Vol. 5. Pp. 1 - 52. Academic Press, London - New York - San Francisco 1977.

33881 - HARNISCHFEGER, G. : Beziehungen zwischen der relativen Größe des Tieftemperaturemissionsmaximums bei 738 nm und der Aktivität von PS I. - Ber. deut. bot. Ges. *91* : 487 - 493, 1978.

33882 - HARNISCHFEGER, G. : Effect of chemical modification of amino groups by fluorescamine on partial reactions of photosynthesis. - Biochim. biophys. Acta *503* : 473 - 479, 1978.

33883 - HARNISCHFEGER, G., CODD, G.A. : Factors affecting energy transfer from phycobilisomes to thylakoids in *Anacystis nidulans*. - Biochim. biophys. Acta *502* : 507 - 513, 1978.

33884 - HARNISCHFEGER, G., HEROLD, B. : Aspects of the regulation of excitation energy transfer in the red alga *Porphyridium aerugineum*. - Ber. deut. bot. Ges. *91* : 477 - 485, 1978.

33885 - HARRIMAN, A., PORTER, G., DUNCAN, I. : Photoredox reactions of manganese. - In : METZNER, H. (ed.) : Photosynthetic Oxygen Evolution. Pp. 393 - 403. Academic Press, London - New York - San Francisco 1978. [Chl.]

33886 - HARRIS, D.A., CROFTS, A.R. : The initial stages of photophosphorylation. Studies using excitation by saturating, short flashes of light. - Biochim. biophys. Acta *502* : 87 - 102, 1978.

33887 - HARRIS, G.C., STERN, A.I. : Stoichiometry of the ribulose-1,5-bisphosphate oxygenase reaction. - J. exp. Bot. *29* : 561 - 566, 1978.

33888 - HARRIS, G.P. : Photosynthesis, productivity and growth : The physiological ecology of phytoplankton. - Ergeb. Limnol. *10* : 1 - 171, 1978.

33889 - HARRIS, J.B. : Development of a tubular apparatus in chloroplasts of ageing *Cyphomandra* leaves. - Cytobios *21* : 151 - 164, 1978.

33890 - HARRIS, P.J.C., WILKINS, M.B. : Evidence of phytochrome involvement in the entrainment of the circadian rhythm of carbon dioxide metabolism in *Bryophyllum*. - Planta *138* : 271 - 278, 1978.

33891 - HARRIS, P.J.C., WILKINS, M.B. : The circadian rhythm in *Bryophyllum* leaves : Phase control by radiant energy. - Planta *143* : 323 - 328, 1978. [Ps.]

33892 - HARRISON, C.R., ARDITTI, J. : Physiological changes during the germination of *Cattleya aurantiaca (Orchidaceae)*. - Bot. Gaz. *139* : 180 - 189, 1978. [Ps, Chl.]

33893 - HARRISON, P.G. : Patterns of uptake and translocation of $^{14}C$ by *Zostera americana* DEN HARTOG in the laboratory. - Aquat. Bot. *5* : 93 - 97, 1978.

33894 - HART, R.H., PEARCE, R.B., CHATTERTON, N.J., CARLSON, G.E., BARNES, D.K., HANSON, C.H. : Alfalfa yield, specific leaf weight, $CO_2$ exchange rate, and morphology. - Crop Sci. *18* : 649 - 653, 1978.

33895 - HARTMAN, F.C., NORTON, I.L., STRINGER, C.D., SCHLOSS, J.V. : Attempts to apply affinity labeling techniques to ribulose bisphosphate carboxylase/oxygenase. - In : SIEGELMAN, H.W., HIND, G. (ed.) : Photosynthetic Carbon Assimilation. Pp. 245 - 269. Plenum Press, New York - London  1978.

33896 - HARTWIG, E.O. : Factors affecting respiration and photosynthesis by the benthic community of a subtidal siliceous sediment. - Mar. Biol. *46* : 283 - 293, 1978.

33897 - HARTZ, T.K., MOORE, F.D. III. : Prediction of potato yield using temperature and insolation data. - Amer. Potato J. *55* : 431 - 436, 1978.

33898 - HARVEY, D.M. : The photosynthetic and respiratory potential of the fruit in relation to seed yield of leafless and semi-leafless mutants of *Pisum sativum* L. - Ann. Bot. *42* : 331 - 336, 1978.

33899 - HARVEY, D.M., GOODWIN, J. : The photosynthetic net carbon dioxide exchange potential in conventional and "leafless" phenotypes of *Pisum sativum* L. in relation to foliage area, dry matter production and seed yield. - Ann. Bot. *42* : 1091 - 1098, 1978.

33900 - HARVEY, D.M.R., FLOWERS, T.J. : Determination of the sodium, potassium and chloride ion concentrations in the chloroplasts of the halophyte *Suaeda maritima* by non-aqueous cell fractionation. - Protoplasma *97* : 337 - 349, 1978.

33901 - HARVEY, G.W., BISHOP, N.I. : Photolability of photosynthesis in two separate mutants of *Scenedesmus obliquus*. Preferential inactivation of photosystem I. - Plant Physiol. *62* : 330 - 336, 1978.

33902 - HASE, T., WAKABAYASHI, S., MATSUBARA, H., EVANS, M.C.W., JENNINGS, J.V. : Amino acid sequence of a ferredoxin from *Chlorobium thiosulfatophilum* strain Tassajara, a photosynthetic green sulfur bacterium. - J. Biochem. (Tokyo) *83* : 1321 - 1325, 1978.

33903 - HASE, T., WAKABAYASHI, S., MATSUBARA, H., KERSCHER, L., OESTERHELT, D., RAO, K.K., HALL, D.O. : Complete amino acid sequence of *Halobacterium halobium* ferredoxin containing an $N^{\epsilon}$- acetyllysine residue. - J. Biochem. (Tokyo) *83* : 1657 - 1670, 1978.

33904 - HASE, T., WAKABAYASHI, S., MATSUBARA, H., RAO, K.K., HALL, D.O., WIDMER, H., GYSI, J., ZUBER, H. : The amino acid sequence of ferredoxin from the alga *Mastigocladus laminosus*. - Phytochemistry *17* : 1863 - 1867, 1978.

33905 - HASE, T., WAKABAYASHI, S., WADA, K., MATSUBARA, H., JÜTTNER, F., RAO, K.K., FRY, I., HALL, D.O. : *Cyanidium caldarium* ferredoxin : a red algal type ? - FEBS Lett. *96* : 41 - 44, 1978.

33906 - HASEBE, H., YAMAZAKI, S., TAMAURA, Y., HAGIWARA, H., INADA, Y. : Inhibition of chloroplast adenosine triphosphatase activity by basic proteins and peptides. - FEBS Lett. *95* : 295 - 298, 1978.

33907 - HASLETT, B.G., BAILEY, C.J., RAMSHAW, J.A.M., SCAWEN, M.D., BOULTER, D. : The amino acid sequence of plastocyanin from *Rumex obtusifolius*. - Phytochemistry *17* : 615 - 617, 1978.

33908 - HASLETT, B.G., EVANS, I.M., BOULTER, D. : Amino acid sequence of plastocyanin from *Solanum crispum* using automatic methods. - Phytochemistry *17* : 735 - 739, 1978.

33909 - HASPELOVÁ-HORVATOVIČOVÁ, A., HORIČKOVÁ, B. : Príjem $^{14}$C zdravými rastlinami jačmeňa a rastlinami napadnutými múčnatkou. [$^{14}$C incorporation by healthy and mildewed barley plants.] - Acta bot. slov. Acad. Sci. slov. Ser. B *2* : 87 - 98, 1978. [In Slovak, ab : E, R.]

33910 - HASUMI, H., NAKAMURA, S. : Studies on the ferredoxin-ferredoxin-NADP reductase complex : kinetic and solvent perturbation studies on the location of sulfhydryl and aromatic amino acid residues. - J. Biochem. (Tokyo) *84* : 707 - - 717, 1978.

33911 - HATANO, S. : Studies on frost hardiness in *Chlorella ellipsoidea* : Effects of antimetabolites, surfactans, hormones, and sugars on the hardening process in the light and dark. - In : LI, P.H., SAKAI, A. (ed.) : Plant Cold Hardiness and Freezing Stress. Mechanisms and Crop Implications. Pp. 175 - 196. Academic Press, New York - San Francisco - London 1978. [Chloroplast.]

33912 - HATANO, S., SADAKANE, H., NAGAYAMA, J., WATANABE, T. : Studies on frost hardiness in *Chlorella ellipsoidea*. III. Changes in $O_2$ uptake and evolution during hardening and after freeze-thawing. - Plant Cell Physiol. *19* : 917 - 926, 1978.

33913 - HATCH, M.D. : Regulation of enzymes in $C_4$ photosynthesis. - In : HORECKER, B.L., STADTMAN, E.R. (ed.) : Current Topics in Cellular Regulation. Vol. 14. Pp. 1 - 27. Academic Press, New York - San Francisco - London 1978.

33914 - HATCH, M.D., OLIVER, I.R. : Activation and inactivation of phosphoenolpyruvate carboxylase in leaf extracts from $C_4$ species. - Aust. J. Plant Physiol. *5* : 571 - 580, 1978.

*33915 - HATCHER, B.G., CHAPMAN, A.R.O., MANN, K.H. : An annual carbon budget for the kelp *Laminaria longicruris*. - Mar. Biol. *44* : 85 - 96, 1977.

33916 - HATFIELD, J.L., CARLSON, R.E. : Photosynthetically active radiation, $CO_2$ uptake, and stomatal diffusive resistance profiles within soybean canopies. - Agron. J. *70* : 592 - 596, 1978.

33917 - HATZIOS, K.K., PENNER, D. : The effect of diflubenzuron [1-(4-chlorophenyl)- -3-(2,6-difluorobenzoyl)urea] on soybean [*Glycine max* (L.) MERR.] photosynthesis, respiration, and leaf ultrastructure. - Pesticide Biochem. Physiol. *9* : 65 - 69, 1978.

33918 - HATZIOS, K.K., PENNER, D. : The effect of fentin hydroxide (triphenyltin hydroxide) on soybean [*Glycine max* (L.) MERR.] and rice (*Oryza sativa* L.) photosynthesis, respiration, and leaf ultrastructure. - Pesticide Biochem. Physiol. *9* : 70 - 74, 1978.

33919 - HAUSKA, G. : Vectorial redox reactions of quinoid compounds and the topography of photosynthetic membranes. - In : HALL, D.O., COOMBS, J., GOODWIN, T.W. (ed.) : Proceedings of the Fourth International Congress on Photosynthesis. Pp. 185 - 196. Biochem. Soc., London 1978.

33920 - HAVRANEK, W.M., BENECKE, U. : The influence of soil moisture on water poten-
        tial, transpiration and photosynthesis of conifer seedlings. - Plant Soil *49* :
        91 - 103, 1978.

*33921 - HAYASHI, K.-I., YAMAMOTO, T., NAKAGAHRA, M. : Genetic control for leaf pho-
        tosynthesis in rice, *Oryza sativa* L. - Ikushugaku Zasshi [Jap. J. Breed.]
        *27* : 49 - 56, 1977.

*33922 - HAYASHIDA, F. : [The effect on photosynthesis in several benthic marine al-
        gae of pollutants in Tagonoura Port, Suruga Bay.] - Environ. Control Biol.
        *13* : 9 - 13, 1975. [In Jap., ab : E.]

33923 - HAYNES, R.C., HAMMER, U.T. : The saline lakes of Saskatchewan IV. Primary
        production by phytoplankton in selected saline ecosystems. - Int. Rev. ges.
        Hydrobiol. *63* : 337 - 351, 1978.

33924 - HAYWARD, S.B., GRANO, D.A., GLAESER, R.M., FISHER, K.A. : Molecular orien-
        tation of bacteriorhodopsin within the purple membrane of *Halobacterium ha-
        lobium*. - Proc. nat. Acad. Sci. USA *75* : 4320 - 4324, 1978.

33925 - HEAL, O.W., LATTER, P.M., HOWSON, G. : A study of the rates of decomposition
        of organic matter. - In : HEAL, O.W., PERKINS, D.F. (ed.) : Production Ecology
        of British Moors and Montane Grasslands. Pp. 136 - 159. Springer-Verlag, Ber-
        lin - Heidelberg - New York 1978.

B33926 - HEAL, O.W., PERKINS, D.F. (ed.) : Production Ecology of British Moors and
        Montane Grasslands.- Springer-Verlag, Berlin-Heidelberg - New York 1978.

33927 - HEANEY, S.I. : Some observations on the use of the *in vivo* fluorescence
        technique to determine chlorophyll-*a* in natural populations and cultures of
        freshwater phytoplankton. - Freshwater Biol. *8* : 115 - 126, 1978.

*33928 - HEARN, A.B. : Crop physiology. - In : ARNOLD, M.H. (ed.) : Agricultural Re-
        search for Development. Pp. 77 - 122. Cambridge University Press, Cambridge
        - New York 1976. [Ps.]

33929 - HEATH, R.L., LEECH, R.M. : The simulation of $CO_2$-supported $O_2$ evolution in
        intact spinach chloroplasts by ammonium ion. - Arch. Biochem. Biophys. *190* :
        221 - 226, 1978.

33930 - HEATHCOTE, P., WILLIAMS-SMITH, D.L., EVANS, M.C.W. : Quantitative electron-
        -paramagnetic-resonance measurements of the electron-transfer components of
        the Photosystem-I reaction centre. The reaction-centre chlorophyll (*P700*),
        the primary electron acceptor X and bound iron-sulphur centre A. - Biochem.
        J. *170* : 373 - 378, 1978.

33931 - HEATHCOTE, P., WILLIAMS-SMITH, D.L., SIHRA, C.K., EVANS, M.C.W. : The role
        of the membrane-bound iron-sulphur centres A and B in the Photosystem I re-
        action centre of spinach chloroplasts. - Biochim. biophys. Acta *503* : 333 -
        342, 1978.

33932 - HEBER, U., EGNEUS, H., HANCK, U., JENSEN, M., KÖSTER, S. : Regulation of
        photosynthetic electron transport and photophosphorylation in intact chlo-
        roplasts and leaves of *Spinacia oleracea* L. - Planta *143* : 41 - 49, 1978.

33933 - HEBER, U., PURCZELD, P. : Substrate and product fluxes across the chloro-
        plast envelope during bicarbonate and nitrite reduction. - In : HALL, D.O.,
        COOMBS, J., GOODWIN; T.W. (ed.) : Proceedings of the Fourth International
        Congress on Photosynthesis. Pp. 107 - 118. Biochem. Soc., London 1978.

33934 - HEDGE, D.M., SARAF, C.S. : Effect of intercropping and phosphorus fertiliza-
        tion on canopy development, drymatter accumulation, AGR, NAR, CCR and NGR
        of pigeon pea [*Cajanus cajan* (L) MILLER] under dryland condition. - In :
        Inter-Disciplinary Symposium on Photosynthesis and Productivity. Abstract
        of Papers. Pp. 57 - 58. Indian nat. Sci. Acad., New Delhi 1978.

33935 - HEGAZI, A.M., KAUSCH, W. : The interaction between salinity and (2-chloro-
        ethyl)-trimethylammonium chloride (CCC) on salt tolerance in maize. - Z.
        Pflanzenphysiol. *88* : 39 - 45, 1978. [Chl.]

*33936 - HEICHEL, G.H., TURNER, N.C. : Phenology and leaf growth of defoliated hard-
        wood trees. - In : ANDERSON, J.F., KAYA, H.K. (ed.) : Perspectives in Forest

Entomology. Pp. 31 - 40. Academic Press, New York - San Francisco - London 1976.

33937 - HEIDE-JØRGENSEN, H.S. : The xeromorphic leaves of *Hakea suaveolens* R.BR. I. Structure of photosynthetic tissue with intercellular pectic strands and tylosoids. - Bot. Tidsskr. *72* : 87 - 103, 1978.

33938 - HEINZ, E., BERTRAMS, M., JOYARD, J., DOUCE, R. : Demonstration of an acyltransferase activity in chloroplast envelopes. - Z. Pflanzenphysiol. *87* : 325 - 331, 1978.

33939 - HEISE, K.-P. : Variation in chloroplast lipid content and its correlation to photosynthetic activities. - Z. Naturforsch. *33 C* : 685 - 687, 1978.

33940 - HEISE, K.-P., HARNISCHFEGER, G. : Correlation between photosynthesis and plant lipid composition. - Z. Naturforsch. *33 C* : 537 - 540, 1978.

33941 - HEITHIER, H., GALLA, H.-J., MÖHWALD, H. : Fluorescence spectroscopic and thermodynamic studies of chlorophyll containing monolayers and vesicles. Part I : Mixed monolayers of pheophytin A and lecithin. - Z. Naturforsch. *33 C* : 382 - 391, 1978.

33942 - HEIZMANN, P., VERDIER, G., YOUMIS, H. : Transcription of nuclear and chloroplast genomes in *Euglena* during greening and in bleached mutants. - In : AKOYUNOGLOU, G., ARGYROUDI-AKOYUNOGLOU, J.H. (ed.) : Chloroplast Development. Pp. 623 - 628. Elsevier/North-Holland Biomedical Press, Amsterdam - New York - Oxford 1978.

33943 - HELDT, H.W., CHON, C.J., LILLEY, R.McC., PORTIS, A. : The role of fructose- - and sedoheptulosebisphosphatase in the control of $CO_2$ fixation. Evidence from the effect of $Mg^{++}$ concentration, pH and $H_2O_2$. - In : HALL, D.O., COOMBS, J., GOODWIN, T.W. (ed.) : Proceedings of the Fourth International Congress on Photosynthesis. Pp. 469 - 478. Biochem. Soc., London 1978.

33944 - HELDT, H.W., CHON, C.J., LORIMER, G.H. : Phosphate requirement for the light activation of ribulose-1,5-bisphosphate carboxylase in intact spinach chloroplasts. - FEBS Lett. *92* : 234 - 240, 1978.

33945 - HELDT, H.W., FLÜGGE, U.I., FLIEGE, R. : The influence of illumination on the transport of 3-phosphoglycerate across the chloroplast envelope. - In : AZZONE, G.F., AVRON, M., METCALFE, J.C., QUAGLIARIELLO, E., SILIPRANDI, N. (ed.) : The Proton and Calcium Pumps. Pp. 105 - 114. Elsevier/North-Holland Biomedical Press, Amsterdam - Oxford - New York 1978.

33946 - HELLER, R.A., PEIRSON, D.R. : The other photosynthesis. - J. chem. Educ. *55* : 233 - 235, 1978.

33947 - HELLINGWERF, K.J., SCHOLTE, B.J., VAN DAM, K. : Bacteriorhodopsin vesicles. An outline of the requirements for light-dependent $H^+$ pumping. - Biochim. biophys. Acta *513* : 66 - 77, 1978.

33948 - HELLINGWERF, K.J., SCHUURMANS, J.J., WESTERHOFF, H.V. : Demonstration of coupling between the protonmotive force across bacteriorhodopsin and the flow throught its photochemical cycle. - FEBS Lett. *92* : 181 - 186, 1978.

33949 - HELLINGWERF, K.J., TEGELAERS, F.P.W., WESTERHOFF, H.V., ARENTS, J.C., VAN DAM, K. : Structural and functional description of membrane vesicles containing the light-driven proton pump from the plasma membrane of the halophilic bacterium *Halobacterium halobium* : Bacteriorhodopsin. - In : CAPLAN, S.R., GINZBURG, M. (ed.) : Energetics and Structure of Halophilic Microorganisms. Pp. 283 - 290. Elsevier/North-Holland Biomedical Press, Amsterdam - New York 1978.

33950 - HELMS, K., WARDLAW, I.F. : Translocation of tobacco mosaic virus and photosynthetic assimilate in *Nicotiana glutinosa*. - Physiol. Plant Pathol. *13* : 23 - 36, 1978.

33951 - HELSEL, D.B., FREY, K.J. : Grain yield variations in oats associated with differences in leaf area duration among oat lines. - Crop Sci. *18* : 765 - 769, 1978.

33952 - HEMPHILL, J.K., VENKETESWARAN, S. : Chlorophyll and carotenoid accumulation
in three chlorophyllous callus phenotypes of *Glycine max*. - Amer. J. Bot.
*65* : 1055 - 1063, 1978.

33953 - HENDRY, G.A.F., STOBART, A.K. : Glycine metabolism in etiolated barley leaves on exposure to light. - Phytochemistry *17* : 69 - 72, 1978. [Chl.]

33954 - HENDRY, G.A.F., STOBART, A.K. : The effect of haem on chlorophyll synthesis
in barley leaves. - Phytochemistry *17* : 73 - 77, 1978.

33955 - HENDRY, G.A.F., STOBART, A.K. : Effect of 2,2'-bipyridyl on porphyrin synthesis in etiolated and light-treated barley leaves. - Phytochemistry *17* : 671 -
674, 1978.

33956 - HENRIQUES, F., PARK, R.B. : Characterization of three new chlorophyll-protein
complexes. - Biochem. biophys. Res. Commun. *81* : 1113 - 1118, 1978.

33957 - HENRIQUES, F., PARK, R.B. : Polypeptide cross-linking in chloroplast membranes. - Arch. Biochem. Biophys. *189* : 44 - 50, 1978.

33958 - HENRIQUES, F., PARK, R.B. : Spectral characterization of five chlorophyll-
-protein complexes. - Plant Physiol. *62* : 856 - 860, 1978.

33959 - HENRY, E.W. : Promellar body development in etioplasts of *Pisum sativum* L.
var. "Alaska" tissue. - Micron *9* : 7 - 8, 1978.

33960 - HERBERT, M., BURKHARD, C., SCHNARRENBERGER, C. : Cell organelles from Cras-
sulacean-Acid-Metabolism (CAM) plants. I. Enzymes in isolated peroxisomes.
- Planta *143* : 279 - 284, 1978.

33961 - HERBERT, S.J. : Plant density and irrigation studies on lupins III. Seed-
-yield relationships of *Lupinus angustifolius* cv. 'Unicrop'. - New Zeal. J.
agr. Res. *21* : 483 - 489, 1978. [Growth analysis.]

33962 - HERBERT, S.J., HILL, G.D. : Plant density and irrigation studies on lupin
I. Growth analysis of *Lupinus angustifolius* cv. 'WAU11B'. - New Zeal. J. agr.
Res. *21* : 467 - 474, 1978.

33963 - HERBERT, S.J., HILL, G.D. : Plant density and irrigation studies on lupin
II. Components of seed yield of *Lupinus angustifolius* cv. WAU11B. - New Zeal.
J. agr. Res. *21* : 475 - 481, 1978.

33964 - HERBETTE, L., SCARPA, A., PACHENCE, J., DUTTON, P.L., BLASIE, J.K., WANG,
C.T., SAITO, A., FLEISCHER, S. : The structure and function of reconstituted
$Ca^{2+}$ ATPase/lipid and photosynthetic reaction center/lipid membranes. - In :
DUTTON, P.L., LEIGH, J.S., SCARPA, A. (ed.) : Frontiers of Biological Ener-
getics. Vol. 2. Pp. 1099 - 1108. Academic Press, New York - San Francisco -
London 1978.

*33965 - HERBLAND, A., VOITURIEZ, B. : Relation chlorophylle *a*-fluorescence *in vivo*
dans l'Atlantique tropical. Influence de la structure hydrologique. - Cah.
ORSTOM, Sér. Océanographie *15* : 67 - 77, 1977.

33966 - HERMANN, T.R., RAYFIELD, G.W. : The electrical response to light of bacteriorhodopsin in planar membranes. - Biophys. J. *21* : 111 - 125, 1978.

33967 - HERRMANN, F.H. : Polypeptide composition of chlorophyll protein complex I
from several plants. - In : AKOYUNOGLOU, G., ARGYROUDI-AKOYUNOGLOU, J.H.
(ed.) : Chloroplast Development. Pp. 221 - 227. Elsevier/North-Holland Biomedical Press, Amsterdam - New York - Oxford 1978.

33968 - HERVEY, A., ROBBINS, W.J. : Development of plants from leaf discs of varie-
gated *Coleus* and its relation to pattern of leaf chlorosis. - In vitro *14* :
294 - 300, 1978.

33969 - HERZOG, H. : Wachstumsverhalten und Kältetoleranz bei Ackerbohnen (*Vicia
faba* L.) unter verschiedenen Testbedingungen II. Assimilationsleistung und
ihre Veränderung nach Gefriertests. - Z. Acker- Pflanzenbau *147* : 111 - 120,
1978.

*33970 - HERZOG, H., GEISLER, G. : Der Einfluß von Cytokininapplikation auf die Assi-
milateinlagerung und die endogene Cytokininaktivität der Karyopsen bei zwei
Sommerweizensorten. - Z. Acker- Pflanzenbau *144* : 230 - 242, 1977.

*33971 - HERZOG, R., HETEŠA, J. :  Primární produkce planktonu jako ukazatel účinnosti hnojení rybníků. [Primary production of plankton as indicator of pond fertilization efficiency.] - Práce výzk. Ústavu rybář. hydrobiol. (Vodňany) 8 : 5 - 25, 1968 (1970). [In Czech.]

33972 - HESS, B., KORENSTEIN, R., KUSCHMITZ, D. : Conformational changes and cooperation of bacteriorhodopsin. - In : SCHÄFER, G., KLINGENBERG, M. (ed.) : Energy Conservation in Biological Membranes. Pp. 152 - 156. Springer-Verlag, Berlin - Heidelberg - New York 1978.

33973 - HESS, B., KORENSTEIN, R., KUSCHMITZ, D. : Coupling of the photocycle to conformational changes in bacteriorhodopsin. - In : CAPLAN, S.R., GINZBURG, M. (ed.) : Energetics and Structure of Halophilic Microorganisms. Pp. 89 - 108. Elsevier/North-Holland Biomedical Press, Amsterdam - New York 1978.

33974 - HESSLER, R., PEVELING, E. : Die Lokalisation von $^{14}$C-Assimilaten in Flechtenthalli von Cladonia incrassata FLOERKE und Hypogymnia physodes (L.) ACH. - Z. Pflanzenphysiol. 86 : 287 - 302, 1978.

33975 - HEYSER, W., EVERT, R.F., FRITZ, E., ESCHRICH, W. : Sucrose in the free space of translocating maize leaf bundles. - Plant Physiol. 62 : 491 - 494, 1978. [Ps.]

33976 - HICKLENTON, P.R., JOLIFFE, P.A. : Effects of greenhouse $CO_2$ enrichment on the yield and photosynthetic physiology of tomato plants. - Can. J. Plant Sci. 58 : 801 - 817, 1978.

33977 - HICKMAN, M., JENKERSON, C.G. : Phytoplankton primary productivity and population efficiency studies in a Prairie-Parkland Lake near Edmonton, Alberta, Canada. - Int. Rev. ges. Hydrobiol. 63 : 1 - 24, 1978.

33978 - HIGGINS, T.J.V., JACOBSEN, J.V. : The influence of plant hormones on selected aspects of cellular metabolism. - In : LETHAM, D.S., GOODWIN, P.B., HIGGINS, T.J.V. (ed.) : Phytohormones and Related Compounds - A Comprehensive Treatise. Vol. 1. Pp. 467 - 514. Elsevier/North-Holland Biomedical Press, Amsterdam - Oxford - New York 1978. [Ps.]

33979 - HIGHFIELD, P.E., ELLIS, R.J. : Synthesis and transport of the small subunit of chloroplast ribulose bisphosphate carboxylase. - Nature 271 : 420 - 424, 1978.

33980 - HIGUCHI, M., KATAYOSE, A. : Preferential inhibition by ethidium bromide of photosynthetic apparatus formation in Rhodopseudomonas spheroides cells. - Plant Cell Physiol. 19 : 963 - 973, 1978.

33981 - HILL, B.C., BOWN, A.W. : Phosphoenolpyruvate carboxylase activity from Avena coleoptile tissue. Regulation by $H^+$ and malate. - Can. J. Bot. 56 : 404 - 407, 1978.

33982 - HILL, R. : Oxygen evolution: Some historical aspects. - Biochem. Soc. Trans. 6 : 901 - 904, 1978.

33983 - HILLER, R.G., PILGER, T.B.G., GENGE, S. : Formation of chlorophyll protein complexes during greening of etiolated barley leaves. - In : AKOYUNOGLOU, G., ARGYROUDI-AKOYUNOGLOU, J.H. (ed.) : Chloroplast Development. Pp. 215 - 220. Elsevier/North-Holland Biomedical Press, Amsterdam - New York - Oxford 1978.

33984 - HINCKLEY, T.M., ASLIN, R.G., AUBUCHON, R.R., METCALF, C.L., ROBERTS, J.E. : Leaf conductance and photosynthesis in four species of the oak-hickory forest type. - Forest Sci. 24 : 73 - 84, 1978.

33985 - HINCKLEY, T.M., LASSOIE, J.P., RUNNING, S.W. : Temporal and spacial variations in the water status of forest trees. - Forest Sci. 24 (Suppl.3) : 1 - 72, 1978.

33986 - HIND, G., MILLS, J.D., SLOVACEK, R.E. : Cyclic electron transport in photosynthesis. - In : HALL, D.O., COOMBS, J., GOODWIN, T.W. (ed.) : Proceedings of the Fourth International Congress on Photosynthesis. Pp. 591 - 600. Biochem. Soc., London 1978.

33987 - HINDMAN, J.C., KUGEL, R., SVIRMICKAS, A., KATZ, J.J. : Stimulated fluorescence and fluorescence quenching in chlorophyll $a$ and bacteriochlorophyll $a$.
- Chem. Phys. Lett. *53* : 197 - 200, 1978.

33988 - HINDMAN, J.C., KUGEL, R., WASIELEWSKI, M.R., KATZ, J.J. : Coherent stimulated light emission (lasing) in covalently linked chlorophyll dimers. - Proc.
nat. Acad. Sci. USA *75* : 2076 - 2079, 1978.

33989 - HINKLE, P.C., McCARTY, R.E. : How cells make ATP. - Sci. Amer.*238* : 104 -
123, 1978.

33990 - HIPKINS, M.F. : Kinetic analysis of the chlorophyll fluorescence inductions
from chloroplasts blocked with 3-(3,4-dichlorophenyl)-1,1-dimethylurea. -
Biochim. biophys. Acta *502* : 514 - 523, 1978.

33991 - HIPKINS, M.F. : The emission yields of delayed and prompt fluorescence from
chloroplasts. - Biochim. biophys. Acta *502* : 161 - 168, 1978.

33992 - HIRAI, A. : [Fraction I protein as a genetic marker of nucleus and chloroplast.] - Tampakushitsu Kakusan Koso *23* : 1023 -   1031, 1978. [In Jap.]

33993 - HIRAKI, K., HAMANAKA, T., MITSUI, T., KITO, Y. : Formation of the two-dimensional hexagonal lattice of bacteriorhodopsin in reconstituted brown membrane. - Biochim. biophys. Acta *536* : 318 - 322, 1978.

33994 - HIRAYAMA, O., NISHIDA, T. : Effects of lipids and their related compounds
on the chloroplast functions. - Agr. biol. Chem. (Tokyo) *42* : 141 - 146,
1978.

33995 - HIRAYAMA, O., NOMOTOBORI, T. : Preparation and characterization of phospho-
lipid-depleted chloroplasts. - Biochim. biophys. Acta *502* : 11 - 16, 1978.

33996 - HIROTA, N. : [Studies on chlorophyll of marine algae - I. Absorption spectra
and chlorophyll contents of dried "Wakame".] - Bull. jap. Soc. sci. Fish.
*44* : 1003 - 1007, 1978. [In Jap., ab : E.]

33997 - HIROTA, N. : [Studies on chlorophyll of marine algae - II. Separation of
chlorophyll and its altered compounds in "Wakame" using TLC.]- Bull. jap.
Soc. sci. Fish. *44* : 1009 - 1014, 1978. [In Jap., ab : E.]

*33998 - HIROTA, O., AKIYAMA, T., TAKEDA, T., MATSUI, T., AIGA, I. : [Studies on the
role of light transmissibility of single leaf and plant type as factors con-
stituting light extinction coefficient K in corn plant populations.] - Proc.
Crop Sci. Soc. Japan *43* : 283 - 288, 1974. [In Jap., ab : E.]

33999 - HIROTA, O., TAKEDA, T. : [Studies on utilization of solar radiation by crop
stands III. Relationships between conversion efficiency of solar radiation
energy and respiration of construction and maintenance in rice and soybean
plant populations.] - Jap. J. Crop Sci. *47* : 336 - 343, 1978. [Ps; In Jap.,
ab : E.]

34000 - HIROTA, O., TAKEDA, T., MURATA, Y., KOBA, M. : [Studies on utilization of
solar radiation by crop stands II. Utilization of short wave and photosynthetically active radiation by rice and soybean plant populations.] - Jap.
J. Crop Sci. *47* : 133 - 140, 1978. [In Jap., ab : E.]

34001 - HIRSCH, R.E., BRODY, S.S. : Spectral properties of chlorophyll $a$ monolayers
in the presence of an exogenous electron donor and acceptor. - Europe. J.
Biochem. *89* : 281 - 286, 1978.

34002 - HITCHCOCK, G.L. : Labelling patterns of carbon-14 in net plankton during a
winter-spring bloom. - J. exp. mar. Biol. Ecol. *31* : 141 - 153, 1978.

34003 - HITZ, W.D., STEWART, R.C. : The steady state pool size of ribulose-1,5-bis-
phosphate during photosynthesis : Effect of $O_2$ and $CO_2$ . - Plant Physiol.
*61* (Suppl.) : 100, 1978.

34004 - HO, K.K., KROGMANN, D.W. : The isolation and characterization of cytochro-
chrome $f$ from cyanobacteria. - Plant Physiol. *61* (Suppl.) : 76, 1978.      ·

34005 - HO, L.C. : The regulation of carbon transport and the carbon balance of mature tomato leaves. - Ann. Bot. *42* : 155 - 164, 1978.

34006 - HO, L.C., THORNLEY, J.H.M. : Energy requirements for assimilate translocation from mature tomato leaves. - Ann. Bot. 42 : 481 - 483, 1978.

34007 - HOARAU, J., REMY, R. : Analysis, at 20 °C, of the Chl a red absorption band
heterogeneity, using the shifting effect of Triton X 100 on chlorophyllous
membranes and chlorophyll protein complexes spectra. - In : AKOYUNOGLOU, G.,
ARGYROUDI-AKOYUNOGLOU, J.H. (ed.) : Chloroplast Development. Pp. 455 - 459.
Elsevier/North-Holland Biomedical Press, Amsterdam - New York - Oxford 1978.

34008 - HOFÄCKER, W. : Untersuchungen zur Photosynthese der Rebe. Einfluß der Ent-
blätterung, der Dekapitierung, der Ringelung und der Entfernung der Traube.
- Vitis 17 : 10 - 22, 1978.

34009 - HOFF, A.J., GORTER DE VRIES, H. : Electron spin resonance in zero magnetic field
of the reaction center triplet of photosynthetic bacteria. - Biochim. biophys. Acta 503 : 94 - 106, 1978.

34010 - HOFF, A.J., MÖBIUS, K. : Nitrogen electron nuclear double resonance and proton triple resonance experiments on the bacteriochlorophyll cation in solution. - Proc. nat. Acad. Sci. USA 75 : 2296 - 2300, 1978.

34011 - HOFFMAN, W.E., DAWES, C.J. : Diurnal photosynthetic rhythms and production
of two Florida benthic algal species, Bostrychia binderi HARVEY and Gracilaria verrucosa (HUDSON) PAPENFUSS. - J. Phycol. 14 (Suppl.) : 39, 1978.

34012 - HOFFMANN, F. : Ein Simulationsmodell für Wasserhaushalt und Stoffbildung
eines vegetativ wachsenden Pflanzenbestandes (Zuckerrüben). - Arch. Acker-
Pflanzenbau Bodenk. 22 : 97 - 107, 1978.

34013 - HOFFMANN, P. : Einheit und Mannigfaltigkeit im Prozeß der Photosynthese
bei Pro- und Eukaryoten. - Biol. Rundschau 16 : 73 - 88, 1978.

34014 - HOFFMANN, P., LEUPOLD, D., HIEKE, B., VOIGT, B. : Laser-spectroscopic characterization of the absorption behaviour of chlorophyll in vitro and in vivo. - Biochem. Physiol. Pflanz. 173 : 460 - 464, 1978.

34015 - HOFFMANN, W., GRACA-MIGUEL, M., BARNARD, P., CHAPMAN, D. : Evidence for conformational transitions in bacteriorhodopsin. - FEBS Lett. 95 : 31 - 34,
1978.

34016 - HOGETSU, K. : Studies on productivity of marine communities. - In : TAMIYA,
H. (ed.) : Summary Report on the Contribution of the Japanese National Com-
mittee for IBP, 1964-1974. JIBP Synthesis. Vol. 20. Pp. 137 - 158. Univ. of
Tokyo Press, Tokyo 1978.

34017 - HOLADAY, S., BOWES, G. : Photosynthetic/photorespiratory variation and dark
fixation in submersed aquatic plants. - Plant Physiol. 61 (Suppl.) : 8,
1978.

34018 - HOLDER, A.A. : Peptide mapping of the large subunit of ribulose bisphospha-
te carboxylase : Application to the enzyme from the genus Oenothera. - In :
SIEGELMAN, H.W., HIND, G. (ed.) : Photosynthetic Carbon Assimilation. P.420.
Plenum Press, New York - London 1978.

34019 - HOLDER, A.A. : Peptide mapping of the ribulose bisphosphate carboxylase large subunit from the genus Oenothera. - Carlsberg Res. Commun. 43 : 391 -
399, 1978.

34020 - HOLMES, N.G., VAN GRONDELLE, R., DUYSENS, L.N.M. : Flash-induced changes
in the in vivo bacteriochlorophyll fluorescence yield at low temperatures
and low redox potentials in carotenoid-containing strains of photosynthetic
bacteria. - Biochim. biophys. Acta 503 : 26 - 36, 1978.

34021 - HOLM-HANSEN, O., RIEMANN, B. : Chlorophyll a determination : improvements
in methodology. - Oikos 30  : 438 - 447, 1978.

34022 - HOLMQVIST, O. : Disappearance of the intracytoplasmic membranes in Rhodopseudomonas sphaeroides : A quantitative ultrastructural study. - Arch.
Microbiol. 117 : 293 - 295, 1978. [Chl.]

34023 - HOLTEN, D., WINDSOR, M.W., PARSON, W.W., GOUTERMAN, M. : Models for bacterial photosynthesis : Electron transfer from photoexcited singlet bacterio-

pheophytin to methyl viologen and $m$-dinitrobenzene. - Photochem. Photobiol. *28* : 951 - 961, 1978.

34024 - HOLTEN, D., WINDSOR, M.W., PARSON, W.W., THORNBER, J.P. : Primary photoche-mical processes in isolated reaction centers of *Rhodopseudomonas viridis*. - Biochim. biophys. Acta *501* : 112 - 126, 1978.

34025 - HOLZAPFEL, C. : Analysis of the prompt fluorescence induction by means of computer simulation of the primary photosynthetic reactions. - Z. Naturfor-sch. *33 C* : 402 - 407, 1978.

34026 - HOMANN, P.H. : Energy conservation in stroma thylakoids. - In : AKOYUNOGLOU, G., ARGYROUDI-AKOYUNOGLOU, J.H. (ed.) : Chloroplast Development. Pp. 461 - 466. Elsevier/North-Holland Biomedical Press, Amsterdam - New York - Oxford 1978.

34027 - HOMANN, P.H. : Oxygen-dependent photooxidations in photosystem II of isola-ted chloroplasts. - In : METZNER, H. (ed.) : Photosynthetic Oxygen Evolu-tion. Pp. 195 - 212. Academic Press, London - New York - San Francisco 1978.

34028 - HONIG, B. : Kinetic and molecular models for proton pumping in bacteriorho-dopsin. - In : CAPLAN, S.R., GINZBURG, M. (ed.) : Energetics and Structure of Halophilic Microorganisms. Pp. 109 - 121. Elsevier/North-Holland Biome-dical Press, Amsterdam - New York 1978.

34029 - HONSELL, E., AVANZINI, A., GHIRARDELLI, L.A. : Two ways of chloroplast deve-lopment in vegetative cells of *Nitophyllum punctatum* (*Rhodophyta*). Ultra-structural study. - J. submicrosc. Cytol. *10* : 227 - 237, 1978.

34030 - HOOPER-REID, N.M., ROBINSON, G.G.C. : Seasonal dynamics of epiphytic algal growth in a marsh pond : productivity, standing crop, and community compo-sition. - Can. J. Bot. *56* : 2434 - 2440, 1978. [Chl.]

34031 - HOOPER-REID, N.M., ROBINSON, G.G.C. : Seasonal dynamics of epiphytic algal growth in a marsh pond : composition, metabolism, and nutrient availability. - Can. J. Bot. *56* : 2441 - 2448, 1978. [Chl.]

34032 - HOPE, A.B. : Analysis of the high-energy state of chloroplasts. - In : DEAMER, D.W. (ed.) : Light Transducing Membranes : Structure, Function, and Evolution. Pp. 289 - 312. Academic Press, New York - San Francisco - London 1978.

*34033 - HOPE, A.B., CHOW, W.S., WAGNER, G. : Ionic exchanges in chloroplasts : Membrane transport and membrane adsorption. - In : THELLIER, M., MONNIER, A., DEMARTY, M., DAINTY, J. (ed.) : Échanges Ioniques Transmembranaires chez les Végétaux. Colloque du CNRS   No. 258. Pp. 577 - 582. Édit. CNRS, Paris 1977.

34034 - HOPKINS, A.W. : The effects of thermal effluent on phytoplankton populations in lake Arlington, Texas. - J. Phycol.*14* (Suppl.) : 24, 1978. [Chl.]

34035 - HOPKINSON, C.S., GOSSELINK, J.G., PARRONDO, R.T. : Aboveground production of seven marsh species in coastal Louisiana. - Ecology *59* : 760 - 769, 1978.

34036 - HORIE, T. : A simulation model for cucumber growth to form basis for mana-ging the plant-environment system. - Acta Hort. *87* : 215 - 224, 1978. [Ps.]

34037 - HORIE, T. : Simulation of the growth of sunflower plant canopy in relation to solar radiation. - In : MONSI, M., SAEKI, T. (ed.) : Ecophysiology of Pho-tosynthetic Productivity. JIBP Synthesis Vol. 19. Pp. 260 - 267. University of Tokyo Press, Tokyo 1978.

34038 - HORIE, T. : Studies on photosynthesis and primary production of rice plants in relation to meteorological environments. 1. Gaseous diffusive resistan-ces, photosynthesis and transpiration in the leaves as influenced by radia-tion intensity and wind speed. - J. agr. Meteorol. *34* : 125 - 136, 1978.

34039 - HORIO, T. : [Capturing light energy photosynthesis.] - Seisan to Gijutsu *30* (1) : 55 - 57, 1978. [In Jap.]

34040 - HORROCKS, R.D., KERBY, T.A., BUXTON, D.R. : Carbon source for developing bolls in normal and superokra leaf cotton. - New Phytol.   *80* : 335 - 340, 1978.

*34041 - HORSTMANN, U. : Application of solar energy bioconversion in developing
countries. - In : MITSUI, A., MIYACHI, S., SAN PIETRO, A., TAMURA, S. (ed.) :
Biological Solar Energy Conversion. Pp. 427 - 436. Academic Press, New York
- San Francisco - London 1977. [Ps $H_2$ production.]

34042 - HORTON, P., CROZE, E., SMUTZER, G. : Interactions between Photosystem II
components in chloroplast membranes. A correlation between the existence of
a low potential species of cytochrome $b$-559 and low chlorophyll fluorescence
in inhibited and developing chloroplasts. - Biochim. biophys. Acta 503 :
274 - 286, 1978.

34043 - HORVÁTH, G., DROPPA, M., JÁNOSSY, A., CSATORDAY, K., FALUDI-DÁNIEL, Á. :
Comparison of the $O_2$ evolving system in granal and agranal chloroplasts of
maize. - In : METZNÉR, H. (ed.) : Photosynthetic Oxygen Evolution. Pp. 117 -
124. Academic Press, London - New York - San Francisco 1978.

34044 - HORVÁTH, G., DROPPA, M., MUSTÁRDY, L.A., FALUDI-DÁNIEL, Á. : Functional
characteristics of intact chloroplasts isolated from mesophyll protoplasts
and bundle sheath cells of maize. - Planta 141 : 239 - 244, 1978.

34045 - HORVÁTH, G., MUSTÁRDY, L.A., FALUDI-DÁNIEL, Á. : Cellular environment affec-
ting granum formation. A working hypothesis. - In : AKOYUNOGLOU, G., ARGY-
ROUDI-AKOYUNOGLOU, J.H. (ed.) : Chloroplast Development. Pp. 863 - 869.
Elsevier/North-Holland Biomedical Press, Amsterdam - New York - Oxford 1978.

34046 - HORVÁTH, I. : Účinok HF na intenzitu fotosyntézy Vicia faba L. [Effect of
HF on photosynthetic rate of Vicia faba L.] - Acta bot. slov. Acad. Sci. slov.,
Ser. B 2 : 145 - 158, 1978. [In Slovak, ab : E, R.]

34047 - HOSHINA, T., KÔZAI, S., ISHIGAKI, K. : [Foliar absorption of $^{15}N$ labeled
urea by tea plant.] - Chagyo Gijutsu Kenkyu[Study of Tea] 54 : 33 - 36,
1978. [Chl; in Jap., ab : E.]

34048 - HOSHINO, T., UZIHARA, K., SHIKATA, S.-I. : [Effects of tall and short isogenic lines among grain sorghum varieties on dry matter production and yield
performance at different planting densities.] - Jap. J. Crop Sci. 47 : 541 -
546, 1978. [Growth analysis; in Jap., ab : E.]

34049 - HOSODA, N., LEE, K.-K., YATAZAWA, M. : Effects of carbon dioxide, oxygen,
and light on nitrogen-fixing activities in Japan clover (Kummerowia striata
S.). - Soil Sci. Plant Nutr. 24 : 113 - 120, 1978. [Ps.]

34050 - HOSSAIN, M., SEN, S. : Radiation induced high fibre yielding chlorophyll
mutant in jute. - Z. Pflanzenzücht. 81 : 77 - 79, 1978.

34051 - HOUCHINS, J.P., BURRIS, R.H. : Hydrogen metabolism in blue-green algae. -
Plant Physiol. 61 (Suppl.) : 2, 1978.

*34052 - HOUGH, R.A., WETZEL, R.G. : Photosynthetic pathways of some aquatic plants.
- Aquatic Bot. 3 : 297 - 313, 1977.

34053 - HOUGH, R.A., WETZEL, R.G. : Photorespiration and $CO_2$ compensation point in
Najas flexilis. - Limnol. Oceanogr. 23 : 719 - 724, 1978.

34054 - HOWELL, S.H. : Cell cycle regulation of messenger RNAs coding for chloroplast proteins in Chlamydomonas reinhardi. - In : AKOYUNOGLOU, G., ARGYROU-
DI-AKOYUNOGLOU, J.H. (ed.) : Chloroplast Development. Pp. 679 - 686. Else-
vier/North-Holland Biomedical Press, Amsterdam - New York - Oxford 1978.

34055 - HOWELL, S.H., GELVIN, S. : The messenger RNAs and genes coding for the
small and large subunits of ribulose 1,5-bisphosphate carboxylase/oxygenase
in Chlamydomonas reinhardi. - In : SIEGELMAN, H.W., HIND, G. (ed.) : Photosynthetic Carbon Assimilation. Pp. 363 - 378. Plenum Press, New York - London 1978.

34056 - HÖXTERMANN, E. : Preparation of water-free chlorophyll-$a$. - Stud. biophys.
72 : 203 - 204, Microfiche 3/1 - 3/13, 1978.

34057 - HOZYO, Y., KATO, S. : [The facility for measuring photosynthetic rate of
crop plants in the controlled environment.] - Misc. Publ. nat. Inst. agr.
Sci., Ser. D 1978 (2) : 1 - 29, 1978. [In Jap., ab : E.]

34058 - HSIAO, S.I., KITTLE, D.W., FOY, M.G. : Effects of crude oils and the oil dispersant Corexit on primary production of arctic marine phytoplankton and seaweed. - Environ. Pollut. *15* : 209 - 221, 1978.

34059 - HUANG, Chuo-Hui, LI, You-Ze, QIU, Guo-Xiong : [The effect of aureomycin on photosynthetic rate of wheat leaf and photophosphorylation reactions of isolated chloroplasts.] - Acta bot. sinica *20* : 330 - 336, 1978. [In Chinese, ab : E.]

*34060 - HUBALD, M., AUGSTEN, H. : The ultrastructure of duckweed chloroplasts (*Lemna gibba* L., G1) influenced by glycine and deficiency conditions. - Acta Biol. Med. exp. *2* : 61 - 64, 1977.

34061 - HUBER, O. : Light compensation point of vascular plants of a tropical cloud forest and an ecological interpretation. - Photosynthetica *12* : 382 - 390, 1978.

34062 - HUBER, S.C. : Regulation of chloroplast photosynthesis by exogenous magnesium. - Plant Physiol. *61* (Suppl.) : 109, 1978.

34063 - HUBER, S.C. : Regulation of chloroplast photosynthetic activity by exogenous magnesium. - Plant Physiol. *62* : 321 - 325, 1978.

34064 - HUBER, S.C. : Substrates and inorganic phosphate control : the light activation of NADP-glyceraldehyde-3-phosphate dehydrogenase and phosphoribulo-kinase in barley (*Hordeum vulgare*) chloroplasts. - FEBS Lett. *92* : 12 - 16, 1978.

34065 - HUFFAKER, R.C., MILLER, B.L. : Reutilization of ribulose bisphosphate carboxylase. - In : SIEGELMAN, H.W., HIND, G. (ed.) : Photosynthetic Carbon Assimilation. Pp. 139 - 152. Plenum Press, New York - London 1978.

34066 - HUISMAN, J.G., MOORMAN, A.F.M., VERKLEY, F.N. : *In vitro* synthesis of chloroplast ferredoxin as a high molecular weight precursor in a cell-free protein synthesizing system from wheat germs. - Biochem. biophys. Res. Commun. *82* : 1121 - 1131, 1978.

34067 - HUISMAN, J.G., STAPEL, S., GEBBINK, M.G.T. : Characterization of ferredoxins on a nanomole scale. - Anal. Biochem. *90* : 501 - 509, 1978.

34068 - HUISMAN, J.G., STAPEL, S., MUIJSERS, A.O. : Two different plant-type ferredoxins in each of two *Petunia* species. - FEBS Lett. *85* : 198 - 202, 1978.

34069 - HUMPHREY, G.F. : The recalculation of marine chlorophyll concentrations with special reference to Australian waters. - Aust. J. mar. Freshwater Res. *29* : 409 - 416, 1978.

34070 - HUNDING, C. : Growth cycle of a freshwater diatom. - Mitt. int. Ver. Limnol. *21* : 136 - 146, 1978. [Ps, Chl.]

34071 - HUNER, N.P.A. : Evidence of a conformational change in ribulose diphosphate carboxylase-oxygenase from Puma rye during cold adaptation. - Plant Physiol. *61* (Suppl.) : 99, 1978.

34072 - HUNER, N.P.A. : The effect of low temperature adaptation of Puma rye on the structure and function of RuBP carboxylase/oxygenase. - In : SIEGELMAN, H.W., HIND, G. (ed.) : Photosynthetic Carbon Assimilation. P. 421. Plenum Press, New York - London 1978.

34073 - HUNER, N.P.A., MACDOWALL, F.D.H. : Evidence for an *in vivo* conformational change in ribulose bisphosphate carboxylase-oxygenase from Puma rye during cold adaptation. - Can. J. Biochem. *56* : 1154 - 1161, 1978.

*34074 - HUNNIUS, W., DROBNY, J. : Der Einfluß unterschiedlicher Bestandesdichten auf verschiedene Leistungsmerkmale von Futterrübensorten.- Z. Acker- Pflanzenbau *144* : 39 - 53, 1977.

B34075 - HUNT, R. : Plant Growth Analysis. (The Institute of Biology's Studies in Biology No.96).- Edward Arnold, London 1978.

34076 - HURDUC, N., TĂNASE, V., JUNCU, A.-M., VELICOGLU, A., DRĂGHICI, L., BUDE, A. : Caracterizarea productivității fotosintetice a soiului de orz de toamnă Miraj. [Characterization of photosynthetic productivity of winter barley cv. Miraj.] - An. Inst. Cercetări Cereale Plante tehnice Fundulea *43* : 441 - 448, 1978. [In Roum., ab : E, R.]

34077 - HURLEY, J.B., BECHER, B., EBREY, T.G. : More evidence that light isomerises the chromophore of purple membrane protein. - Nature *272* : 87 - 88, 1978.

34078 - HURLEY, J.B., EBREY, T.G. : Energy transfer in the purple membrane of *Halobacterium halobium*. - Biophys. J. *22* : 49 - 66, 1978.

34079 - HUSKISSON, N.S., WARD, P.F.V. : A reliable method for scintillation counting of $^{14}CO_2$ trapped in solutions of sodium hydroxide, using a scintillant suitable for general use. - Int. J. appl. Radiat. Isotop. *29* : 729.- 734, 1978.

34080 - HUTBER, G.N., SMITH, A.J., ROGERS, L.J. : Comparative biological activities of two ferredoxins and a flavodoxin from the cyanobacterium *Nostoc* strain MAC. - FEMS Microbiol. Lett. *4* : 11 - 14, 1978.

34081 - HUTSON, K.G., ROGERS, L.J., HASLETT, B.G., BOULTER, D., CAMMACK, R. : Comparative studies on two ferredoxins from the cyanobacterium *Nostoc* strain MAC. - Biochem. J. *172* : 465 - 477, 1978.

34082 - HUTTER, K.-J., EIPEL, H.E. : Flow cytomeric determinations of cellular substances in algae, bacteria, moulds and yeasts. - Antonie van Leeuwenhoek *44* : 269 - 282, 1978. [Chl.]

*34083 - HWANG, S.B., KORENBROT, J.I., STOECKENIUS, W. : Light-dependent proton transport by bacteriorhodopsin incorporated in an interface film. - In : HALL, Z., KELLY, R., FRED-FOX, C. (ed.) : Progress in Clinical and Biological Research. Vol. 15. Cellular Neurobiology. Pp. 81 - 93. A.R. Liss, Inc., New York 1977.

34084 - HWANG, S.-B., KORENBROT, J.I., STOECKENIUS, W. : Transient photovoltages in purple membrane multilayers. Charge displacement in bacteriorhodopsin and its photointermediates. - Biochim. biophys. Acta *509* : 300 - 317, 1978.

*34085 - HWANG, S.-B., STOECKENIUS, W. : Purple membrane vesicles : Morphology and proton translocation.-J. Membr. Biol. *33* : 325 - 350, 1977.

*34086 - HYNNINEN, P.H. : Chlorophylls. V. Isolation of chlorophylls *a* and *b* using an improved two-phase extraction method followed by a precipitation and a separation on a sucrose column. - Acta chem. scand. *B 31* : 829 - 835, 1977.


34087 - IBRAHIM, R.K., PHAN, C.T. : Phenolic synthesis in relation to chloroplast ultrastructure in flax callus and cell suspension cultures. - Biochem. Physiol. Pflanzen *172* : 199 - 212, 1978. [Chl.]

34088 - ICHII, M. : [A new method for tracing of internode length in wheat.]-Jap. J. Crop Sci. *47* : 243 - 248, 1978. [In Jap., ab : E.]

*34089 - IDA, S. : An improved method for the purification of ferredoxin from spinach leaves. - Bull. Res. Inst. Food Sci., Kyoto Univ. *40* : 7 - 9, 1977.

34090 - IDSO, S.B., HATFIELD, J.L., REGINATO, R.J., JACKSON, R.D. : Wheat yield estimation by albedo measurement. - Remote Sens. Environ. *7* : 273 - 276, 1978.

34091 - IFENKWE, O.P., ALLEN, E.J. : Effects of row width and planting density on growth and yield of two maincrop potato varieties. 1. Plant morphology and dry-matter accumulation. - J. agr. Sci. *91* : 265 - 278, 1978.

34092 - IFENKWE, O.P., ALLEN, E.J. : Effects of row width and planting density on growth and yield of two maincrop potato varieties. 2. Number of tubers, total and graded yields and their relationships with above-ground stem densities. - J. agr. Sci. *91* : 279 - 289, 1978.

34093 - IINO, M., HASHIMOTO, T., HEBER, U. : Inhibition of photosynthesis and respiration by batatasins. - Planta *138* : 167 - 172, 1978.

*34094 - IL'INA, L.P., SIMONOVA, E.I., OKUNTSOV, M.M. : Zaklyuchitel'nyÏ étap biosinteza khlorofilla v étiolirovannykh prorostkakh yachmenya raznogo vozrasta, osveshchennykh monokhromaticheskimi uchastkami spektra. [Terminal step in the biosynthesis of chlorophyll in etiolated barley seedlings of different ages illuminated with the monochromatic region of spectrum.] - In : Fiziologicheskie i Biokhimicheskie Osnovy Adaptatsii Rastenii k Usloviyam Severa. Pp. 67 - 71. Izd. Yakutsk. Fil. sibir. Otd. Akad. Nauk SSSR, Yakutsk 1976. [In R.]

34095 - **IMAI, K., MURATA, Y.** : Effect of carbon dioxide concentration on growth and dry matter production of crop plants. III. Relationship between $CO_2$ concentration and nitrogen nutrition in some $C_3$- and $C_4$-species. - Jap. J. Crop Sci. *47* : 118 - 123, 1978.

34096 - **IMAI, K., MURATA, Y.** : Effect of carbon dioxide concentration on growth and dry matter production of crop plants. IV. After-effects of carbon dioxide-treatments on the apparent photosynthesis, dark respiration and dry matter production. - Jap. J. Crop Sci. *47* : 330 - 335, 1978.

34097 - **IMAI, K., MURATA, Y.** : Effect of carbon dioxide concentration on growth and dry matter production of crop plants. V. Analysis of after-effect of carbon dioxide-treatment on apparent photosynthesis. - Jap. J. Crop Sci. *47* : 587 - 595, 1978.

34098 - **INADA, K.** : Photosynthetic enhancement spectra in higher plants. - Plant Cell Physiol. *19* : 1007 - 1017, 1978.

34099 - **INCROPERA, F.P., THOMAS, J.F.** : A model for solar radiation conversion to algae in a shallow pond. - Solar Energy *20* : 157 - 165, 1978.

*34100 - **INDIATI, R., PIERANDREI, F., ROSSI, C. DE, CALÈ, M.T.** : Intensità fotosintetica e respiratoria; composizione chimica e attività enzimatiche di alcune colture foraggere in rapporto ai trattamenti fertilizzanti. [Photosynthetic and respiratory rates, chemical composition and enzymatic activities of some fodder cultures in connection with fertilizing treatments.] - Ann. Ist. sper. Nutr. Piante *8* (5) : 1 - 26, 1977. [In Ital., ab : E.]

34101 - **INO, Y., OSHIMA, Y.** : Effect of fluctuating light on photosynthesis of tree leaves. - In : MONSI, M., SAEKI, T. (ed.) : Ecophysiology of Photosynthetic Productivity. JIBP Synthesis. Vol. 19. Pp. 19 - 25. University of Tokyo Press, Tokyo 1978.

34102 - **INOUÉ, H.** : Break points in Arrhenius plots of the Hill reaction of spinach chloroplast fragments in the temperature range from -25 to 25 °C. - Plant Cell Physiol. *19* : 355 - 363, 1978.

34103 - **INOUE, K., NISHIMURA, M., AKAZAWA, T.** : Effect of α-hydroxy-2-pyridinemethanesulfonate on glycolate metabolism in spinach leaf protoplasts. - Plant Cell Physiol. *19* : 317 - 325, 1978.

34104 - **INOUE, Y., KOBAYASHI, Y., SHIBATA, K., HEBER, U.** : Synthesis and hydrolysis of ATP by intact chloroplasts under flash illumination and in darkness. - Biochim. biophys. Acta *504* : 142 - 152, 1978.

34105 - **INOUE, Y., SHIBATA, K.** : Oscillation of thermoluminescence at medium-low temperature. - FEBS Lett. *85* : 193 - 197, 1978.

34106 - **INOUE, Y., SHIBATA, K.** : Thermoluminescence bands of chloroplasts as characterized by flash excitation. - In : HALL, D.O., COOMBS, J., GOODWIN, T.W. (ed.) : Proceedings of the Fourth International Congress on Photosynthesis. Pp. 211 - 221. Biochem. Soc., London 1978.

*34107 - **INTERESSE, F.S., RUGGIERO, P.** : L'auto-oxydation de l'huile d'olive : influence des pigments chlorophylliens. - Inform. oléicoles int. *58-59* : 125 - 138, 1972.

*34108 - **INTERESSE, F.S., RUGGIERO, P., VITAGLIANO, M.** : L'autossidazione dell'olio di oliva : influenza dei pigmenti clorofillici. [Autooxidation of olive oil. Effects of chlorophyll pigments.] - Ind. Agr. *9* : 318 - 324, 1971. [In Ital., ab : E, F.]

34109 - **INUYAMA, S.** : Varietal differences in leaf water potential, leaf diffusive resistance and grain yield of grain sorghum affected by drought stress. - Jap. J. Crop Sci. *47* : 255 - 261, 1978.

34110 - INUYAMA, S. : [Effect of plant densities under two irradiation regimes on leaf water potential, leaf diffusive resistance during drought stress period and grain yield of grain sorghum.] - Jap. J. Crop Sci. *47* : 596 - 601, 1978. [In Jap., ab : E.]

34111 - ÏORDANOV, I., ZDRAVCHEVA, V., VELINOVA, E. : Vliyanie na khloramfenikola, tetratsiklina i tsiklokheksimida v"rkhu s"d"rzhanieto i razpredelenieto na plastidnite pigmenti, spektralnite kharakteristiki na pigment-belt"chnite kompleksi i s"stava na lamelarnite belt"tsi na khloroplasti ot fasul (*Phaseolus vulgaris* L.). [Effect of chloramphenicol, tetracycline and cycloheximide on the plastid pigment content and distribution, pigment-protein complexes, spectral characteristics and the lamellar proteins of bean chloroplasts.] - Fiziol. Rast. (Sofia) *4* (2) : 11 - 22, 1978. [In Bulg., ab : E, R.]

34112 - IPPEN, E.P., SHANK, C.V., LEWIS, A., MARCUS, M.A. : Subpicosecond spectroscopy of bacteriorhodopsin. - Science *200* : 1279 - 1281, 1978.

34113 - IRIYAMA, K. : A rapid and convenient method for purification and isolation of chlorophyll-*a* from *Porphyra vezoensis*. - Biochem. biophys. Res. Commun. *83* : 501 - 505, 1978.

34114 - IRIYAMA, K., YOSHIURA, M., SHIRAKI, M. : Micro-method for the qualitative and quantitative analysis of photosynthetic pigments using high-performance liquid chromatography. - J. Chromatogr. *154* : 302 - 305, 1978.

34115 - IRIYAMA, K., YOSHIURA, M., SHIRAKI, M., SUGI, M., IIJIMA, S. : [Photosynthesis and Langmuir-Blodget films.] - Maku [Membrane] *3* (4) : 256 - 264, 1978. [In Jap.]

34116 - ISARANGKURA, R., PEASLEE, D., LOCKARD, R. : Utilization and redistribution of Zn during vegetative growth of corn. - Agron. J. *70* : 243 - 246, 1978. [Ps.]

34117 - ISHIBASHI, H., YUN, S.-J., HYEON, S.-B., SUZUKI, A., TAMURA, S. : A simple method for screening photorespiration inhibitors using isolated spinach cells and an oxygen electrode. - Agr. biol. Chem. *42* : 1807 - 1809, 1978.

34118 - ISHIHARA, K., EBARA, H., HIRASAWA, T., OGURA, T. : [The relationship between environmental factors and behaviour of stomata in rice plants. VII. The relation between nitrogen content in leaf blades and stomatal aperture.] - Jap. J. Crop Sci. *47* : 664 - 673, 1978. [In Jap., ab : E.]

34119 - ISHIHARA, K., IIDA, O., HIRASAWA, T., OGURA, T. : [Relationship between potassium content in leaf blades and stomatal aperture in rice plants.] - Jap. J. Crop Sci. *47* : 719 - 720, 1978. [In Jap.]

34120 - ISHIHARA, K., SAGO, R., OGURA, T. : [The relationship between environmental factors and behaviour of stomata in rice plants. V. Effects of partial excision of root system on diurnal course of stomatal aperture.] - Jap. J. Crop Sci. *47* : 499 - 505, 1978. [In Jap., ab : E.]

34121 - ISHIHARA, K., SAGO, R., OGURA, T. : [The relationship between environmental factors and behaviour of stomata in rice plants. VI. Comparisons between the diurnal course of stomatal aperture of rice plants grown in the border and interior of paddy fields.] - Jap. J. Crop Sci. *47* : 515 - 528, 1978. [In Jap., ab : E. ]

34122 - ISHII, R., MURATA, Y. : [Photosynthesis in $C_3$- and $C_4$-plants.] - Jap. J. Crop Sci. *47* : 165 - 188, 1978. [In Jap.]

34123 - ISHII, R., MURATA, Y. : Further evidence of the Kok effects in $C_3$ plants and the effects of environmental factors on it. - Jap. J. Crop Sci. *47* : 547 - 550, 1978.

34124 - ITAI, C. : Response of *Eucalyptus occidentalis* to water stress induced by NaCl. - Physiol. Plant. *43* : 377 - 379, 1978. [Ps.]

34125 - ITAI, C., BENZIONI, A., MUNZ, S. : Heat stress : Effects of abscisic acid and kinetin on response and recovery of tobacco leaves. - Plant Cell Physiol. *19* : 453 - 460, 1978.

34126 - ITOH, S. : Electrostatic state of the membrane surface and the reactivity
         between ferricyanide and electron transport components inside chloroplast
         membrane. - Plant Cell Physiol. *19* : 149 - 166, 1978.

34127 - ITOH, S. : Membrane surface potential and the reactivity of the system II
         primary electron acceptor to charged electron carriers in the medium. - Bio-
         chim. biophys. Acta *504* : 324 - 340, 1978.

34128 - IVANCHANKA, V.M., GANCHARYK, M.M. : Fotasintèz i strukturna-funktsyyanal'ny
         stan sfarmiravanykh khlaraplastaŭ. [Photosynthesis and structural and func-
         tional state of formed chloroplasts.] - Vestsi Akad. Navuk belarus. SSR,
         Ser. biyal. Navuk *1978* (6) : 30 - 36, 140, 1978. [In Belorus., ab : E, R.]

34129 - IVANCHENKO, V.M., LEGENCHENKO, B.I., KRUCHININA, S.S. : Vodnyǐ rezhim i èner-
         geticheskiǐ obmen rasteniǐ v svyazi s ikh gomeostazom. [Water relations and
         energy balance in plants as related to their homeostasis.] - In : Vodnyǐ
         Rezhim Rasteniǐ v Svyazi s Raznymi Ekologicheskimi Usloviyami. Pp. 236 - 244.
         Izdatel'stvo Kazanskogo Universiteta, Kazan' 1978. [In R.]

*34130 - IVANOV, I.D., DEMINA, N.S. : Azotfiksatsiya i vydelenie molekulyarnogo vodo-
         roda fotosinteziruyushchimi bakteriyami. [Nitrogen fixation and molecular
         hydrogen evolution by photosynthetic bacteria.] - Uspekhi Mikrobiol. *5* :
         50 - 61, 1968. [In R.]

34131 - IVANOV, M.V., ZYAKUN, A.M., GOGOTOVA, G.I., BONDAR', V.A. : Razdelenie izo-
         topov ugleroda fotosinteziruyushchimi bakteriyami, rastushchimi na bikarbo-
         nate, obogashchennom izotopom $^{13}$C. [Carbon isotope separation by photosyn-
         thetic bacteria growing on bicarbonate enriched with the $^{13}$C isotope.] -
         Dokl. Akad. Nauk SSSR *242* : 1417 - 1420, 1978. [In R.]

34132 - IVANOVA, N.A. : Vliyanie defoliatsii na stroenie ust'ichnogo apparata i fo-
         tosinteticheskuyu aktivnost' list'ev. [Influence of defoliation on the struc-
         ture of stoma apparatus and leaf photosynthetic activity.] - In : Mezostruk-
         tura i Funktsional'naya Aktivnost' Fotosinteticheskogo Apparata. Pp. 132 -
         136. Ural'skiǐ Gosudarstvennyǐ Universitet, Sverdlovsk 1978. [In R.]

*34133 - IVANOVSKIǏ, R.N. : Fotoindutsirovannoe vosstanovlenie NAD(F) v kletkakh ze-
         lenykh bakteriǐ. [Photoinduced reduction of NAD(P) in the cells of green
         bacteria.] - Mikrobiologiya *44* : 965 - 969, 1975. [In R, ab : E.]

34134 - IVORY, D.A., WHITEMAN, P.C. : Effect of temperature on growth of five sub-
         tropical grasses. I Effect of day and night temperature on growth and mor-
         phological development. - Aust. J. Plant Physiol. *5* : 131 - 148, 1978.
         [Growth analysis.]

34135 - IVORY, D.A., WHITEMAN, P.C. : Effect of temperature on growth of five sub-
         tropical grasses. II Effect of low  night temperature. - Aust. J. Plant Phy-
         siol. *5* : 149 - 157, 1978. [Growth analysis.]

*34136 - IWAKIRI, S. : [Regional difference in characteristics of soybean plant growth
         in Japan.] - J. agr. Meteorol. *31* : 83 - 87, 1975. [Growth analysis; in Jap.]

34137 - IWANAGA, M., MUKAI, Y., PANAYOTOV, I., TSUNEWAKI, K. : Genetic diversity of
         the cytoplasm in *Triticum* and *Aegilops*. VII. Cytoplasmic effects on respi-
         ratory and photosynthetic rates. - Jap. J. Genet. *53* : 387 - 396, 1978.

34138 - IZAWA, S., PAN, R.L. : Photosystem I electron transport and phosphorylation
         supported by electron donation to the plastoquinone region. - Biochem. bio-
         phys. Res. Commun. *83* : 1171 - 1177, 1978.

34139 - IZMEST'EVA, L.R., BELYAEV, A.A. : O pervichnoǐ produktivnosti pelagiali
         ozera Baǐkal. [Primary productivity of the pelagic zone of the lake Baikal.]
         - In : Gidrobiologicheskie i Ikhtiologicheskie Issledovaniya v Vostochnoǐ
         Sibiri. Vol. 2. Pp. 36 - 46. Irkutsk 1978. [Chl; in R.]

34140 - JABBEN, M., DEITZER, G.F. : A method for measuring phytochrome in plants
         grown in white light. - Photochem. Photobiol. *27* : 799 - 802, 1978. [Chl.]

34141 - JABLONSKI, P.P., ANDERSON, J.W. : Light-dependent reduction of oxidized glu-
         tathione by ruptured chloroplasts. - Plant Physiol. *61* : 221 - 225, 1978.

34142 - JACKSON, C., DENCH, J., MOORE, A.L., HALLIWELL, B., FOYER, C.H., HALL, D.O. :
Subcellular localisation and identification of superoxide dismutase in the
leaves of higher plants. - Europe. J. Biochem. *91* : 339 - 344, 1978. [Chloroplast.]

*34143 - JACKSON, G.A. : Biological constraints on seaweed culture. - In : MITSUI, A.,
MIYACHI, S., SAN PIETRO, A., TAMURA, S. (ed.) : Biological Solar Energy Conversion. Pp. 437 - 448. Academic Press, New York - San Francisco - London
1977.

34144 - JACOBSEN, T.R. : A quantitative method for the separation of chlorophylls *a*
and *b* from phytoplankton pigments by high pressure liquid chromatography. -
Mar. Sci. Commun. *4* : 33 - 47, 1978.

*34145 - JACQUET, J., BOUTIBONNES, P. : Effets des flavacoumarines (aflatoxines) sur
quelques animaux et végétaux. - Compt. rend. Séances Soc. Biol. *164* : 2239 -
- 2244, 1970. [Chl.]

*34146 - JACQUET, J., BOUTIBONNES, P., SAINT, S. : Effets biologiques des flavacoumarines d'*Aspergilus parasiticus* A.T.C.C. 15517. II. Végétaux. - Rev. Immunol.
*35* : 219 - 240, 1971. [Chl.]

34147 - JACQUOT, J.-P., VIDAL, J., GADAL, P., SCHÜRMANN, P. : Evidence for the existence of several enzyme-specific thioredoxins in plants. - FEBS Lett. *96* :
243 - 246, 1978.

34148 - JAHNKE, H., SCHÖNBORN, M., ZIMMERMANN, G. : Cathodic reduction of oxygen on
chelates. - In : METZNER, H. (ed.) : Photosynthetic Oxygen Evolution. Pp.
439 - 452. Academic Press, London - New York - San Francisco 1978.

34149 - JAHNKE, L.S., SOULEN, T.K. : Effect of manganese on growth and restoration
of photosynthesis in manganese deficient algae. - Z. Pflanzenphysiol. *88* :
83 - 93, 1978.

34150 - JAMES, T.D.W., SMITH, D.W. : Seasonal changes in the caloric value of the
leaves and twigs of *Populus tremuloides*. - Can. J. Bot. *56* : 1804 - 1805,
1978.

*34151 - JANA, B.L., CHAUDHURI, B.B. : Effect of direction of planting on the yield
of jute. - Indian J. agr. Sci. *46* : 403 - 406, 1976. [Growth analysis.]

34152 - JANARDHAN, K.V., MURTY, K.S. : Association of photosynthetic efficiency with
various growth parameters and yield in rice. - In : Inter-Disciplinary Symposium on Photosynthesis and Productivity. Abstract of Papers. Pp. 40 - 42.
Indian nat. Sci. Acad., New Delhi 1978.

34153 - JANARDHAN, K.V., MURTY, K.S. : Photosynthetic carbon dioxide fixation by rice ears. - Curr. Sci. *47* : 810 - 811, 1978.

34154 - JANARDHAN, K.V., MURTY, K.S. : Variability in photorespiration and photosynthesis in rice varieties. - Indian J. exp. Biol. *16* : 116 - 117, 1978.

*34155 - JÁNOSSY, A.G.S., MUSTÁRDY, L.A., FALUDI-DÁNIEL, Á. : X-ray microanalytical
study of Mn and Fe compartmentation in maize chloroplasts. - Acta histochem.
*58* : 317 - 323, 1977.

34156 - JAQUIÉRY, R., KELLER, E.R. : Beeinflussung des Fruchtansatzes bei der Ackerbohne (*Vicia faba* L.) durch die Verteilung der Assimilate. Teil I. - Angew.
Bot. *52* : 261 - 276, 1978.

34157 - JASSBY, A.D.: Polarographic measurements of photosynthesis and respiration. -
In : HELLEBUST, J.A., CRAIGIE, J.S. (ed.) : Handbook of Phycological Methods.
Physiological and Biochemical Methods. Pp. 285 - 296. Cambridge University
Press, London 1978.

34158 - JASSBY, A.D. : Polarographic measurements of respiration following light--dark transitions. - In : HELLEBUST, J.A., CRAIGIE, J.S. (ed.) : Handbook
of Phycological Methods. Physiological and Biochemical Methods. Pp. 297 -
- 303. Cambridge University Press, London 1978.

34159 - JEANNEAU, Y. : Sur la présence de phytoferritine dans les chloroplastes des
cellules des feuilles de *Brassica oleracea* L. porteuses de tumeurs. - Compt.
rend. Acad. Sci. Paris, Sér. D *287* : 619 - 622, 1978.

34160 - JEFFREY, S.W., VESK, M. : Chloroplast structural changes induced by white
light in the marine diatom *Stephanopyxis turris*. - J. Phycol. *14* : 238 - 240,
1978.

34161 - JEFFRIES, T.W., LEACH, K.L. : Intermittent illumination increases biophoto-
lytic hydrogen yield by *Anabaena cylindrica*. - Appl. environ. Microbiol.
*35* : 1228 - 1230, 1978.

34162 - JEFFRIES, T.W., TIMOURIAN, H., WARD, R.L. : Hydrogen production by *Anabaena
cylindrica* : Effects of varying ammonium and ferric ions, pH, and light. -
Appl. environ. Microbiol. *35* : 704 - 710, 1978.

34163 - JEN, J.J., THOMAS, R.L. : Antagonistic effect of CPTA and far-red light on
the carotenogenesis in lutescent tomatoes. - J. Food Biochem. *2* : 23 - 27,
1978.

34164 - JENNI, B., STUTZ, E. : Physical mapping of the ribosomal DNA region of *Euglena gracilis* chloroplast DNA. - Europe. J. Biochem. *88* : 127 - 134, 1978.

34165 - JENNINGS, R.C., FORTI, G. : Interaction of cations with the thylakoid mem-
branes. - In : AZZONE, G.F., AVRON, M., METCALFE, J.C., QUAGLIARIELLO, E.,
SILIPRANDI, N. (ed.) : The Proton and Calcium Pumps. Pp. 93 - 103. Elsevier/
North-Holland Biomedical Press, Amsterdam - Oxford - New York 1978.

34166 - JENNINGS, R.C., FORTI, G., GEROLA, P.D., GARLASCHI, F.M. : Studies on cation-
-induced thylakoid membrane stacking, fluorescence yield, and photochemical
efficiency. - Plant Physiol. *62* : 879 - 884, 1978.

34167 - JENNINGS, R.C., GARLASCHI, F.M., FORTI, G., GEROLA, P. : Cations and energy
spillover in chloroplasts. - Plant Sci. Lett. *13* : 1 - 8, 1978.

34168 - JENNINGS, R.C., GARLASCHI, F.M., GEROLA, P.D., FORTI, G. : Partition zone
penetration by chymotrypsin and the localization of the chloroplast flavo-
protein and photosystem II. - In : AKOYUNOGLOU, G., ARGYROUDI-AKOYUNOGLOU,
J.H. (ed.) : Chloroplast Development. Pp. 467 - 474. Elsevier/North-Holland
Biomedical Press, Amsterdam - New York - Oxford 1978.

34169 - JENSEN, A. : Chlorophylls and carotenoids. - In : HELLEBUST, J.A., CRAIGIE,
J.S. (ed.) : Handbook of Phycological Methods. Physiological and Biochemical
Methods. Pp. 59 - 70. Cambridge University Press, London 1978.

34170 - JENSEN, C.R. : Effects of salinity in the root medium. V. Alteration in diurnal rhythmic activity of top and root $CO_2$-exchange, nutrient uptake, and
water balance at onset of high $KNO_3$-concentrations. - Acta Agr. scand. *28* :
303 - 312, 1978.

34171 - JENSEN, K.G., HULBARY, R.L. : Chloroplast development during sporogenesis
in six species of mosses. - Amer. J. Bot. *65* : 823 - 833, 1978.

34172 - JENSEN, L.H. : Structure and function of iron-sulfur proteins. - In : SCHÄ-
FER, G., KLINGENBERG, M. (ed.) : Energy Conservation in Biological Membranes.
Pp. 74 - 83. Springer-Verlag, Berlin - Heidelberg - New York 1978.

34173 - JENSEN, R.G., SICHER, R.C.,Jr., BAHR, J.T. : Regulation of ribulose 1,5-bis-
phosphate carboxylase in the chloroplast. - In : SIEGELMAN, H.W., HIND, G.
(ed.) : Photosynthetic Carbon Assimilation. Pp. 95 - 112. Plenum Press,
New York - London 1978.

34174 - JESKE, C., SENGER, H. : Development of pigments and activity of the photo-
synthetic apparatus in some higher plants. - In : AKOYUNOGLOU, G., ARGYROUDI-
AKOYUNOGLOU, J.H. (ed.) : Chloroplast Development. Pp. 475 - 480. Elsevier/
North-Holland Biomedical Press, Amsterdam - New York - Oxford 1978.

34175 - JESKE, H., WERZ, G. : The influence of light intensity on pigment composition
and ultrastructure of plastids in leaves of diseased *Abutilon sellowianum*
REG. (*Malvaceae*). - Phytopathol. Z. *91* : 1 - 13, 1978.

34176 - JEŠKO, T., TROUGHTON, A. : Root tips removal in relation to net photosynthesis and growth in *Lolium perenne* L. - Biológia(Bratislava) *33* : 65 - 71,1978.

34177 - JEWSON, D.H., TAYLOR, J.A. : The influence of turbidity on net phytoplankton
photosynthesis in some Irish lakes. - Freshwater Biol. *8* : 573 - 584, 1978.

34178 - JOHAL, S., BOURQUE, D.P. : Enzymatically active crystalline RuBP carboxylase from spinach. - In : SIEGELMAN, H.W., HIND, G. (ed.) : Photosynthetic Carbon Assimilation. Pp. 421 - 422. Plenum Press, New York - London 1978.

34179 - JOHANNES, B., BUDZIKIEWICZ, H. : Zur Photosynthese grüner Pflanzen, VIII [1] : *in vitro*-Sauerstoffaustausch zwischen Carotinoid-Epoxiden und Wasser ? - Z. Naturforsch. *33 C* : 116 - 119, 1978.

34180 - JOHANSSON, G., WESTRIN, H. : Specific extraction of intact chloroplasts using aqueous biphasic systems. - Plant Sci. Lett. *13* : 201 - 212, 1978.

34181 - JOHNS, G.G. : Transpirational, leaf area, stomatal and photosynthetic responses to gradually induced water stress in four temperate herbage species. - Aust. J. Plant Physiol. *5* : 113 - 125, 1978.

34182 - JOHNS, S.R., LESLIE, D.R., WILLING, R.I., BISHOP, D.G. : Studies on chloroplast membranes. III. $^{13}C$ chemical shifts and longitudinal relaxation times of 1,2-diacyl-3-(6-sulpho-α-quinovosyl)-*sn*-glycerol. - Aust. J. Chem. *31* : 65 - 72, 1978.

34183 - JOHNSTONE, I.M. : Phenotypic plasticity in *Draparmaldia (Chlorophyta : Chaetophoraceae)* . I. Effects of the chemical environment. - J. Phycol. *14* : 302 - 308, 1978. [Chl.]

34184 - JOINT, I.R. : Microbial production of an estuarine mudflat. - Estuarine coastal mar. Sci. *7* : 185 - 195, 1978. [Chl.]

34185 - JOLIOT, P. : The photosynthetic intramembrane electric field. - In : PULLMAN, B. (ed.) : Frontiers in Physicochemical Biology. Pp. 485 - 497. Academic Press, New York - San Francisco - London 1978.

34186 - JOLIOT, P., JOLIOT, A. : La photosynthèse. - Recherche *9* : 331 - 338, 1978.

34187 - JONES, C.E., MACKAY, R.A. : Reactions in microemulsions. 3. Photodegradation of chlorophyll. - J. phys. Chem. *82* : 63 - 65, 1978.

34188 - JONES, C.R., COOK, J.R. : Culture pH, $CO_2$ tension, and cell division in *Euglena gracilis* Z. - J. cell.Physiol. *96* : 253 - 259, 1978. [Ps.]

34189 - JONES, E.P., WARD, T.V., ZWICK, H.H. : A fast response atmospheric $CO_2$ sensor for eddy correlation flux measurements. - Atmos. Environ. *12* : 845 - 851, 1978.

34190 - JONES, H.E., GORE, A.J.P. : A simulation of production and decay in blanket bog. - In : HEAL, O.W., PERKINS, D.F. (ed.) : Production Ecology of British Moors and Montane Grasslands. Pp. 160 - 186. Springer-Verlag, Berlin - Heidelberg - New York 1978.

34191 - JONES, K.J., TETT, P., WALLIS, A.C., WOOD, B.J.B. : Investigation of a nutrient-growth model using a continuous culture of natural phytoplankton. - J. mar. biol. Assoc. U K *58* : 923 - 941, 1978.

34192 - JONES, K.J., TETT, P., WALLIS, A.C., WOOD, B.J.B. : The use of small, continuous and multispecies cultures to investigate the ecology of phytoplankton in a Scottish sea-loch. - Mitt. int. Ver. Limnol. *21* : 398 - 412, 1978. [Chl.]

34193 - JONES, M.B., MILBURN, T.R. : Photosynthesis in papyrus (*Cyperus papyrus* L.). - Photosynthetica *12* : 197 - 199, 1978.

34194 - JONES, O.T.G. : Biosynthesis of porphyrins, hemes, and chlorophylls. - In : CLAYTON, R.K., SISTROM, W.R. (ed.) : The Photosynthetic Bacteria. Pp. 751 - 777. Plenum Press, New York - London 1978.

34195 - JONES, P.G., LAING, D.R. : The effects of phenological and meteorological factors on soybean yield. - Agr. Meteorol. *19* : 485 - 495, 1978.

34196 - JONES, R., WILKINS, M.B., COGGINS, J.R., FEWSON, C.A., MALCOLM, A.D.B. : Phosphoenolpyruvate carboxylase from the crassulacean plant *Bryophyllum fedtschenkoi* HAMET *et* PERRIER. Purification, molecular and kinetic properties. - Biochem. J. *175* : 391 - 406, 1978.

34197 - JONES, R.C. : Algal biomass dynamics during colonization of artificial is-
lands : experimental results and a model. - Hydrobiologia 59 : 165 - 180,
1978.

34198 - JOSE, A.M., SCHÄFER, E. : Distorted phytochrome action spectra in green
plants. - Planta 138 : 25 - 28, 1978. [Chl.]

34199 - JOSET-ESPARDELLIER, F., ASTIER, C., EVANS, E.H., CARR, N.G. : Cyanobacteria
grown under photoautotrophic, photoheterotrophic, and heterotrophic regimes : sugar metabolism and carbon dioxide fixation. - FEMS Microbiol. Lett.
4 : 261 - 264, 1978.

34200 - JOSHI, G.V. : Photsynthesis and productivity in coastal ecosystems. - In :
Inter-Disciplinary Symposium on Photosynthesis and Productivity. Abstract
of Papers. Pp. 31 - 34. Ind. nat. Sci. Acad., New Delhi 1978.

34201 - JOURNEAUX, R., VIOVY, R. : Orientation of chlorophylls in liquid crystals. -
Photochem. Photobiol. 28 : 243 - 248, 1978.

34202 - JUNGE, W., AUSLÄNDER, W. : Proton release during photosynthetic water oxidation : kinetics under flashing light. - In : METZNER, H.(ed.) : Photosynthetic Oxygen Evolution. Pp. 213 - 228. Academic Press, London - New York -
San Francisco 1978.

34203 - JUNGE, W., McGEER, A.J., AUSLÄNDER, W., KOLLIA, J. : Proton pumping across
the thylakoid membrane resolved in time and space. - In : SCHÄFER, G.,
KLINGENBERG, M. (ed.) : Energy Conservation in Biological Membranes. Pp.
113 - 127. Springer-Verlag, Berlin - Heidelberg - New York 1978.

34204 - JUNGE, W., SCHAFFERNICHT, H. : On the internal structure of photosystem I
in green plants. - In : HALL, D.O., COOMBS, J., GOODWIN, T.W. (ed.) : Proceedings of the Fourth International Congress on Photosynthesis. Pp. 21 -
32. Biochem. Soc., London 1978.

34205 - JUPIN, H. : Les transformations d'énergie dans le chloroplast. - Année biol.
17 : 297 - 319, 1978.

34206 - JURGENS, S.K., JOHNSON, R.R., BOYER, J.S. : Dry matter production and translocation in maize subjected to drought during grain fill. - Agron. J. 70 :
678 - 682, 1978. [Growth analysis.]

34207 - JURSINIC, P., GOVINDJEE, WRAIGHT, C.A. : Membrane potential and microsecond
to millisecond delayed light emission after a single excitation flash in etiolated chloroplasts. - Photochem. Photobiol. 27 : 61 - 71, 1978.

34208 - JURSINIC, P.A. : Flash polarographic detection of superoxide production as
a means of monitoring electron flow between photosystems I and II. - FEBS
Lett. 90 : 15 - 20, 1978.

34209 - JYUNG, W.H. : The characteristics of mineral deficiency induced chlorophyll
loss. - Plant Physiol. 61 (Suppl.) : 78, 1978.


34210 - KACHAN, A.A., LITSOV, N.I., NIKOLAEVSKAYA, V.I. : Dvukhstupenchataya model'
protsessa fotorazlozheniya vody pri fotosinteze. [A two-stage model of water
photodecomposition during photosynthesis.] - Dokl. Akad. Nauk ukr. SSR, Ser.
B 1978 : 456 - 459, 1978. [In R, ab : E.]

34211 - KACHAN, A.A., NEGIEVICH, L.A. : O fotosensibilizirovannom khlorofillom a
okislenii vody metilovym krasnym v adsorbirovannom na aérosile sostoyanii.
[Chlorophyll a-photosensitized oxidation of water with methyl red in the
state of adsorption on aerosil.] - Dokl. Akad. Nauk SSSR 241 : 1204 - 1206,
1978. [In R.]

34212 - KADER, A.A., MORRIS, L.L. : Tomato fruit color measured with an Agtron E5-W
reflectance spectrophotometer. - HortScience 13 : 577 - 578, 1978.

34213 - KADONO, Y. : Effect of oxygen deficit on the photosynthetic and respiratory
activities of submerged plants. - Jap. J. Ecol. 28 : 319 - 323, 1978.

34214 - KAFALIEVA-BOEVA, D.N., VAKLINOVA, S.G. : Differences in the content of the
pigments and Hill's test in mutant forms in peas. - Dokl. bolg. Akad. Nauk
*31* : 461 - 464, 1978.

34215 - KAGEYAMA, A., YOKOHAMA, Y. : The function of siphonein in a siphonous green
alga *Dichotomosiphon tuberosus*. - Jap. J. Phycol. *26* : 151 - 155, 1978.

34216 - KAJAK, Z. : The characteristics of a temperate eutrophic, dimictic lake
(Lake Mikolajskie, Northern Poland). - Int. Rev. ges. Hydrobiol. *63* : 451 -
480, 1978.

34217 - KAKUNO, T., HIURA, H., YAMASHITA, J., BARTSCH, R.G., HORIO, T. : Complete
stabilization of water-soluble hydrogenase from *Rhodospirillum rubrum* under
air atmosphere with a high concentration of chloride ions. - J. Biochem.
(Tokyo) *84* : 1649 - 1651, 1978.

34218 - KALER, V.L., FRIDLYAND, L.E. : Kharakteristika aktivnosti fotosinteticheskogo
apparata kak funktsiya kontsentratsiǐ ego osnovnykh élementov. Matematicheska-
ya model'. [The characteristics of the activity of the photosynthetic apparatus as a function of concentrations of its main elements : a mathematical
model.] - Fiziol. Rast. *25* : 484 - 491, 1978. [In R, ab : E.]

34219 - KALER, V.L., FRIDLYAND, L.E. : Ontogeneticheskaya adaptatsiya fotosinteza
rasteniǐ kak sledstvie kolichestvennykh izmeneniǐ osnovnykh élementov fotosinteticheskogo apparata. Teoreticheskoe rassmotrenie. [Ontogenetic adaptation of plant photosynthesis as a result of quantitative changes in the main
elements of the photosynthetic apparatus : a theoretical consideration.] -
Fiziol. Rast. *25* : 664 - 670, 1978. [In R, ab : E.]

34220 - KALER, V.L., FRIDLYAND, L.E. : Vozmozhnyǐ mekhanizm kooperativnogo upravle-
niya v polifermentnykh sistemakh. [The possible mechanism of cooperative
control in multienzyme systems.] - Mol. Biol. (Moskva) *12* : 421 - 428, 1978.
[Chl; in R, ab : E.]

34221 - KALISKY, O., LACHISH, U., OTTOLENGHI, M. : Time resolution of a back photo-
reaction in bacteriorhodopsin. - Photochem. Photobiol. *28* : 261 - 263, 1978.

34222 - KÄLLQVIST, T., MEADOWS, B.S. : The toxic effect of copper on algae and roti-
fers from a soda lake (Lake Nakuru, East Africa). - Water Res. *12* : 771 -
775, 1978. [Ps.]

34223 - KAMEKE, E. von, WEGMANN, K. : Properties and function of two manganese-con-
taining proteins from *Dunaliella* chloroplasts. - In : METZNER, H. (ed.) :
Photosynthetic Oxygen Evolution. Pp. 371 - 380. Academic Press, London - New
York - San Francisco 1978.

34224 - KAMIMURA, Y., MATSUZAKI, E. : Cytochrome components of green alga, *Bryopsis
maxima* : Purification and properties of cytochrome *f* from membrane fragments.
- Plant Cell Physiol. *19* : 1175 - 1183, 1978.

34225 - KAMÍNEK, M., LUŠTINEC, J. : Sensitivity of oat chlorophyll retention bioassay to natural and synthetic cytokinins. - Biol. Plant. *20* : 377 - 382,
1978.

34226 - KANDEL, M., GORNALL, A.G., CYBULSKY, D.L., KANDEL, S.I. : Carbonic anhydrase from spinach leaves. Isolation and some chemical properties. - J. biol.
Chem. *253* : 679 - 685, 1978.

*34227 - KANDELAKI, A.A., OKROPIRIDZE, T.D., CHKHUBIANISHVILI, R.I. : Nekotorye ana-
tomo-morfologicheskie i fiziologicheskie izmeneniya v bukovykh drevostoyakh
v svyazi s kompleksno-vyborochnymi rubkami. [Some anatomical, morphological
and physiological changes in beech stands in connection with voluntary-se-
lected cuttings.] - In : Voprosy Gornogo Lesovedeniya i Lesovodstva v Gruzii.
Vol. 25. Pp. 64 - 70. Sabchota Adzhara, Batumi 1976. [Ps; in R, ab : E, Ge-
orgian.]

B34228 - KANEMASU, E.T., RASMUSSEN, V.P., BAGLEY, J. : Estimating Water Requirements
for Corn with a "Pocket" Calculator. - Agricultural Experiment Station, Kansas State University, Manhattan 1978. [Growth analysis.]

34229 - KANEMATSU, S., ASADA, K. : Crystalline ferric superoxide dismutase from an
anaerobic green sulfur bacterium, *Chlorobium thiosulfatophilum*. - FEBS Lett.
*91* : 94 - 98, 1978.

34230 - **KANEMATSU, S., ASADA, K.** : Superoxide dismutase from an anaerobic photosynthetic bacterium, *Chromatium vinosum*. - Arch. Biochem. Biophys. *185* : 473 -
482, 1978.

34231 - **KANETI, J., KARANOV, E.N.** : Dependence between chemical structure and inhibition of Hill's reaction for certain heterocyclic derivatives of urea and
acylamides. - Dokl. bolg. Akad. Nauk *31* : 473 - 476, 1978.

34232 - **KANIUGA, Z., MICHALSKI, W.** : Photosynthetic apparatus in chilling-sensitive
plants II. Changes in free fatty acid composition and photoperoxidation in
chloroplastsfollowing cold storage and illumination of leaves in relation
to Hill reaction activity. - Planta *140* : 129 - 136, 1978.

34233 - **KANIUGA, Z., SOCHANOWICZ, B., ZĄBEK, J., KRZYSTYNIAK, K.** : Photosynthetic
apparatus in chilling-sensitive plants I. Reactivation of Hill reaction activity inhibited on the cold and dark storage of detached leaves and intact
plants. - Planta *140* : 121 - 128 , 1978.

34234 - **KANIUGA, Z., ZĄBEK, J., SOCHANOWICZ, B.** : Photosynthetic apparatus in chilling-sensitive plants III. Contribution of loosely bound manganese to the
mechanism of reversible inactivation of Hill reaction activity following
cold and dark storage and illumination of leaves. - Planta *144* :49-56, 1978.

34235 - **KANNANGARA, C.G., GOUGH, S.P.** : Biosynthesis of Δ-aminolevulinate in greening barley leaves : glutamate 1-semialdehyde aminotransferase. - Carlsberg
Res. Commun. *43* : 185 - 194, 1978.

34236 - **KANNANGARA, C.G., GOUGH, S.P., WETTSTEIN, D. von** : The biosynthesis of
Δ-aminolevulinate and chlorophyll and its genetic regulation. - In :
**AKOYUNOGLOU, G., ARGYROUDI-AKOYUNOGLOU, J.H.** (ed.) : Chloroplast Develop-
ment. Pp. 147 - 160. Elsevier/North-Holland Biomedical Press, Amsterdam -
New York - Oxford 1978.

34237 - **KAPLAN, S.** : Control and kinetics of photosynthetic membrane development. -
In : **CLAYTON, R.K., SISTROM, W.R.** (ed.) : The Photosynthetic Bacteria. Pp.
809 - 839. Plenum Press, New York - London 1978.

34238 - **KAPLANOVÁ, M., SOCHA, J.** : The effect of some herbicides on the photooxidation of chlorophyll-*a* in solution. - In : **FIALA, J.** (ed.) : Third International Seminar on Energy Transfer in Condensed Matter. Pp. 152 - 157. Univerzita Karlova, Praha 1978.

*34239 - **KAPPEN, L., LANGE, O.L., SCHULZE, E.-D., EVENARI, M., BUSCHBOM, U.** : Photosynthese und Wasserhaushalt von Wild- und Kulturpflanzen in der Negev-Wüste.
- Verhandl. Ges. Ökol. *1973* : 77 - 85, 1973.

34240 - **KARABASHEV, G.S., SOLOV'EV, A.N.** : O svyazi maksimumov intensivnosti fluorestsentsii pigmentov fitoplanktona s polozheniem sezonnogo piknoklina.
[Relationship between fluorescence intensity maxima of phytoplankton pigments and the location of the seasonal pycnokline.] - Okenologiya *18* : 709 -
715, 1978. [In R, ab : E.]

*34241 - **KARAMAN, I.P.** : Ispol'zovanie ènergii solnechnoĭ radiatsii na formirovanie
urozhaya tomatov v zavisimosti ot ikh mineral'nogo pitaniya. [Utilization
of solar radiant energy for the formation of tomato yield in dependence on
mineral nutrition.] - In : Pitanie Rasteniĭ i Primenenie Udobreniĭ. Pp. 59 -
62, 102. Kishinev. sel'skokhoz. Inst. Im. M.V.Frunze, Kishinev 1977. [In R.]

*34242 - **KARAMAN, I.P.** : Vliyanie udobreniĭ i gustoty stoyaniya rasteniĭ na soderzhanie khlorofilla v list'yakh tomatov. [Effect of mineral nutrition and plant
density on chlorophyll content in tomato leaves.] - In : Pitanie Rasteniĭ i
Primenenie Udobreniĭ. Pp. 62 - 64, 102. Kishinev. sel'skokhoz. Inst. Im.
M.V.Frunze, Kishinev 1977. [In R.]

34243 - **KARAMANOS, A.J.** : Water stress and leaf growth of field beans (*Vicia faba*
L.) in the field : leaf number and total leaf area. - Ann. Bot. *42* : 1393 -
1402, 1978.

34244 - **KARAPETYAN, N.V., RAKHIMBERDIEVA, M.G., BUKHOV, N.G.** : Priroda izmeneniĭ
vykhoda fluorestsentsii fragmentov khloroplastov, obogashchennykh P700.

[Nature of fluorescence yield changes of P700-enriched fragments.] - Biokhimiya *43* : 1319 - 1327, 1978. [In R, ab : E.]

34245 - **KARATAGLIS, S.S.** : Effect of EDTA on chlorophyll synthesis and root elongation of *Anthoxanthum odoratum.* - Ber. deut. bot. Ges. *91* : 297 - 304, 1978.

34246 - **KARAVAEV, V.A., PAVLOVA, I.E., KUKUSHKIN, A.K.** : Vliyanie usloviĭ osveshcheniya
pri vyrashchivanii na induktsiyu fluorestsentsii list'ev drevesnykh rasteniĭ.
[The effect of illuminance during cultivation on the induction of fluorescence in tree leaves.] - Fiziol. Rast. *25* : 798 - 802, 1978. [Chl; in R,
ab : E.]

34247 - **KARL, D.M., HAUGHSNESS, J.A., CAMBELL, L., HOLM-HANSEN, O.** : Adenine nucleotide extraction from multicellular organisms and beach sand : ATP recovery,
energy charge ratios and determination of carbon/ATP ratios. - J. exp. mar.
Biol. Ecol. *34* : 163 - 181, 1978.

B34248 - **KARNAUKHOV, V.N.** : Rol' Mollyuskov s Vysokim Soderzhaniem Karotinoidov
v Okhrane Vodnoĭ Sredy ot Zagryazneniya. [Role of Molluscs with High Content
of Carotenoids in Prevention of Water Environment from Pollution.] - Nauch.
Tsentr biol. Issled. Akad. Nauk SSSR, Inst. biol. Fiz. Akad. Nauk SSSR,
Pushchino 1978. [Car, Chl.]

34249 - **KARNOK, K.J., BEARD, J.B.** : The effect of chilling on the chloroplast ultrastructure of *Cynodon dactylon* PERS. as affected by gibberellic acid. - Plant
Physiol. *61* (Suppl.) : 94, 1978.

34250 - **KARPILOV, Yu.S., KARPILOVA, I.F., KERIMOV, S.** : Rol' piruvatkinazy v nespet-
sificheskikh izmeneniyakh fotosinteticheskogo metabolizma ugleroda, vyzyva-
emykh podavleniem fotofosforilirovaniya. [Role of pyruvate kinase in non-
-specific changes of carbon photosynthetic metabolism, caused by the inhibition of photophosphorylation.] - Biokhimiya *43* : 290 - 295, 1978. [In R,
ab : E.]

34251 - **KARPILOV, Yu.S., KARTASHOVA, R.I., TITLYANOV, E.A.** : Sostav produktov fotosinteza u nekotorykh mnogokletochnykh vodorosleĭ yaponskogo morya. [Compo-
sition of photosynthesis products in some multicellular algae of the sea
of Japan.] - Bot. Zh. *63* : 434 - 437, 1978. [In R.]

34252 - **KARPILOV, Yu.S., KERIMOV, S.Kh., MASLOV, A.N., BELOBRODSKAYA, L.K., KARPI-
LOVA, I.F., NOVITSKAYA, I.L., GERUS, E.V., KARTASHOVA, R.I.** : Prevrashcheniya serina i glitsina v list'yakh $C_3$ i $C_4$-rasteniĭ i ikh uchastie v mito-
khondrial'nom dykhanii. [Transformations of serine and glycine in leaves
of $C_3$ and $C_4$ plants and their participation in mitochondrial respiration.]
- In : **KARPILOV, Yu.S., ROMANOVA, A.K.** (ed.) : Mekhanizm Fotodykhaniya i
ego Osobennosti u Rasteniĭ Razlichnykh Tipov. Pp. 187 - 224. Pushchino
1978. [In R.]

34253 - **KARPILOV, Yu.S., KERIMOV, S.Kh., NOVITSKAYA, I.L., KARPILOVA, I.F.** : Metabolizm alanina i glyukozy v list'yakh rasteniĭ raznykh tipov v atmosfere
bez $CO_2$ . [Metabolism of alanine and glucose in leaves of different type
of plants in the $CO_2$-free atmosphere.] - Fiziol. Biokhim. kul'tur. Rast.
*10* : 499 - 506, 1978. [In R, ab : E.]

34254 - **KARPILOV, Yu.S., KUZ'MIN, A.N., BIL', K.Ya.** : Raspredelenie fermentov glikoliza v assimilyatsionnykh tkanyakh list'ev $C_4$-rasteniĭ i ikh svyaz' s oso-
bennostyami reaktsiĭ fotosinteza i fotodykhaniya. [Glycolytic enzymes in
assimilatory tissues of $C_4$-plant leaves and their relation to the peculiarities of the reactions of photosynthesis and photorespiration.] - Fiziol.
Rast. *25* : 1129 - 1135, 1978. [In R, ab : E.]

34255 - **KARPILOV, Yu.S., LYUBIMOV, V.Yu., CHERMNYKH, R.M., KOSOBRYUKHOV, A.A.** :
Priroda svetoindutsirovannogo okisleniya organicheskikh kislot i ego voz-
mozhnaya rol' v regulyatsii psevdotsiklicheskogo fotofosforilirovaniya v
khloroplastakh. [Nature of light-induced oxidation of organic acids and its
possible role in the regulation of pseudocyclic photophosphorylation in
chloroplasts.] - In : **KARPILOV, Yu.S., ROMANOVA, A.K.** (ed.) : Mekhanizm
Fotodykhaniya i ego Osobennosti u Rasteniĭ Razlichnykh Tipov. Pp. 74 - 89.
Pushchino 1978. [In R.]

34256 - KARPILOV, Yu.S., NOVITSKAYA, I.L., BELOBRODSKAYA, L.K., BIL', K.Ya., MASLOV, A.I., KUZ'MIN, A.N., KARPILOVA, I.F., KERIMOV, S.Kh., POPOVA, E.I., PETRUK-HIN, Yu.A., GERTS, S.M. : Reaktsii glikoliza v avtotrofnoĭ kletke. Rol' v fotosinteticheskom metabolizme i svetovom dykhanii. [Glycolysis reactions in an autotrophic cell. Role in the photosynthetic metabolism and photorespiration.] - In : KARPILOV, Yu.S., ROMANOVA, A.K. (ed.) : Mekhanizm Foto-dykhaniya i ego Osobennosti u Rasteniĭ Razlichnykh Tipov. Pp. 90 - 187. Pushchino 1978. [In R.]

B34257 - KARPILOV, Yu.S., ROMANOVA, A.K. (ed.) : Mekhanizm Fotodykhaniya i ego Oso-bennosti u Rasteniĭ Razlichnykh Tipov. [Mechanism of Photorespiration and its Peculiarities in Various Plant Types.] - Akad. Nauk SSSR, Pushchino 1978. [In R.]

34258 - KARPOV, E.A., MEDYANNIKOV, V.M. : Raspredelenie radioaktivnykh izotopov ugleroda i fosfora v tallomakh morskoĭ vodorosli *Sargassum pallidum*. [Distri-bution of radioisotopes of carbon and phosphorus in thalli of marine alga *Sargassum pallidum*.] - Sb. Rabot Akad. Nauk SSSR, dal'nevost. nauch. Tsentr, Inst. Biol. Morya (Vladivostok) *11* (Ėkologicheskie Aspekty Fotosinteza Mor-skikh Makrovodorosleĭ) : 175 - 182, 187 - 188, 1978. [In R, ab : E.]

34259 - KARPOVA, R.N., PERSANOV, V.M., KARPILOV, Yu.S. : Biosintez NADP-malatdegid-rogenaz pri zelenenii ėtiolirovannykh list'ev kukuruzy. [Biosynthesis of NADP-malate dehydrogenases in greening etiolated maize leaves.] - Biokhi-miya *43* : 1636 - 1639, 1978. [In R, ab : E.]

34260 - KARTUSCH, B. : Unterschiedliches Photosyntheseverhalten immergrüner Pflan-zen in Abhängigkeit von den klimatischen Faktoren. - Phyton *19* : 61 - 69, 1978.

*34261 - KARUNAKARAN, K., KISS, I.S. : Chlorophyll mutation yields by ethyl methane-sulfonate, gamma rays and fast neutrons in rice. - Riso *19* : 287 - 292, 1970.

34262 - KASIMOV, I. : Vliyanie na nyakoi agrotekhnichni faktori v"rkhu listnata ploshch na tsentralniya brat na pshenitsata. [Effect of some agrotechnical factors on the leaf area of the central tiller in wheat.] - Rasteniev. Nau-ki *15* (4) : 86 - 96, 1978. [In Bulg., ab : E, R.]

34263 - KATES, M., KUSHWAHA, S.C. : Biochemistry of the lipids of extremely halo-philic bacteria. - In : CAPLAN, S.R., GINZBURG, M. (ed.) : Energetics and Structure of Halophilic Microorganisms. Pp. 461 - 480. Elsevier/North-Hol-land Biomedical Press, Amsterdam - New York 1978. [Bacteriorhodopsin.]

34264 - KATO, S., HOZYO, Y. : [The speed and coeficient of $^{14}C$-photosynthates trans-location in the stem of grafts between improved variety and wild type plant in *Ipomoea*.] - Bull. nat. Inst. agr.  Sci., Ser. D *29* : 113 - 131, 1978. [In Jap., ab : E.]

34265 - KATO, Y. : The involvement of photosynthesis in inducing bud formation on excised leaf segments of *Heloniopsis orientalis* (*Liliaceae*). - Plant Cell Physiol. *19* : 791 - 799, 1978.

34266 - KATSUMI, M., KAZAMA, H. : Gibberellin control of cell elongation in cucumber hypocotyl sections. - Bot. Mag. (Tokyo), spec. Issue *1*: 141 - 158, 1978. [Ps inhibitors.]

34267 - KATZ, A., DEHAN, K., ITAI, C. : Kinetic reversal of NaCl effects. - Plant Physiol. *62* : 836 - 837, 1978. [Ps, Chl.]

34268 - KAUFMANN, K.J., SUNDSTROM, V., YAMANE, T., RENTZEPIS, P.M. : Kinetics of the 580-nm ultrafast bacteriorhodopsin transient. - Biophys. J. *22* : 121 - 124, 1978.

34269 - KAWANABE, S., OKUBO, T. : Comparison of net assimilation rate and crop growth rate between tropical and temperate grasses. - In : MONSI, M., SAEKI, T. (ed.) : Ecophysiology of Photosynthetic Productivity. JIBP Synthe-sis. Vol. 19. Pp. 185 - 194. Univ. of Tokyo Press, Tokyo 1978.

34270 - **KAWASAKI, H., TOMURA, K.** : [Bacteria using photoenergy, bacteriorhodopsin.] - Hakko to Kogyo *36* : 118 - 126, 1978. [In Jap.]

34271 - **KAWASE, M.** : Aeration and waterlogging damages. - HortScience *13* : 370, 1978. [Chl.]

*34272 - **KAYSER, H.** : Waste-water assay with continuous algal cultures : The effect of mercuric acetate on the growth of some marine dinoflagellates. - Mar. Biol. *36* : 61 - 72, 1976. [Chl.]

*34273 - **KAYSER, H.** : Effect of zinc sulphate on the growth of mono- and multispecies cultures of some marine plankton algae. - Helgoländer wiss. Meeresunters. *30* : 682 - 696, 1977. [Chl.]

34274 - **KAYUMOV, A.K., TARAN, A.A., DENISYUK, R.V., LISITSKIĬ, V.V., KOKLYUKOV, A.M.** : Uskorennyĭ metod opredeleniya soderzhaniya karotina v abrikosakh I produktakh ikh pererabotki. [Rapid method for determination of carotene in apricots and products of their processing.]-Konservn. ovoshchesush. Promyshlennost' *1978* (3) : 36 - 37, 48, 1978. [In R.]

34275 - **KAZAMA, H., KATSUMI, M.** : Effects of light on auxin-induced elongation of light-grown cucumber hypocotyl sections. - Plant Cell Physiol. *19* : 1137 - 1144, 1978. [Ps inhibitors.]

34276 - **KAZAMA, H., KATSUMI, M., UEDA, K.** : Light-controlled sugar-starch interconversion in epidermal chloroplasts of light-grown cucumber hypocotyl sections and its relationships to cell elongation. - Bot. Mag. (Tokyo) *91* : 121 - 130, 1978.

34277 - **KE, B.** : The primary electron acceptors in green-plant Photosystem I and photosynthetic bacteria. - In : SANADI, D.R., VERNON, L.P. (ed.) : Current Topics in Bioenergetics. Vol. 7. Pp. 76 - 138. Academic Press, New York - San Francisco - London 1978.

34278 - **KEELEY, J.E.** : Malic acid accumulation in roots in response to flooding : Evidence contrary to its role as an alternative to ethanol. - J. exp. Bot. *29* : 1345 - 1349, 1978.

34279 - **KEELEY, P.E., THULLEN, R.J.** : Light requirements of yellow nutsedge (*Cyperus-esculentus*) and light interception by crops. - Weed Sci. *26* : 10 - 16, 1978.

34280 - **KEIFER, D.W., SPANSWICK, R.M.** : Activity of the electrogenic pump in *Chara corallina* as inferred from measurements of the membrane potential, conductance, and potassium permeability. - Plant Physiol. *62* : 653 - 661, 1978.

34281 - **KEISTER, D.L.** : Respiration vs. photosynthesis. - In : CLAYTON, R.K., SISTROM, W.R. (ed.) : The Photosynthetic Bacteria. Pp. 849 - 856. Plenum Press, New York - London 1978.

34282 - **KELL, D.B., FERGUSON, S.J., JOHN, P.** : Measurement by a flow dialysis technique of the steady-state proton-motive force in chromatophores from *Rhodospirillum rubrum*. Comparison with phosphorylation potential. - Biochim. biophys. Acta *502* : 111 - 126, 1978.

34283 - **KELLER, T.** : Der Einfluss einer $SO_2$-Belastung zu verschiedenen Jahreszeiten auf $CO_2$-Aufnahme und Jahrringbau der Fichte. - Schweiz. Z. Forstwesen *129* : 381 - 393, 1978. [Ps.]

34284 - **KELLER, T.** : Einfluss niedriger $SO_2$-Konzentrationen auf die $CO_2$-Aufnahme von Fichte und Tanne. - Photosynthetica *12* : 316 - 322, 1978.

34285 - **KELLER, T.** : Wintertime atmospheric pollutants - do they affect the performance of deciduous trees in the ensuing growing season? - Environ. Pollut. *16* : 243 - 247, 1978. [Production.]

34286 - **KELLEY, P.M., IZAWA, S.** : The role of chloride ion in photosystem II. I.Effects of chloride ion on photosystem II electron transport and on hydroxylamine inhibition. - Biochim. biophys. Acta *502* : 198 - 210, 1978.

34287 - **KELLOMÄKI, S.** : Typpilannoituksen vaikutus havupuiden fotosynteesikapateettiin. [Effect of some nitrogen fertilizers on photosynthetic capacity of coniferous trees.] - Silva fenn. *12* : 231 - 239, 1978. [In Finn., ab : E.]

*34288 - KELLOMÄKI, S., HARI, P. : Rate of photosynthesis of some forest mosses as function of temperature and light intensity and effect of water content of moss cushion on photosynthetic rate. - Silva fenn. *10* : 288 - 295, 1976.

*34289 - KELLOMÄKI, S., HARI, P., KOPONEN, T. : Ecology of photosynthesis in *Dicranum* and its taxonomic significance. - In : Congrès International de Bryologie. Bryophytorum Bibliotheca 13. Pp. 485 - 507. Bordeaux 1977.

*34290 - KELLOMÄKI, S., HARI, P., VÄISÄNEN, E. : Annual production of some forest mosses as a function of light available for photosynthesis. - Silva fenn. *11* : 81 - 86, 1977. [Ps.]

*34291 - KELLOMÄKI, S., HARI, P., VUOKKO, R., VÄISÄNEN, E., KANNINEN, M. : Above ground growth rate of a dwarf shrub community. - Oikos *29* : 143 - 149, 1977.

34292 - KELLY, G.J. : Aspects of enzyme regulation illustrated by the properties of two phosphofructokinases from spinach leaves. - In : HALL, D.O., COOMBS, J., GOODWIN, T.W. (ed.) : Proceedings of the Fourth International Congress on Photosynthesis. Pp. 437 - 446. Biochem. Soc., London 1978.

34293 - KELLY, M.H., FITZPATRICK, L.C., PEARSON, W.D. : Phytoplankton dynamics, primary productivity and community metabolism in a north-central Texas pond. - Hydrobiologia *58* : 245 - 260, 1978.

34294 - KEMP, G.A. : Growth of primary leaves of beans (*Phaseolus vulgaris*) under suboptimal temperatures. - Can. J. Plant Sci. *58* : 169 - 174, 1978. [Growth analysis.]

34295 - KENNEDY, R.A. : Relationship between the stage of leaf development, photosynthetic rates and water-use efficiency in $C_3$ and $C_4$ plants. - Plant Physiol. *61* (Suppl.) : 86, 1978.

34296 - KENNER, G.W., RIMMER, J., SMITH, K.M., UNSWORTH, J.F. : Pyrroles and related compounds. Part 39. Structural and biosynthetic studies of the *Chlorobium* chlorophylls-660 (bacteriochlorophylls *c*). Incorporations of methionine and porphobilinogen. - J. chem. Soc. Perkin Trans. I *1978* : 845 - 852, 1978.

34297 - KENYON, C.N. : Complex lipids and fatty acids of photosynthetic bacteria. - In : CLAYTON, R.K., SISTROM, W.R. (ed.) : The Photosynthetic Bacteria. Pp. 281 - 313. Plenum Press, New York - London 1978. [Chromatophores.]

34298 - KENYON, W.H., KRINGSTAD, R., BLACK, C.C. : Diurnal changes in the malic acid content of vacuoles isolated from leaves of the Crassulacean Acid Metabolism plant, *Sedum telephium*. - FEBS Lett. *94* : 281 - 283, 1978.

34299 - KERBER, N.L., PUCHEU, N.L., GARCIA, A.F. : Phenazine methosulfate mediated photoinactivation of some energy linked reactions in *Rhodospirillum rubrum*. - Biochem. biophys. Res. Commun. *81* : 667 - 671, 1978.

34300 - KERBER, N.L., PUCHEU, N.L., GARCIA, A.F. : PMS photo-inhibition in *Rhodospirillum rubrum* membranes in the presence of permeant entities affecting either the $\Delta\psi$ or the $\Delta$ pH components of the protonmotive force. - FEBS Lett. *94* : 265 - 268, 1978.

34301 - KERR, J.P., McPHERSON, H.G. : Evapotranspiration and physiological response to water stress of several pasture and crop species. - Proc. N. Zeal. Grassland Assoc. *39* (1) : 70 - 78, 1978. [Growth analysis.]

34302 - KERSHAW, K.A., SMITH, M.M. : Studies on lichen-dominated systems. XXI. The control of seasonal rates of net photosynthesis by moisture, light, and temperature in *Stereocaulon paschale*. - Can. J. Bot. *56* : 2825 - 2830, 1978.

34303 - KESSLER, E. : Hydrogenase in green algae. - In : Hydrogenases : Their Catalytic Activity, Structure and Function. Pp. 415 - 422. Erich Goltze K.G., Göttingen 1978.

34304 - KHAĬLOV, K.M. : Ob otsenke funktsional'nogo sostoyaniya rasteniĭ po analizu statisticheskoĭ svyazi morfologicheskikh i fiziologicheskikh parametrov. [Evaluation of the functional state of plants by the analysis of the specific relationship of morphological and physiological parameters.] - Fiziol. Rast. *25* : 1262 - 1269, 1978. [Ps; in R, ab E.]

34305 - KHAKIMOV, Ya.I., KVITKO, K.V. : Geneticheskiǐ kontrol' chuvstvitel'nosti k gerbitsidu diuronu u *Chlamydomonas reinhardii*. [Genetic control of herbicide diuron sensitivity in *Chlamydomonas reinhardii*.] - Genetika *14* : 1319 - 1327, 1978. [In R, ab E.]

34306 - KHAN, A., SOLTANPOUR, P.N. : Factors associated with Zn chlorosis in dryland beans. - Agron. J. *70* : 1022 - 1026, 1978.

34307 - KHAN, A.A., MALHOTRA, S.S. : Biosynthesis of lipids in chloroplasts isolated from Jack pine needles. - Phytochemistry *17* : 1107 - 1110, 1978.

*34308 - KHAN, M.A., TSUNODA, S. : Growth analysis of cultivated wheat species and their wild relatives with special reference to dry matter distribution among different plant organs and to leaf area expansion. - Tohoku J. agr. Res. *21* : 47 - 59, 1970.

*34309 - KHAN, M.A., TSUNODA, S. : Growth analysis of six commercially cultivated wheats of West Pakistan with special reference to a semi-dwarf modern wheat variety, Mexi-Pak. - Tohoku J. agr. Res. *21* : 60 - 72, 1970.

*34310 - KHAN, M.A., TSUNODA, S. : Comparative leaf anatomy of cultivated wheats and wild relatives with reference to their leaf photosynthetic rates. - Jap. J. Breed. *21* : 143 - 150, 1971. [Ps.]

34311 - KHANGULOV, S.V., GOL'DFEL'D, M.G. : Svetozavisimye reaktsii $Mn^{2+}$ i $P700^+$ v khloroplastakh v usloviyakh impul'snogo osveshcheniya. [Light-dependent reactions of $Mn^{2+}$ ions and $P700^+$ in flash excited chloroplasts.] - Biofizika *23* : 272 - 278, 1978. [In R, ab : E.]

34312 - KHANNA CHOPRA, R. : Carbon-dioxide assimilation in some reproductive plant organs. - In : Inter-Disciplinary Symposium on Photosynthesis and Productivity. Abstract of Papers. Pp. 51 - 53. Indian nat. Sci. Acad., New Delhi 1978.

34313 - KHANOVA, L.A., TARASEVICH, M.R. : Élektrokhimicheskoe povedenie khlorofilla v vodnykh rastvorakh. Katodnoe vosstanovlenie. [Electrochemical behavior of chlorophyll in aqueous solutions. Cathodic reduction.] - Élektrokhimiya *14* : 168 - 171, 1978. [In R.]

34314 - KHARANYAN, N.N. : Vodnyǐ rezhim i aktivnost' nekotorykh fermentov rasteniǐ pshenitsy, obrabotannykh retardantom khlorkholinkhloridom (CCC) pri razlichnoǐ vodoobespechennosti. [Water regime and activity of some enzymes in wheat plants affected by CCC under different water supply.] - In : Vodnyǐ Rezhim Rasteniǐ v Svyazi s Raznymi Ékologicheskimi Usloviyami. Pp. 103 - 107. Izdatel'stvo Kazanskogo Universiteta, Kazan' 1978. [Ps; In R.]

34315 - KHAVARI-NEJAD, R.A. : Inactivation of ribulose-1,5-bisphosphate carboxylase/oxygenase by light in the presence of FMN. - Plant Physiol. *61* (Suppl.) : 99, 1978.

34316 - KHISAMUTDINOVA, V.I., KUZ'MINA, G.G., VASIL'EVA, I.M. : Nekotorye vzaimosvyazi mezhdu ovodnennost'yu khloroplastov i ikh fotokhimicheskoǐ aktivnost'yu v usloviyakh zakalivaniya. [Some relationships between chloroplast hydration and photochemical activity during hardening.] - In : Vodnyǐ Rezhim Rasteniǐ v Svyazi s Raznymi Ékologicheskimi Usloviyami. Pp. 257 - 262. Izdatel'stvo Kazanskogo Universiteta, Kazan' 1978. [In R.]

34317 - KHMARA, L.A., ZAKRZHEVSKIǏ, D.A., ROZONOVA, L.N., KALASHNIKOV, Yu.E. : O vliyanii defitsita margantsa na vydelenie kisloroda i strukturu khloroplastov u gorokha. [Effect of manganese deficiency on oxygen elimination and structure of chloroplasts in pea.] - Fiziol. Biokhim. kul't. Rast. *10* : 416 - 421, 1978. [In R, ab : E.]

34318 - KHODASEVICH, É.V., ARNAUTOVA, A.I., MYSHKOVETS, E.N. : Ul'trastrukturnaya organizatsiya khloroplastov v svyazi s obratimoǐ degradatsieǐ fonda pigmentov u khvoǐnykh. [The structural organization of chloroplasts related to reversible degradation of the pigment pool in conifers.] - Fiziol. Rast. *25* : 810 - 814, 1978. [In R, ab : E.]

34319 - KHODZHAEV, A.S., MEĬSTRIK, I.A., RAKHMANKULOVA, M.E. : O vzaimosvyazi chisla
        khloroplastov v kletke s ikh aktivnost'yu v ontogeneze lista khlopchatnika.
        [Correlation between the number of chloroplasts in the cell and their activi-
        ty in the course of cotton leaf ontogenesis.] - Fiziol. Rast. 25 : 541 -
        546, 1978. [In R, ab : E.]

34320 - KHOKHLOVA, V.A., KARAVAĬKO, N.N., PODERGINA, T.A., KULAEVA, O.N. : Antagonizm
        v deĭstvii abstsizovoĭ kisloty i tsitokinina na strukturnuyu i biokhimiches-
        kuyu differentsiatsiyu khloroplastov v izolirovannykh semyadolyakh tykvy.
        [The antagonistic effect of abscisic acid and cytokinin on the structural
        and biochemical differentiation of chloroplasts in isolated pumpkin cotyle-
        dons.] - Tsitologiya 20 : 1033 - 1039, 1978. [In R, ab : E.]

34321 - KHRISTIN, M.S., AKULOVA, E.A. : Kompleksoobrazovanie ferredoksina i ferre-
        doksin-NADF-reduktazy v model'noĭ sisteme i khloroplastakh. [Complexing of
        ferredoxin and ferredoxin-NADP-reductase in a model system and chloroplasts.]
        - Dokl. Akad. Nauk SSSR 238 : 992 - 995, 1978. [In R.]

*34322 - KHRYANIN, V.N., KUPCHININ, Yu.I. : Osobennosti deĭstviya gibberellovoĭ kis-
        loty i "gibrelata" na soderzhanie khlorofilla i svyaz' ego s belkom. [Peculi-
        arities of action of gibberellic acid and "gibrelate" on chlorophyll content
        and its binding to protein.] - In : Regulirovanie Rosta i Razvitiya Rasteniĭ
        s Pomoshchyu Khimicheskikh Sredstv i Usloviĭ Vneshneĭ Sredy. Vol. 2. Pp. 64 -
        68. Kalininskiĭ gosudarstvennyĭ Universitet, Kalinin 1973. [In R.]

34323 - KILPATRICK, D.J. : Growth models for unthinned stands of Sitka spruce in Nor-
        thern Ireland. - Forestry 51 : 47 - 56, 1978.

34324 - KIM, C.S., MOROSS, G., MOYER, W., KAUFMAN, G., MacCOLL, R., EDWARDS, M.R. :
        Automated acquisition and analysis of data from the photoelectric scanner
        of the model E analytical ultracentrifuge. - Anal. Biochem. 86 : 371 - 377,
        1978. [Biliproteins.]

34325 - KIMURA, M. : Measurement of constructive and maintenance respiration in grow-
        ing Helianthus tuberosus leaves. - In : MONSI, M., SAEKI, T. (ed.) : JIBP
        Synthesis. Vol. 19. Pp. 237 - 240. University of Tokyo Press, Tokyo 1978.

34326 - KIMURA, M., YOKOI, Y., HOGETSU, K. : Quantitative relationships between
        growth and respiration. II. Evaluation of constructive and maintenance re-
        spiration in growing Helianthus tuberosus leaves. - Bot. Mag. (Tokyo) 91 :
        43 - 56, 1978.

34327 - KINDMAN, L.A., COHEN, C.E., ZELDIN, M.H., BEN-SHAUL, Y., SCHIFF, J.A. :
        Events surrounding the early development of Euglena chloroplasts. 12. Spec-
        troscopic examination of the protochlorophyll(ide) phototransformation in
        intact cells. - Photochem. Photobiol. 27 : 787 - 794, 1978.

*34328 - KING, D., ERBES, D.L., BEN-AMOTZ, A., GIBBS, M. : The mechanism of hydrogen
        photoevolution in photosynthetic organisms. - In : MITSUI, A., MIYACHI, S.,
        SAN PIETRO, A., TAMURA, S. (ed.) : Biological Solar Energy Conversion. Pp.
        69 - 75. Academic Press, New York - San Francisco - London 1977.

34329 - KIPE-NOLT, J.A., STEVENS, S.E., Jr., STEVENS, C.L.R. : Biosynthesis of δ-ami-
        nolevulinic acid by blue-green algae (Cyanobacteria). - J. Bacteriol. 135 :
        286 - 288, 1978.

34330 - KIRA, T. : Studies on biological production in forest and freshwater ecosys-
        tems in West Malaysia. - In : TAMIYA, H. (ed. ) : Summary Report on the Con-
        tribution of the Japanese National Committee for IBP, 1964 - 1974. JIBP Syn-
        thesis. Vol. 20. Pp. 225 - 234. University of Tokyo Press, Tokyo 1978.
        [Growth analysis.]

34331 - KIRICHENKO, A.B., KIRICHENKO, E.B., CHEBOTAR', A.A. : Plastidnyĭ apparat
        razvivayushchikhsya semyapochek Hordeum vulgare L. [Plastid apparatus of
        developing ovules of Hordeum vulgare L.] - Fiziol. Rast. 25 : 113 - 117,
        1978. [In R, ab : E.]

34332 - KIRKHAM, M.B. : Water relations of cadmium-treated plants. - J. Environ.
        Qual. 7 : 334 - 336, 1978. [Stomatal resistance.]

34333 - KIRKHAM, M.B., AHRING, R.M. : Leaf temperature and internal water status of
        wheat grown at different root temperatures. - Agron. J. 70 : 657 - 662, 1978.
        [Stomatal resistance.]

34334 - KIRKHAM, M.B., SMITH, E.L. : Water relations of tall and short cultivars of
        winter wheat. - Crop Sci. 18 : 227 - 230, 1978. [Stomatal resistance.]

34335 - KIS, P. : Improved method for preparation of microcrystalline chlorophyll a
        with Anacystis nidulans as a source. - Experientia 34 : 1289, 1978.

34336 - KISELEV, B.A., EVSTIGNEEV, V.B. : Pigment-pigment electron transfer and se-
        paration of charges in photoexcised chlorophyll aggregates. - In : FIALA, J.
        (ed.) : Third International Seminar on Energy Transfer in Condensed Matter.
        Pp. 120 - 128. Univerzita Karlova, Praha 1978.

34337 - KLEIBEUKER, J.F., PLATENKAMP, R.J., SCHAAFSMA, T.J. : The triplet state of
        photosynthetic pigments. I. Pheophytins. - Chem. Phys. 27 : 51 - 64, 1978.

34338 - KLEIBEUKER, J.F., VAN DER BENT, S.J., SCHAAFSMA, T.J. : The properties of
        the triplet state of photosynthetic pigments. - In : FIALA, J. (ed.) : Third
        International Seminar on Energy Transfer in Condensed Matter. Pp. 52 - 66.
        Univerzita Karlova, Praha 1978.

34339 - KLEIN, G., RÜDIGER, W. : Über die Bindungen zwischen Chromophor und Protein
        in Biliproteiden, V. Stereochemie von Modell-Imiden. - Justus Liebigs Ann.
        Chem. 1978 : 267 - 279, 1978.

34340 - KLEIN, O., DÖRNEMANN, D., SENGER, H. : Two pathways for the biosynthesis of
        δ-aminolevulinic acid in Scenedesmus obliquus mutant C-2A'. - In : AKOYUNO-
        GLOU, G., ARGYROUDI-AKOYUNOGLOU, J.H. (ed.) : Chloroplast Development. Pp.
        45 - 50. Elsevier/North-Holland Biomedical Press, Amsterdam - New York -
        Oxford 1978.

34341 - KLEIN, O., SENGER, H. : Biosynthetic pathways to δ-aminolevulinic acid indu-
        ced by blue light in the pigment mutant C-2A' of Scenedesmus obliquus. -
        Photochem. Photobiol. 27 : 203 - 208, 1978.

34342 - KLEIN, O., SENGER, H. : Two biosynthetic pathways to δ-aminolevulinic acid
        in a pigment mutant of the green alga, Scenedesmus obliquus. - Plant Phy-
        siol. 62 : 10 - 13, 1978.

34343 - KLEIN, U., BETZ, A. : Induced protein synthesis during the adaptation to $H_2$
        production in Chlamydomonas moewusii. - Physiol. Plant. 42 : 1 - 4, 1978.

34344 - KLEINEN HAMMANS, J.W. : P750 sensitized photooxidations in Anacystis nidu-
        lans. - Plant Cell Physiol. 19 : 1457 - 1463, 1978.

34345 - KLEMPERER, G., EISENBACH, M., GARTY, H., CAPLAN, S.R. : The effect of salt
        on the light-induced pH changes in purple membrane from Halobacterium halo-
        bium. - In : CAPLAN, S.R., GINZBURG, M. (ed.) : Energetics and Structure of
        Halophilic Microorganisms. Pp. 291 - 296. Elsevier/North-Holland Biomedical
        Press, Amsterdam - New York 1978.

34346 - KLENOVSKÁ, S. : Dependence of some physiological processes in tobacco ex-
        plants upon the carrier of the experimental material. - Acta Fac. Rerum nat.
        Univ. comenianae, Physiol. Plant. 15 : 17 - 23, 1978. [Chl.]

34347 - KLEO, J., LUTZ, M. : Resonance Raman scattering of cation radical of bacteri-
        ochlorophyll in solution and in reaction centers of photosynthetic bacteria.
        - In : SCHMID, E.D., KRISHNAN, R.S., KIEFER, W. (ed.) : Proceedings of the
        VIth International Conference on Raman Spectroscopy. Vol. 6(2). Pp. 160 -
        - 161. Heyden, London 1978.

34348 - KLIMOV, S.V., BOCHAROV, E.A., DZHANUMOV, D.A. : Svyaz' formativnykh protses-
        sov s fotosintezom i poslesvecheniem u prorostkov ozimoĭ pshenitsy. [Corre-
        lation between formative processes, photosynthesis and delayed light emission
        in winter wheat seedlings.] - Fiziol. Rast. 25 : 106 - 112, 1978. [In R, ab :
        E.]

34349 - KLIMOV, V.V., ALLAKHVERDIEV, S.I., PASHCHENKO, V.Z. : Izmerenie ënergii akti-
        vatsii i vremeni zhizni fluorestsentsii khlorofilla fotosistemy 2. [Measuring
        in activation energy and life-time of fluorescence of photosystem 2 chloro-
        phyll.] - Dokl. Akad. Nauk SSSR 242 : 1204 - 1207, 1978. [In R.]

34350 - **KLIMOV, V.V., KLEVANIK, A.V., SHUVALOV, V.A., KRASNOVSKY, A.A.** : Reduction of pheophytin in the primary light reaction of photosystem II. - In : METZNER, H. (ed.) : Photosynthetic Oxygen Evolution. Pp. 147 - 155. Academic Press, London - New York - San Francisco 1978.

34351 - **KLOPATEK, J.M., STEARNS, F.W.** : Primary productivity of emergent macrophytes in a Wisconsin freshwater marsh ecosystem. - Amer. Midland Naturalist *100* : 320 - 332, 1978.

34352 - **KLUGE, M.** : Ecological aspects of Crassulacean Acid Metabolism (CAM). - In : HALL, D.O., COOMBS, J., GOODWIN, T.W. (ed.) : Proceedings of the Fourth International Congress on Photosynthesis. Pp. 335 - 345. Biochem. Soc., London 1978.

34353 - **KLUGE, M.** : Metabolism of organic acids. - Progr. Bot. *40* : 119 - 125, 1978. [Ps.]

B34354 - **KLUGE, M., TING, I.P.** : Crassulacean Acid Metabolism. Analysis of an Ecological Adaptation. - Springer-Verlag, Berlin - Heidelberg - New York 1978.

34355 - **KNAFF, D.B.** : Active transport in the photosynthetic bacterium *Chromatium vinosum*. - Arch. Biochem. Biophys. *189* : 225 - 230, 1978.

34356 - **KNAFF, D.B.** : Reducing potentials and the pathway of $NAD^+$ reduction. - In : CLAYTON, R.K., SISTROM, W.R. (ed.) : The Photosynthetic Bacteria. Pp. 629 - 640. Plenum Press, New York - London 1978.

34357 - **KNAFF, D.B., MALKIN, R.** : The primary reaction of chloroplast photosystem II. - In : SANADI, D.R., VERNON, L.P. (ed.) : Current Topics in Bioenergetics. Vol.7. Pp. 139 - 172. Academic Press, New York - San Francisco - London 1978.

34358 - **KNAFF, D.B., SMITH, J.M., MALKIN, R.** : Complex formation between ferredoxin and nitrite reductase. - FEBS Lett. *90* : 195 - 197, 1978.

34359 - **KNOECHEL, R., KALFF, J.** : An *in situ* study of the productivity and population dynamics of five freshwater planktonic diatom species. - Limnol. Oceanogr. *23* : 195 - 218, 1978.

34360 - **KNOPF, U.C., STUTZ, E.** : Molecular cloning of the gene region coding for the chloroplast rRNA of *Euglena gracilis*. - Mol. gen. Genet. *163* : 1 - 6, 1978.

34361 - **KNOX, R.S., DAVIDOVICH, M.A.** : Theory of fluorescence polarization in magnetically oriented photosynthetic systems. - Biophys. J. *24* : 689 - 712, 1978.

34362 - **KOBAYASHI, M., FUJII, K., SHIMAMOTO, I., MAKI, T.** : Treatment and re-use of industrial waste water by phototrophic bacteria. - Prog. Water Technol. *11* : 279 - 284, 1978.

34363 - **KOBAYASHI, M., KURATA, S.** : The mass culture and cell utilization of photosynthetic bacteria. - Process Biochem. *13* (9) : 27 - 30, 1978.

34364 - **KOBAYASHI, T.** : [Role and use of photosynthetic bacteria in higher plants.] - Hakko To Kogyo *36* : 574 - 583, 1978. [In Jap.]

34365 - **KOBAYASHI, T., NISHIMURA, S., TANAKA, S.** : [Growth of tropical and subtropical grasses in the southwestern area of Japan, as influenced by air temperature. 1. Effect of seeding time, growth stage and cutting frequency on the growth and yield.] - Sci. Bull. Fac. Agr. Kyushu Univ. *32* : 169 - 175, 1978. [In Jap., ab : E.]

34366 - **KOBAYASHI, T., NISHIMURA, S., TANAKA, S.** : [Growth of tropical and subtropical grasses in the Southwestern area of Japan, as influenced by air temperature. 2. Effect of seeding time and cutting management on the winter survival and yield in the second year.] - Sci. Bull. Fac. Agr. Kyushu Univ. *32* : 177 - 182, 1978. [In Jap., ab : E.]

34367 - **KOBAYASHI, Y., INOUE, Y., SHIBATA, K.** : Light-dependent binding of *p*-nitrothiophenol to photosystem II : oscillatory reactivity under illumination with flashes. - In : METZNER, H. (ed.) : Photosynthetic Oxygen Evolution. Pp. 157 - 170. Academic Press, London - New York - San Francisco 1978.

34368 - KOCH, K., KENNEDY, R.A. : Characteristics of crassulacean acid metabolism
         (CAM) in the succulent $C_4$ dicot, *Portulaca oleracea* L. - Plant Physiol. *61*
         (Suppl.) : 37, 1978.

34369 - KOCH, K., KENNEDY, R.A. : Effect of seasonal changes in the midwest on Cras-
         sulacean acid metabolism (CAM) in *Opuntia humifusa* RAF. - Plant Physiol. *61*
         (Suppl.) : 100, 1978.

34370 - KOCH, R., ALLEWELDT, G. : Der Gaswechsel reifender Weinbeeren. - Vitis *17* :
         30 - 44, 1978.

34371 - KOCHUBEĬ, S.M., SHADCHINA, T.M., YAKOVENKO, A.M., MANUIL'SKAYA, S.V.,
         OSTROVSKAYA, L.K. : O roli lipidov v organizatsii blizhaĭshego okruzheniya
         tsentrov fotosistemy 1. [The role of some lipids in organization of the nea-
         rest surroundings of Photosystem 1 reaction centres.] - Mol. Biol. (Moskva)
         *12* : 47 - 54, 1978. [In R, ab : E.]

34372 - KOENIG, F., RADUNZ, A., SCHMID, G.H., MENKE, W. : Antisera to the coupling
         factor of photophosphorylation and its subunits. - Z. Naturforsch. *33 C* :
         529 - 536, 1978.

34373 - KOH, S., KUMURA, A. : [Studies on matter production in wheat plant III. Chan-
         ges with growth in photosynthetic capacity  and respiratory capacity of
         wheat stand.] - Jap. J. Crop Sci. *47* : 63 - 68, 1978. [In Jap., ab : E.]

34374 - KOH, S., KUMURA, A., MURATA, Y. : [Studies on matter production in wheat
         plant IV. After-effects  of low night temperature on daytime photosynthesis
         examined under controlled conditions.] - Jap. J. Crop Sci. *47* : 69 - 74,
         1978. [In Jap., ab : E.]

34375 - KOH, S., KUMURA, A., MURATA, Y. : [Studies on matter production in wheat
         plant V. The mechanism involved in an after-effect of low night temperature.]
         - Jap. J. Crop Sci. *47* : 75 - 81, 1978. [In Jap., ab : E.]

34376 - KOH, S., KUMURA, A., MURATA, Y. : [Studies on matter production in wheat
         plant VI. Quantitative estimate of photosynthetic depression in the field.]-
         Jap. J. Crop Sci. *47* : 293 - 299, 1978. [In Jap., ab : E.]

34377 - KOIKE, H., SATOH, K., KATOH, S. : Effects of dibromothymoquinone and batho-
         phenanthroline on flash-induced cytochrome $f$ oxidation in spinach chloro-
         plasts. - Plant Cell Physiol. *19* : 1371 - 1380, 1978.

34378 - KOIKE, H., SATOH, K., KATOH, S. : Effects of electron transport inhibitors
         on the magnitude of flash-induced cytochrome $F$ oxidation in spinach chloro-
         plasts. - In : HALL, D.O., COOMBS, J., GOODWIN, T.W. (ed.) : Proceedings of
         the Fourth International Congress on Photosynthesis. Pp. 765 - 769. Biochem.
         Soc., London 1978.

34379 - KOIVUNIEMI, P., TOLBERT, N.E., CARLSON, J. : Ribulose-1,5-bisphosphate car-
         boxylase/oxygenase and polyphenol oxidase in the tobacco mutant SU/su. -
         Plant Physiol. *61* (Suppl.)  : 99, 1978.

34380 - KOKA, P., SONG, P.-S. : Protection of chlorophyll $a$ by carotenoid from pho-
         todynamic decomposition. - Photochem. Photobiol. *28* : 509 - 515, 1978.

34381 - KOLESNIKOV, M.P. : Tetrapirrol'nye pigmenty i karotinoidy v pochvakh. [Te-
         trapyrrolic pigments and carotenoids in soils.] - Pochvovedenie *1978* (9) :
         116 - 124, 1978. [In R, ab : E.]

34382 - KOLESNIKOV, M.P., EGOROV, I.A. : Porphyrins and phycobilins in precambrian
         rocks. - In : NODA, H. (ed.) : Origin of Life. Pp. 533 - 539. Bus. Centre
         Acad. Soc. Japan 1978.

34383 - KOLESNIKOV, P.A., PETROCHENKO, E.I., ZORĖ, S.V. : Regulyatory obrazovaniya
         glioksalevoĭ kisloty iz ribozo-5-fosfata v ėkstraktakh iz khloroplastov.
         [Compounds regulating synthesis of glyoxalic acid from ribose-5-phosphate
         in extracts from chloroplasts.] - Fiziol. Rast. *25* : 350 - 355, 1978. [In R,
         ab : E.]

34384 - KOLLER, H.R., THORNE, J.H. : Soybean pod removal alters leaf diffusion re-
         sistance and leaflet orientation. - Crop Sci. *18* : 305 - 307, 1978. [Stoma-
         tal resistance.]

34385 - **KOLMAKOV, P.V., LAVIN, P.I., TITLYANOV, É.A.** : Temnovaya fiksatsiya neorga-
nicheskogo ugleroda nekotorymi vidami morskikh prikreplennykh zelenykh vodorosleĭ. [Dark fixation of inorganic carbon by some species of marine attached green algae.] - Nauch. Soobshch. Inst. Biol. Morya, dal'nevostoch.
nauch. Tsentr Akad. Nauk SSSR *3* (Biologicheskie Issledovaniya Dal'nevostochnykh Moreĭ) : 39 - 41, 1978. [Ps; in R.]

34386 - **KOLMAKOV, P.V., TARANKOVA, Z.A.** : Opredelenie potentsial'noĭ intensivnosti
fotosinteza u morskikh tallomnykh vodorosleĭ. [Determination of the photosynthetic capacity of seaweeds.] - Sb. Rabot Akad. Nauk SSSR, dal'nevost.
nauch. Tsentr, Inst. Biol. Morya (Vladivostok) *11* (Ékologicheskie Aspekty
Fotosinteza Morskikh Makrovodorosleĭ) : 21 - 27, 1978. [In R, ab : E.]

34387 - **KOMÁRKOVÁ, J., MARVAN, P.** : Primary production and functioning of algae in
the fishpond littoral.- In : DYKYJOVÁ, D., KVĚT, J. (ed.) : Pond Littoral
Ecosystems. (Ecological Studies Vol. 28.) Pp. 321 - 337. Springer-Verlag,
Berlin - Heidelberg - New York  1978.

34388 - **KOMARNYTS'KYĬ, I.K., KOSAKOVS'KA, I.V.** : D-rybulozo-1,5-dyfosfatkarboksyla-
za *Nicotiana tabacum* L. [D-ribulose-1,5-bisphosphate carboxylase of *Nicotiana tabacum* L.] - Ukr. bot. Zh. *35* : 154 - 157, 223, 1978. [In Ukr., ab :
E, R.]

34389 - **KONG, J.L.Y., LOACH, P.A.** : Covalently-linked porphyrin quinone complexes
as RC models. - In : DUTTON, P.L., LEIGH, J.S., SCARPA, A. (ed.) : Frontiers
of Biological Energetics : Electrons to Tissues. Vol. 1. Pp. 73 - 82. Academic Press, New York - San Francisco - London 1978.

34390 - **KONIS, Y., KLEIN, S., OHAD, I.** : The effect of levulinic acid on the light
induced development of Photosystem I and II activities in greening maize
leaves. - Photochem. Photobiol. *27* : 177 - 182, 1978.

34391 - **KONISHI, T., PACKER, L.** : A proton channel in bacteriorhodopsin. - FEBS
Lett. *89* : 333 - 336, 1978.

34392 - **KONISHI, T., PACKER, L.** : The role of tyrosine in the proton pump of bacteriorhodopsin. - FEBS Lett. *92* : 1 - 4, 1978.

34393 - **KONONENKO, A.A., RUBIN, A.B.** : Funktionelle Organisation des Elektronentrans-
portsystems und damit verbundener Photosyntheseprozesse in Bakterien. -
Biol. Rundschau *16* : 213 - 222, 1978.

34394 - **KONOPKA, A., BROCK, T.D.** : Changes in photosynthetic rate and pigment content of blue-green algae in Lake Mendota. - Appl. environ. Microbiol. *35* :
527 - 532, 1978.

34395 - **KONOPKA, A., BROCK, T.D., WALSBY, A.E.** : Buoyancy regulation by planktonic
blue-green algae in Lake Mendota, Wisconsin. - Arch. Hydrobiol. *83* : 524 -
537, 1978. [Ps.]

34396 - **KOOP, H.-U., SCHMID, R., HEUNERT, H.-H., MILTHALER, B.** : Chloroplast migration : a new circadian rhythm in *Acetabularia*. - Protoplasma *97* : 301 - 310,
1978.

34397 - **KORDAN, H.A.** : Effect of oxygen on greening of coleoptiles in light-germi-
nated rice seedlings. - Ann. Bot. *42* : 259 - 261, 1978.

34398 - **KORENSTEIN, R., HESS, B., KUSCHMITZ, D.** : Branching reactions in the photocycle of bacteriorhodopsin.-FEBS Lett. *93* : 266 - 270, 1978.

34399 - **KORNYUSHENKO, G.A., EVDOKIMOVA, I.V., PSURTSEVA, N.V.** : Kharakter fotoindutsirovannykh prevrashcheniĭ ksantofillov v izolirovannykh khloroplastakh
rasteniĭ, vyrosshikh pri raznoĭ intensivnosti sveta. [Characteristics of
photoinduced transformations of xanthophylls in isolated chloroplasts of
plants grown under different irradiance.]-Bot. Zh. *63* : 580 - 585, 1978.
[In R.]

34400 - **KORNYUSHENKO, G.A., EVDOKIMOVA, I.V., SAPOZHNIKOV, D.I.** : Vliyanie $NAD.H_2$
na prevrashcheniya ksantofillov v izolirovannykh khloroplastakh. [The effect
of $NADH_2$ on xanthophyll transformations in isolated chloroplasts.] - Fiziol.
Rast. *25* : 510 - 517, 1978. [In R, ab : E.]

34401 - **KORONA, V.V.** : Khloroplastogenez. Aksiomaticheskiĭ podkhod. [Chloroplasto-genesis. Axiomatical approach.] - In : Mezostruktura i Funktsional'naya Aktivnost' Fotosinteticheskogo Apparata. Pp. 74 - 78. Ural'skiĭ gosudarst-vennyĭ Universitet, Sverdlovsk 1978. [In R.]

34402 - **KORPILAHTI, E., HARI, P.** : A method for approximating the effect of shading on total amount of $CO_2$ fixed by branches of different species during the growing season. - Flora *167* : 257 - 264, 1978.

34403 - **KORZEŃ, A., PERUCKA, I.** : Zawartość witaminy C, sumy α- i β-karotenów oraz suchej masy kilku odmian papryki słodkiej. [Content of vitamin C, total α- and β-carotenes and dry matter in several varieties of sweet pepper.] - Rocz. Nauk roln. A *103* (2) : 19 - 24, 1978. [In Pol., ab : E, R.]

34404 - **KOSTLAN, N.V.** : Fiziologo-biokhimichni osnovy pidvyshchennya efektyvnosti masovogo vyroshchuvannya khlorely. [Physiological and biochemical bases for increasing efficiency of *Chlorella* mass cultivation.] - Ukr. bot. Zh. *35* : 620 - 624, 671, 1978. [Ps; in Ukr., ab : E, R.]

34405 - **KOSYAK, A.V., GOGOTOV, I.N., KULAKOVA, S.M.** : Fotovydelenie vodoroda tsiano-bakteriyami *Anabaena cylindrica*. [Hydrogen photoevolution by the cyanobac-terium *Anabaena cylindrica*.] - Mikrobiologiya *47* : 605 - 610, 1978. [In R, ab : E.]

34406 - **KOVÁČ, J., HENSELOVÁ, M., VARKONDA, Š.** : A rapid method for detection of synergic "*in vitro*" effect of Hill reaction inhibitor mixtures. - Photosyn-thetica *12* : 87 - 88, 1978.

*34407 - **KOZHOVA, O.M., ZAGORENKO, G.F., MAKSIMOV, V.N.** : Velichiny fotosinteza fito-planktona, opredelyayushchie moshchnost' trofogennogo sloya v ozere Khubsu-gul. [Photosynthetic rates of phytoplankton determining the thickness of trophogenic layer in the lake Khubsugul.] - In : Prirodnye Usloviya i Resur-sy Nekotorykh Raĭonov MNR. Pp. 57 - 58. Irkutsk 1977. [In R.]

34408 - **KOZIOL, M.J., COWLING, D.W.** : Growth of ryegrass (*Lolium perenne* L.), expo-sed to $SO_2$. II. Changes in the distribution of photoassimilated $^{14}C$. - J. exp. Bot. *29* : 1431 - 1439, 1978.

34409 - **KOZIOL, M.J., JORDAN, C.F.** : Changes in carbohydrate levels in red kidney bean (*Phaseolus vulgaris* L.) exposed to sulphur dioxide. - J. exp. Bot. *29* : 1037 - 1043, 1978.

34410 - **KPODAR, M.P., PIQUEMAL, M., CALMÈS, J., LATCHÉ, J.-C.** : Relations entre nu-trition azotée et métabolisme photorespiratoire chez une plante à oxalate, *Fagopyrum esculentum* M. - Physiol. vég. *16* : 117 - 130, 1978.

*34411 - **KRÁĽOVIČ, J.** : Reakcia viniča na niektoré pesticídy. [Response of grape-vine on some pesticides.] - Acta Inst. bot. Acad. Sci. slov. Ser. B1 *1975* : 349 - 358, 1975. [Ps; in Slovak, ab : E, R.]

34412 - **KRÁĽOVIČ, J., KRÁLOVÁ, V., KAULOVÁ, J.** : Reakcia cukrovej repy na ošetrova-nie oxychloridom meďnatým a napadnutie cerkosporiózou. [Response of sugar beet to Cu-oxychloride and cercosporosis.] - Acta bot. slov. Acad. Sci. slov. Ser. B 2 : 115 - 131, 1978. [Ps; in Slovak, ab : E, R.]

34413 - **KRÁĽOVIČ, J., PAPÁNEK, D., SEKERA, D.** : Účinok antiperonospórových fungicí-dov na asimilačnú produkciu, pučanie očiek a napadnutie viniča múčnatkou. [Effect of fungicides active against downy mildew of grape-vine on photo-synthetic production, bud opening, and mildew attack.] - Acta bot. slov. Acad. Sci. slov. Ser. B 2 : 133 - 143, 1978. [In Slovak, ab : E, R.]

34414 - **KRAMER, D., FINDENEGG, G.R.** : Variations in the ultrastructure of *Scenedesmus obliquus* during adaptation to low $CO_2$ level. - Z. Pflanzenphysiol. *89* : 407 - 410, 1978. [Chloroplast.]

*34415 - **KRÁMER, M.** : Karotin és A-vitamin meghatározása élelmiszerekben. [Determi-nation of carotene and vitamin A in foods.] - Élelmezési Ipar *25* (6) : 169 - 176, 1971. [In Hung., ab : E, G, R.]

34416 - **KRAMER, P.J.** : The use of controlled environments in research. - HortScience *13* : 447 - 451, 1978. [Ps.]

34417 - **KRASNA, A.I.** : Photoreduction of hydrogen from water in algae and in a coupled system of chloroplasts and hydrogenase. - In : SCHLEGEL, H.G., SCHNEIDER, K. (ed.) : Hydrogenases : Their Catalytic Activity, Structure and Function. Pp. 423 - 437. Erich Goltze KG, Göttingen 1978.

34418 - **KRASNOVSKIĬ, A.A. ml., KOVALEV, Yu.V.** : Fosforestsentsiya khlorofilla v list'yakh i kletkakh vodorosleĭ. [Phosphorescence of chlorophyll in leaves and cells of algae.] - Biofizika *23* : 920 - 922, 1978. [In R, ab : E.]

34419 - **KRASNOVSKY, A.A., BRIN, G.P.** : Photosensitization by titanium dioxide and zinc oxide : oxygen and hydrogen evolution. - In : METZNER, H. (ed.) : Photosynthetic Oxygen Evolution. Pp. 405 - 410. Academic Press, London - New York - San Francisco 1978.

34420 - **KRASNOVSKY, A.A. Jr., LEBEDEV, N.N., LITVIN, F.F.** : Phosphorescence of photosynthetic pigments in model systems and living structures. - In : FIALA, J. (ed.) : Third International Seminar on Energy Transfer in Condensed Matter. Pp. 111 - 119. Univerzita Karlova, Praha 1978.

*34421 - **KRAUSCH, H.-D.** : Die Makrophyten der mittleren Saale und ihre Biomasse. - Limnologica (Berlin) *10* : 57 - 72, 1976.

34422 - **KRAUSE, G.H.** : Effects of uncouplers on $Mg^{2+}$-dependent fluorescence quenching in isolated chloroplasts. - Planta *138* : 73 - 78, 1978.

34423 - **KRAUSE, G.H., KIRK, M., HEBER, U., OSMOND, C.B.** : $O_2$-dependent inhibition of photosynthetic capacity in intact isolated chloroplasts and isolated cells from spinach leaves illuminated in the absence of $CO_2$. - Planta *142* : 229 - 233, 1978.

34424 - **KRAUSE, G.H., LORIMER, G.H., HEBER, U., KIRK, M.R.** : Photorespiratory energy dissipation in leaves and chloroplasts. - In : HALL, D.O., COOMBS, J., GOODWIN, T.W. (ed.) : Proceedings of the Fourth International Congress on Photosynthesis. Pp. 299 - 310. Biochem. Soc., London 1978.

*34425 - **KRAUZE, A.** : Zmiany jakościowe i ilościowe karotenoidów w wybranych odmianach pomidorów oraz w koncentratach pomidorowych. [Qualitative and quantitative changes of carotenoids in some cultivars of tomato and tomato puree.] - Pracy Zakresu Towarozn. Chem., wyzsza Szk. ekon. Poznaniu, Zesz. nauk. Ser. I, *25* : 75 - 100, 1966. [In Pol.]

34426 - **KREĬTSBERG, O.Ė., PAVULINYA, D.A.** : Vzaimosvyaz' izmeneniĭ soderzhaniya zhirnykh kislot polyarnykh lipidov i khlorofilla v period adaptatsii rasteniĭ ogurtsa k povyshennoĭ intensivnosti sveta. [Interrelationships of changes in the level of fatty acids, polar lipids and chlorophyll during the cucumber plant adaptation to high illumination.] - In : Fiziologo-biokhimicheskie Issledovaniya Rasteniĭ. Pp. 38 - 48, 156. Zinatne, Riga 1978. [In R.]

34427 - **KREMER, B.P.** : Aspects of $CO_2$-fixation in some freshwater *Rhodophyceae*. - Phycologia *17* : 430 - 434, 1978.

34428 - **KREMER, B.P.** : Patterns of photoassimilatory products in Pacific *Rhodophyceae*. - Can. J. Bot. *56* : 1655 - 1659, 1978.

34429 - **KREMER, B.P., BERKS, R.** : Photosynthesis and carbon metabolism in marine and freshwater diatoms. - Z. Pflanzenphysiol. *87* : 149 - 165, 1978.

34430 - **KREMER, B.P., FEIGE, G.B., SCHNEIDER, Hj.A.W.** : A new proposal for the systematic position of *Cyanidium caldarium*. - Naturwissenschaften *65* : 157, 1978. [Chl, photosynthates.]

34431 - **KRENDELEVA, T.E., GALYNINA, N.I., KONONENKO, A.A., TIMOFEEV, K.N., KAĬRIS, A.V., RUBIN, A.B.** : Fotoindutsirovannyĭ transport ėlektronov i sopryazhennye protsessy v obogashchennykh fotosistemoĭ I digitoninovykh fragmentakh khloroplastov gorokha. [Light-induced electron transport and coupled processes in pea subchloroplast particles enriched in photosystem I.] - Biokhimiya *43* : 1251 - 1259, 1978. [In R, ab : E.]

34432 - **KRENDELEVA, T.E., KONONENKO, A.A., TCHERNOVA, N.A., RUBIN, A.B.** : Light-induced electron transport and coupled processes in digitonin-fractionated pea subchloroplast particles enriched in photosystem I. I. Experimental study. - Plant Sci. Lett. *11* : 11 - 18, 1978.

34433 - KREUTZ, W. : On the structural arrangement of light reaction centres I and II in the photosynthetic membrane. - In : METZNER, H. (ed.) : Photosynthetic Oxygen Evolution. Pp. 77 - 90. Academic Press, London - New York - San Francisco 1978.

34434 - KRIETSCH, W.K.G., KUNTZ, G.W.K. : Specific enzymatic determination of adenosine triphosphate. - Anal. Biochem. *90* : 829 - 831, 1978.

34435 - KRINSKY, N.I. : Non-photosynthetic functions of carotenoids. - Phil. Trans. roy. Soc. London, Ser. B *284* : 581 - 590, 1978. [Chl.]

34436 - KRISHNAMOORTHI, K.P. : Biological productivity. - In : Inter-Disciplinary Symposium on Photosynthesis and Productivity. Abstract of Papers. Pp. 25 - 28. Indian nat. Sci. Acad., New Delhi 1978.

34437 - KRISHNA MURTHY, K., SINGH, M. : Studies on photosynthesis and translocation in wheat genotypes. - In : Inter-Disciplinary Symposium on Photosynthesis and Productivity. Abstract of Papers. Pp. 50 - 51. Indian nat. Sci. Acad., New Delhi 1978.

*34438 - KRISTKALNE, S.Kh. : Dinamika soderzhaniya khlorofilla v usloviyakh kratkosrochnogo zateneniya. [Dynamics of chlorophyll content under conditions of short-term shading.] - In : Adaptatsiya Fiziologo-Biokhimicheskikh Sistem Rasteniya k Peremene Osveshcheniya. Part 1. Pp. 56 - 67. Zinatne, Riga 1977. [In R.]

*34439 - KRISTKALNE, S.Kh., GUBAR', G.D., VITOLA, A.K., KREĬTSBERG, O.E., SELGA, M.P. : Dinamika sukhoĭ biomassy pri oslablenii intensivnosti osveshcheniya. [Dynamics of dry biomass at the weakening of light. ] - In : Adaptatsiya Fiziologo-Biokhimicheskikh Sistem Rasteniya k Peremene Osveshcheniya. Part 1. Pp. 13 - 31. Zinatne, Riga 1977. [In R.]

34440 - KROCHKO, J.E., BEWLEY, J.D., PACEY, J. : The effects of rapid and very slow speeds of drying on the ultrastructure and metabolism of the desiccation-sensitive moss *Cratoneuron filicinum* (HEDW.) SPRUCE. - J. exp. Bot. *29* : 905 - 917, 1978. [Chloroplast.]

34441 - KRÓL, M. : The relationship between photosynthetic activities and the polypeptide pattern of pine seedling chloroplasts. - Z. Pflanzenphysiol. *86* : 379 - 387, 1978.

34442 - KRÓLIKOWSKA, J. : The transpiration of helophytes. - Ekol. pol. *26* : 193 - 212, 1978. [Growth analysis.]

*34443 - KR"STEV, K.K., DIMOV, A. : Dishane, fotosinteza i plastidni pigmenti pri trite osnovni tipa na infektsiya ot kafyava r"zhda po pshenitsata (*Puccinia recondita* ROB. ex DESM. f.sp. *tritici* ERIKSS.). [Respiration, photosynthesis and plastid pigments in the three basic types of infection with leaf brown rust of wheat (*Puccinia recondita* ROB. ex DESM. f. sp. *tritici* ERIKSS.).] - Rasteniev. Nauki *12* (1) : 120 - 128, 1975. [In Bulg., ab : E, R.]

34444 - KRUCZYNSKI, W.L., SUBRAHMANYAM, C.B., DRAKE, S.H. : Studies on the plant community of a North Florida salt marsh. Part I. Primary production. - Bull. mar. Sci. *28* : 316 - 334, 1978.

34445 - KRUPA, J. : Photosynthesis rate in moss leaves of various anatomical structure. - Acta Soc. Bot. Pol. *47* : 391 - 402, 1978.

34446 - KRUPA, Z., BASZYŃSKI, T. : The effect of exogenous lipids on reconstitution of heptane-extracted spinach chloroplasts. - Bull. Acad. pol. Sci., Sér. Sci. biol. *26* : 885 - 889, 1978.

34447 - KRUPENKO, A.N., GANAGO, I.B., SHUVALOVA, N.P., BELL, L.N. : Rol' fotookislitel'nykh protsessov v énergetike fotosinteziruyushchikh kletok khlorelly. [The role of photooxidative processes in the energetics of photosynthesizing *Chlorella* cells.] - Fiziol. Rast. *25* : 471 - 476, 1978. [In R, ab : E.]

34448 - KRYŃSKA, W., MARTYNIAK, B. : Wartość odżywcza kapusty wczesnej i pomidorów uprawianych na terenie falistym. [Feeding values of early cabbage and tomatoes cultivated in a hilly area.] - Roczniki Nauk roln. Ser. A *103* (4): 79 - 92, 1978. [Car; In Pol., ab : E, R.]

34449 - KRYUKOV, P.G., LAZAREV, Yu.A., LETOKHOV, V.S., MATVEETS, Yu.A., TERPUGOV, E.L., CHEKULAEVA, L.N., SHARKOV, A.V. : Pikosekundnyĭ flesh-fotoliz bakteriorodopsina iz *Halobacterium halobium* pri komnatnoĭ i nizkikh temperaturakh. [Picosecond flash-photolysis of bacteriorhodopsin from *Halobacterium halobium* at room- and low temperatures.] - Biofizika *23* : 171 - 173, 1978. [In R, ab : E.]

34450 - KU, H.S., FINDAK, D. : The mechanism of chilling injury in cotton and cucumber. - Plant Physiol. *61* (Suppl.) : 55, 1978. [Ps.]

34451 - KU, S.-B., EDWARDS, G.E. : Oxygen inhibition of photosynthesis. III. Temperature dependence of quantum yield and its relation to $O_2/CO_2$ solubility ratio. - Planta *140* : 1 - 6, 1978.

34452 - KU, S.B., EDWARDS, G.E. : Photosynthetic efficiency of *Panicum hians* and *Panicum milioides* in relation to $C_3$ and $C_4$ plants. - Plant Cell Physiol. *19* : 665 - 675, 1978.

34453 - KU, S.B., EDWARDS, G.E., SMITH, D. : Photosynthesis and nonstructural carbohydrate concentration in leaf blades of *Panicum virgatum* as affected by night temperature. - Can. J. Bot. *56* : 63 - 68, 1978.

*34454 - KUDREV, T.G., MANOLOVA, N.I., BUCHVAROV, P.Z. : Influence of magnet-treated nutrient solution on the growth of young maize plants. - Dokl. bolg. Akad. Nauk *30* : 1617 - 1620, 1977. [Ps.]

34455 - KUEHN, G.D., HSU, T.-C. : Preparative-scale enzymic synthesis of D-[$^{14}$C]ribulose 1,5-bisphosphate. - Biochem. J. *175* : 909 - 912, 1978.

34456 - KUEN, M., VAKLINOVA, S. : Intenzivnost na fotosintezata i aktivnostta na FEP i RDF karboksilazata pri ligulna i bezligulna tsarevitsa. [Photosynthetic rate and activities of PEP and RuBP carboxylases in ligula-possessing and ligula-less maize.] - Fiziol. Rast. (Sofia) *4* (1) : 19 - 23, 1978. [In Bulg., ab : E, R.]

34457 - KUFNER, R., CZYGAN, F.-C., SCHNEIDER, L. : Veränderungen des Pigmentgehalts und der Ultrastruktur bei den Plastiden der Nadelblätter von *Taxus baccata* (L.) während ihrer Entwicklung. - Ber. deut. bot. Ges. *91* : 325 - 337, 1978.

34458 - KÜHBAUCH, W. : Die Nichtstrukturkohlenhydrate in Gräsern des gemäßigten Klimabereiches, ihre Variationsmöglichkeiten und mikrobielle Verwertung. - Landwirtsch. Forschung *31* : 251 - 268, 1978. [Photosynthates.]

34459 - KUJIRA, Y., KANDA, M. : Competition among individual plants in crop population. IV. Growth analysis from the viewpoint of root behavior. - Jap. J. Crop Sci. *47* : 221 - 227, 1978.

34460 - KUKHTIN, V.V., PETROV, É.G., KHARKYANEN, V.N., KHRISTOFOROV, L.N. : Razdelenie zaryadov v fotosinteticheskikh reaktsionnykh tsentrakh bakteriĭ. [The charge separation in bacterial photosynthetic reaction centres.] - Mol. Biol. (Moskva) *12* : 1246 - 1255, 1978. [In R, ab : E.]

34461 - KUKUSHKIN, A.K., TIKHONOV, A.N. : Teoreticheskoe issledovanie vliyaniya nekotorykh obratnykh reaktsiĭ na kinetiku pervichnykh protsessov fotosinteza. [Theoretical investigation of the influence of back reactions on the kinetics of primary processes of photosynthesis.] - Biofizika *23* : 620 - 623, 1978. [In R, ab : E.]

34462 - KULANDAIVELU, G. : Assessing the photosynthetic efficiency *in vivo* by the fluorescence technique. - In : Inter-Disciplinary Symposium on Photosynthesis and Productivity. Abstract of Papers. Pp. 11 - 12. Indian nat. Sci. Acad., New Delhi 1978.

*34463 - KUL'TEBAEV, É. T. : Soderzhanie khlorofilla i intensivnost' fotosinteza v list'yakh razlichnykh sortov yabloni v Zailiĭskom Alatau. [Chlorophyll level and photosynthetic rate in leaves of different cultivars of apple tree in the Trans-Ili Ala-Tau.] - Tr. kazan. gos. sel'skokhoz. Inst. *13* (3) : 86 - 90, 1970. [In R.]

*34464 - KUL'TEBAEV, É.T. : Vliyanie semennykh podvoev na nakoplenie khlorofillov
         *a* i *b* v list'yakh Aporta Aleksandra. [Effect of stocks on the accumulation
         of chlorophylls *a* and *b* in leaves of "Aport Aleksandra".] - Izv. Akad. Nauk
         kaz. SSR, Ser. biol. *13* (4) : 21 - 23, 1975. [In R, ab : Kaz.]

34465 - KUMAKHOVA, T.A. : Vzaimosvyaz' vodnogo rezhima i komponentov pigmentnoĭ sis-
         temy v list'yakh ozimoĭ pshenitsy v zavisimosti ot usloviĭ pitaniya i vlazh-
         nosti pochvy. [Relationship between water regime and components of pigment
         system in winter wheat leaves as related to nutrition and soil moisture.] -
         In : Vodnyĭ Rezhim Rasteniĭ v Svyazi s Raznymi Ékologicheskimi Usloviyami.
         Pp. 35 - 40. Izdatel'stvo Kazanskogo Universiteta, Kazan' 1978. [In R.]

34466 - KUMAZAWA, S., MITSUI, A. : Hydrogenase and nitrogenase participation in the
         photoproduction of hydrogen by marine blue-green algae. - Plant Physiol.
         *61* (Suppl.) : 77, 1978.

34467 - KUMURA, A., KOH, S., FUKAI, S., KAGANO, J. : Diurnal variation in $CO_2$ ex-
         change in stands of various field-crops. - In : MONSI, M., SAEKI, T. (ed.) :
         Ecophysiology of Photosynthetic Productivity. JIBP Synthesis Vol. 19. Pp.
         82 - 90. University of Tokyo Press, Tokyo 1978.

34468 - KUNERT, K.J., BÖGER, P. : Action of EMD-IT 5914 on chloroplasts. - Weed Sci.
         *26* : 292 - 296, 1978. [Ps, Chl, Car.]

34469 - KUNG, M.C., DEVAULT, D. : High-order fluorescence and exciton interaction
         in photosynthetic bacteria. - Biochim. biophys. Acta *501* : 217 - 231, 1978.

*34470 - KUNG, S.D. : Isoelectric points of the polypeptide components of tobacco
         Fraction 1 protein. - Bot. Bull. Acad. sin. *17* : 185 - 191, 1976.

34471 - KÜNSTLE, E., MITSCHERLICH, G. : Photosynthese, Transpiration und Respira-
         tion in einem jungen Mischbestand. - Angew. Bot. *52* : 233 - 252, 1978.

34472 - KUO, J. : Morphology, anatomy and histochemistry of the Australian seagras-
         ses of the genus *Posidonia* KÖNIG (*Posidoniaceae*). I. Leaf blade and leaf
         sheath of *Posidonia australis* HOOK. F. - Aquat. Bot. *5* : 171 - 190, 1978.
         [Chloroplast.]

34473 - KÜPPERS, U., KREMER, B.P. : Longitudinal profiles of carbon dioxide fixa-
         tion capacities in marine macroalgae. - Plant Physiol. *62* : 49 - 53, 1978.

*34474 - KURATA, A., SARACENI, C., RUGGIU, D., NAKANISHI, M., MELCHIORRI-SANTOLINI,
         U., KADOTA, H. : Relationship between B group vitamins and primary produc-
         tion and phytoplankton population in Lake Mergozzo (Northern Haly). -
         Mem. Ist. ital. Idrobiol. "Dott. Marco de Marchi" *33* : 257 - 284, 1976.

34475 - KURGANOVA, L.N., ANISIMOV, A.A. : Sopryazhennoe deĭstvie γ-radiatsii i
         azotnogo pitaniya na fotosintez. [Combined effect of γ-radiation and ni-
         trogen nutrition on photosynthesis.] - Fiziol. Biokhim. kul't. Rast. *10* :
         507 - 510, 1978. [In R.]

*34476 - KURITA, S., TOYADA, K., ENDO, T., MOCHIZUKI, N., HONYA, M., ONAMI, T. :
         Hydrogen photoproduction from water. - In : MITSUI, A., MIYACHI, S.,
         SAN PIETRO, A., TAMURA, S. (ed.) : Biological Solar Energy Conversion.
         Pp. 87 - 100. Academic Press, New York - San Francisco - London 1977.

34477 - KUROSAKI, T., IZUMI, K. : [Studies on the histochemical of carrot.- I.
         Seasonal fluctuation in the distribution of carotinoid and starch during
         the growth of carrot.] - Hiroshima Daigaku Gakko Kyoikugakubu Kiyo, Dai-
         -2-bu. [Bull. Fac. School Educ., Hiroshima Univ. Part 2] *1* (2) : 187 - 197,
         1978. [In Jap., ab : E.]

34478 - KURSAR, T.A., WOOD, N.B., ALBERTE, R.S. : The composition of the photo-
         synthetic unit in the red alga *Gracilaria* sp. and its mutants. - Plant Phy-
         siol. *61* (Suppl.) : 87, 1978.

34479 - KUSANAGI, Y., ICHII, M., FUKUDA, K. : Design of experiments for parameter
         estimation in a growth curve model based on the internode length of wheat. -
         Tech. Bull. Fac. Agr., Kagawa Univ. *30* : 61 - 73, 1978.

34480 - **KUSHNIRENKO, M.D.** : Reaktsia khloroplastov rasteniĭ razlichnoĭ ustoĭchivosti k zasukhe na vodnyĭ stress. [Response of chloroplasts in plants with
different drought resistance to water stress.] - In : Problemy Zasukhoustoĭchivosti Rasteniĭ. Pp. 165 - 182. Nauka, Moskva 1978. [In R.]

*34481 - **KUSHNIRENKO, M.D., KRYUKOVA, E.V., PECHERSKAYA, S.N., KANASH, E.V.** : Vliyanie vodnogo stressa na sostoyanie khloroplastov rasteniĭ razlichnykh èko-
logicheskikh grupp. [The effect of water stress on the chloroplasts state in
plants of different ecological types.] - Izv. Akad. Nauk mold. SSR, Ser.
biol. khim. Nauk *1977* (3) : 17 - 24, 1977. [In R.]

*34482 - **KUSHNIRENKO, M.D., KURCHATOVA, G.P., KRYUKOVA, E.V., MEDVEDEVA, T.N.** :
Zavisimost' sostoyaniya vody, pigmentnoĭ sistemy i èlektricheskogo sopro-
tivleniya tkaneĭ ot stepeni ustoĭchivosti rasteniĭ k zavyadaniyu. [Depen-
dence of state of water, pigment system and electric resistance of tissues
on the degree of plant resistance to wilting.] - Izv. Akad. Nauk mold. SSR,
Ser. biol. khim. Nauk *1970* (2) : 43 - 47, 95, 1970. [In R.]

*34483 - **KUSHNIRENKO, M.D., MEDVEDEVA, T.N., KRYUKOVA, E.V., SEMENCHENKO, P.P.** :
Izmenenie pigmentnoĭ sistemy list'ev rasteniĭ v zavisimosti ot ikh vodnogo
rezhima. [Changes in the pigment system of plant leaves as dependent on
their water relations.] - Izv. Akad. Nauk mold. SSR *1967* (9) : 69 - 80,
1967. [In R.]

34484 - **KUSK, K.O.** : Effects of crude oil and aromatic hydrocarbons on the photosynthesis of the diatom *Nitzschia palea*. - Physiol. Plant. *43* : 1 - 6, 1978.

34485 - **KUSUMOTO, T.** : Photosynthesis and respiration in leaves of main component
species. - In : **KIRA, T., ONO, Y., HOSOKAWA, T.** (ed.) : Biological Production in a Warm-Temperate Evergreen Oak Forest in Japan. JIBP Synthesis Vol.
18. Pp. 88 - 98. University of Tokyo Press, Tokyo 1978.

34486 - **KVĚT, J.** : Growth analysis of fishpond littoral communities. - In : **DYKYJO-
VÁ, D., KVĚT, J.** (ed.) : Pond Littoral Ecosystems. (Ecological Studies.
Vol. 28.) Pp. 198-206. Springer-Verlag, Berlin - Heidelberg - New York 1978.

34487 - **KVĚT, J., HUSÁK, S.** : Primary data on biomass and production estimates in
typical stands of fishpond littoral plant communities. - In : **DYKYJOVÁ, D.,
KVĚT, J.** (ed.) : Pond Littoral Ecosystems. (Ecological Studies. Vol. 28.)
Pp. 211 - 216. Springer-Verlag, Berlin - Heidelberg - New York 1978.

34488 - **KWANYUEN, P., WILDMAN, S.G.** : Evidence that genetic information for chloroplast coupling factor I is shared by nuclear and chloroplast DNA. - Biochim. biophys. Acta *502* : 269 - 275, 1978.


*34489 - **LaBAUGH, J.** : Phytoplankton primary production, chlorophyll *a* concentrations, and assimilation numbers in Spruce Knob Lake, West Virginia. - Proc.
West Virginia Acad. Sci. *47* : 190 - 197, 1975.

34490 - **LADYGIN, V.G., KOSTIKOV, A.P.** : Spektral'nye svoĭstva khlorofilla i fotokhimicheskaya aktivnost' fotosistem pigmentnych mutantov *Chlamydomonas*.
[Spectral characteristics of chlorophyll and photochemical activity of photosystems of *Chlamydomonas* pigment mutants.]- Fiziol. Rast. *25* : 500 - 509,
1978. [In R, ab : E.]

34491 - **LADYGIN, V.G., SEMENOVA, G.A., TAGEEVA, S.V.** : Variabel'nost' v nakoplenii
khlorofilla i struktura khloroplastov u dochernikh shtammov fenotipicheski
zheltogo mutanta Zh-4 *Chlamydomonas reinhardii*. [Variability of chlorophyll
accumulation and structure of chloroplasts in daughter strains of phenoty-
pically yellow mutant Y-4 of *Chlamydomonas reinhardii*.] - Tsitologiya *20* :
998 - 1004, 1978. [In R, ab : E.]

34492 - **LAGOUTTE, B., DURANTON, J.** : A new enzymatic activity associated with thy-
lakoid membranes. - In : **AKOYUNOGLOU, G., ARGYROUDI-AKOYUNOGLOU, J.H.** (ed.)
: Chloroplast Development. Pp. 229 - 234. Elsevier/North-Holland Biomedical
Press, Amsterdam - New York - Oxford 1978.

34493 - **LAKHANOV, A.P.** : Ustoĭchivost' pigmentnogo kompleksa fasoli k nizkim polo-
zhitel'nym temperaturam v protsesse ontogeneza rasteniĭ. [Resistance of
the bean pigment complex to low positive temperatures in the process of
plant ontogenesis.] - Fiziol.Rast. 25 : 12 - 17, 1978. [In R, ab : E.]

34494 - **LAKSO, A.N.** : Seasonal changes in stomatal response to leaf water potential
in apple. - HortScience 13 : 376, 1978. [Ps, stomatal resistance.]

34495 - **LAKSO, A.N., BARNES, J.E.** : Apple leaf photosynthesis in alternating light. -
HortScience 13 : 473 - 474, 1978.

34496 - **LAKSO, A.N., KLIEWER, W.M.** : The influence of temperature on malic acid metabolism in grape berries. II. Temperature responses of net dark $CO_2$ fixation and malic acid pools. - Amer. J. Enol. Viticult. 29 : 145 - 149, 1978.

34497 - **LAKSO, A.N., SEELEY, E.J.** : Environmentally induced responses of apple tree
photosynthesis. - HortScience 13 : 646 - 650, 1978.

34498 - **LAL, B., AMBASHT, R.S.** : Growth of *Chrozophora rottleri* A. JUSS in relation
to different watering levels. - Indian J. Ecol. 5 : 172 - 180, 1978. [Growth
analysis.]

34499 - **LAL, P., REDDY, G.G., MODI, M.S.** : Accumulation and redistribution pattern
of dry matter and N in triticale and wheat varieties under water stress
condition. - Agron. J. 70 : 623 - 626, 1978.

34500 - **LAL, R., MAURYA, P.R., OSEI-YEBOAH, S.** : Effects of no-tillage and ploughing
on efficiency of water use in maize and cowpea. - Exp. Agr. 14 : 113 - 120,
1978. [Growth analysis.]

*34501 - **LALLYETT, C.I.K.** : Ultrastructural changes in trifoliate leaves of bean
(*Phaseolus vulgaris*) following inoculation of monofoliate leaves with *Pseudomonas phaseolicola*. - Physiol. Plant Pathol. 10 : 229 - 236, 1977.[ Chl.]

34502 - **LAL THAKORE, B.B., CHAKRAVARTI, B.P., PRASAD, B.** : Biochemical changes in
maize leaves after infection by *Physoderma maydis*. - Biovigyanam 4 : 7 - 11,
1978. [Chl.]

34503 - **LAMBERT, R.J., JOHNSON, R.R.** : Leaf angle, tassel morphology, and the performance of maize hybrids. - Crop Sci. 18 : 499 - 502, 1978. [Canopy architecture.]

34504 - **LAMOREAUX, R.J., CHANEY, W.R.** : The effect of cadmium on net photosynthesis,
transpiration, and dark respiration of excised silver maple leaves. - Physiol. Plant. 43 : 231 - 236, 1978.

34505 - **LAMOREAUX, R.J., CHANEY, W.R.** : Photosynthesis and transpiration of excised
silver maple leaves exposed to cadmium and sulphur dioxide. - Environ. Pollut. 17 : 259 - 267, 1978.

34506 - **LAMOREAUX, R.J., CHANEY, W.R., BROWN, K.M.** : The plastochron index : A review after two decades of use. - Amer. J. Bot. 65 : 586 - 593, 1978.

34507 - **LAMOREAUX, R.J., STRICKLAND, R.C., CHANEY, W.R.** : The plastochron index
as applied to a cadmium toxicity study. - Environ. Pollut. 16 : 311 - 317,
1978. [Primary production.]

34508 - **LANE, H.C., THOMPSON, A.C.** : Morphological and physiological differences
between the cotyledons of intact and debudded cotton plants. - Bot. Gaz.
139 : 207 - 210, 1978. [Chl.]

34509 - **LANE, M.D., MIZIORKO, H.M.** : Mechanism of action of ribulose bisphosphate
carboxylase/oxygenase. - In : SIEGELMAN, H.W., HIND, G. (ed.) : Photosynthetic Carbon Assimilation. Pp. 19 - 40. Plenum Press, New York - London
1978.

34510 - **LANG, A.** : Interactions between source, path and sink in determining phloem translocation rate. - Aust. J. Plant Physiol. 5 : 665 - 674, 1978. [Pho-
tosynthates.]

34511 - **LANGE, O.L.** : Ökologische Untersuchungen - Grundlagen für den Kulturpflan-
zenanbau in Wüstengebieten. - Jahresber. Bayer. Julius-Maximilians-Univ.
Würzburgsakad. Jahr. 1976/77 : 37 - 53, 1978. [Ps.]

34512 - LANGE, O.L., SCHULZE, E.-D., EVENARI, M., KAPPEN, L., BUSCHBOM, U. : The temperature-related photosynthetic capacity of plants under desert conditions. III. Ecological significance of the seasonal changes of the photosynthetic response to temperature. - Oecologia *34* : 89 - 100, 1978.

34513 - LANGENBERG, W.G. : Relative speed of fixation of glutaraldehyde and osmic acid in plant cells measured by grana appearance in chloroplasts. - Protoplasma *94* : 167 - 173, 1978.

34514 - LÄNNERGREN, C. : Net- and nanoplankton : effects of an oil spill in the North Sea. - Bot. mar. *21* : 353 - 356, 1978. [Chl.]

34515 - LÄNNERGREN, C. : Phytoplankton production at two stations in Lindåspollene, a Norwegian Land-locked Fjord, and limiting nutrients studied by two kinds of bio-assays. - Int. Rev. ges. Hydrobiol. *63* : 57 - 76, 1978. [Chl.]

34516 - LANYI, J.K. : Light energy conversion in *Halobacterium halobium*. - Microbiol. Rev. *42* : 682 - 706, 1978.

34517 - LAROW, E.J., McNAUGHT, D.C. : Systems and organismal aspects of phosphorus remineralization. - Hydrobiologia *59* : 151 - 154, 1978. [Ps.]

34518 - LARSON, D.W. : Patterns of lichen photosynthesis and respiration following prolonged frozen storage. - Can. J. Bot. *56* : 2119 - 2123, 1978.

34519 - LARSON, D.W. : Possible misestimates of lake primary productivity due to vertical migrations by dinoflagellates. - Arch. Hydrobiol. *81* : 296 - 303, 1978.

34520 - LARSSON, C.-M., TILLBERG, J.-E. : Effects of phosphate readdition on ATP levels and $O_2$ exchange in phosphorus-starved *Scenedesmus*. - Z. Pflanzenphysiol. *90* : 21 - 31, 1978.

34521 - LARSSON, C.-M., TILLBERG, J.-E., HALLMÉN, G. : Light-induced phosphate binding in relation to photophosphorylation and levels of ATP, ADP and AMP in the green alga *Scenedesmus*. - Physiol. Plant. *44* : 115 - 121, 1978.

34522 - LASCELLES, J. : Regulation of pyrrole synthesis. - In : CLAYTON, R.K., SISTROM, W.R.(ed.) : The Photosynthetic Bacteria. Pp. 795 - 808. Plenum Press, New York - London 1978. [Chl.]

34523 - LASLEY, S.E., GARBER, M.P. : Photosynthetic contribution of cotyledons to early development of cucumber. - HortScience *13* : 191 - 193, 1978.

34524 - LATCHÉ, J.C., VIALA, G., CALMÉS, J., CAVALIÉ, G. : Étude comparative du métabolisme photorespiratoire chez différentes varietés de Soja. - Ann. Amélior. Plant. *28* : 77 - 87, 1978.

34525 - LATZKO, E., KELLY, G.J. : Control of carbon metabolism through enzyme regulation and membrane-mediated metabolite transport. - Progr. Bot. *40* : 99 - 118, 1978.

34526 - LÄUFER, A., INOUE, Y., SHIBATA, K. : Enhanced charging of thermoluminescence A-band by a combination of flash and continuous light excitation. - In : AKOYUNOGLOU, G., ARGYROUDI-AKOYUNOGLOU, J.H. (ed.) : Chloroplast Development. Pp. 379 - 387. Elsevier/North-Holland Biomedical Press, Amsterdam - New York - Oxford 1978.

34527 - LAVERYCHEVA, I.G. : K izucheniyu funktsii fermenta v reaktsii Khilla po ee temperaturnoĭ zavisimosti. [Studying the enzyme function in Hill reaction with its temperature dependence.] - Vestn. leningrad. Univ. *1978* (9) : 100 - 111, 166, 1978. [In R, ab : E.]

34528 - LAVIN, P.I., TITLYANOV, É. A., KALUGINA, V.M. : Sezonnaya dinamika fotosinteza, temnovogo dykhaniya i soderzhaniya agara u morskoĭ vodorosli *Ahnfeltia tobuchiensis*. [Seasonal changes of photosynthesis, dark respiration and agar content in the marine alga *Ahnfeltia tobuchiensis*.] - Sb. Rabot Akad. Nauk SSSR, dal'nevost. nauch. Tsentr, Inst. Biol. Morya (Vladivostok) *11* (Ékologicheskie Aspekty Fotosinteza Morskikh Makrovodorosleĭ) : 150 - 157, 186 - 187, 1978. [In R, ab : E.]

34529 - **LAVINTMAN, N., GALLING, G., OHAD, I.** : Repair of photosynthetic activity in thylakoids formed at the non-permissive temperature in a *ts* mutant of *Chlorella*. - In : **AKOYUNOGLOU, G., ARGYROUDI-AKOYUNOGLOU, J.H.** (ed.) : Chloroplast Development. Pp. 875 - 884. Elsevier/North-Holland Biomedical Press, Amsterdam - New York - Oxford 1978.

34530 - **LAVOREL, J.** : Photosynthesis and molecular organization. - CNRS Res. *8* : 17 - 25, 1978.

34531 - **LAVOREL, J.** : On the origin of damping of the oxygen yield in sequences of flashes. - In : **METZNER, H.** (ed.) : Photosynthetic Oxygen Evolution. Pp. 249 - 268. Academic Press, London - New York - San Francisco 1978.

34532 - **LAW, K.N., LO, S.N., KORAN, Z.** : Utilization of spruce foliage. Extraction of protein and chlorophyll-carotene. - Wood Sci. *11* (2) : 91 - 96, 1978.

34533 - **LAWANSON, A.O., FANIMOKUN, V.O., ADELUSI, S.A.** : Heat-induced quantitative changes in carotenoids and protochlorophyll precursors in *Zea mays*. - Z. Pflanzenphysiol. *86* : 423 - 431, 1978.

34534 - **LAWLIS, V.B., McFADDEN, B.A.** : Modification of ribulose bisphosphate carboxylase by 2,3-butadione. - Biochem. biophys. Res. Commun. *80* : 580 - 585, 1978.

34535 - **LAWLOR, D.W., FOCK, H.** : Photosynthesis, respiration, and carbon assimilation in water-stressed maize at two oxygen concentrations. - J. exp. Bot. *29* : 579 - 593, 1978.

34536 - **LAWRENCE, B.A., LEWIS, M.C., MILLER, P.C.** : A simulation model of population processes of arctic tundra graminoids. - In : **TIESZEN, L.L.** (ed.) : Vegetation and Production Ecology of an Alaskan Arctic Tundra. Pp. 599 - 619. Springer-Verlag, New York - Heidelberg - Berlin 1978.

34537 - **LAWRENCE, J.R., HAYNES, R.C., HAMMER, U.T.** : Contribution of photosynthetic green sulphur bacteria to total primary production in a meromictic saline lake. - Verh. int. Verein. Limnol. *20* : 201 - 207, 1978.

34538 - **LAWS, E.A., WONG, D.C.L.** : Studies of carbon and nitrogen metabolism by three marine phytoplankton species in nitrate-limited continuous culture. - J. Phycol. *14* : 406 - 416, 1978.

34539 - **LAWYER, A.L., ZELITCH, I.** : Inhibition of glutamate : Glyoxalate aminotransferase activity in tobacco leaves and callus by glycidate, an inhibitor of photorespiration. - Plant Physiol. *61* : 242 - 247, 1978.

34540 - **LAZǍR-KEUL, G., CRǍCIUN, C., SORAN, V.** : Wheat chloroplast development under lindane treatment. - Biochem. Physiol. Pflanzen *172* : 191 - 194, 1978.

34541 - **LEACH, G.J.** : The ecology of lucerne pastures. - In : **WILSON, J.R.** (ed.) : Plant Relations in Pastures. Pp. 290 - 308. CSIRO, Melbourne 1978. [Stomatal resistance.]

34542 - **LEAMER, R.W., NORIEGA, J.R., WIEGAND, C.L.** : Seasonal changes in reflectance of two wheat cultivars. - Agron. J. *70* : 113 - 118, 1978.

34543 - **LEBEDEV, N.N., KRASNOVSKIĬ, A.A.** ml. : O svyazi mezhdu vremenem zhizni i énergieĭ tripletnogo sostoyaniya khlorofillov. Otsenka énergii tripletnogo sostoyaniya bakteriokhlorofilla i bakteriofeofitina *a*. [Relation between the lifetime and energy of chlorophyll triplet state. Estimation of the energy of bacteriochlorophyll *a* and bacteriopheophytin *a* triplet states.] - Biofizika *23* : 1095 - 1096, 1978. [In R, ab : E.]

34544 - **LEBEDEV, S.I., SAVCHENKO, N.P., KRIVORUCHKO, L.G.** : Aktivnost' lipoksigenazy i soderzhanie karotinoidov v otdel'nykh organakh rasteniĭ. [The activity of lipoxygenase and the content of carotenoids in plant organs.] - Fiziol. Rast. *25* : 597 - 600, 1978. [In R.]

34545 - **LEBEDEVA, E.V., VIL'YAMS, M.V., TSVETKOVA, I.V.** : Vliyanie dlitel'nosti oblucheniya na rost i produktivnost' rasteniĭ stolovoĭ svekly. [The effect of the duration of illumination on the growth and productivity of *Beta vulgaris* L.] - Fiziol. Rast. *25* : 191 - 195, 1978. [In R, ab : E.]

34546 - **LEBEL, D., POIRIER, G.G., BEAUDOIN, A.R.** : A convenient method for the
ATPase assay. - Anal. Biochem. *85* : 86 - 89, 1978.

34547 - **LECHOWICZ, M.J.** : Carbon dioxide exchange in *Cladina* lichens from subarctic
and temperate habitats. - Oecologia *32* : 225 - 237, 1978.

34548 - **LECHOWSKI, Z.** : Fotosynteza typu pośredniego $C_3$ - $C_4$. [Photosynthesis
of $C_3$ - $C_4$ intermediate type.] - Wiadom. bot. *22* : 89 - 97, 1978. [In Pol.]

34549 - **LECHOWSKI, Z.** : Fotosynteza typu pośredniego $C_3$ - $C_4$. Aspekty biochemiczne
i fizjologiczne. [Photosynthesis of the intermediate type $C_3$ - $C_4$. Biochemical and physiological aspects.] - Wiadom. bot. *22* : 183 - 198, 1978. [In
Pol.]

34550 - **LEDENT, J.F.** : Beam light interception by leaves with undulating edges -
a simulation of maize leaf section. - Agr. Meteorol. *19* : 399 - 410, 1978.

34551 - **LEDENT, J.F.** : Changes in the angle and curvature of the uppermost leaves
of winter wheat. - J. agr. Sci. *90* : 319 - 323, 1978.

34552 - **LEDENT, J.F.** : Mechanisms determining leaf movement and leaf angle in wheat
(*Triticum aestivum* L.). - Ann. Bot. *42* : 345 - 351, 1978.

34553 - **LeDREW, E.F., WELLER, G.** : A comparison of the radiation and energy balance
during the growing season for an arctic and alpine tundra. - Arctic alpine
Res. *10* : 665 - 678, 1978. [Canopy.]

34554 - **LEE, J.W.** : Influence of nitrogen source on nitrogen metabolism in detached wheat heads. - Aust. J. Plant Physiol. *5* : 779 - 785, 1978. [Ps.]

34555 - **LEE, K., NALEWAJKO, C.** : Photosynthesis, extracellular release and glycollic
acid uptake by plankton : Fractionation studies. - Verh. int. Verein.Lim-
nol. *20* : 257 - 262, 1978.

34556 - **LEEGOOD, R.C., AP REES, T.** : Dark fixation of $CO_2$ during gluconeogenesis
by the cotyledons of *Cucurbita pepo* L. - Planta *140* : 275 - 282, 1978.

34557 - **LEHMAN, J.L., VASCONCELOS, A.C.** : Isolation of intact chloroplasts from
the marine diatom *Cylindrotheca closterium*. - Plant Physiol. *61* (Suppl.) :
114, 1978.

34558 - **LEHMAN, J.T.** : Enhanced transport of inorganic carbon into algal cells and
its implications for the biological fixation of carbon. - J. Phycol. *14* :
33 - 42, 1978.

34559 - **LEHMANN, M., WÖBER, G.** : Enzymes of glycogen mobilization in the photosynthetic procaryote, *Anacystis nidulans*. - Planta *143* : 63 - 65, 1978. [Photo-
synthates.]

34560 - **LEHMANN, M., WÖBER, G.** : Glykogen-Stoffwechsel bei *Anacystis nidulans*. *In
vivo*- und *in vitro*-Versuche zur Regulation durch Licht. - Ber. deut. bot.
Ges. *91* : 527 - 538, 1978.

34561 - **LEHMANN, M., WÖBER, G.** : Light dark modulation of glycogen phosphorylase
activity in the blue-green alga, *Anacystis nidulans*. - Plant Cell Environ.
*1* : 155 - 160, 1978. [Ps.]

34562 - **LEHNBERG, W.** : Die Wirkung eines Licht-Temperatur-Salzgehalt Komplexes auf
den Gaswechsel von *Delesseria sanguinea (Rhodophyta)* aus der westlichen
Ostsee. - Bot. mar. *21* : 485 - 497, 1978.

34563 - **LEHNER, K., HELDT, H.W.** : Dicarboxylate transport across the inner membrane
of the chloroplast envelope. - Biochim. biophys. Acta *501* : 531 - 544, 1978.

34564 - **LEICKNAM, J.-P., HENRY, M., KLÉO, J.** : Autoassociation de la pyrochlorophylle
*a* en solution. I.- Structure des complexes, comparaison aves la chlorophylle
*a*. - J. Chim. phys. *75* : 529 - 534, 1978.

34565 - **LEICKNAM, J.-P., HENRY, M., KLÉO, J.** : Autoassociation de la pyrochlorophylle *a* en solution. II.- Grandeurs thermodynamiques. - J. Chim. phys. *75* : 535
- 543, 1978.

34566 - LEIGH, J.S., Jr.: EPR Studies of primary events in bacterial photosynthesis.
- In : CLAYTON, R.K., SISTROM, W.R. (ed.) : The Photosynthetic Bacteria.
Pp. 431 - 438. Plenum Press, New York - London 1978.

34567 - LEKAN, J.F., WILSON, R.E. : Spatial variability of phytoplankton biomass
in the surface waters of Long Island. - Estuar. coast. mar. Sci. 6 : 239 -
251, 1978.

34568 - LELĔTKIN, V.A. : Ustanovka dlya izmereniya spektral'nykh i kineticheskikh
kharakteristik vydeleniya fotosinteticheskogo kisloroda morskimi vodoros-
lyami. [Apparatus for spectral and kinetic measurements of photosynthetic
oxygen evolution in seaweeds.] - Sb. Rabot Akad. Nauk SSSR, dal'nevost.
nauch. Tsentr, Inst. Biol. Morya (Vladivostok) 11 (Ėkologicheskie Aspekty
Fotosinteza Morskikh Makrovodorosleĭ) : 55 - 63, 1978. [In R, ab : E.]

34569 - LELĔTKIN, V.A. : Izuchenie ėffekta Emersona u simbioticheskikh zooksantell.
[Study of Emerson enhancement effect in symbiotic Zooxanthella.]- Sb. Rabot
Akad. Nauk SSSR, dal'nevost. nauch. Tsentr, Inst. Biol. Morya (Vladivostok)
12 [Biologiya Korallovykh Rifov (Fotosintez Zooksantell i Vodorosleĭ-Makrofitov)] : 65 - 74, 1978. [In R, ab : E.]

34570 - LELĔTKIN, V.A., ZVALINSKIĬ, V.I. : Kinetika vydeleniya kisloroda pri mono-
khromaticheskom vozbuzhdenii vodorosli Ulva fenestrata. [Kinetics of oxygen
evolution in green alga Ulva fenestrata subjected to monochromatic light
action.] - Sb. Rabot Akad. Nauk SSSR, dal'nevost. nauch. Tsentr, Inst. Biol.
Morya (Vladivostok) 11 (Ėkologicheskie Aspekty Fotosinteza Morskikh Makrovodorosleĭ) : 121 - 135, 186, 1978. [In R, ab : E.]

34571 - LEMOALLE, J. : Une solution graphique d'integration de la production primaire sur la profondeur et dans temps. - Cah. O.R.S.T.O.M., sér. Hydrobiol.
12 : 181 - 185, 1978.

34572 - LEMOINE, Y., JUPIN, H. : Analyse cinétique et spectroscopique de la fluorescence chez un mutant photosensible de Tabac. - Photosynthetica 12 : 35 -.
50, 1978.

34573 - LENDZIAN, K.J. : Interactions between magnesium ions, pH, glucose-6-phos-
phate, and NADPH/NADP$^+$ ratios in the modulation of chloroplast glucose-6-phos-
phate dehydrogenase in vitro. - Planta 141 : 105 - 110, 1978.

34574 - LENDZIAN, K.J. : Activation of ribulose-1,5-bisphosphate carboxylase by
chloroplast metabolites in a reconstituted spinach chloroplast system. -
Planta 143 : 291 - 296, 1978.

34575 - LENDZIAN, K.J., ZIEGLER, H. : Activation of NADP$^+$-linked glyceraldehyde-3-
-phosphate dehydrogenase in a reconstituted spinach chloroplast system
without functioning photophosphorylation. - Biochem. Physiol. Pflanzen
173 : 500 - 504, 1978.

34576 - LENDZIAN, K.J., ZIEGLER, H., SANKHLA, N. : Effect of phosphon-D on photosynthetic light reactions and on reactions of the oxidative and reductive
pentose phosphate cycle in a reconstituted spinach (Spinacia oleracea L.)
chloroplast system. - Planta 141 : 199 - 204, 1978.

34577 - LEONARDI, S., LINSER-BOURDELLON, A. : Valeurs énergétiques observeés dans
le Quercus ilex L., station de Monte Minardo (Etna). - Flora 167 : 35 - 39,
1978.

34578 - LEONG, T.-Y., SCHWEIGER, H.-G. : The role of chloroplast membrane protein
synthesis in the circadian clock. Occurrence of a polypeptide which tentati-
vely is involved in the clock. - In : AKOYUNOGLOU, G., ARGYROUDI-AKOYUNO-
GLOU, J.H. (ed.) : Chloroplast Development. Pp. 323 - 332. Elsevier/North-
-Holland Biomedical Press, Amsterdam - New York - Oxford 1978.

34579 - LESCURE, A.M. : Chloroplast differentiation in cultured tobacco cells :
in vitro protein synthesis efficiency of plastids at various stages of their
evolution. - Cell Differentiation 7 : 139 - 152, 1978.

34580 - LESCURE, A.-M., SEYER, P. : Intérêt des cultures de cellules in vitro pour
l'étude des organites chez les végétaux supérieurs : exemple des chloroplastes. - Physiol. vég. 16 : 449 - 450, 1978.

34581 - LESSERTOIS, D., MONÉGER, R. : Evolution des pigments pendant la croissance et la maturation du fruit de *Prunus persica*. - Phytochemistry *17* : 411 - 415, 1978.

34582 - LEUPOLD, D., VOIGT, B., MORY, S., HOFFMANN, P., HIEKE, B. : Low-intensity two-step absorption of chlorophyll-A *in vivo*. - Biophys. J. *21* : 177 - 180, 1978.

34583 - LEUTWILER, L.S., CHAPMAN, D.J. : Biosynthesis of carotenoids in *Rhodomicrobium vannielii*. - FEBS Lett. *89* : 248 - 252, 1978.

34584 - LEVANON, H., NORRIS, J.R. : The photoexcited triplet state and photosynthesis. - Chem. Rev. *78* : 185 - 198, 1978.

34585 - LEVI, C., PERCHOROWICZ, J.T., GIBBS, M. : Malate synthesis by dark carbon dioxide fixation in leaves. - Plant Physiol. *61* : 477 - 480, 1978.

34586 - LEVI, C., PREISS, J. : Amylopectin degradation in pea chloroplast extracts. - Plant Physiol. *61* : 218 - 220, 1978. [Ps.]

34587 - LEVITT, J. : Role of SH and SS groups in damage to biological systems at low water activities. - In : CROWE, J.H., CLEGG, J.S. (ed.) : Dry Biological Systems. Pp. 243 - 256. Academic Press, New York - San Francisco - London 1978. [Ps.]

34588 - LEWANDOWSKA, M., JARVIS, P.G. : Quantum requirements of photosynthetic electron transport in Sitka spruce from different light environments. - Physiol. Plant. *42* : 277 - 282, 1978.

34589 - LEWENSTEIN, A., BACHOFEN, R. : $CO_2$-fixation and ATP synthesis in continuous cultures of *Chlorella fusca*. - Arch. Microbiol. *116* : 169 - 173, 1978.

34590 - LEWIS, A. : The molecular mechanism of excitation in visual transduction and bacteriorhodopsin. - Proc. nat. Acad. Sci. USA *75* : 549 - 553, 1978.

34591 - LEWIS, A., MARCUS, M.A., EHRENBERG, B., CRESPI, H. : Experimental evidence for secondary protein-chromatophore interactions at the Schiff base linkage in bacteriorhodopsin : Molecular mechanism for proton pumping. - Proc. nat. Acad. Sci. USA *75* : 4642 - 4646, 1978.

34592 - LEWIS, A.G. : Concentrations of nutrients and chlorophyll on a cross-channel transect in Juan de Fuca Strait, British Columbia. - J. Fish. Res. Board Can. *35* : 305 - 314, 1978.

34593 - LEWIS, W.M. Jr. : Analysis of succession in a tropical phytoplankton community and a new measure of succession rate. - Amer. Naturalist *112* : 401 - 414, 1978. [Ps.]

34594 - LI, B.D. : Razdelenie, identifikatsiya i kolichestvennoe opredelenie fotosinteticheskikh pigmentov makrobentosnykh vodorosleĭ. [Separation, identification and quantitative determination of photosynthetic pigments in seaweeds.] - Sb. Rabot Akad. Nauk SSSR, dal'nevost. nauch. Tsentr, Inst. Biol. Morya (Vladivostok) *11* (Ėkologicheskie Aspekty Fotosinteza Morskikh Makrovodorosleĭ) : 38 - 54, 1978. [In R, ab : E.]

34595 - LI, B.D. : Sezonnaya dinamika pigmentov fotosinteza u morskikh zelenykh vodorosleĭ. [Seasonal variations of photosynthetic pigment contents in marine green algae.] - Sb. Rabot Akad. Nauk SSSR, dal'nevost. nauch. Tsentr, Inst. Biol. Morya (Vladivostok) *11* (Ėkologicheskie Aspekty Fotosinteza Morskikh Makrovodorosleĭ) : 158 - 166, 187, 1978. [In R, ab : E.]

34596 - LI, B.D., TITLYANOV, É.A. : Adaptatsiya benticheskikh rasteniĭ k svetu. III. Soderzhanie fotosinteticheskikh pigmentov v morskikh makrofitakh iz razlichnykh po osveshchennosti mest obitaniya. [Adaptation of benthic plants to light. III. Content of photosynthetic pigments in marine macrophytes from differently illuminated habitats.] - Biol. Morya *1978* (2) : 47 - 55, 1978. [In R, ab : E.]

34597 - LI, P.H., PALTA, J.P. : Frost hardening and freezing stress in tuber-bearing *Solanum* species. - In : LI, P.H., SAKAI, A. (ed.) : Plant Cold Hardiness and Freezing Stress. Mechanisms and Crop Implications. Pp. 49 - 71. Academic Press, New York - San Francisco - London 1978. [Chloroplast.]

34598 - **LI, Shu-Jun, CAI, Jian-Ping, WANG, Guo-Qing, WANG, Mei-Qi, ZHAO, Hai-Ying** :
[Studies on structure and function of chloroplasts II. Isolation and inter-
changeability of pure chloroplast coupling factors.] - Acta bot. sin. *20* :
103 - 107, 1978. [In Chin., ab : E.]

34599 - **LI, W.K.W.** : Cellular composition and physiology of *Thalassiosira fluviati-
lis* accomodated to cadmium stress. - J. Phycol. *21* (Suppl.) : 21, 1978. [Ps,
Chl, Car.]

*34600 - **LI, Y.-S.** : Step-wise stabilization of charge separation at photosystem II
reaction center. - A hypothesis. - Bot. Bull. Acad. sin. *18* : 169 - 178,
1977.

34601 - **LI, Y.-S.** : Derivation of an electron-transport rate equation, energy-con-
servation equations and a luminescence-flux equation of algal and plant photosynthesis. - Biochem. J. *174* : 569 - 577, 1978.

34602 - **LI, Y.-S.** : Redox of Q regulated slow fluorescence induction of isolated
chloroplasts. - Bot. Bull. Acad. sin. *19* : 33 - 40, 1978.

34603 - **LIAAEN-JENSEN, S.** : Chemistry of carotenoid pigments. - In : **CLAYTON, R.K.,
SISTROM, W.R.** (ed.) : The Photosynthetic Bacteria. Pp. 233 - 247. Plenum
Press, New York - London 1978.

34604 - **LIAAEN-JENSEN, S.** : Marine carotenoids. - In : **SCHEUER, P.J.** (ed.) : Marine
Natural Products. Vol. 2. Pp. 1 - 73. Academic Press, New York - London -
San Francisco 1978.

34605 - **LIBERMAN, E.A.** : Bioênergetika i protonno-êlektronnye sistemy membran. [Bio-
energetic and proton-electron systems of membranes.] - Biofizika *12* : 174 -
179, 1978. [Ps; in R, ab : E.]

34606 - **LICHTENTHALER, H., BUSCHMANN, C.** : Control of chloroplast development by red
light, blue light and phytohormones. - In : **AKOYUNOGLOU, G., ARGYROUDI-AKO-
YUNOGLOU, J.H.** (ed.) : Chloroplast Development. Pp. 801 - 816. Elsevier/
North-Holland Biomedical Press, Amsterdam - New York - Oxford 1978.

34607 - **LICHTENTHALER, H.K., PFISTER, K.** : New aspects on the function of naphtho-
quinones in photosynthesis. - In : **METZNER, H.** (ed.) : Photosynthetic Oxygen Evolution. Pp. 171 - 193. Academic Press, London - New York - San Francisco 1978.

*34608 - **LICHTENTHALER, H.K., STRAUB, V.** : Die Bildung von Lipochinonen in Gewebe-
kulturen. - Planta med. *1975* (Suppl.): 198 - 213, 1975.

34609 - **LIEBERMAN, J.R., ARNTZEN, C.J.** : Chlorophyll fluorescence *in vivo* : A tool
for detecting environmental stress. - Plant Physiol. *61* (Suppl.) : 31, 1978.

34610 - **LIEBERMAN, J.R., BOSE, S., ARNTZEN, C.J.** : Requirement of the light-harvesting pigment·protein complex for magnesium ion regulation of excitation energy distribution in chloroplasts. - Biochim. biophys. Acta *502* : 417 - 429,
1978.

*34611 - **LIETH, H.** : Primary production : terrestrial ecosystems. - Hum. Ecol. *1* :
303 - 332, 1973.

34612 - **LIETH, H.** : Die pflanzliche Primärproduktivität. - Bürger + Univ. Informa-
tionsbl. Universitätsges. Osnabrück e.V. *3* : 38 - 51, 1978.

34613 - **LIETH, H.** : Vegetation and $CO_2$ changes. - In : **WILLIAMS, J.** (ed.) : Carbon
Dioxide, Climate and Society. Pp. 103 - 109. Pergamon Press, Oxford 1978.
[Production.]

34614 - **LILLEY, R.M.** : Carbon traffic between chloroplast and cytoplasm during photosynthesis and its regulation by cytoplasmic orthophosphate. - Proc. aust.
biochem. Soc. *11* : Q1 - Q2, 1978.

*34615 - **LIN, A.C.** : [Studies on tillering and photosynthetic activities of 1st and
2nd crops of rice.] - J. agr. Assoc. China-Taiwan [Chung-Hua Nung Hsueh Hui
Pao] *100* : 87 - 97, 1977. [In Chin.]

34616 - **LIN, C.H., STOCKING, C.R.** : Influence of leaf age, light, dark, and iron deficiency on polyribosome levels in maize leaves. - Plant Cell Physiol. *19* : 461 - 470, 1978. [Chl.]

34617 - **LINEK, V., BENEŠ, P.** : Comparison of regression and moment methods of evaluation of oxygen probe responses. - Biotechnol. Bioeng. *20* : 903 - 912, 1978.

34618 - **LINK, G., BEDBROOK, J.R., BOGORAD, L., COEN, D.M., RICH, A.** : The expression of the gene for the large subunit of ribulose 1,5-bisphosphate carboxylase in maize. - In : SIEGELMAN, H.W., HIND, G. (ed.) : Photosynthetic Carbon Assimilation. Pp. 349 - 362. Plenum Press, New York - London 1978.

34619 - **LINK, G., COEN, D.M., BOGORAD, L.** : Differential expression of the gene coding for the large subunit of ribulose-1,5-bisphosphate carboxylase in *Zea mays*. - In : AKOYUNOGLOU, G., ARGYROUDI-AKOYUNOGLOU, J.H. (ed.) : Chloroplast Development. Pp. 559 - 564. Elsevier/North-Holland Biomedical Press, Amsterdam - New York - Oxford 1978.

34620 - **LINK, G., COEN, D.M., BOGORAD, L.** : Differential expression of the gene for the large subunit of ribulose bisphosphate carboxylase in maize leaf cell types. - Cell *15* : 725 - 731, 1978.

34621 - **LINTHURST, R.A., REIMOLD, R.J.** : Estimated net aerial primary productivity for selected estuarine angiosperms in Maine, Delaware, and Georgia. - Ecology *59* : 945 - 955, 1978.

34622 - **LINVILL, D.E., DALE, R.F., HODGES, H.F.** : Solar radiation weighting for weather and corn growth models. - Agron. J. *70* : 257 - 263, 1978.

*34623 - **LIPINSKY, E.S., McCLURE, T.A.** : Using sugar crops to capture solar energy. - In : MITSUI, A., MIYACHI, S., SAN PIETRO, A., TAMURA, S. (ed.) : Biological Solar Energy Conversion. Pp. 397 - 410. Academic Press, New York - San Francisco - London 1977.

*34624 - **LIPPERT, L.F., HALL, M.O.** : Determination of total carotenoids in *Capsicum*.- HortScience *8* : 38 - 40, 1973.

*34625 - **LISTER, R., LEMON, E.** : Interactions of atmospheric carbon dioxide, diffuse light, plant productivity and climate processes - model predictions. - In : ENGELMAN, R.J., SEHMEL, G.A. (ed.) : Atmosphere-Surface Exchange of Particulate and Gaseous Pollutans. Pp. 112 - 135. National Technical Information Service, Springfield, Va. 1976.

34626 - **LITTLER, M.M., MURRAY, S.N.** : Influence of domestic wastes on energetic pathways in rocky intertidal communities. - J. appl. Ecol. *15* : 583 - 595, 1978. [Energy content.]

34627 - **LITVIN, F.F., IGNATOV, N.V., EFIMTSEV, E.I., BELYAEVA, O.B.** : Two successive photochemical reactions in protochlorophyll(ide) reduction into chlorophyll-(ide). - Photosynthetica *12* : 375 - 381, 1978.

34628 - **LITVIN, F.F., STADNICHUK, I.N., KRUGLOV, V.P.** : Razlozhenie na komponenty spektrov fluorestsentsii i pogloshcheniya khlorofilla *a* v kletke. [Computer curve analysis of chlorophyll *a* absorption and fluorescence *in vivo*.] - Biofizika *23* : 450 - 455, 1978. [In R, ab : E.]

*34629 - **LITVIN, F.F., STADNICHUK, I.N., SHUBIN, V.V.** : Vtoraya proizvodnaya spektrov fluorestsentsii i vozbuzhdeniya fluorestsentsii khlorofilla *a* i soprovozhdayushchikh pigmentov v vysshikh rasteniyakh i vodoroslyakh. [Second derivative of the fluorescence and excitation fluorescence spectra of chlorophyll *a* and accompanying pigments in higher plants and algae.] - Nauch. Dokl. vyssh. Shkoly, biol. Nauki *19* (9) : 36 - 46, 1976. [In R.]

34630 - **LITVINENKO, L.G., LEBEDEV, S.I.** : Vplyv azotnogo zhyvlennya na aktyvnist' fotosyntetychnogo aparatu kukurudzy. [Effect of nitrogenous nutrition on the activity of maize photosynthetic apparatus.] - Ukr. bot. Zh. *35*(1) : 56 - 58, 111, 1978. [In Ukr., ab : E,R.]

34631 - **LIU, E.H., SHARITZ, R.R., SMITH, M.H.** : Thermal sensitivities of malate dehydrogenase isozymes in *Typha*. - Amer. J. Bot. *65* : 214 - 220, 1978.

34632 - LIU, W.T.,  POOL, R., WENKERT, W., KRIEDEMANN, P.E. : Changes in photosynthesis, stomatal resistance and abscisic acid of *Vitis labruscana* through
drought and irrigation cycles. - Amer. J. Enol. Viticult. *29* : 239 - 246,
1978.

34633 - LIU, W.T., WENKERT , W., ALLEN, L.H.,Jr., LEMON, E.R. : Soil-plant water relations in a New York vineyard : Resistances to water movement. - J. amer.
Soc. hort. Sci. *103* : 226 - 230, 1978. [Stomatal resistance.]

34634 - LIUM, B.W., SHOAF, W.T. : The use of magnesium carbonate in chlorophyll determinations. - Water Resour. Bull. *14* : 190 - 194, 1978.

34635 - LIVDANE, B.A. : Nakoplenie krakhmala i belkov v sozrevayushchikh zernovkakh
yachmenya pri defitsite medi. [Accumulation of starch and proteins in ripening barley caryopses under copper deficit.] - In : Fiziologo-Biokhimicheskie
Issledovaniya Rasteniĭ. Pp. 92 - 99, 158. Zinatne, Riga 1978. [Ps; in R.]

34636 - LLOYD, N.D.H., WOOLHOUSE, H.W. : Leaf resistances in different populations
of *Sesleria caerulea* (L.) ARD. - New Phytol. *80* : 79 - 85, 1978.

34637 - LOBBAN, C.S. : Translocation of ${}^{14}C$ in *Macrocystis integrifolia (Phaeophyceae)*. - J. Phycol. *14* : 178 - 182, 1978.

34638 - LOBBAN, C.S. : Translocation of ${}^{14}C$ in *Macrocystis pyrifera* (Gigant kelp). -
Plant Physiol. *61* : 585 - 589, 1978.

34639 - LOCKWOOD, D.W., SPARKS, D. : Translocation of ${}^{14}C$ in "Stuart" pecan in the
spring following assimilation of ${}^{14}CO_2$ during the previous growing season. -
J. amer. Soc. hort. Sci. *103* : 38 - 45, 1978.

34640 - LOCKWOOD, D.W., SPARKS, D. : Translocation of ${}^{14}C$ from tops and roots of pecan in the spring following assimilation of ${}^{14}CO_2$ during the previous growing season. - J. amer. Soc. hort. Sci. *103* : 45 - 49, 1978.

34641 - LOMMEN, M.A.J., TAKEMOTO, J. : Comparison, by freeze-fracture electron microscopy, of chromatophores, spheroplast-derived membrane vesicles, and whole
cells of *Rhodopseudomonas sphaeroides*. - J. Bacteriol. *136* : 730 - 741,
1978.

34642 - LOMMEN, M.A.J., TAKEMOTO, J. : Ultrastructure of carotenoid mutant strain
R-26 of *Rhodopseudomonas sphaeroides*. - Arch. Microbiol. *118* : 305 - 308,
1978.

34643 - LONERGAN, T.A., SARGENT, M.L. : Effects of acetazolamide (Diamox), ethoxzo-
lamide and high levels of $CO_2$ on carbonic anhydrase, photosystem activity,
and oxygen evolution in *Euglena gracilis*. - Physiol. Plant. *43* : 55 - 61,
1978.

34644 - LONERGAN, T.A., SARGENT, M.L. : Regulation of the photosynthesis rhythm in
*Euglena gracilis*. I. Carbonic anhydrase and glyceraldehyde-3-phosphate dehydrogenase do not regulate the photosynthesis rhythm. - Plant Physiol. *61* :
150 - 153, 1978.

34645 - LONG, S.P. : Solar energy conversion in biology - a report of the symposium.
- In : HALL, D.O., COOMBS, J., GOODWIN, T.W. (ed.) : Proceedings of the
Fourth International Congress on Photosynthesis. Pp. 677 - 682. Biochem.
Soc., London 1978.

34646 - LONG, S.P., WOOLHOUSE, H.W. : The responses of net photosynthesis to vapour
pressure deficit and $CO_2$ concentration in *Spartina townsendii (sensu lato)*,
a $C_4$ species from a cool temperature climate. - J. exp. Bot. *29* : 567 - 577,
1978.

34647 - LONG, S.P., WOOLHOUSE, H.W. : The responses of net photosynthesis to light
and temperature in *Spartina townsendii (sensu lato)*, a $C_4$ species from a
cool temperate climate. - J. exp. Bot. *29* : 803 - 814, 1978.

34648 - LONGSTRETH, D.J., NOBEL, P.S. : Salinity-induced changes in leaf anatomy
and photosynthesis. - Plant Physiol. *61* (Suppl.) : 94, 1978.

34649 - LONGUENESSE, J.J. : Température nocturne et photosynthèse I.-Étude biblio-
graphique. - Ann. agron. *29* : 525 - 539, 1978.

34650 - **LOOMIS, R.S., NG, E.** : Influences of climate on photosynthetic productivity of sugar beet. - In : **HALL, D.O., COOMBS, J., GOODWIN, T.W.** (ed.) : Proceedings of the Fourth International Congress on Photosynthesis. Pp. 259 - 268. Biochem. Soc., London 1978.

*34651 - **LOPUSHINSKY, W.** : Water relations and photosynthesis in lodgepole pine. - In : **BAUMGARTNER, D.M.** (ed.) : Management of Lodgepole Pine Ecosystems. Symp. Proc. Pp. 135 - 153. Washington State University Cooperative Extension Service, Pullman 1975.

34652 - **LOR, K.-L., COSSINS, E.A.** : Relationships between glycollate and folate metabolism in *Euglena gracilis*. - Phytochemistry *17* : 659 - 665, 1978.

34653 - **LORD, J.M., BOWDEN, L.** : Evidence that glyoxysomal malate synthase is segregated by the endoplasmic reticulum. - Plant Physiol. *61* : 266 - 270, 1978.

34654 - **LORENC-PLUCIŃSKA, G.** : The effect of $SO_2$ on the photosynthesis and dark respiration of larch and pine differing in resistance to this gas. - Arboretum kórnickie *23* : 121 - 132, 1978.

34655 - **LORENC-PLUCIŃSKA, G.** : Effect of sulphur dioxide on photosynthesis, photorespiration and dark respiration of Scots pines differing in resistance to this gas. - Arboretum kórnickie *23* : 133 - 144, 1978.

34656 - **LORIMER, G.H.** : Retention of the oxygen atoms at carbon-2 and carbon-3 during the carboxylation of ribulose 1,5-bisphosphate. - Europe.J. Biochem. *89* : 43 - 50, 1978.

34657 - **LORIMER, G.H., BADGER, M.R., HELDT, H.W.** : The activation of ribulose 1,5-bisphosphate carboxylase/oxygenase. - In : **SIEGELMAN, H.W., HIND, G.** (ed.) : Photosynthetic Carbon Assimilation. Pp. 283 - 306. Plenum Press, New York - London 1978.

34658 - **LORIMER, G.H., OSMOND, C.B., AKAZAWA, T., ASAMI, S.** : On the mechanism of glycolate synthesis by *Chromatium* and *Chlorella*. - Arch. Biochem. Biophys. *185* : 49 - 56, 1978.

34659 - **LORIMER, G.H., WOO, K.C., BERRY, J.A., OSMOND, C.B.** : The $C_2$ photorespiratory carbon oxidation cycle in leaves of higher plants: pathway and consequences. - In : **HALL, D.O., COOMBS, J., GOODWIN, T.W.** (ed.) : Proceedings of the Fourth International Congress on Photosynthesis. Pp. 311 - 322. Biochem. Soc., London 1978.

34660 - **LOS', S.I.** : Porivnyal'na kharakterystyka biliproteïdiv sin'ozelenykh vodorosteï (na pidstavi vikovoï minlyvosti). [Comparative characteristic of the blue-green algae biliproteins with regard to age variability.] - Ukr. bot. Zh. *35* : 614 - 619, 671, 1978. [In Ukr., ab : E, R.]

34661 - **LOSADA, M.** : Energy-transducing redox systems and the mechanism of oxidative phosphorylation. - Bioelectrochem. Bioenerg. *5* : 296 - 310, 1978. [Ps.]

34662 - **LOSEV, A.P.** : Chlorophyll energetics in concentrated solutions and associates. - In : **FIALA, J.** (ed.) : Third International Seminar on Energy Transfer in Condensed Matter. Pp. 91 - 110. Univerzita Karlova, Praha 1978.

34663 - **LOSEVA, N.L., RYBKINA, G.V., BEL'KOVICH, T.M.** : K voprosu o svyazi vodnogo rezhima s fotokhimicheskoï aktivnost'yu khloroplastov. [Relationship between water regime and photochemical activity of chloroplasts.] - In : Vodnyï Rezhim Rasteniï v Svyazi s Raznymi Ékologicheskimi Usloviyami. Pp. 325 - 330. Izdatel'stvo Kazanskogo Universiteta, Kazan' 1978. [In R.]

34664 - **LOSKUTNIKOV, A.I., PETROV, V.E.** : Rol' sveta v reparatsii énergeticheskogo obmena assimiliruyushcheï kletki posle teplovogo povrezhdeniya. [Role of light in reparation of energy metabolism of assimilating cell after heat injury.] - Fiziol. Biokhim. kul't. Rast. *10* : 204 - 210, 1978. [In R, ab : E.]

34665 - **LOVE, J.M., BARDEN, J.A.** : The effect of a pinching agent on the net photosynthesis of "Golden Delicious" and "Delicious" apple leaves. - HortScience *13* : 281 - 282, 1978.

34666 - **LOWER, R.L., PHARR, D.M., SMITH, O.S., SOX, H.N.** : Cotyledon angle as a predictor of growth rate of cucumber plant. - Euphytica *27* : 701 - 706, 1978.

34667 - **LOZIER, R.H., CHAE, Q., MOWERY, P.C., STOECKENIUS, W.** : Flash photolysis studies of bacteriorhodopsin species formed at extreme low pH. - In : **CAPLAN, S.R., GINZBURG, M.** (ed.) : Energetics and Structure of Halophilic Microorganisms. Pp. 297 - 301. Elsevier/North-Holland Biomedical Press, Amsterdam - New York 1978.

34668 - **LOZIER, R.H., NIEDERBERGER, W., OTTOLENGHI, M., SIVORINOVSKY, G., STOECKENIUS, W.** : On the photocycles of light- and dark-adapted bacteriorhodopsin. - In : **CAPLAN, S.R., GINZBURG, M.** (ed.) : Energetics and Structure of Halophilic Microorganisms. Pp. 123 - 141. Elsevier/North-Holland Biomedical Press, Amsterdam - New York 1978.

34669 - **LOZOVA, G.I.** : Évolyutsiya pigmentnogo aparatu roslyn. [Evolution of pigment apparatus in plants.] - Ukr. bot. Zh. *35* : 575 - 584, 670, 1978. [In Ukr., ab : E, R.]

34670 - **LOZOVA, G.I., L'VIVS'KA, N.R.** : Osoblyvosti pigmentvmisnykh kompleksiv pry porushenni struktury khloroplastiv. [Peculiarities of pigment-containing complexes with disturbance in chloroplast structure.] - Ukr. bot. Zh. *35* : 640 - 645, 672, 1978. [In Ukr., ab : E, R.]

*34671 - **LUCAS, W.J., SMITH, F.A.** : The formation of alkaline and acid regions at the surface of *Chara corallina* cells. - J. exp. Bot. *24* : 1 - 14, 1973. [Ps.]

34672 - **LUCAS, W.J., TYREE, M.T.** : $HCO_3^-$ transport in *Potamogeton lucens* and *Chara corallina*. - Plant Physiol. *61* (Suppl.) : 107, 1978.

34673 - **LUCAS, W.J., TYREE, M.T., PETROV, A.** : Characterization of photosynthetic $^{14}$carbon assimilation by *Potamogeton lucens* L. - J. exp. Bot. *29* : 1409 - 1421, 1978.

34674 - **LUCERO, H., LESCANO, W.I.M., VALLEJOS, R.H.** : Inhibition of energy conservation reactions in chromatophores of *Rhodospirillum rubrum* by antibiotics. - Arch. Biochem. Biophys. *186* : 9 - 14, 1978.

*34675 - **LÜDERITZ, R., KLEMME, J.-H.** : Isolierung und Charakterisierung eines membrangebundenen Pyruvatdehydrogenase-Komplexes aus dem phototrophen Bakterium *Rhodospirillum rubrum*. - Z. Naturforsch. *32C* : 351 - 361, 1977.

34676 - **LUDLOW, M.M.** : Light relations of pasture plants. - In : **WILSON, J.R.** (ed.) : Plant Relations in Pastures. Pp. 35 - 49. CSIRO, Melbourne 1978. [Ps.]

34677 - **LUGO, A.E., GONZALES-LIBOY, J.A., CINTRON, B., DUGGER, K.** : Structure, productivity, and transpiration of a subtropical dry forest in Puerto Rico. - Biotropica *10* : 278 - 291, 1978.

34678 - **LUHADIYA, A.P., KULKARNI, P.R.** : Dehydration of green chillies (*Capsicum frutescens*). - J. Food Sci. Technol. *15* : 139 - 142, 1978. [Chl.]

34679 - **LUISETTI, J., GALLA, H.-J., MÖHWALD, H.** : Energy transfer and fluorescence quenching in chlorophyll containing vesicles. - Ber. Bunsenges. phys. Chem. *82* : 911 - 916, 1978.

34680 - **LUNDIN, A., BALTSCHEFFSKY, M.** : Measurement of photophosphorylation and ATPase using purified firefly luciferase. - In : **COLOWICK,S.P.,KAPLAN, N.O.** (ed.): Methods in Enzymology. Vol.57. Pp. 50 - 56. Academic Press, New York - San Francisco - London 1978.

34681 - **LUPATOV, V.M.** : Issledovanie skorosti final'noĭ stadii obrazovaniya kisloroda pri fotosinteze. [Kinetics of the final stage in oxygen formation during photosynthesis.] - Biofizika *23* : 462 - 466, 1978. [In R, ab : E.]

34682 - **LURIE, S.** : The effect of wavelength of light on stomatal opening. - Planta *140* : 245 - 249, 1978. [Ps.]

34683 - **LURIE, S., PAZ, N., STRUCK, N., BRAVDO, B.-A.** : RuBP carboxylase/oxygenase activity and gas exchange rates during leaf and fruit development. - In : **SIEGELMAN, H.W., HIND, G.** (ed.) : Photosynthetic Carbon Assimilation. P. 422. Plenum Press, New York - London 1978.

34684 - LÜTTGE, U., BALL, E. : Free running oscillations of transpiration and $CO_2$ exchange in CAM plants without a concomitant rhythm of malate levels. - Z. Pflanzenphysiol. *90* : 69 - 77, 1978.

34685 - LÜTZ, C. : Separation and comparison of prolamellar bodies and prothyla-koids of etioplasts from *Avena sativa* L. - In : AKOYUNOGLOU, G., ARGYROUDI--AKOYUNOGLOU, J.H. (ed.) : Chloroplast Development. Pp. 481 - 488. Elsevier/ North-Holland Biomedical Press, Amsterdam - New York - Oxford 1978.

34686 - LUTZ, M., AGALIDIS, I. : Resonance Raman scattering of cis conformers of C-40 carotenoids. Evidence for cis carotenoids in bacterial reaction cen-ters. - In : SCHMID, E.D., KRISHNAN, R.S., KIEFER, W. (ed.) : Proceedings of the 6th International Conference on Raman Spectroscopy. Vol. *6* (2). Pp. 162 - 163. Heyden, London 1978.

34687 - LUTZ, M., AGALIDIS, I., HERVO, G., COGDELL, R.J., REISS-HUSSON, F. : On the state of carotenoids bound to reaction centers of photosynthetic bacteria : a resonance Raman study. - Biochim. biophys. Acta *503* : 287 - 303, 1978.

34688 - LUTZ, M., BROWN, J.S., REMY, R. : Resonance Raman scattering of chlorophyll--protein complexes. In : SCHMID, E.D., KRISHNAN, R.S., KIEFER, W. (ed.) : Proceedings of the 6th International Conference on Raman Spectroscopy. Vol. *6*(2).Pp.158 - 159.Heyden, London 1978.

34689 - LUUKKANEN, O. : Investigations on factors affecting net photosynthesis in trees : Gas exchange in clones of *Picea abies* (L.) KARST. - Acta forest. fenn. *162* : 1 - 63, 1978.

34690 - LUXMOORE, R.J., HUFF, D.D., McCONATHY, R.K., DINGER, B.E. : Some measured and simulated plant water relations of yellow-poplar. - Forest Sci. *24* : 327 - 341, 1978.

34691 - LYAKHNOVICH, Ya.P., GONTAREVA, T.V. : O soderzhanii khlorofillida i gidro-liticheskoĭ aktivnosti khlorofillazy v svyazi s vozrastom kletki khlorelly. [Chlorophyllide content and hydrolytic activity of chlorophyllase in rela-tion to *Chlorella* cell age.] - Vestsi Akad. Navuk belarus. SSR, Ser. biyal. Navuk *1978* (1) : 119, 1978. [In R.]

34692 - LYMAN, H., SRINIVAS, U.K. : Regulation of chloroplast DNA synthesis : Pos-sible role of chloroplast nucleases in *Euglena*. - In : AKOYUNOGLOU, G., ARGYROUDI-AKOYUNOGLOU, J.H. (ed.) : Chloroplast Development. Pp. 593 - 607. Elsevier/North-Holland Biomedical Press, Amsterdam - New York - Oxford 1978.

34693 - LYTVYNENKO, L.G., LEBEDEV, S.I. : Vplyv azotnogo zhyvlennya na aktyvnist' fotosyntetychnogo aparatu kukuruzy. [Effect of nitrogen nutrition on acti-vity of maize photosynthetic apparatus.] - Ukr. bot. Zh. *35* : 56 - 58, 1978. [In Ukr., ab : E.]

34694 - LYUBIMOV, V.Yu. : Psevdotsiklicheskiĭ transport ėlektronov v khloroplas-takh. [Pseudocyclic electron transport in chloroplasts.] - In : KARPILOV, Yu.S., ROMANOVA, A.K. (ed.) : Mekhanizm Fotodykhaniya i Ego Osobennosti u Rasteniĭ Razlichnykh Tipov. Pp. 17 - 32. Pushchino 1978. [In R.]

34695 - LYUBIMOV, V.Yu., KARPILOV, Yu.S. : Svetozavisimyĭ okislitel'nyĭ metabolizm organicheskikh kislot v mezofil'nykh khloroplastakh kukuruzy. [Light-depen-dent oxidative metabolism of organic acids in mesophyll chloroplasts of maize.] - In : KARPILOV, Yu.S., ROMANOVA, A.K. (ed.) : Mekhanizm Fotodykha-niya i ego Osobennosti u Rasteniĭ Razlichnykh Tipov. Pp. 58 - 74. Pushchino 1978. [In R.]

34696 - MacCOLL, R. : Excitation energy transfer studies of phycoerythrins. - Trends in Fluorescence *1* (2) : 31 - 33, 1978.

34697 - MacCOLL, R. , BERNS, D.S. : Energy transfer studies on cryptomonad bili-proteins. - Photochem. Photobiol. *27* : 343 - 349, 1978.

34698 - MacCOLL, R., EDWARDS, M.R., HAAKSMA, C. : Some properties of allophycocya-nin from a thermophilic blue-green alga. - Biophys. Chem. *8* : 369 - 376, 1978.

34699 - MacDONALD, I.R., GORDON, D.C. : Geostimulation of root growth and pigment synthesis in mustard seedlings. - Nature *272* : 48 - 49, 1978.

*34700 - MACHE, R., LOISEAUX, S. : Modifications des chloroplastes de *Marchantia polymorpha* dues à un excès de lumière. - Bull. Soc. bot. Fr. *121* : 191 - 193, 1974.

34701 - MÄCHLER, F., NÖSBERGER, J. : The adaptation to temperature of photorespiration and of the photosynthetic carbon metabolism of altitudinal ecotypes of *Trifolium repens* L. - Oecologia *35* : 267 - 276, 1978.

34702 - MÄCHLER, F., NÖSBERGER, J. : Die Beziehung zwischen der Photosyntheserate und dem Chlorophyllgehalt bei Stark- und Schwachlichtpflanzen von *Trifolium repens* L. - Angew. Bot. *52* : 155 - 159, 1978.

34703 - MACHOWICZ, E., WIĘCKOWSKI, S. : The photochemical activity of chloroplasts and subchloroplast particles isolated from etiolated bean leaves exposed to continuous light. - J. exp. Bot. *29* : 1383 - 1389, 1978.

34704 - MACKENDER, R.O. : The amino acid composition of the chloroplast envelope membranes of *Vicia faba* L. - Plant Sci. Lett. *12* : 279 - 285, 1978.

34705 - MACKENDER, R.O. : Etioplast development in dark-grown leaves of *Zea mays* L. - Plant Physiol. *62* : 499 - 505, 1978.

B34706 - MADER, P., NAUŠ, J., MAKOVEC, P., KUPKA, J., NOVÁK, V., GRÉC, L. : Effect of Nitrogen Nutrition on the Photophysical, Photochemical and Photosynthetic Activities of the Spring Barley Autotrophic Apparatus. - Praha 1978.

34707 - MADER, P., NOVÁK, V., NAUŠ, J., VACEK, K., PAVELEK, M., VOSPĚLOVÁ, V., CHLAD, F., JEŽKOVÁ, J. : Luminiscence, fotosyntetická aktivita a pigment-proteinové složení thylakoidů u rostlin ječmene během ontogenese. [Luminescence, photosynthetic activity, and pigment-protein composition of thylakoids in barley plants in the course of ontogenesis.] - Sborník vysoké školy zemědělské Praze,Fak. agron., Ser. A *1* : 133 - 148, 1978. [In Czech, ab : E, R.]

34708 - MAGOMEDOV, I.M., TISHCHENKO, N.N. : Metodika issledovaniya glavnykh fermentov $C_4$ fotosinteza. [Methods for studies of main enzymes of $C_4$ photosynthesis.] - Tr. priklad. Bot. Genet. Selektsii *61* (3) : 105 - 110, 1978. [In R, ab : E.]

34709 - MAGOMEDOV, I.M., TISHCHENKO, N.N., AGAEV, M.G. : Biokhimicheskie osobennosti fotosinteza $C_4$ rastenii v svyazi s ikh évolyutsiei.[Biochemical characteristics of $C_4$ photosynthesis of plants in relation to their evolution.] - Tr. petergof. biol.Inst. *27* : 142 - 146, 1978. [In R, ab : E.]

34710 - MAILLARD-SEVHONKIAN, S., PILET, P.-E. : Effect of abscisic acid on differentiation and ribulose diphosphate carboxylase of chloroplasts. - Plant Cell Physiol. *19* : 811 - 817, 1978.

*34711 - MAJOR, D.J. : Seasonal dry-weight distribution of single-stalked and multi--tillered corn hybrids grown at three population densities in southern Alberta. - Can. J. Plant Sci. *57* : 1041 - 1047, 1977.

34712 - MAJOR, D.J., BOLE, J.B., CHARNETSKI, W.A. : Distribution of photosynthates after $^{14}CO_2$ assimilation by stems, leaves, and pods of rape plants. - Can. J. Plant Sci. *58* : 783 - 787, 1978.

*34713 - MAKI, T. : [Interrelationships between zero-plane displacement aerodynamic roughness length and plant canopy height.] - J. agr. Meteorol. *31* : 7 - 15, 1975. [Growth analysis; in Jap., ab : E.]

*34714 - MAKI, T. : [Wind profile parameters of various canopies as influenced by wind velocity and stability.]- J. agr. Meteorol. *31* : 61 - 70, 1975. [Growth analysis; in Jap., ab : E.]

34715 - MÄKIRINTA, U. : Spektrale Lichtmessungen im freien Wasser und in der Wasservegetation des Sees Kukkia, Südfinnland, unter besonderer Berücksichtigung der Zonation. - Aquilo Ser. bot. *16* : 39 - 53, 1978.

34716 - MAKOVCOVÁ, O., ŠINDELÁŘ, L. : Changes in phosphoenolpyruvate carboxylase and ribulosebisphosphate carboxylase activities in tobacco plants infected with tobacco mosaic virus. - Biol. Plant. *20* : 135 - 137, 1978.

34717 - **MAKOVKINA, L.E., MESTECHKINA, N.M., KUZNETSOV, B.A., MUTUSKIN, A.A., PSHE-NOVA, K.V.** : Okislitel'no-vosstanovitel'nye potentsialy nekotorykh metalloproteidov. [Redox potentials of some metalloproteins.] - Biokhimiya *43* : 564 - 567, 1978. [Cyt, plastocyanin; In R, ab : E.]

34718 - **MAKSIMOVA, I.V., KAMAEVA, S.S.** : Vliyanie kontsentratsii kisloroda i intensivnosti sveta na fotosintez i vydelenie organicheskikh veshchestv kletkami *Chlorella pyrenoidosa* shtamm 82. [Effect of oxygen concentration and illuminance on photosynthesis and excretion of organic substances by *Chlorella pyrenoidosa* strain 82 cells.] - Nauch. Dokl. vyssh. Shkoly, biol. Nauki *1978* (3) : 113 - 119, 1978. [In R.]

34719 - **MAKSIMOVA, I.V., KAMAEVA, S.S., MAKSIMOV, V.N.** : Zavisimost' éffekta Varburga u *Chlorella pyrenoidosa* shtamm 82 ot intensivnosti sveta. [Relationship between Warburg effect and illuminance in *Chlorella pyrenoidosa* strain 82.] - Nauch. Dokl. vyssh. Shkoly, biol. Nauki *1978* (8) : 107 - 116, 1978. [In R.]

34720 - **MAKSIMOVA, L.A.** : Vzaimosvyaz' mezhdu osnovnymi pokazatelyami fotosinteticheskoĭ deyatel'nosti rasteniĭ kukuruzy v usloviyakh orosheniya. [Relationship between basic characteristics of photosynthetic activity of irrigated maize plants.] - Byull. vsesoyuz. nauch.-issl. Inst. Kukuruzy *1978* (2-3) : 32 - 36, 1978. [In R.]

34721 - **MAKUNGA, O.H.D., PEARMAN, I., THOMAS, S.M., THORNE, G.N.** : Distribution of photosynthate produced before and after anthesis in tall and semi-dwarf winter wheat, as affected by nitrogen fertiliser. - Ann. appl. Biol. *88* : 429 - 437, 1978.

*34722 - **MALCHEV, E.** : Spektralna kharakteristika na feoforbidi i tsvetna kharakteristika na khlorofili, feofitini i feoforbidi. [Spectral characteristic of pheophorbides and colour characteristics of chlorophylls, pheophytins and pheophorbides.] - Nauch. Tr. vissh. Inst. khranit. vkus. Prom. (Plovdiv) *19* (3) : 199 - 207, 1972. [In Bulg., ab : E, R.]

34723 - **MALKIN, R.** : Oxidation-reduction potential dependence of the flash-induced 518 nm absorbance change in chloroplasts. - FEBS Lett. *87* : 329 - 333, 1978.

34724 - **MALKIN, R., BARBER, J.** : New insights on the primary electron-acceptor complex of Photosystem II. - Biochem. Soc. Trans. *6* : 909 - 913, 1978.

34725 - **MALKIN, R., BEARDEN, A.J.** : Electron paramagnetic resonance studies of plastocyanin in the photosynthetic electron transport chain. - In : **HALL, D.O., COOMBS, J., GOODWIN, T.W.** (ed.) : Proceedings of the Fourth International Congress on Photosynthesis. Pp. 787 - 791. Biochem. Soc., London 1978.

34726 - **MALKIN, R., BEARDEN, A.J.** : Membrane-bound iron-sulfur centers in photosynthetic systems. - Biochim. biophys. Acta *505* : 147 - 181, 1978.

34727 - **MALKIN, R., POSNER, H.B.** : On the site of function of the Rieske iron-sulfur center in the chloroplast electron transport chain. - Biochim. biophys. Acta *501* : 552 - 554, 1978.

34728 - **MALKIN, S., BARBER, J.** : Induction patterns of delayed luminescence from isolated chloroplasts. I. Response of delayed luminescence to changes in the prompt fluorescence yield. - Biochim. biophys. Acta *502* : 524 - 541, 1978.

34729 - **MALKIN, S., FORK, D.C., ARMOND, P.A.** : Probing photosynthetic unit sizes of leaves by fluorescence-induction measurements. - Carnegie Inst. Year Book *77* : 237 - 240, 1978.

34730 - **MALKINA, I.S.** : Opredelenie intensivnosti fotosinteza v krone vzroslykh derev'ev. [Determination of photosynthetic rate in the crown of mature oak trees.] - Fiziol. Rast. *25* : 792 - 797, 1978. [In R, ab : E.]

34731 - **MALKINA, I.S.** : Fotosintez v krone vzroslogo dereva. [Photosynthesis in the crown of a mature tree.] - Lesovedenie *1978*(1) : 78 - 85, 1978. [In R, ab : E.]

34732 - **MALLOTT, P.G., DAVY, A.J.** : Analysis of effects of the bird cherry-oat aphid on the growth of barley: Unrestricted infestation. - New Phytol. *80* : 209 - 218, 1978. [Гs.]

*34733 - MALONE, T.C. : Phytoplankton productivity in the apex of the New York Bight:
Environmental regulation by productivity/chlorophyll $a$. - Amer. Soc. Limnol.
Oceanogr. Spec. Symp. 2 : 260 - 272, 1976.

34734 - MAL'YAN, A.N., AKULOVA, E.A. : O mekhanizme stimulyatsii rastvorimoĭ ATPazy
khloroplastov (CF$_1$). [The stimulation mechanism of soluble ATPase from chloroplasts (CF$_1$).] - Biokhimiya 43 : 1206 - 1211, 1978. [In R, ab : E.]

*34735 - MAMYTOV, A.M., KHASANOVA, M.R. : Vliyanie gerbitsidov na biologicheskuyu
aktivnost' pochv i produktivnost' rasteniĭ sakharnoĭ svekly. [Effect of herbicides on soil biological activity and sugar beet plant productivity.] -
Izv. Akad. Nauk SSSR, Ser. biol. 1977 (4) : 525 - 533, 1977. [In R, ab : E.]

34736 - MANCINELLI, A.L., RABINO, I. : The "high irradiance responses" of plant
photomorphogesis. - Bot. Rev. 44 : 129 - 180, 1978. [Ps, Chl.]

*34737 - MANDAHAR, C.L., GARG, I.D. : Note on a statistical method for the determina-
tion of the leaf area of potato. - Indian J. agr. Sci. 46 : 436 - 438, 1976.

34738 - MANDAL, R.K., KUNDU, A.B. : Role of phosphoenolpyruvate carboxylase in a C$_3$
plant. - In : Inter-Disciplinary Symposium on Photosynthesis and Productivity. Abstract of Papers. Pp. 45 - 46. Indian nat. Sci. Acad., New Delhi
1978.

34739 - MANETAS, Y., AKOYUNOGLOU, G. : Studies on the fate of δ-aminolevulinic acid
induced-protochlorophyllide. - In : AKOYUNOGLOU, G., ARGYROUDI-AKOYUNOGLOU,
J.H. (ed.) : Chloroplast Development. Pp. 105 - 110. Elsevier/North-Holland
Biomedical Press, Amsterdam - New York - Oxford 1978.

34740 - MANIKOWSKI, H., BAUMAN, D., MARTYŃSKI, T., FRĄCKOWIAK, D. : Influence of
dye-dye interaction on lifetime of photosynthetic pigments. - In : FIALA, J.
(ed.) : Third International Seminar on Energy Transfer in Condensed Matter.
Pp. 129 - 135. Univerzita Karlova, Praha 1978.

34741 - MANN, K.H. : Some adaptations for high productivity in seaweeds. - In :
HALL, D.O., COOMBS, J., GOODWIN, T.W. (ed.) : Proceedings of the Fourth Congress on Photosynthesis. Pp. 235 - 244. Biochem. Soc., London 1978. [Ps.]

34742 - MANOLOV, P., RANGELOV, B., BORICHENKO, N. : Ekstraktsiya i radiokhromato-
grafsko razdelyane na rannite produkti na fotosintezata pri visshite C$_3$
rasteniya. [Extraction and radiochromatographic separation of early products
of photosynthesis in higher C$_3$ plants.] - Fiziol. Rast. (Sofia) 4 (2) : 23 -
35, 1978. [In Bulg., ab : E, R.]

34743 - MANSFIELD, T.A., WELLBURN, A.R., MOREIRA, T.J.S. : The rôle of abscisic
acid and farnesol in the alleviation of water stress. - Phil. Trans. roy.
Soc. London B 284 : 471 - 482, 1978. [Ps.]

34744 - MANWARING, J., PULLIN, C.A., EVANS, E.H. : Alteration in concentration  of
carotenoids in Rhodopseudomonas capsulata transferred from dark to light
growth. - Biochem. Soc. Trans. 6 : 1041 - 1043, 1978.

34745 - MANZANO, C., CANDAU, P., GUERRERO, M.G. : Affinity chromatography of Anacystis nidulans ferredoxin-nitrate reductase and NADP reductase on reduced
ferredoxin-Sepharose. - Anal. Biochem. 90 : 408 - 412, 1978.

34746 - MAPLESTON, R.E., GRIFFITHS, W.T. : Effects of illumination of etiolated
leaves on the redox state of NADP in the plastids. - FEBS Lett. 92 : 168 -
172, 1978.

34747 - MARCHAND, P.J., CHABOT, B.F. : Winter water relations of tree-line plant
species on Mt. Washington, New Hampshire. - Arctic alpine Res. 10 : 105 -
116, 1978.

34748 - MARCUS, M.A., LEWIS, A. : Resonance Raman spectroscopy of the retinylidene
chromophore in bacteriorhodopsin (bR$_{570}$), bR$_{560}$, M$_{412}$ and other intermediates : Structural conclusions based on kinetics, analogues, models, and
isotopically labeled membranes. - Biochemistry 17 : 4722 - 4735, 1978.

34749 - MARES, D.J., HAWKER, J.S., POSSINGHAM, J.V. : Starch synthesizing enzymes
in chloroplasts of developing leaves of spinach (*Spinacea oleracea* L.). -
J. exp. Bot. *29* : 829 - 835, 1978.

34750 - MARGULIES, M.M. : Thylakoid subfractions containing ribosomes. - Plant Physiol. *61* (Suppl.) : 114, 1978.

34751 - MARIANI COLOMBO, P. : An ultrastructural study of thallus organization in
*Udotea petiolata*. - Phycologia *17* : 227 - 235, 1978. [Chloroplast.]

34752 - MARKOVSKAYA, E.F. : Karotinoidy raznykh organov *Pinus sylvestris* L. (*Pinaceae*). [Carotenoids of different organs of *Pinus sylvestris* L. (*Pinaceae*).]-
Bot. Zh. *63* : 437 - 441, 1978. [In R.]

34753 - MARKS, S.B., WYDRZYNSKI, T., GOVINDJEE, SCHMIDT, P.G., GUTOWSKY, H.S. :
An NMR study of manganese in chloroplast membranes. - In : AGRIS, P.F.,
LOEPPKY, R.N., SYKES, B.D. (ed.) : Biomolecular Structure and Function. Pp.
95 - 100. Academic Press, New York - San Francisco - London 1978.

34754 - MARKS, T.C. : The carbon economy of *Rubus chamaemorus* L. II. Respiration. -
Ann. Bot. *42* : 181 - 190, 1978.

34755 - MARKS, T.C., TAYLOR, K. : The carbon economy of *Rubus chamaemorus* L. I.
Photosynthesis. - Ann. Bot. *42* : 165 - 179, 1978.

34756 - MARKVART, T. : Exciton transport in the photosynthetic unit. - J. theor.
Biol. *72* : 91 - 116, 1978.

34757 - MARKVART, T. : Infinite model of connected photosynthetic units. - J. theor.
Biol. *72* : 117 - 130, 1978.

34758 - MARKWELL, J.P., REINMAN, S., THORNBER, J.P. : Chlorophyll-protein complexes
from higher plants : a procedure for improved stability and fractionation. -
Arch. Biochem. Biophys. *190* : 136 - 141, 1978.

34759 - MARÓTI, P., RINGLER, A., LACZKÓ, G., SZALAY, L. : Kinetic analysis of the
fast phase of the delayed fluorescence excited by nanosecond laser pulses
in *Chlorella*. - Acta phys. pol. *A 54* : 789 - 796, 1978.

*34760 - MORÓTI, P., RINGLER, A., VIZE, L., SZALAY, L. : The effect of time-depen-
dent coherence of excitation on the primary processes of photosynthesis. -
Acta phys. chem. *23* : 155 - 160, 1977.

34761 - MARRS, B.L. : Genetics and bacteriphage. - In : CLAYTON, R.K., SISTROM,
W.R. (ed.) : The Photosynthetic Bacteria. Pp. 873 - 883. Plenum Press,
New York - London 1978. [Chl, Car.]

34762 - MARRS, B.L.: Mutations and genetic manipulations as probes of bacterial
photosynthesis. - In : SANADI, D.R., VERNON, L.P. (ed.) : Current Topics
in Bioenergetics. Vol. 8. Pp. 261 - 294. Academic Press, New York - San
Francisco - London 1978.

34763 - MARSHALL, J.W. : An experimental study of benthic algal standing crops in
the Leeston Drain, Canterbury. - Mauri Ora *6* : 33 - 39, 1978. [Chl.]

34764 - MARSHO, T.V., SOKOLOVE, P.M. : Regulation of electron flow pathways in chloroplasts. - Plant Physiol. *61* (Suppl.) : 76, 1978.

34765 - MÅRTENSSON, O., ÖSTERDAHL, B.-G. : A note on the chorophylls of *Andreaea
rupestris* HEDW. - Lindbergia *4* : 207 - 208, 1978.

34766 - MARTIN, B., MÅRTENSSON, O., ÖQUIST, G. : Effects of frost hardening and
dehardening on photosynthetic electron transport and fluorescence properties in isolated chloroplasts of *Pinus silvestris*. - Physiol. Plant. *43* :
297 - 305, 1978.

34767 - MARTIN, B., MÅRTENSSON, O., ÖQUIST, G. : Seasonal effects on photosynthetic electron transport and fluorescence properties in isolated chloroplasts
of *Pinus silvestris*. - Physiol. Plant. *44* : 102 - 109, 1978.

34768 - MARVAN, P., KOMÁREK, J., ETTL, H., KOMÁRKOVÁ, J. : Structural elements.
Principal populations of algae. Spatial distribution. - In : DYKYJOVÁ, D.,
KVĚT, J. (ed.) : Pond Littoral Ecosystems. Pp. 295 - 313. Springer-Verlag,
Berlin - Heidelberg - New York 1978.

34769 - **MASAMOTO, K., NISHIMURA, M.** : Effects of ethanol on the interaction of photosynthetic processes in spinach chloroplasts. - Plant Cell Physiol. *19* : 1543 - 1552, 1978.

34770 - **MASCIA, P.** : An analysis of precursors accumulated by several chlorophyll biosynthetic mutants of maize. - Mol. gen. Genet. *161* : 237 - 244, 1978.

34771 - **MASCIA, P.N., ROBERTSON, D.S.** : Studies of chloroplast development in four maize mutants defective in chlorophyll biosynthesis. - Planta *143* : 207 - 211, 1978.

34772 - **MASINOVSKÝ, Z., LIEBL, V.** : Natural selection and some aspects of the evolution of photosynthesis. - In : NOVÁK, V.J.A., LEONOVICH, V.V., PACLTOVÁ, B. (ed.) : Natural Selection. Pp. 195 - 203. Czechoslovak Acad. Sci., Praha 1978.

34773 - **MASLOVA, T.G., KOROLĚVA, O.Ya., ZELENSKIĬ, M.I., GABR, M.A., SAPOZHNIKOV, D.I.** : Tormozhenie vydeleniya kisloroda v protsesse fotosinteza pri narushenii violaksantinovogo tsikla. [Inhibition of oxygen evolution in photosynthesis upon damage of the violaxanthin cycle.] - Fiziol. Rast. *25* : 91 - 96, 1978. [In R, ab : E.]

34774 - **MASTERS, B.R., MAUZERALL, D.** : Effect of quinones on the photoelectric properties of chlorophyll *a*-containing lipid bilayers. - J. Membrane Biol. *41* : 377 - 388, 1978.

34775 - **MATHIS, P.** : Studies of the donor side of photosystem II by absorption spectroscopy. - In : METZNER, H. (ed.) : Photosynthetic Oxygen Evolution. Pp. 125 - 134. Academic Press, London - New York - San Francisco 1978.

34776 - **MATHIS, P., SAUER, K., REMY, R.** : Rapidly reversible flash-induced electron transfer in a *P*-700 chlorophyll-protein complex isolated with SDS. - FEBS Lett. *88* : 275 - 278, 1978.

34777 - **MATHIS, P., VAN BEST, J.A.** : A study of photosystem-2 reactions by flash absorption spectroscopy. - In : HALL, D.O., COOMBS, J., GOODWIN, T.W. (ed.) : Proceedings of the Fourth International Congress on Photosynthesis. Pp. 387 - 396. Biochem. Soc., London 1978.

34778 - **MATORIN, D.N., MARENKOV, V.S., DOBRYNIN, S.A., ORTOIDZE, T.V., VENEDIKTOV, P.S.** : Ustanovka dlya registratsii zamedlennoĭ flyuorestsentsii fotosintezi-ruyushchikh organizmov s impul'snym rezhimom osveshcheniya. [Device for recording delayed fluorescence of photosynthesizing organisms with the impulse regime of irradiation.] - Nauch. Dokl. vyssh. Shkoly, biol. Nauki *21* (11) : 127 - 130, 1978. [In R.]

34779 - **MATORIN, D.N., VENEDIKTOV, P.S., GOSTIMSKIĬ, S.A., TIMOFEEV, K.N.** : Delayed fluorescence and ESR signal of pea mutant with blocked electron flow from Photosystem 1. - Photosynthetica *12* : 274 - 279, 1978.

34780 - **MATORIN, D.N., VENEDIKTOV, P.S., TIMOFEEV, K.N., RUBIN, A.B.** : Issledovanie induktsionnykh krivykh zamedlennoĭ fluorestsentsii zelenykh rasteniĭ. [Induction curves of delayed fluorescence of green plants.] - Nauch. Dokl. vyssh. Shkoly, biol. Nauki *21* (2) : 35 - 41, 1978. [In R.]

34781 - **MATSUDA, T.** : [Studies on the breeding of high yield variety in air-cured tobacco. I. Relations among plant type, growth habit and yield in some bred lines.] - Ikushugaku Zasshi [Jap. J. Breed.] *28* (1) : 1 -12, 1978. [Growth analysis; in Jap., ab : E.]

34782 - **MATSUDA, T.** : [Studies on the breeding of high yield variety of air-cured tobacco. II. Relations among plant type, growth habit and assimilation rate in some bred lines.] - Ikushugaku Zasshi [Jap. J. Breed.] *28* (2) : 106 - 112, 1978. [In Jap., ab : E.]

34783 - **MATSUDA, T., OHASHI, Y.** : [Studies on the breeding of high yield variety of air-cured tobacco. III. Varietal differences in apparent photosynthetic rate of Japanese domestic tobacco cultivars.] - Ikushugaku Zasshi [Jap. J. Breed.] *28* (3) : 177 - 185, 1978. [In Jap., ab : E.]

34784 - MATSUI,T.,EGUCHI, H. : Image processing of plants for evaluation of growth in relation to environment control. - Acta Hort. *87* (Potential Productivity in Protected Cultivation) : 283 - 290, 1978.

34785 - MATSUNO, K. : Evolution of dissipative system : A theoretical basis of Margalef's principle on ecosystem. - J. theor. Biol. *70* : 23 - 31, 1978. [Ps production.]

34786 - MATSUOKA, Y. : [Experimental studies of sulfur dioxide injury on rice plant and its mechanism.] - Spec. Bull. Chiba-Ken agr. exp. Sta. *7* : 1 - 63, 1978. [In Jap., ab : E.]

34787 - MATSUOKA, Y. : Injury of rice plants caused by sulfur dioxide and its mechanism. - Jap. agr. Res. quart. *12* : 183 - 186, 1978. [Ps.]

34788 - MATSUSHIMA, J., KOYAMA, H. : [Influence of the order of nitrogen dioxide and sulfur dioxide on visible injury and photosynthesis in morning glory.] - Mie Daigaku Kankyo Kagaku Kenkyu Kiyo [Rep. Environ. Sci. Mie Univ.] *1978* (3) : 63 - 69, 1978. [In Jap., ab : E.]

34789 - MATSUURA, K., NISHIMURA, M. : Diffusion-potential-induced oxidation and reduction of cytochromes in chromatophores from *Rhodopseudomonas sphaeroides*. - J. Biochem. (Tokyo) *84* : 539 - 546, 1978.

34790 - MATTHEIS, J.R., REBEIZ, C.A. : Chloroplast biogenesis. XXIII. The conversion of exogenous protochlorophyllide into phototransformable protochlorophyllide *in vitro*. - Photochem. Photobiol. *28* : 55 - 60, 1978.

*34791 - MAUDINAS, B., HERBER, R., VILLOUTREIX, J. : Action de différentes substances chimiques sur la carotenogenèse de *Rhodopseudomonas spheroides*. - Chem.-biol. Interact. *5* : 341 - 350, 1972.

34792 - MAUER, J., MAYO, J.M., DENFORD, K. : Comparative ecophysiology of the chromosome races in *Viola adunca* J.E. SMITH. - Oecologia *35* : 91 - 104, 1978.

34793 - MAULOOD, B.K., HINTON, G.C.F., BONEY, A.D. : Diurnal variation of phytoplankton in Loch Lomond. - Hydrobiologia *58* : 99 - 117, 1978. [Chl.]

34794 - MAUNEY, J.R., FRY, K.E., GUINN, G. : Relationship of photosynthetic rate to growth and fruiting of cotton, soybean, sorghum, and sunflower. - Crop Sci. *18* : 259 - 263, 1978.

34795 - MAUZERALL, D. : Bacteriochlorophyll and photosynthetic evolution. - In : CLAYTON, R.K., SISTROM, W.R. (ed.) : The Photosynthetic Bacteria. Pp. 223 - 231. Plenum Press, New York - London 1978.

34796 - MAUZERALL, D. : Photoredox reactions of porphyrins and the origins of photosynthesis. - In : VAN TAMELEN, E.E. (ed.) : Bioorganic Chemistry. Vol. IV. Pp. 303 - 314. Academic Press, New York - San Francisco - London 1978.

34797 - MAUZERALL, D. : Multiple excitations and the yield of chlorophyll *a* fluorescence in photosynthetic systems. - Photochem. Photobiol. *28* : 991 - 998, 1978.

34798 - MAYNE, B.C., RAY, T.B., PETERS, G.A., TOIA, R.E., Jr. : Photosynthesis and acetylene reduction in the *Azolla-Anabaena* symbiosis. - Plant Physiol. *61* (Suppl.) : 102, 1978.

34799 - MAZLIAK, P., DUBACQ, J.-P., DECOTTE, A.-M. : Effets de concentrations croissantes de détergents sur les transferts d'électrons dans les chloroplastes. - Compt. rend. Acad. Sci. Paris, Sér. D *286* : 749 - 752, 1978.

34800 - MBAKU, S.B., FRITZ, G.J., BOWES, G. : Photosynthetic and carbohydrate metabolism in isolated leaf cells of *Digitaria pentzii*. - Plant Physiol. *62* : 510 - 515, 1978.

*34801 - McBRIDE, A.C., LIEN, S., TOGASAKI, R.K., SAN PIETRO, A. : Mutational analysis of *Chlamydomonas reinhardi* : Application to biological solar energy conversion. - In : MITSUI, A., MIYACHI, S., SAN PIETRO, A., TAMURA, S. (ed.) : Biological Solar Energy Conversion. Pp. 77 - 86. Academic Press, New York - San Francisco - London 1977.

34802 - McCARTY, R.E. : AMP is converted to ADP and ATP in the medium before it is bound to coupling factor 1 in illuminated spinach chloroplasts thylakoids. - FEBS Lett. *95* : 299 - 302, 1978.

34803 - McCARTY, R.E. : The ATPase complex of chloroplasts and chromatophores. - In : SANADI, D.R., VERNON, L.P. (ed.) : Current Topics in Bioenergetics. Vol.7. Pp. 245 - 278. Academic Press, New York - San Francisco - London 1978.

34804 - McCARTY, R.E. : The reaction of coupling factor 1 in chloroplast thylakoids with N-substituted maleimides. - In : HALL, D.O., COOMBS, J., GOODWIN, T.W. (ed.) : Proceedings of the Fourth International Congress on Photosynthesis. Pp. 571 - 580. Biochem. Soc., London 1978.

34805 - McCARTY, R.E. : The stoichiometry of the proton-translocating ATPase of chloroplast thylakoids. - In : AZZONE, G.F., AVRON, M., METCALFE, J.C., QUAGLI-ARIELLO, E., SILIPRANDI, N. (ed.) : The Proton and Calcium Pumps. Pp. 65 - 70. Elsevier/North-Holland Biomedical Press, Amsterdam - Oxford - New York 1978.

34806 - McCAUGHEY, J.H. : Estimation of net radiation for a coniferous forest, and the effects of logging of net radiation and the reflection coefficient. - Can. J. Forest. Res. *8* : 450 - 455, 1978.

34807 - McCLELLAND, M.R., OLIVER, L.R., MATHIS, W.D., FRANS, R.E. : Responses of six morningglory *(Ipomoea)* species to bentazon. - Weed Sci. *26* : 459 - 464, 1978. [Ps.]

34808 - McCOLLUM, R.E. : Analysis of potato growth under different P regimes. II. Time by P-status interactions for growth and leaf efficiency. - Agron. J. *70* : 58 - 67, 1978.

34809 - McCREE, K.J., KRESOVICH, S. : Growth and maintenance requirements of white clover as a function of daylength. - Crop Sci. *18* : 22 - 25, 1978.

34810 - McCREE, K.J., SILSBURY, J.H. : Growth and maintenance requirements of subterranean clover. - Crop Sci. *18* : 13 - 18, 1978. [Ps.]

34811 - McCULLOUGH, J.D. : A study of phytoplankton primary productivity and nutrient concentrations in Livingston reservoir, Texas. - Texas J. Sci. *30* : 377 - 387, 1978.

34812 - McCURRY, S.D., HALL, N.P., PIERCE, J., PAECH, C., TOLBERT, N.E. : Ribulose--1,5-bisphosphate carboxylase/oxygenase from parsley. - Biochem. biophys. Res. Commun. *84* : 895 - 900, 1978.

34813 - McCURRY, S.D., PAECH, C., PIERCE, J., TOLBERT, N.E. : Substrate inhibition of ribulose-1,5-bisphosphate carboxylase. - Plant Physiol. *61* (Suppl.) : 99, 1978.

34814 - McDANIEL, G.L., BRESENHAM, G.L. : Use of antitranspirants to improve water relations of cineraria. - HortScience *13* : 466 - 467, 1978. [Stomatal resistance.]

34815 - McDANIEL, M.E., DUNPHY, D.J. : Differential iron chlorosis of oat cultivars. - Crop Sci. *18* : 136 - 138, 1978.

*34816 - McEVOY, F.A., LYNN, W.S. : Isolation of the chloroplast membrane proteins. - Biochem. Soc. Trans. *1* : 894 - 896, 1973.

34817 - McFADDEN, B.A., PUROHIT, K. : Chemosynthetic, photosynthetic, and cyanobacterial ribulose bisphosphate carboxylase. - In : SIEGELMAN, H.W., HIND, G. (ed.) : Photosynthetic Carbon Assimilation. Pp. 179 - 207. Plenum Press, New York - London 1978.

34818 - McINTIRE, C.D. : Marine plant production and utilization - a systems perspective. - In : KRAUSS, R. (ed.) : The Marine Plant Biomass of the Pacific Northwest Coast. Pp. 315 - 333. Oregon State University Press, 1978.

34819 - McINTOSH, R.P., DAWES, C.J. : Low salinity tolerance and the effect of spring water on the distribution of *Boatrychia binderi (Rhodophyta)* in Floridian estuaries. - J. Phycol. *14* (Suppl.) : 38, 1978.

34820 - McKEE, J.W.A., HAWKE, J.C. : The incorporation of [$^{14}$C]bicarbonate and $^{14}CO_2$ into the constituent fatty acids of monogalactosyldiacylglycerol by spinach chloroplasts and leaves. - FEBS Lett. *94* : 273 - 276, 1978.

34821 - McKERSIE, B.D., THOMPSON, J.E. : Phase behavior of chloroplast and microsomal membranes during leaf senescence. - Plant Physiol. *61* : 639 - 643, 1978.

34822 - McKINNEY, D.W., BUCHANAN, B.B., WOLOSIUK, R.A. : Activation of chloroplast ATPase by reduced thioredoxin. - Phytochemistry *17* : 794 - 795, 1978.

34823 - McKINNEY, H.H., MENSER, H.A. : Chlorotic leaf markings in Betzes barley grown in a controlled environment chamber. - Plant Dis. Rep. *60* : 1017 - 1019, 1976. [Chl.]

34824 - McLACHLAN, J., BIDWELL, R.G.S. : Photosynthesis of eggs, sperm, zygotes, and embryos of *Fucus serratus*. - Can. J. Bot. *56* : 371 - 373, 1978.

34825 - McLAUGHLIN, S.P. : Determining understory production in southwestern Ponderosa pine forests. - Bull. Torey bot. Club *105* : 224 - 229, 1978.

34826 - McMILLEN, G.G., McCLENDON, J.H. : Biochemical differentiation in sun and shade leaves of trees. - Plant Physiol. *61* (Suppl.) : 9, 1978.

34827 - McMILLEN, G.G., McCLENDON, J.H. : Leaf angle: An adaptive feature of sun and shade leaves. - Plant Physiol. *61* (Suppl.) : 52, 1978.

34828 - McMULLEN, C.R., GARDNER, W.S., MYERS, G.A. : Aberrant plastids in barley leaf tissue infected with barley stripe mosaic virus. - Phytopathology *68* : 317 - 325, 1978.

34829 - MEDINA, E., SOBRADO, M., HERRERA, R. : Significance of leaf orientation for leaf temperature in an Amazonian sclerophyll vegetation. - Radiat. Environ. Biophys. *15* : 131 - 140, 1978.

34830 - MEDVEDEV, A.M. : Luchshie nizkostebel'nye yarovye pshenitsy mirovoǐ kollektsii VIR i perspektivy ikh vnedreniya na polivnykh zemlyakh Srednego Povolzh'ya. [The best low culm spring wheats of the world collection of the All-union Institute of Plant Production and the perspectives of their growing on irrigated soils of Central Povolzh'e.] - Dokl. VASKHNIL *1978* (10) : 9 - 11, 1978. [Growth analysis; in R.]

34831 - MEDYANNIKOV, V.M., KHOLUPENKO, I.P. : Postuplenie $^{14}$C-assimilyatov v list'ya soi v zavisimosti ot udel'noǐ radioaktivnosti $^{14}CO_2$ i zavyadaniya rasteniǐ. [Transport of $^{14}$C-photosynthates into soybean leaves in dependence on specific radioactivity of $^{14}CO_2$ and plant wilting.] - In : Pogloshchenie i Peredvizhenie Veshchestv u Rasteniǐ. Pp. 35 - 38. Dal'nevostoch. nauch. Tsentr Akad. Nauk SSSR, biol.-pochv. Inst., Vladivostok 1978. [In R.]

34832 - MEEKS, J.C., CASTENHOLZ, R.W. : Photosynthetic properties of the extreme thermophile *Synechococcus lividus* - I. Effect of temperature on fluorescence and enhancement of $CO_2$ assimilation. - J. therm. Biol. *3* : 11 - 18, 1978.

34833 - MEEKS, J.C., CASTENHOLZ, R.W. : Photosynthetic properties of the extreme thermophile *Synechococcus lividus* - II. Stoichiometry between oxygen evolution and $CO_2$ assimilation. - J. therm. Biol. *3* : 19 - 24, 1978.

34834 - MEHROTRA, O.N., SAXENA, H.K., LAL, K.B., SINGH, R., YADAV, S.S. : Varietal differences in photosynthesis and productivity of Bengal gram (*Cicer arietinum*, L.). - In : Inter-Disciplinary Symposium on Photosynthesis and Productivity. Abstract of Papers. Pp. 47 - 49. Indian nat. Sci. Acad., New Delhi 1978.

34835 - MEINL, G., MOLL, A. : Ertragsphysiologische Untersuchungen an Kopfkohl unter Einbeziehung seiner Photosyntheserate. - Arch. Züchtungsforsch. *8* : 221 - 229, 1978.

34836 - MEINZER, F.C. : Observaciones sobre la distribución taxonómica y ecológica de la fotosíntesis $C_4$ en la vegetación del noroeste de Centroamérica. [Observations on the taxonomic and ecologic distribution of $C_4$ photosynthesis in northwestern Central America.] - Rev. Biol. trop. *26* : 359 - 369, 1978. [In Span., ab : E.]

34837 - MEISCH, H.-U. : The role of vanadium in chlorophyll biosynthesis of green algae. - In : AKOYUNOGLOU, G., ARGYROUDI-AKOYUNOGLOU, J.H. (ed.) : Chloroplast Development. Pp. 141 - 146. Elsevier/North-Holland Biomedical Press, Amsterdam - New York - Oxford 1978.

34838 - MEISCH, H.-U., BAUER, J. : The role of vanadium in green plants. IV. Influence on the formation of δ-aminolevulinic acid in *Chlorella*. - Arch. Microbiol. *117* : 49 - 52, 1978.

34839 - MEKHTI-ZADE, É.R., BEZUGLOV, V.K. : Dinamika pokazateleĭ vodnogo rezhima
list'ev i énergeticheskogo obmena khloroplastov ozimoĭ pshenitsy. [Dynamics
of leaf water regime characteristics and energy exchange of chloroplasts
in winter wheat.] - In : Vodnyĭ Rezhim Rasteniĭ v Svyazi s Raznymi Ékologicheskimi Usloviyami. Pp. 269 - 273. Izdatel'stvo Kazanskogo Universiteta,
Kazan' 1978. [Chl; in R.]

*34840 - MELACK, J.M. : Primary productivity and fish yields in tropical lakes. -
Trans. amer. Fish. Soc. *105* : 575 - 580, 1976. [Ps.]

34841 - MELANDRI, B.A., CASADIO, R., BACCARINI MELANDRI, A. : Energy levels and
rates of photophosphorylation in bacterial chromatophores. - In : HALL, D.O.,
COOMBS, J., GOODWIN, T.W. (ed.) : Proceedings of the Fourth International
Congress on Photosynthesis. Pp. 601 - 609. Biochem. Soc., London 1978.

34842 - MELANDRI, B.A., DE SANTIS, A., VENTUROLI, G., BACCARINI-MELANDRI, A. :
The rates of onset of photophosphorylation and of the protonic electrochemical potential difference in bacterial chromatophores. - FEBS Lett. *95* :
130 - 134, 1978.

34843 - MELIS, A. : Oxidation-reduction potential dependence of the two kinetic
components in chloroplast system II primary photochemistry. - FEBS Lett.
*95* : 202 - 206, 1978.

34844 - MELIS, A., HOMANN, P.H. : A selective effect of $Mg^{2+}$ on the photochemistry
at one type of reaction center in photosystem II of chloroplasts. - Arch.
Biochem. Biophys. *190* : 523 - 530, 1978.

34845 - MELL, V., SENGER, H. : Photochemical activities. Pigment distribution and
photosynthetic unit size of subchloroplast particles isolated from synchronized cells of *Scenedesmus obliquus*. - Planta *143* : 315 - 322, 1978.

34846 - MELLER, E., HAREL, E. : The pathway of 5-aminolevulinic acid synthesis in
*Chlorella vulgaris* and in *Fremyella diplosiphon*. - In : AKOYUNOGLOU, G.,
ARGYROUDI-AKOYUNOGLOU, J.H. (ed.) : Chloroplast Development. Pp. 51 - 57.
Elsevier/North-Holland Biomedical Press, Amsterdam - New York - Oxford
1978.

34847 - MELOUK, H.A. : Determination of leaf necrosis caused by *Cercospora arachi-
dicola* HORI in peanut as measured by loss in total chlorophyll. - Peanut
Sci. *5* (1) : 17 - 18, 1978.

34848 - MENDE, D., NIEMEYER, H., HECKER, S., WIESSNER, K. : *In vivo* regulation of
the photosynthetic electron transport. - Photosynthetica *12* : 440 - 448,
1978.

B34849 - MENGEL, K., KIRKBY, E.A. : Principles of Plant Nutrition. - International
Potash Institute, Berne 1978. [Ps.]

34850 - MENKE, W., KOENIG, F., SCHMID, G.H., RADUNZ, A. : Functional characterization of four thylakoid membrane polypeptides with apparent molecular weights between 40 000 and 48 000. - Z. Naturforsch. *33 C* : 280 - 289, 1978.

34851 - MEON, S., WALLACE, H.R., FISHER, J.M. : Water relations of tomato (*Lycopersicon esculentum* MILL. cv. Early Dwarf Red) infected with *Meloidogyne
javanica* (TREUB), CHITWOOD. - Physiol. Plant Pathol. *13* : 275 - 281, 1978.

34852 - MERRICK, J.M. : Metabolism of reserve materials. - In : CLAYTON, R.K.,
SISTROM, W.R. (ed.) : The Photosynthetic Bacteria. Pp. 199 - 219. Plenum
Press, New York - London 1978. [Chromatophores.]

34853 - MERRILL, J.E., WAALAND, J.R. : An investigation of photosynthetic rates
in a fast growing strain of the red alga *Gigartina exasperata*. - J. Phycol. *14* (Suppl.) : 38, 1978.

34854 - MERTENS, W.C., WRIGHT, R.D. : Root and shoot growth rate relationships of
two cultivars of Japanese Holly. - J. amer. Soc. hort. Sci. *103* : 722 -
724, 1978. [Dry-matter production.]

34855 - MERZLYAK, M.N. : Densitometricheskoe opredelenie karotinoidov rasteniĭ v
tonkom sloe na plastinakh "Silufol". [Densitometric determination of plant
carotenoids in the thin layer on "Silufol" plates.] - Nauch. Dokl. vyssh.
Shkoly, biol. Nauki 21 (1) : 134 - 138, 1978. [In R.]

34856 - MERZLYAK, M.N., POGOSYAN, S.I., YUFEROVA, S.G., SHEVYRËVA, V.V. : Ispol'-
zovanie 2-tiobarbiturovoĭ kisloty pri issledovanii pereokisleniya lipidov
v tkanyakh rasteniĭ. [Use of 2-thiobarbituric acid in studies of lipid
peroxidation in plant tissues.] - Nauch. Dokl. vyssh. Shkoly, biol. Nauki
21 (9) : 86 - 94, 1978. [Chloroplast; in R.]

34857 - MESSIER, J., GUCKERT, A. : Contribution à la mesure de la photosynthèse au
champ par la méthode utilisant le $^{14}CO_2$. - Bull. École nat. super. Agron.
Ind. aliment. (Nancy) 20 (1-2) : 31 - 45, 1978.

34858 - METZ, J., BISHOP, N.I. : Characterization of low fluorescent photosynthetic mutants of Scenedesmus obliquus blocked on the oxidizing side of PS-II.
- Plant Physiol. 61 (Suppl.) : 88, 1978.

34859 - METZNER, H. : Oxygen evolution as energetic problem. - In : METZNER, H.
(ed.) : Photosynthetic Oxygen Evolution. Pp. 60 - 76. Academic Press, London - New York - San Francisco 1978.

34860 - METZNER, H. : Solar energy conversion by plant photosynthesis. - Festkör-
perprobleme 18 : 33 - 52, 1978.

34861 - MEYER, J., KELLEY, B.C., VIGNAIS, P.M. : Aerobic nitrogen fixation by Rhodopseudomonas capsulata. - FEBS Lett. 85 : 224 - 228, 1978. [Ps.]

34862 - MEYER, J., KELLEY, B.C., VIGNAIS, P.M. : Effect of light on nitrogenase
function and synthesis in Rhodopseudomonas capsulata. - J. Bacteriol. 136 :
201 - 208, 1978. [Ps, Chl.]

34863 - MEYER, J., KELLEY, B.C., VIGNAIS, P.M. : Nitrogen fixation and hydrogen
metabolism in photosynthetic bacteria. - Biochimie 60 : 245 - 260, 1978.

34864 - MEZENTSEV, V.V., MOLCHANOV, M.I., TRUSOVA, V.M. : K izucheniyu stroeniya
lamellyarnoĭ sistemy khloroplastov. [Structure of chloroplast lamellar
system.] - Dokl. Akad. Nauk SSSR 240 : 213 - 216, 1978. [In R.]

B34865 - Mezostruktura i Funktsional'naya Aktivnost' Fotosinteticheskogo Apparata.
[Mesostructure and Functional Activity of the Photosynthetic Apparatus.] -
Ural'skiĭ gosudarstvennyĭ Universitet, Sverdlovsk 1978. [In R.]

34866 - MGALOBLISHVILI, M.P., KHETSURIANI, N.D., KALANDADZE, A.N., SANADZE, G.A. :
O lokalizatsii biosinteza izoprena v khloroplastakh list'ev topolya.
[Localization of isoprene biosynthesis in poplar leaf chloroplasts.] -
Fiziol. Rast. 25 : 1055 - 1061, 1978. [Ps; in R, ab : E.]

34867 - MICHEL, J.-M. : Dependence of the early steps of chlorophyll(ide) synthesis on respiration in dark-grown Euglena gracilis. - Planta 141 : 45 - 50,
1978.

34868 - MICHELS, P.A.M., KONINGS, W.N. : The electrochemical proton gradient generated by light in membrane vesicles and chromatophores from Rhodopseudomonas sphaeroides. - Europe. J. Biochem. 85 : 147 - 155, 1978.

34869 - MIKHAĬLOV, O.F., BESSONOVA, V.P., KORYTOVA, A.I. : Vliyanie kinetina i rent-
genovskogo oblucheniya na rostovye protsessy i nakoplenie pigmentov v prorostkakh gorokha. [Effect of kinetin and X-ray irradiation on growth processes and pigment accumulation in pea seedlings.] - Fiziol. Biokhim.
kul't. Rast. 10 : 70 - 76, 99, 112, 1978. [In R, ab : E.]

*34870 - MIKHEEVA, T.M. : O pokazatelyakh udel'noĭ aktivnosti fitoplanktona i neko-
torykh prichinakh, ikh opredelyayushchikh. [Phytoplankton specific activity
indices and some factors determining them.] - Gidrobiol. Zh. 13 (3) : 11 -
16, 1977. [Ps; in R, ab : E.]

34871 - MIKULOVICH, T.P. : Synthesis of plastid and cytoplasmic ribosomal RNAs in
isolated pumpkin cotyledons. I. Effect of detachment and light. - Biochem.
Physiol. Pflanz. 172 : 93 - 100, 1978.

34872 - MIKULOVICH, T.P., WOLLGIEHN, R., KHOKHLOVA, W.A., NEUMANN, D., KULAEVA, O.N.:
Synthesis of plastid and cytoplasmic ribosomal RNAs in isolated pumpkin cotyledons. 2. Effect of cytokinin and light. - Biochem. Physiol. Pflanzen
*172* : 101 - 110, 1978. [Chl.]

*34873 - MILBOURN, G.M. : Yield potential of maize in different regions of the world.
- Ann. appl. Biol. *87* : 242 - 245, 1977. [Ps.]

34874 - MILICĂ, C.I. : Fiziologia porumbului in condiții dirijate de nutriție și
umiditate a solului. [Maize physiology under controlled nutrition and soil
moisture.] - Lucrări științ.,Ser.Agron. *1978* : 45 - 46, 1978. [ Dry-matter
accumulation; in Roum., ab : E.]

34875 - MILICĂ, C.I., MATEI, V., COSMULESCU, E. : Die Dynamik der Wachstumsprozessen
und Synthese des Zuckers bei den Ruben an verschiedenen ernährenden Verhält-
nissen. - Lucrări științ.,Ser.Agron. *1978* : 77 - 80, 1978.

34876 - MILICĂ, C.I., POPESCU, I., TEȘU, V., AFUSOAIE, I. : Dynamics of growth, development and metabolic processes in some corn hybrids. - Lucrări științ.,Ser.
Agron. *1978*: 47 - 50, 1978. [Dry-matter accumulation.]

34877 - MILICĂ, C.I., TEȘU, V., AIRINEI, A. : Particularități fiziologice ale unor
hibrizi de floarea soarelui. [Physiological   peculiarities differen-
tiating   the productivity of new sunflower hybrids.] - Lucrări științ.,Ser.
Agron. *1978* : 75 - 76, 1978. [Chl; in Roum., ab : E.]

34878 - MILIUS, A., KYVASK, V. : Sezonnye izmeneniya biomassy, soderzhaniya khlorofilla *a* i fosfataznoĭ aktivnosti fitoplanktona v ozere Viĭtna-Linayarv.
[Seasonal variation of phytoplankton biomass, chlorophyll *a* content, and
phosphatase activity in Lake Viitna-Linajarv.] - Eesti NSV Tead. Akad. Toim.,
Biol. *27* : 306 - 313, 1978. [In R, ab : E, Est.]

34879 - MILLER, B.F. : Effects of sprouting on nutritional value of wheat. - Proc.
10th nat. Conf. Wheat Util. Res. *ARM-W-4* : 144 - 147, 1978. [Car.]

34880 - MILLER, G.R., WATSON, A. : Heather productivity and its relevance to the
regulation of red grouse populations. - In : HEAL, O.W., PERKINS, D.F. (ed.):
Production Ecology of British Moors and Montane Grasslands. Pp. 277 - 285.
Springer-Verlag, Berlin-Heidelberg-New York 1978.

34881 - MILLER, J.E. : A proposal for the genetic manipulation of plant metabolism.
- J. theor. Biol. *74* : 153 - 154, 1978. [Ps, chloroplast.]

34882 - MILLER, K.R. : Structural organization in the photosynthetic membrane. - In :
AKOYUNOGLOU, G., ARGYROUDI-AKOYUNOGLOU, J.H. (ed.) : Chloroplast Development.
Pp. 17 - 30. Elsevier/North-Holland Biomedical Press, Amsterdam - New York -
Oxford 1978.

34883 - MILLER, K.R., OHAD, I. : Chloroplast membrane biogenesis in *Chlamydomonas* :
correlation between the formation of membrane components and membrane structure. - Cell Biol. Int. Rep. *2* : 537 - 549, 1978.

34884 - MILLER, M.C., ALEXANDER, V., BARSDATE, R.J. : The effects of oil spills on
phytoplankton in an arctic lake and ponds. - Arctic *31* : 192 - 218, 1978.
[Chl.]

34885 - MILLER, M.C., HATER, G.R., VESTAL, J.R. : Effect of Prudhoe crude oil on
carbon assimilation by planktonic algae in an arctic pond. - In : ADRIANO,
D.C., BRISBIN, I.L.,Jr. (ed.) : Environmental Chemistry and Cycling Processes. Pp. 833 - 850. N.T.I.S., Springfield, Va. 1978.

34886 - MILLER, P.C., OECHEL, W.C., STONER, W.A., SVEINBJÖRNSSON, B. : Simulation
of $CO_2$ uptake and water relations of four arctic bryophytes at Point Barrow,
Alaska. - Photosynthetica *12* : 7 - 20, 1978.

*34887 - MILLER, P.C., STONER, W.A., HOM, J., POOLE, D.K. : Potential influence of
thermal effluents on the production and water-use efficiency of mangrove
species in south Florida. - In : ESCH, G.W., McFARLANE, R.W. (ed.) : Thermal
Ecology. II. Proceedings of a Symposium Held at Augusta, Georgia, 1975.
Pp. 39 - 45. Tech. Inform. Center, Energy Res. Develop. Administr., Washington, D.C. 1976.

34888 - MILLER, P.C., STONER, W.A., TIESZEN, L.L., ALESSIO, M., McCOWN, B., CHAPIN,
F.S., SHAVER, G. : A model of carbohydrate,nitrogen, phosphorus allocation
and growth in tundra production. - In : TIESZEN, L.L. (ed.) : Vegetation and
Production Ecology of an Alaskan Arctic Tundra. Pp. 577 - 598. Springer-Verlag, New York - Heidelberg - Berlin 1978.

34889 - MILLS, J.D., BARBER, J. : Fluorescence changes in isolated broken chloroplasts and the electrical double layer. - Biophys. J. 21 : 257 - 272, 1978.

34890 - MILLS, J.D., HIND, G. : Use of the fluorescent lanthanide $Tb^{3+}$ as a probe
for cation-binding sites associated with isolated chloroplast thylakoid membranes. - Photochem. Photobiol. 28 : 67 - 73, 1978.

34891 - MILLS, J.D., SLOVACEK, R.E., HIND, G. :  Cyclic electron transport in isolated intact chloroplasts. Further studies with antimycin. - Biochim. biophys. Acta 504 : 298 - 309, 1978.

34892 - MILLS, W.R., WILSON, K.G. : Amino acid biosynthesis in isolated pea chloroplasts: metabolism of labeled aspartate and sulfate. - FEBS Lett. 92 : 129 -
132, 1978.

34893 - MIMURO, M., FUJITA, Y. : Excitation energy transfer between pigment system
II units in blue-green algae. - Biochim. biophys. Acta 504  : 406 - 412,
1978.

34894 - MINARČIC, P., PAULECH, C. : Vplyv múčnatky trávnej na chloroplasty listov
jačmeňa, ich štruktúru a funkciu. [Influence of powdery mildew on the chloroplasts of barley leaves, their structure and function.] - Acta bot. slov.
Acad. Sci. slov., Ser. B 2 : 35 - 52, 1978. [In Slov., ab : E, R.]

34895 - MINCHIN, F.R., SUMMERFIELD, R.J., EAGLESHAM, A.R.J., STEWART, K.A. : Effects
of short-term waterlogging on growth and yield of cowpea (Vigna unguiculata).
- J. agr. Sci. 90 : 355 - 366, 1978. [Primary production.]

34896 - MINCHIN, P.E.H. : Analysis of tracer profiles with applications to phloem
transport. - J. exp. Bot. 29 : 1441 - 1450, 1978. [Photosynthates.]

34897 - MIRŽINSKI-STEFANOVIČ, L. : Proučavanje dejstva atrazina na fotohemijske pro-
cese u biljke kukuruza. [Effects of atrazine on photochemical processes in
maize.] - Arhiv poljopr. Nauke 31 : 13 - 28, 1978. [In Croat., ab : E.]

34898 - MISAGHI, I.J., DeVAY, J.E., DUNIWAY, J.M. :  Relationship between occlusion
of xylem elements and disease symptoms in leaves of cotton infected with
Verticillium dahliae. - Can. J. Bot. 56 : 339 - 342, 1978. [Chl.]

34899 - MISHKIND, M., MAUZERALL, D. : Diurnal variation of photosynthesis in Ulva. -
Plant Physiol. 61 (Suppl.) : 87, 1978.

34900 - MITSNER, B.I., KARNAUKHOVA, E.N., ZVONKOVA, E.N., EVSTIGNEEVA, R.P : Podkhod
k napravlennomu vvedeniyu retinalya v lys-lys-fragment "khromofornogo tsentra" bakteriorodopsina. [An approach to specific introduction of retinal
into Lys-Lys fragment of bacteriorhodopsin chromophoric centre.] - Bioorg.
Khim. 4 : 1684 - 1686, 1978. [In R, ab : E.]

34901 - MITSUI, A. : Bio-solar hydrogen production. - In : VEZIROGLU, T.N., SEIFRITZ,
W. (ed.) : Hydrogen Energy System. Vol.3. Pp. 1267 - 1291. Pergamon Press,
Oxford - New York 1978. [Ps.]

34902 - MITSUI, A. : Marine photosynthetic microorganisms as potential energy resources : Research on nitrogen fixation and hydrogen production. - In : Proceedings of the 5th International Ocean Development Conference. Pp. B1-29-52.
IODC Organizing Committee, Tokyo 1978.

34903 - MITSUI, A. : Marine algae and aquatic plants as food components. - In :
OVERMIRE, T.G. (ed.) : Microbial Conversion for Food and Fodder Production
and Waste Management. Pp. 3 -31. 1978.  [Ps.]

*34904 - MITSUI, A., KUMAZAWA, S. : Hydrogen production by marine photosynthetic or-
ganisms as a potential energy resource. - In : MITSUI, A., MIYACHI, S., SAN
PIETRO, A., TAMURA, S. (ed.) : Biological Solar Energy Conversion. Pp. 23 -
51. Academic Press, New York - San Francisco - London 1977.

*34905 - **MIYACHI, S., KAMIYA, A., MIYACHI, S.** : Wavelength-effects of incident light on carbon metabolism in *Chlorella* cells. - In : MITSUI, A., MIYACHI, S., SAN PIETRO, A., TAMURA, S. (ed.) : Biological Solar Energy Conversion. Pp. 167 - 182. Academic Press, New York - San Francisco - London 1977.

34906 - **MIYACHI, S., MIYACHI, S., KAMIYA, A.** : Wavelength effects on photosynthetic carbon metabolism in *Chlorella*. - Plant Cell Physiol. *19* : 277 - 288, 1978.

34907 - **MIYAKE, H., MAEDA, E.** : Starch accumulation in bundle sheath chloroplasts during the leaf development of C3 and C4 plants of *Gramineae*. - Can. J. Bot. *56* : 880 - 882, 1978.

34908 - **MIYAKE, J., OCHIAI-YANAGI, S., KASUMI, T., TAKAGI, T.** : Isolation of a membrane protein from *R. rubrum* chromatophores and its abnormal behavior in SDS-polyacrylamide gel electrophoresis due to a high binding capacity for SDS. - J. Biochem. (Tokyo) *83* : 1679 - 1686, 1978.

34909 - **MIYASAKA, T., WATANABE, T., FUJISHIMA, A., HONDA, K.** : Light energy conversion with chlorophyll monolayer electrodes. *In vitro* electrochemical simulation of photosynthetic primary processes. - J. amer. chem. Soc. *100* : 6657 - 6665, 1978.

34910 - **MIYAZAKI, T., MURAKAMI, S., MORITA, S.** : Average distance between bacteriochlorophyll molecules on chromatophore membranes of *Rhodopseudomonas sphaeroides*. - Plant Cell Physiol. *19* : 1333 - 1338, 1978.

34911 - **MIZIORKO, H.M.** : Magnetic resonance studies on ribulose bisphosphate carboxylase. - In : SIEGELMAN, H.W., HIND, G. (ed.) : Photosynthetic Carbon Assimilation. Pp. 41 - 42. Plenum Press, New York - London 1978.

34912 - **MIZUSAWA, M., KAGEYAMA, A., YOKOHAMA, Y.** : Physiology of benthic algae in tide pools I. Photosynthesis-temperature relationships in summer. - Jap. J. Phycol. *26* : 109 - 114, 1978.

34913 - **MO, M., TJØRNHOM, T.** : Losses of carbon-containing substances during dry matter determination by oven-drying. - Acta Agr. scand. *28* : 196 - 202, 1978.

34914 - **MOED, J.R., HALLEGRAEFF, G.M.** : Some problems in the estimation of chlorophyll-*a* and phaeopigments from pre- and post-acidification spectrophotometric measurements. - Int. Rev. ges. Hydrobiol. *63* : 787 - 800, 1978.

34915 - **MOELLER, R.E.** : Carbon-uptake by the submerged hydrophyte *Utricularia purpurea*. - Aquatic Bot. *5* : 209 - 216, 1978.

34916 - **MOELLER, R.E.** : Seasonal changes in biomass, tissue chemistry, and net production of the evergreen hydrophyte, *Lobelia dortmanna*. - Can. J. Bot. *56* : 1425 - 1433, 1978.

34917 - **MOESTRUP, Ø.** : On the phylogenetic validity of the flagellar apparatus in green algae and other chlorophyll *a* and *b* containing plants. - BioSystems *10* : 117 - 144, 1978.

34918 - **MOGILEVA, G.A., SAKHAROVA, O.V., ZELENSKIĬ, M.I.** : Metodika massovykh opredeleniĭ fotokhimicheskoĭ aktivnosti i fotofosforilirovaniya na pshenitse. [Methods for mass determinations of photochemical activity and photophosphorylation in wheat.]-Tr. priklad. Bot. Genet. Selektsii *61* (3) : 111 - 118, 1978. [In R, ab : E.]

34919 - **MOHANTY, P.** : Primary process in photosynthesis and plant productivity. - In : Inter-Disciplinary Symposium on Photosynthesis and Productivity. Abstract of Papers. Pp. 10 -11. Indian nat. Sci. Acad., New Delhi 1978.

34920 - **MOHR, H.** : Photomorphogenese, erläutert am Beispiel der Plastidenentwicklung. - Freiburger Universitätsblätter *59* : 41 - 47, 1978. [Chl.]

34921 - **MOHR, H.** : Pattern specification and realization in photomorphogenesis. - Bot. Mag. (Tokyo) Special Issue *1* (Controlling Factors in Plant Development) : 199 - 217, 1978. [Chl.]

34922 - **MOHR, H., OELZE-KAROW, H.** : Phytochrome and chloroplast development. - In : AKOYUNOGLOU, G., ARGYROUDI-AKOYUNOGLOU, J.H. (ed.) : Chloroplast Develop-

ment. Pp. 769 - 779. Elsevier/North-Holland Biomedical Press, Amsterdam -
New York - Oxford 1978.

34923 - **MOKRONOSOV, A.T.** : Éndogennaya regulyatsia fotosinteza v tselom rastenii.
[Endogenous control of photosynthesis in the whole plant.] - Fiziol. Rast.
*25* : 938 - 951, 1978. [In R, ab : E.]

34924 - **MOKRONOSOV, A.T.** : Mezostruktura i funktsional'naya aktivnost' fotosinteticheskogo apparata. [Mesostructure and functional activity of the photosynthetic apparatus.] - In : Mezostruktura i Funktsional'naya Aktivnost' Fotosinteticheskogo Apparata. Pp. 5 - 30. Ural'skiĭ gosudarstvennyĭ Universitet, Sverdlovsk 1978. [In R.]

34925 - **MOKRONOSOV, A.T., BORZENKOVA, R.A.** : Metodika kolichestvennoĭ otsenki struktury i funktsional'noĭ aktivnosti fotosinteziruyushchikh tkaneĭ i organov.
[Procedure of quantitative estimation of the structure and functional activity of photosynthesizing tissues and organs.] - Tr. priklad. Bot. Genet.
Selektsii  *61* (3) : 119 - 133, 1978. [In R, ab : E.]

34926 - **MOKRONOSOV, A.T., SHMAKOVA, T.V.** : Sravnitel'nyĭ analiz mezostruktury fotosinteticheskogo apparata u mezofitnykh i kserofitnykh rasteniĭ. [Comparati-
ve analysis of mesostructure of photosynthetic apparatus of mesophytic and
xerophytic plants.] - In : Mezostruktura i Funktsional'naya Aktivnost'
Fotosinteticheskogo Apparata. Pp. 103 - 107. Ural'skiĭ gosudarstvennyĭ Universitet, Sverdlovsk 1978. [In R.]

34927 - **MOLCHANOV, M.I.** : Aminoatsilfosfatidilglitseriny membrannoĭ sistemy plastid
pri biogeneze khloroplastov. [The aminoacyl phosphatidyl glycerols of the
plastid membrane system in the process of chloroplast biogenesis.] - Biokhimiya *43* : 312 - 320, 1978. [In R, ab : E.]

*34928 - **MOLCHANOV, M.I., BALAUR, N.S., TRUSOVA, V.M.** : Ob izmeneniyakh fosfolipidov membrannoĭ sistemy plastid pri formirovanii ul'trastruktury khloroplastov pod deĭstviem sveta. [The changes of phospholipids in the membrane
system of plastids at the formation of chloroplast ultrastructure under the
action of light.] - Bioorg. Khim. *1* : 702 - 708, 1975. [In R, ab : E.]

34929 - **MOLCHANOV, M.I., MEZENTSEV, V.V.** : K izucheniyu funktsional'noĭ roli amino-
atsilfosfatidilglitserinov v lamellyarnoĭ sisteme khloroplastov. [Functi-
onal role of aminoacyl phosphatidyl glycerols in chloroplast lamellar system.] - Biokhimiya *43* : 1429 - 1437, 1978. [In R, ab : E.]

*34930 - **MOLCHANOV, M.I., TRUSOVA, V.M.** : O biosinteze belkov membrannoĭ sistemy
plastid pri biogeneze khloroplastov. [Biosynthesis of proteins of the plastid membrane system during chloroplast biogenesis.] - Dokl. Akad. Nauk SSSR
*224* : 479 - 482, 1975. [In R.]

*34931 - **MOLCHANOV, M.I., TRUSOVA, V.M., KSENZENKO, S.M., KOTOVSKAYA, A.P.** : Biosintez i fraktsionnyĭ sostav lipoproteidov lamellyarnoĭ sistemy khloroplastov.
[Biosynthesis and fractional composition of lipoproteins of the chloroplast
lamellar system.] - Dokl. Akad. Nauk SSSR *235* : 236 - 239, 1977. [In R.]

34932 - **MOLL, A., HEERKLOSS, B.** : Produktions- und Attraktionsleistung während der
Alterung der Kartoffelpflanze. - Biochem. Physiol. Pflanzen *173* : 213 - 219,
1978.

34933 - **MOLL, A., HENNIGER, W.** : Genotypische Photosyntheserate von Kartoffeln und
ihre mögliche Rolle für die Ertragsbildung. - Photosynthetica *12* : 51 - 61,
1978.

34934 - **MOLL, W.A.W., DE WIT, B., LUTTER, R.** : Chlorophyllase activity in developing
leaves of *Phaseolus vulgaris* L. - Planta *139* : 79 - 83, 1978.

34935 - **MOLL, W.A.W., STEGWEE, D.** : The activity of Triton X-100 soluble chlorophyllase in liposomes. - Planta *140* : 75 - 80, 1978.

34936 - **MONDAL, M.H., BRUN, W.A., BRENNER, M.L.** : IAA levels and photosynthesis
in leaves of control and depodded soybean plants. - Plant Physiol. *61*
(Suppl.) : 8, 1978.

34937 - MONDAL, M.H., BRUN, W.A., BRENNER, M.L. : Effects of sink removal on photosynthesis and senescence in leaves of soybean (*Glycine max* L.) plants. -
Plant Physiol. *61* : 394 - 397, 1978.

34938 - MONDOVÌ, B., GRAZIANI, M.T., MORPURGO, L. : Properties of the copper site
of plastocyanin. - In : HALL, D.O., COOMBS, J., GOODWIN, T.W. (ed.) : Proceedings of the Fourth International Congress on Photosynthesis. Pp. 799 -
803. Biochem. Soc., London 1978.

34939 - MONSELISE, S.P., VARGA, A., BRUINSMA, J. : Growth analysis of the tomato
fruit, *Lycopersicon esculentum* MILL. - Ann. Bot. *42* : 1245 - 1247, 1978.

34940 - MONSI, M. : Studies on production processes. - In : TAMIYA, H. (ed.) :
Summary Report on the Contribution of the Japanese National Committee for
IBP, 1964 - 1974. JIBP Synthesis. Vol. 20. Pp. 57 - 95. Univ. of Tokyo
Press, Tokyo 1978.

34941 - MONTALBINI, P., KOCH, F., BURBA, M., ELSTNER, E.F. : Increase in lipid-de-
pendent carotene destruction as compared to ethylene formation and chlorophyllase activity following mixed infection of sugar beet (*Beta-vulgaris*
L.) with beet yellows virus and beet mild yellowing virus. - Physiol. Plant
Pathol. *12* : 211 - 223, 1978.

34942 - MONTEITH, J.L. : Reassessment of maximum growth rates for $C_3$ and $C_4$ crops.
- Exp. Agr. *14* : 1- 5, 1978.

34943 - MONTENY, B., GOSSE, G. : Variations du rayonnement photosynthétiquement
actif en région tropicale humide. - Arch. Meteorol. Geophys. Bioklimatol.,
Ser. B *25* : 371 - 382, 1978.

34944 - MONTENY, B., GOSSE, G. : Trouble atmosphérique et rayonnement solaire en
basse Côte d'Ivoire. - Agr. Meteorol. *19* : 121 - 136, 1978. [Radiation in
canopy.]

34945 - MOONEY, H.A., BJÖRKMAN, O., COLLATZ, G.J. : Photosynthetic acclimation to
temperature in the desert shrub, *Larrea divaricata*. I. Carbon dioxide exchange characteristics of intact leaves. - Plant Physiol. *61* : 406 - 410,
1978.

34946 - MOONEY, H.A., FERRAR, P.J., SLATYER, R.O. : Photosynthetic capacity and
carbon allocation patterns in diverse growth forms of *Eucalyptus*. - Oecologia *36* : 103 - 111, 1978.

34947 - MOORE, D.P. : When $C_4$ plants do best. - Nature *272* : 400 - 401, 1978.

34948 - MOORE, F.D., SHEPHARD, D.C. : Chloroplast autonomy in pigment synthesis. -
Protoplasma *94* : 1 - 17, 1978.

34949 - MOORE, P.H., MARETZKI, A. : Biochemical responses of *Saccharum* sp. to
drought. - Plant Physiol. *61* (Suppl.) : 81, 1978. [Ps.]

34950 - MORADSHAHI, A., VINES, H.M., BLACK, C.C. : Phosphoenolpyruvate carboxylase
and the diurnal regulation of Crassulacean acid metabolism. - Plant Physiol.
*61* (Suppl.) : 109, 1978.

34951 - MOREL, A. : Available, usable, and stored radiant energy in relation to marine photosynthesis. - Deep-Sea Res. *25* : 673 - 688, 1978.

34952 - MOREL, C., QUEIROZ, O. : Dawn signal as a rhythmical timer for the seasonal
adaptive variation of CAM : a model. - Plant Cell Environment *1* : 141 - 149,
1978.

34953 - MOREL, N.M.L., RUETER, J.G., MOREL, F.M.M. : Copper toxicity to *Skeletonema
costatum* (*Bacillariophyceae*). - J. Phycol. *14* : 43 - 48, 1978. [Chl.]

34954 - MORESHET, S., FALKENFLUG, V. : A krypton diffusion porometer for the direct
field measurement of stomatal resistance. - J. exp. Bot. *28* : 267 - 275,
1978.

34955 - **MORI, S.** : Studies on productivity of freshwater communities. - In :
TAMIYA, H. (ed.) : Summary Report on the Contribution of the Japanese Natio-
nal Committee for IBP, 1964 - 1974. JIBP Synthesis. Vol. 20. Pp. 117 - 135.
Univ. of Tokyo Press, Tokyo 1978.

34956 - **MORITA, K.** : A physiological study on the dynamic status of leaf nitrogen
in rice plants. - Bull. Hokuriku nat. agr. exp. Sta. *21* : 1 - 61, 1978. [Ps,
Chl, chloroplast.]

34957 - **MORITA, S., MIYAZAKI, T.** : Dichroism of bacteriochlorophyll in chromatophores of photosynthetic bacteria. - J. Biochem. (Tokyo) *83* : 1715 - 1720,
1978.

34958 - **MOROT-GAUDRY, J.-F., FARINEAU, J.** : Étude comparée des réactions de carbo-
xylations photosynthétiques chez un Maïs normal (W64A) et chez un Maïs
mutant opaque 2 (W64 o2) : mise en évidence de deviations métaboliques chez
le mutant.- Physiol. vég. *16* : 451 - 467, 1978.

34959 - **MOROT-GAUDRY, J.F., THOMAS, D.A., DEROCHE, M.E., CHARTIER, P.** : Growth,
leaf optical properties, chlorophyll content and net assimilation rate in
maize seedlings with and without the gene *opaque 2*. - Photosynthetica *12* :
284 - 289, 1978.

34960 - **MOROZOV, V.L.** : Struktura assimilyatsionnogo apparata dominantov kamchatskogo krupnotravya. [Structure of the assimilatory apparatus of tall herb
dominants in Kamchatka.] - Izvest. sib. Otd. Akad. Nauk SSSR, Ser. biol.
Nauk *1978* (2) : 36 - 42, 1978. [In R, ab : E.]

34961 - **MOROZOV, V.L.** : Fotosinteticheskaya deyatel'nost' krupnotravnykh soobshchestv na Kamchatke. [Photosynthetic activity of tall herb communities in
Kamchatka.] - Izv. sib. Otd. Akad. Nauk SSSR, Ser. biol. Nauk *1978* (2) :
42 - 48, 1978. [In R, ab : E.]

34962 - **MOROZOV, V.L.** : Sutochnaya i sezonnaya dinamika fotosinteza dominantov
Kamchatskogo krupnotrav'ya. [Diurnal and seasonal dynamics of photosynthesis in dominants of Kamchatka high-grasses.] - Bot. Zh. *63* : 682 - 689,
1978. [In R, ab : E.]

34963 - **MOROZOV, V.L., BELAYA, G.A.** : Akkumulyatsiya solnechnoĭ ėnergii kamchatskim
krupnotrav'em v razlichnykh ėkologicheskikh usloviyakh. [Accumulation of
solar energy by tall herbaceous vegetation of the Kamchatka peninsula in
different ecological conditions.] - Ėkologiya *1978* (1) : 34 - 41, 1978.
[In R.]

34964 - **MORRILL, L.C., LOEBLICH, A.R. III.** : Photoheterotrophy in the *Pyrrhophyta*.
- J. Phycol. *14* (Suppl.) : 39, 1978.

34965 - **MORRISON, L.E., LOACH, P.A.** : Complex charge recombination kinetics of the
phototrap in *Rhodospirillum rubrum*. - Photochem. Photobiol. *27* : 751 - 757,
1978.

34966 - **MOSER, T.J., NASH, T.H. III.** : Photosynthetic patterns of *Cetraria cuculla-
ta* (BELL.) ACH. at Anaktuvuk pass, Alaska. - Oecologia *34* : 37 - 44, 1978.

34967 - **MOSIER, A.R.** : Inhibition of photosynthesis and nitrogen fixation in algae
by volatile nitrogen bases. - J. environ. Qual. *7* : 237 - 240, 1978.

34968 - **MOSKALENKO, A.A., EROKHIN, Yu.E.** : Investigation of light-harvesting complex *Rhodopseudomonas sphaeroides*. - FEBS Lett. *87* : 254 - 256, 1978.

34969 - **MOSKOWITZ, E., MALLEY, M.M.** : Energy transfer and photooxidation kinetics
in reaction centers on the picosecond time scale. - Photochem. Photobiol.
*27* : 55 - 59, 1978.

34970 - **MOSS, M.R.** : The potential primary productivity of Peninsular Malaysia. -
J. environ. Manage. *6* : 171 - 183, 1978.

*34971 - **MOSSER, J.L., BROCK, T.D.** : Temperature optima for algae inhabiting cold
mountain streams. - Arct. alp. Res. *8* : 111 - 114, 1976. [Ps.]

34972 - **MOTTA, N., MEDINA, E.** : Early growth and photosynthesis of tomato (*Lycoper-
sicum esculentum* L.) under nutritional deficiencies. - Turrialba *28* : 135 -
141, 1978.

34973 - **MOUNTZ, J.M., TIEN, H.T.** : Photoeffects of pigmented lipid membranes in a microporous filter. - Photochem. Photobiol. *28* : 395 - 400, 1978.

34974 - **MOURA, J.J.G., XAVIER, A.V., HATCHIKIAN, E.C., LE GALL, J.** : Structural control of the redox potentials and of the physiological activity by oligomerization of ferredoxin. - FEBS Lett. *89* : 177 - 179, 1978.

34975 - **MOURIOUX, G., DOUCE, R.** : Transport spécifique du sulfate à travers l'enveloppe des chloroplastes d'Épinard. - Compt. rend. Acad. Sci. Paris, Sér. D *286* : 277 - 280, 1978.

*34976 - **MUCKLE, G., RÜDIGER, W.** : Chromophore content of C-phycoerythrin from various cyanobacteria. - Z. Naturforsch. *32 C* : 957 - 962, 1977.

34977 - **MUELLER, U., CHERRY, R.J., HEYN, M.P.** : Bacteriorhodopsin : rotational mobility and self-assocoation in membranes and determination of chromophore orientation. - In : CAPLAN, S.R., GINZBURG, M. (ed.) : Energetics and Structure of Halophilic Microorganisms. Pp. 303 - 308. Elsevier/North-Holland Biomedical Press, Amsterdam - New York 1978.

34978 - **MUKERJI, D., GLOVER, H.E., MORRIS, I.** : Diversity in the mechanism of carbon dioxide fixation in *Dunaliella tertiolecta (Chlorophyceae)*. - J. Phycol. *14* : 137 - 142, 1978.

34979 - **MUKERJI, D., MORRIS, I.** : Measurement of carboxylases (RuDPCase and PEPCase) in cell suspensions of *Phaedactylum tricornutum* treated with organic solvents. - Z. Pflanzenphysiol. *90* : 95 - 99, 1978.

34980 - **MUKHERJEE, D., AFRIA, B.S.** : Studies on the metabolism of some $C_3$ and $C_4$ plants. - In : Inter-Disciplinary Symposium on Photosynthesis and Productivity. Abstract of Papers. Pp. 24 - 25. Indian nat. Sci. Acad., New Delhi 1978.

34981 - **MUKHERJEE, T., SAPRE, A.V., MITTAL, J.P.** : On the nature of chlorophyll-*a* in aqueous micellar systems. - Photochem. Photobiol. *28* : 95 - 96, 1978.

34982 - **MUKHIN, E.N., CHERMNYKH, R.M.** : Kataliziruemoe ferredoksinom fotofosforilirovanie v khloroplastakh gorokha. [Ferredoxin-catalyzed photophosphorylation in pea chloroplasts.] - Dokl. Akad. Nauk SSSR *235* : 1207 - 1210, 1977. [In R.]

34983 - **MUKHIN, E.N., GINS, V.K., RED'KO, T.P.** : K kharakteristike plastotsianina iz list'ev kukuruzy (*Zea mays* L.) i gorokha (*Pisum sativum* L.). [Characteristics of plastocyanin from maize (*Zea mays* L.) and pea (*Pisum sativum* L.) leaves.]- Izv. Akad. Nauk SSSR, Ser. biol. *1978* : 399 - 404, 1978. [In R, ab : E.]

34984 - **MUKOHATA, Y., NAKABAYASHI, S., HIGASHIDA, M.** : Quercetin, an energy transfer inhibitor in photophosphorylation. - FEBS Lett. *85* : 215 - 218, 1978.

34985 - **MÜLLER, C.** : On the productivity and chemical composition of some benthic algae in hard-water streams. - Verh. int. Verein. Limnol. *20* : 1457 - 1462, 1978.

34986 - **MÜLLER, M., SANTARIUS, K.A.** : Changes in chloroplast membrane lipids during adaptation of barley to extreme salinity. - Plant Physiol. *62* : 326 - 329, 1978.

34987 - **MULLER, R.N.** : The phenology, growth and ecosystem dynamics of *Erythronium americanum* in the northern hardwood forest. - Ecol. Monogr. *48* : 1 - 20, 1978. [Ps.]

34988 - **MÜLLER, W., WEGMANN, K.** : Sucrose biosynthesis in *Dunaliella*. I. Thermic and osmotic regulation. - Planta *141* : 155 - 158, 1978. [Ps.]

34989 - **MÜLLER, W., WEGMANN, K.** : Sucrose biosynthesis in *Dunaliella*. III. Regulation by a membrane change. - Planta *141* : 165 - 167, 1978. [Chloroplast membrane.]

34990 - **MULLET, J.E., ARNTZEN, C.J.** : Structure and characterization of "native" photosystem I complexes. - Plant Physiol. *61* (Suppl.) : 76, 1978.

*34991 - **MUNAWAR, M., MUNAWAR, I.F.** : Some observations on the growth of diatoms in Lake Ontario with emphasis on *Melosira binderana* KÜTZ during thermal bar conditions. - Arch. Hydrobiol. *75* : 490 - 499, 1975. [Chl.]

*34992 - **MUNAWAR, M., MUNAWAR, I.F.** : The abundance and significance of phytoflagellates and nannoplankton in the St. Lawrence Great Lakes. 1. Phytoflagellates. - Verh. int. Verein. Limnol. *19* : 705 - 723, 1975. [Ps production.]

*34993 - **MUNAWAR,M.,MUNAWAR,I.F.** : A lakewide study of phytoplankton biomass and its species composition in Lake Erie, April - December 1970. - J. Fish. Res. Board Canada *33* : 581 - 600, 1976. [Primary productivity.]

*34994 - **MUNAWAR, M., STADELMANN, P., MUNAWAR, I.F.** : Phytoplankton biomass, species composition and primary production at a nearshore and a midlake station of Lake Ontario during IFYGL (IFYGL). - Proc. 17th Conf. Great Lakes Res. *17* (2) : 629 - 652, 1974. [Ps.]

 34995 - **MURADO, M.A.** : Interacciones fitoplancton-hidrocarburos aromáticos policlorados. I. Efectos de tres Hallowax sobre el crecimiento, pigmentos fotosintéticos y respiración endógena de *Nitzschia* sp. [Interaction phytoplankton-polychlorinated aromatic hydrocarbons. I. Effects of three Hallowax on the growth, photosynthetic pigments and dark respiration of *Nitzschia* sp.] - Invest. pesquera *42* : 273 - 292, 1978.    [In Span., ab : E.]

 34996 - **MURAKAMI, S., STROTMANN, H.** : Adenylate kinase bound to the envelope membranes of spinach chloroplasts. - Arch. Biochem. Biophys. *185* : 30 - 38, 1978.

 34997 - **MURAKAMI, T.** : [Studies on the photosynthetic rate of mulberry plant : I. The variation of photosynthetic rates of leaves with leaf order in various growing periods.] - Bull. sericult. Exp. Station (Tokyo) *27* : 353 - 368, 1978. [In Jap., ab : E.]

*34998 - **MURATA, N.** : [Chlorophyll fluorescence and its temperature dependence in a blue-green alga *Anacystis nidulans*.]- Kagaku no Ryoiki, Zokan *114* : 89 - 95, 1976. [In Jap.]

 34999 - **MURATA, N., FORK, D.C.** : Temperature dependence of the turnover of P700 in photosynthesis of the blue-green alga *Anacystis nidulans*. - Carnegie Inst. Year Book *77* : 289 - 291, 1978.

 35000 - **MURATA, N., SATO, N.** : Studies on the absorption spectra of chlorophyll *a* in aqueous dispersions of lipids from the photosynthetic membranes. - Plant Cell Physiol. *19* : 401 - 410, 1978.

 35001 - **MUREĬ, I.A., SHUL'GIN, I.A.** : Fiziologicheskiĭ analiz prikhodyashcheĭ FAR k rasteniyu. [Physiological analysis of PhAR reaching the plant.] - Bot. Zh. *63* : 962 - 973, 1978. [In R, ab : E.]

 35002 - **MUREĬ, I.A., SHUL'GIN, I.A.** : Éffektivnost' ispol'zovaniya FAR na istinnyĭ fotosintez i obrazovanie biomassy rasteniĭ. [The effectivity of PhAR utilization in net photosynthesis and plant biomass formation.] - Bot. Zh. *63* : 1731 - 1743, 1978. [In R, ab : E.]

 35003 - **MUREĬ, I.A., SHUL'GIN, I.A.** : Uvelichenie éffektivnosti ispol'zovaniya FAR na fotosintez v poseve po mere zateneniya list'ev. [An increase in the efficiency of PhAR utilization in canopy photosynthesis in the course of shadowing of leaves.] - Fiziol. Rast. *25* : 492 - 499, 1978. [In R, ab : E.]

 35004 - **MUREĬ, I.A., SHUL'GIN, I.A.** : Kolichestvennyĭ fiziologicheskiĭ analiz pogloshcheniya FAR rasteniem. [Quantitative physiological analysis of PhAR absorption by plants.] - Fiziol. Rast. *25* : 671 - 680, 1978. [In R, ab : E.]

 35005 - **MURPHY, D.J., LEECH, R.M.** : The pathway of [$^{14}$C]bicarbonate incorporation into lipids in isolated photosynthesising spinach chloroplasts. - FEBS Lett. *88* : 192 - 196, 1978.

 35006 - **MURPHY, T.M.** : Immunochemical comparisons of ribulosebisphosphate carboxylases using antisera to tobacco and spinach enzymes. - Phytochemistry *17* : 439 - 443, 1978.

35007 - **MURTAGH, G.J.** : The effect of evaporative demand on the growth of well-watered kikuyu. - Agr. Meteorol. *19* : 379 - 389, 1978.

35008 - **MURTHY, M.S.** : Relationships between vegetation and climate in the upper catchment area of the Narmada river, Central India. - Vegetatio *36* : 53 - 60, 1978. [Growth analysis.]

35009 - **MURZAEVA, S.V., AKULOVA, E.A.** : Rol' katalazy i peroksidazy v regulyatsii fotokhimicheskikh reaktsiĭ khloroplastov, svyazannykh s metabolizmom perekisi vodoroda. [Role of catalase and peroxidase in regulation of photochemical reactions in chloroplasts connected with metabolism of $H_2O_2$.] - In : **KARPILOV, Yu.S., ROMANOVA, A.K.** (ed.) : Mekhanizm Fotodykhaniya i ego Osobennosti u Rasteniĭ Razlichnykh Tipov. Pp. 33 - 58. Pushchino 1978. [In R.]

35010 - **MUSSELMAN, R.C., KENDER, W.J., CROWE, D.E.** : Determining air pollutant effects on the growth and productivity of 'Concord' grapevines using open-top chambers. - J. amer. Soc. hort. Sci. *103* : 645 - 648, 1978. [Chl, stomatal resistance.]

35011 - **MUSTARDY, L.A., BRANGEON, J.** : 3-dimensional chloroplast infrastructure : Developmental aspects. - In : **AKOYUNOGLOU, G., ARGYROUDI-AKOYUNOGLOU, J.H.** (ed.) : Chloroplast Development. Pp. 489 - 494. Elsevier/North-Holland Biomedical Press, Amsterdam - New York - Oxford 1978.

35012 - **MUZAFAROV, E.N., AKULOVA, E.A., IVANOV, B.N., RUZIEVA, R.Kh.** : Ingibirovanie kvertsetinom èlektronnogo transporta i fotofosforilirovaniya v khloroplastakh. [Inhibition of photophosphorylation and of electron transport in isolated chloroplasts by quercetin.] - Mol. Biol. (Moskva) *12* : 100 - 107, 1978. [In R, ab : E.]

35013 - **MUZAFAROV, E.N., ROSHCHINA, V.V., KHRISTIN, M.S.** : O rabotakh E.A.Akulovoĭ ob osobennostyakh mekhanizmov èlektronnogo transporta pri fotosinteze. [E.A.Akulova's works on the specific features of electron transport mechanism in photosynthesis.] - Fiziol. Rast. *25* : 843 - 847, 1978. [In R.]

35014 - **M'YAKUSHKO, V.K.** : Sosnovi lisy Malogo Polissya. [Pine forests of Lesser Polesye.] - Ukr. bot. Zh. *35* : 284 - 290, 1978. [Primary production; in Ukr., ab : E.]

35015 - **MYERS, J., GRAHAM, J.-R., WANG, R.T.** : On spectral control of pigmentation in *Anacystis nidulans* (*Cyanophyceae*). - J.Phycol. *14* : 513 - 518, 1978.

35016 - **NAGARAJAH, S.** : Some differences in the responses of stomata of the two leaf surfaces in cotton. - Ann. Bot. *42* : 1141 - 1147, 1978. [Resistances.]

35017 - **NAGARAJAH, S., RATNASURIYA, G.B.** : The effect of phosphorus and potassium deficiencies on transpiration in tea (*Camellia sinensis.*).-Physiol. Plant. *42* : 103 - 108, 1978. [Chl.]

*35018 - **NAGL, W.** : Elektronenmikroskopischer Vergleich der Nukleolen und Plastiden im Gametophyten und Sporophyten des Lebermooses *Sphaerocarpos donnellii*. - Phyton (Austria) *16* : 159 - 163, 1974.

35019 - **NAIM, R., NEUMANN, P.M.** : Mechanism of senescence induction by silicone oil. - Physiol. Plant. *42* : 57 - 60, 1978. [Ps, Chl.]

35020 - **NAIMAN, R.J., SIBERT, J.R.** : Transport of nutrients and carbon from the Nanaimo River to its estuary. - Limnol. Oceanogr. *23* : 1183 - 1193, 1978.[Chl.]

35021 - **NAITO, K., TSUJI, H., HATAKEYAMA, I.** : Effect of benzyladenine on DNA, RNA, protein, and chlorophyll contents in intact bean leaves : Differential responses to benzyladenine according to leaf age. - Physiol. Plant. *43* : 367 - 371, 1978.

35022 - **NAKAMURA, H., OTA, Y.** : An injury to rice plants caused by photochemical oxidants in Japan. - Jap. agr. Res. quart. *12* (2) : 69 - 73, 1978. [Ps, Chl.]

35023 - **NAKAMURA, H., SAKA, H.** : [Photochemical oxidants injury in rice plants. III. Effect of ozone on physiological activities in rice plants.] - Jap. J. Crop Sci. *47* : 707 - 714, 1978. [Chl; in Jap., ab : E.]

*35024 - NAKAMURA, Y., MIYACHI, S. : Conversion of glucose-polymer to sucrose in
         *Chlorella* induced by high temperature. - In : MITSUI, A., MIYACHI, S.,
         SAN PIETRO, A., TAMURA, S. (ed.) : Biological Solar Energy Conversion. Pp.
         203 - 212. Academic Press, New York - San Francisco - London 1977. [Photo-
         synthates.]

 35025 - NAKASEKO, K., GOTOH, K., SATO, H. : Physio-ecological studies on prolifica-
         cy in maize III. The relationships between expression of second ear and
         some physio-ecological characteristics. - Jap. J. Crop Sci. *47* : 212 - 220,
         1978. [Growth analysis.]

 35026 - NAKATANI, H.Y., BARBER, J., FORRESTER, J.A. : Surface charges on chloro-
         plast membranes as studied by particle electrophoresis. - Biochim. biophys.
         Acta *504* : 215 - 225, 1978.

 35027 - NAKAYAMA, O., OHNO, M., YASUI, T. : [Effect of enrichment with digested
         night soil on the growth of marine plankton.] - Bull. Jap. Soc. Fish. *44* :
         1099 - 1103, 1978. [Chl; in Jap., ab : E.]

*35028 - NALBORCZYK, E. : Determination of the photosynthesis/light response curves -
         a radiometric method. - Newslet. Appl. nucl. Meth. Biol. Agr. *1976* (6) :
         8 - 11, 1976.

 35029 - NALBORCZYK, E. : Dark carboxylation and its possible effect on the value
         of $\delta^{13}C$ in $C_3$ plants. - Acta Physiol. Plant. *1* : 53 - 58, 1978.

*35030 - NALBORCZYK, T., NALBORCZYK, E. : A method for simultaneous measurement of
         photosynthesis of different plant organs. - Newslett. Appl. nucl. Meth. Biol.
         Agr. *1976* (7) : 21 - 25, 1976.

 35031 - NAMBUDIRI, E.M.V., TIDWELL, W.D., SMITH, B.N., HEBBERT, N.P. : A $C_4$ plant
         from the Pliocene. - Nature *276* : 816 - 817, 1978.

 35032 - NANDI, D.L. : Studies on $\delta$-aminolevulinic acid synthase of *Rhodopseudomonas
         spheroides*. Reversibility of the reaction, kinetic, spectral, and other
         studies related to the mechanism of action. - J. biol. Chem. *253* : 8872 -
         8877, 1978.

 35033 - NASSERY, H., OGATA, G., NIEMAN, R.H., MAAS, E.V. : Growth, phosphate pools,
         and phosphate mobilization of saltstressed sesame and pepper. - Plant Phy-
         siol. *62* : 229 - 231, 1978. [Photosynthates.]

 35034 - NASYROV, Yu.S. : Genetic control of photosynthesis and improving of crop
         productivity. - Annu. Rev. Plant Physiol. *29* : 215 - 237, 1978.

 35035 - NATO, A., MATHIEU, Y. : Studies of some tobacco mutant photosynthetic acti-
         vities and plastid differentiation. - In : AKOYUNOGLOU, G., ARGYROUDI-AKOY-
         UNOGLOU, J.H. (ed.) : Chloroplast Development. Pp. 733 - 738. Elsevier/
         North-Holland Biomedical Press, Amsterdam - New York - Oxford 1978.

 35036 - NATO, A., MATHIEU, Y. : Changes in phosphoenolpyruvate carboxylase and
         ribulose-biphosphate carboxylase activities during the photoheterotrophic
         growth of *Nicotiana tabacum* (cv. *Xanthi*) cell suspensions. - Plant Sci. Lett.
         *13* : 49 - 56, 1978.

 35037 - NÁTR, L. : Programování, modelování a prognózy výnosů obilovin. [Program-
         ming, modelling and prognoses of cereal yields.] - Stud. Inf. ÚVTIZ, Základ.
         Vědy Zeměd. *1978* (2) : 1 - 84, 1978. [In Czech, ab : E, R.]

 35038 - NÁTR, L. : Vztah mezi obsahem dusíku v listech a rychlostí fotosyntézy.
         [Relation between the nitrogen content in leaves and the rate of photosyn-
         thesis.] - Agrochémia *18*(4): 98 - 100, 1978. [In Czech.]

 35039 - NÁTR, L., KOUSALOVÁ, I. : Vliv sponu rostlin na výnos a strukturu porostu
         jarního ječmene. [Effect of spacing of plants on the yield and structure
         of spring barley stands.] - Rostl. Výr. *24* : 1301 - 1311, 1978. [In Czech,
         ab : E, G, R.]

 35040 - NAUMOV, A.V. : O roli dykhatel'nogo gazoobmena v produktivnosti estestven-
         nykh i kul'turnykh fitotsenozov. [The role of respiratory gas exchange in
         the productivity of natural and cultivated phytocenoses.]-Ékologiya *1978*(1):
         19 - 26, 1978. [In R.]

35041 - **NAUŠ, J.** : Fluorescence characteristics of chlorophyll *a* in polymer matrix of cellulose triacetate. - In : FIALA, J. (ed.) : Third International Seminar on Energy Transfer in Condensed Matter. Pp. 300 - 306. Univerzita Karlova, Praha 1978.

35042 - **NAUŠ, J., MADER, P., FOJTÍK, L.** : Fluorescence of chlorophyll on several structural levels of the spring barley under the different nutrition conditions and different water state. - In : FIALA, J. (ed.) : Third International Seminar on Energy Transfer in Condensed Matter. Pp. 158 - 166. Univerzita Karlova, Praha 1978.

35043 - **NAUWERCK, A., PECHLANER, R., SACHERER, F.A., WEISS, A.** : Die Phytoplanktonentwicklung in einem eutrophen Baggersee bei Lampertheim (Hessen, B.R.D.) im Jahresgang. - Ber. naturwiss.-med. Ver. Innsbruck *65* : 23 - 54, 1978.

35044 - **NAYAK, S.K., JANARDHAN, K.V., MURTY, K.S.** : Photosynthetic efficiency of rice as influenced by light intensity and quality. - Indian J. Plant Physiol. *21* : 48 - 52, 1978.

35045 - **NAZAROV, S.K.** : Strukturno-funktsional'nye osobennosti lista u trekh ěkotipov *Rubus Chamaemorus* L. [Structural and functional peculiarities of the leaf of three ecotypes of *Rubus chamaemorus* L.] - In : Mezostruktura i Funktsional'naya Aktivnost' Fotosinteticheskogo Apparata. Pp. 108 - 111. Ural'skiǐ gosudarstvennyǐ Universitet, Sverdlovsk 1978. [Ps, Chl; in R.]

35046 - **NAZAROVA, I.G., EVSTIGNEEV, V.B.** : O mekhanizme aktivirovaniya fotosensibiliziruyushchego deǐstviya vodorastvorimykh analogov khlorofilla pri svyazyvanii ikh polimerom. [Activation mechanism of the photosensitizing effect of water-soluble analogs of chlorophyll while binding them with a polymer.] -Biofizika *23* : 419 - 423, 1978. [In R, ab : E.]

35047 - **NEALES, T.F., NICHOLLS, A.O.** : Growth responses of young wheat plants to a range of ambient $CO_2$ levels. - Aust. J. Plant Physiol. *5* : 45 - 59, 1978.

35048 - **NEAMŢU, G., ILLYES, G.** : Cercetǎri chemotaxonomice la plantele superioare X. Pigmenţii carotenoidici şi clorofilieni din plante superioare acvatice. [Chemotaxonomic studies on higher plants. X. Carotenoid and chlorophyll pigments in aquatic higher plants.] - Stud. Cercet. Biochim. *21* : 63 - 67, 1978. [In Roum., ab : E.]

35049 - **NEDYALKOV, N., STOǏNOVA, E.** : Vliyanie na atrazina v"rkhu ovodnenostta i s"stoyanieto na vodata v listata na tsaravitsata, kolichestvoto na plastidnite pigmenti i intenzivnostta na transpiratsiyata i fotosintezata v zavisimost ot pochvenata vlazhnost. [Effect of atrazine on the water content and state in maize leaves, plastid pigment quantity and rates of transpiration and photosynthesis in relation to soil moisture.] - Fiziol. Rast. (Sofia) *4* (1) : 43 - 51, 1978. [In Bulg., ab : E, R.]

35050 - **NEKRASOV, L.I., MAMLEEVA, N.A.** : Adsorbtsionnye sloi khlorofilla i ikh fiziko-khimicheskie svoǐstva. [Adsorption layers of chlorophyll and their physico-chemical properties.] - Zh. fiz. Khim. *52* : 2721 - 2735, 1978. [In R.]

35051 - **NEKRASOVA, G.F.** : Formirovanie struktury i fotosinteticheskoǐ funktsii v protsesse rosta lista. [Formation of structure and photosynthetic function during leaf growth.] - In : Mezostruktura i Funktsional'naya Aktivnost' Fotosinteticheskogo Apparata. Pp. 61 - 73. Ural'skiǐ gosudarstvennyǐ Universitet, Sverdlovsk 1978. [In R.]

35052 - **NELSON, C.J., TREHARNE, K.J., COOPER, J.P.** : Influence of temperature on leaf growth of diverse populations of tall fescue. - Crop Sci. *18* : 217 - 220, 1978.

35053 - **NELSON,N., EYTAN, E.** : Preparation of chloroplast highly depleted of $CF_1$ with high degree of reconstitution. - Plant Physiol. *61* (Suppl.) : 89, 1978.

35054 - **NELSON, N., EYTAN, E., JULIAN, C.** : The function of individual polypeptides in photosynthetic energy transduction. - In : HALL, D.O., COOMBS, J., GOODWIN, T.W. (ed.) : Proceedings of the Fourth International Congress on Photosynthesis. Pp. 559 - 570. Biochem. Soc., London 1978.

35055 - NELSON, W.L., DALE, R.F. : A methodology for testing the accuracy of yield predictions from weather-yield regression models for corn. - Agron.J. *70* : 734 - 740, 1978.

35056 - NEUBAUER, H. : Der Einfluß von Stickstoffdüngung, Schnitthäufigkeit und umbruchloser Regeneration der Grasnarbe auf den Nährstoffgehalt und-ertrag von Dauerwiesen. - Bodenkultur *29* : 40 - 58, 1978.

35057 - NEUMANN, K.-H., PERTZSCH, C., MOOS, M., KRÖMMELBEIN, G. : Untersuchungen zur Charakterisierung des Ernährungssystems von Karottengewebekulturen : Photosynthese, Kohlenstoff- und Stickstoffernährung, Mineralstoffverbrauch. - Z. Pflanzenernähr. Bodenk. *141* : 299 - 311, 1978.

35058 - NEWMAN, P.J., SHERMAN, L.A. : Isolation and characterization of Photosystem I and II membrane particles from the blue-green alga, *Synechococcus cedrorum*. - Biochim. biophys. Acta *503* : 343 - 361, 1978.

35059 - NEYRA, C.A., HAGEMAN, R.H. : Pathway for nitrate assimilation in corn (*Zea mays* L.) leaves. Cellular distribution of enzymes and energy sources for nitrate reduction. - Plant Physiol. *62* : 618 - 621, 1978. [Ps, Chl.]

35060 - NEZHNEV, Yu.N., ZUBANOVA, L.S. : Vliyanie margantsa na urozhaǐ i kachestvo tomatov. [Effect of manganese on the yield and quality of tomatoes.]- Agrokhimiya *1978* (4) : 104 - 107, 1978. [Chl; in R.]

35061 - NGUYEN HUU THUOC, LE VAN QUY, DANG XUYEN NHU, HUYNH NGOC THACH : The effect of 2-pyridylhydroxymethanesulfonic acid (PHMSA) on glycolate oxidase activity and photosynthesis of soybean and *Azolla pinnata* R. BR. - Photosynthetica *12* : 449 - 451, 1978.

35062 - NICHIPOROVICH, A.A. : Énergeticheskaya éffektivnost' i produktivnost' fotosinteziruyushchikh sistem kak integral'naya problema. [Energetic efficiency and productivity of photosynthesizing systems as an integral problem.] - Fiziol. Rast. *25* : 922 - 937, 1978. [in R, ab : E.]

35063 - NICHOLLS, K.H., DILLON, P.J. : An evaluation of phosphorus-chlorophyll-phytoplankton relationships for lakes. - Int. Rev. ges. Hydrobiol. *63* : 141 - 154, 1978.

35064 - NICOL, W.M. : Production and food uses of plant carbohydrates. - In : HALL, D.O., COOMBS, J., GOODWIN, T.W. (ed.) : Proceedings of the Fourth International Congress on Photosynthesis. Pp. 735 - 744. Biochem. Soc., London 1978.

35065 - NIEDEN, U. ZUR, NEUMANN, D. : Effects of (2-chloroethyl)-trimethyl-ammonium chloride (CCC) on chlorophyll content and ultrastructure of plastids of *Pisum sativum*. - Biochem. Physiol. Pflanzen *173* : 202 - 212, 1978.

35066 - NIEDERMAN, R.A., GIBSON, K.D. : Isolation and physicochemical properties of membranes from purple photosynthetic bacteria. - In : CLAYTON, R.K., SISTROM, W.R. (ed.) : The Photosynthetic Bacteria. Pp. 79 - 118. Plenum Press, New York - London 1978.

35067 - NIELSEN, N.C. : Light induction of acetyl CoA carboxylase activity during chloroplast development. - In : AKOYUNOGLOU, G., ARGYROUDI-AKOYUNOGLOU, J.H. (ed.) : Chloroplast Development. Pp. 315 - 322. Elsevier/North-Holland Biomedical Press, Amsterdam - New York - Oxford 1978.

35068 - NIELSEN, N.C., SMILLIE, R.M. : The effect of p-phenylenediamine and dibromothymoquinone on the photosynthetic electron transport and proton pump activities of isolated barley chloroplasts. - Arch. Biochem. Biophys. *186* : 52 - 59, 1978.

35069 - NIEMECK, R.A., MATHIESON, A.C. : Physiological studies of intertidal fucoid algae. - Bot. mar. *21* : 221 - 227, 1978. [Ps.]

35070 - NIEMI, Å. : Ecology of phytoplankton in the Twärminne area, SW coast of Finland. III. Environmental conditions and primary production in Pojoviken in the 1970s. - Acta bot. fenn. *106* : 1 - 28, 1978.

35071 - **NIGON, V., HEIZMANN, P.** : Morphology, biochemistry, and genetics of plastid development in *Euglena gracilis*. - Int. Rev. Cytol. *53* : 211 - 290, 1978.

35072 - **NIGON, V., VERDIER, G., SALVADOR, G., HEIZMANN, P., RAVEL-CHAPUIS, P., FREYSSINET, G.** : Biochemical sequences during the greening of dark-grown *Euglena gracilis*. - In : AKOYUNOGLOU, G., ARGYROUDI-AKOYUNOGLOU, J.H. (ed.) : Chloroplast Development. Pp. 629 - 640. Elsevier/North-Holland Biomedical Press, Amsterdam - New York - Oxford 1978.

35073 - **NII, N.** : [Effects of leaf age and defoliation on growth at the early stage of fruit development in satsuma mandarin.]- J. jap. Soc. hort. Sci. *47* : 172 - 180, 1978. [In Jap., ab : E.]

35074 - **NIIMI, Y., TORIKATA, H.** : Changes in cytokinin activities, photosynthesis and respiration of the grape flower clusters during their development. - J. jap. Soc. hort. Sci. *47* : 301 - 307, 1978.

35075 - **NIKANDROV, V.V., BRIN, G.P., KRASNOVSKIĬ, A.A.** : Svetovaya aktivatsiya NADH i NADPH. [Light-induced activation of NADH and NADPH.] - Biokhimiya *43* : 636 - 645, 1978. [In R, ab : E.]

35076 - **NIKANDROV, V.V., CHAN VAN NI, BRIN, G.P., KRASNOVSKIĬ, A.A.** : Issledovanie usloviĭ fotovosstanovleniya metilviologena khloroplastami. [Methylviologen photoreduction by chloroplasts.] - Mol. Biol. (Moskva) *12* : 1278 - 1287, 1978. [In R, ab : E.]

35077 - **NIKANDROV, V.V., KRASNOVSKIĬ, A.A.** : Fosforestsentsiya vosstanovlennykh nikotinamidnykh kofermentov. [Phosphorescence of reduced nicotinamide coenzymes.] - Biofizika *23* : 721 - 723, 1978. [In R, ab : E.]

35078 - **NIKITINA, K.A., BARSKIĬ, E.L., GUSEV, M.V.** : Avtotrofnye komponenty vodnykh ėkosistem okrestnosteĭ g. Sochi v zimniĭ period. [Autotrophic components of aqueous ecosystems near Sochi in winter.] - Mikrobiologiya *47* : 510 - 520, 1978. [In R, ab : E.]

35079 - **NIKOLOV, G.** : Arkhitektonika na pamukoviya posev v zavisimost ot vodoobezpechenostta mu. [Architecture of cotton crop as dependent on its water supply.] - Rasteniev"dni Nauki *15* (9-10) : 97 - 107, 1978. [In Bulg., ab : E, R.]

35080 - **NIKOLOV, G.** : Biologichna produktivnost na pamuka. [The biological productivity of cotton.] - Rasteniev"dni Nauki *15* (1) : 49 - 59, 1978. [Energy utilization; in Bulg., ab : E, R.]

35081 - **NILSEN, S., MORTENSEN, L.** : Spectral response of photorespiration effect of plant age and chlorophyll content in spruce. - Z. Pflanzenphysiol. *89* : 433 - 441, 1978.

*35082 - **NIL'SON, T.** : Teoriya propuskaniya radiatsii neodnorodnym rastitel'nym pokrovom. [A theory of radiation penetration into non-homogeneous plant canopies.] - In : Propuskanie Solnechnoĭ Radiatsii Rastitel'nym Pokrovom. Pp. 5 - 70. Akad. Nauk ėst. SSR, Inst. Astrofiz. Fiz. Atmosfery, Tartu 1977. [In R.]

*35083 - **NIL'SON, T., ROSS, V., ROSS, Yu.** : Nekotorye voprosy arkhitektoniki rasteniĭ i rastitel'nogo pokrova. [Some problems of the architecture of plants and plant canopies.] - In : Propuskanie Solnechnoĭ Radiatsii Rastitel'nym Pokrovom. Pp. 71 - 144. Akad. Nauk ėst. SSR, Inst. Astrofiz. Fiz. Atmosfery, Tartu 1977. [In R.]

*35084 - **NIMBALKAR, J.D., JOSHI, G.V.**: Effect of salt stress on photosynthesis in sugarcane var. Co. 740. - Biovigyanam *2* : 137 - 144, 1976.

35085 - **NISHI, N., SCHMIDT, J., HOFF, A.J., VAN DER WAALS, J.H.** : Fluorescence detection of electron spin echoes in the triplet state of nonphosphorescent molecules and a photosynthetic bacterium. - Chem. Phys. Lett. *58* : 205 - 207, 1978.

35086 - **NISHI, N.N., HOFF, A.J., SCHMIDT, J., VAN DER WAALS, J.H.** : Electron spin echo studies of complex biological systems. An application to "signal II" of plant photosynthesis. - Chem. Phys. Lett.*58* : 164 - 170, 1978.

35087 - **NISHIDA, K.** : Effect of leaf age on light and dark $^{14}CO_2$ fixation in a CAM plant, *Bryophyllum calycinum*. - Plant Cell Physiol. *19* : 935 - 941, 1978.

35088 - **NISHIMURA, M., AKAZAWA, T.** : Biosynthesis of ribulose-1,5-bisphosphate carboxylase in spinach leaf protoplasts. - Plant Physiol. *62* : 97 - 100, 1978.

35089 - **NISHIMURA, M., BEEVERS, H.** : Isolation and characterization of plastids from protoplasts from castor bean endosperm. - Plant Physiol. *61* (Suppl.) : 114, 1978.

35090 - **NISHIMURA, M., BEEVERS, H.** : Isolation of intact plastids from protoplasts from castor bean endosperm. - Plant Physiol. *62* : 40 - 43, 1978.

35091 - **NISHIO, T., FUJII, S.** : [Studies on the physical characteristics of rice seedling. VI. Effects of light intensity and air temperature.]- Jap. J. Crop Sci. *47* : 431 - 437, 1978. [Dry-matter accumulation ; in Jap., ab : E.]

35092 - **NISHIOKA, M., HOZUMI, K., KIRITA, H., NAGANO, M.** : Estimation of canopy photosynthesis and respiration. - In : **KIRA, T., ONO, Y., HOSOKAWA, T.** (ed.) : Biological Production in a Warm-Temperate Evergreen Oak Forest of Japan. JIBP Synthesis Vol. 18. Pp. 99 - 111. Univ. of Tokyo Press, Tokyo 1978.

35093 - **NISHIZAKI, Y.** : Synergism of triggered luminescence by simultaneous treatment of pH transition and potassium addition in chloroplasts. - Biochim. biophys. Acta *503* : 170 - 177, 1978.

35094 - **NOAILLES, M.-C.** : Étude ultrastructurale de la récupération hydrique après une période de sécheresse chez une hypnobryale : *Pleurozium schreberi* (WILLD.) MITT. - Ann. Sci. nat. Bot. Paris, Sér. 12,*19* : 249 - 265, 1978. [Chloroplast.]

35095 - **NOBEL, P.S.** : Surface temperatures of cacti - influences of environmental and morphological factors. - Ecology *59* : 986 - 996, 1978. [Ps.]

35096 - **NOBEL, P.S.** : Microhabitat, water relations, and photosynthesis of a desert fern, *Notholaena parryi*. - Oecologia *31* : 293 - 309, 1978.

35097 - **NOBEL, P.S., HARTSOCK, T.L.** : Resistance analysis of nocturnal carbon dioxide uptake by a Crassulacean acid metabolism succulent, *Agave deserti*. - Plant Physiol. *61* : 510 - 514, 1978.

35098 - **NOBEL, P.S., LONGSTRETH, D.J., HARTSOCK, T.L.** : Effect of water stress on the temperature optima of net $CO_2$ exchange for two desert species. - Physiol. Plant. *44* : 97 - 101, 1978.

35099 - **NOBEL, W., KOHLER, A.** : Die Wirkung der Salzbelastung auf submerse Weichwasser-Makrophyten unter Laborbedingungen. - Verh. Ges. Ökol. (7. Jahresversammlung Kiel 1977) : 273 - 279, 1978.

35100 - **NOGUCHI, H., MAEKAWA, T., FUJIMOTO, S., SATAKE, I., SAKAKIBARA, M.** : Physico-chemical studies on fraction I protein from alfalfa. - Agr. biol. Chem. *42* : 1553 - 1558, 1978.

35101 - **NOGUCHI, K., NAKAYAMA, K.** : [Studies on competition between upland crops and weeds. II. Comparison of early growth of crops and weeds.] - Jap. J. Crop Sci. *47* : 48 - 55, 1978. [Growth analysis ; in Jap., ab : E.]

35102 - **NOGUCHI, K., NAKAYAMA, K.** : [Studies on competition between upland crops and weeds. III. Effect of shade on growth of weeds.] - Jap. J. Crop Sci. *47* : 56 - 62, 1978. [In Jap., ab : E.]

35103 - **NOGUCHI, K., NAKAYAMA, K.** : [Studies on competition between upland crops and weeds. IV. Changes of light environment in crop canopies and hypothesis about the period for weed-free maintenance.] - Jap. J. Crop Sci. *47* : 381 - 387, 1978. [Growth analysis ; in Jap., ab : E.]

35104 - **NOLAN, W.G., BISHOP, D.G.** : Effect of amphotericin B on membrane-associated photosynthetic reactions in maize chloroplasts. - Arch. Biochem. Biophys. *190* : 473 - 482, 1978.

35105 - **NOMOTO, N.** : Photosynthetic matter productivity in the canopy of stands measured by an improved half-leaf method. - In : **MONSI, M., SAEKI, T.** (ed.) : Ecophysiology of Photosynthetic Productivity. JIBP Synthesis Vol. 19. Pp. 77 - 82. Univ. of Tokyo Press, Tokyo 1978.

35106 - NOODÉN, L.D., LEOPOLD, A.C. : Phytohormones and the endogenous regulation of senescence and abscission.- In : LETHAM, D.S., GOODWIN, P.B., HIGGINS, T.J.V. (ed.) : Phytohormones and Related Compounds - a Comprehensive Treatise. Vol. II. Pp. 329 - 369. Elsevier/North-Holland Biomedical Press, Amsterdam - New York - Oxford 1978. [Ps.]

35107 - NORRIS, J.R., KATZ, J.J. : Oxidized bacteriochlorophyll as photoproduct. - In : CLAYTON, R.K., SISTROM, W.R. (ed.) : The Photosynthetic Bacteria. Pp. 397 - 418. Plenum Press, New York - London 1978.

35108 - NORRIS, J.R., THURNAUER, M.C. : Magnetic resonance spectroscopy of the primary donor of photosynthesis. - In : FIALA, J. (ed.) : Third International Seminar on Energy Transfer in Condensed Matter. Pp. 72 - 82. Univerzita Karlova, Praha 1978.

*35109 - NORTH, W.J. : Possibilities of biomass from the ocean. The marine farm project. - In : MITSUI, A., MIYACHI, S., SAN PIETRO, A., TAMURA, S. (ed.) : Biological Solar Energy Conversion. Pp. 347 - 361. Academic Press, New York - San Francisco - London 1977.

35110 - NOVACKY, A., FISCHER, E., ULLRICH-EBERIUS, C.I., LÜTTGE, U., ULLRICH, W.R. : Membrane potential changes during transport of glycine as a neutral amino acid and nitrate in *Lemna gibba* G 1. - FEBS Lett. *88* : 264 - 267, 1978.

35111 - NOVACKY, A., ULLRICH-EBERIUS, C.I., LÜTTGE, U. : Membrane potential changes during transport of hexoses in *Lemna gibba* G 1. - Planta *138* : 263 - 270, 1978. [Ps.]

*35112 - NOVÁK, V. : Fotosyntetická kapacita listů jarního ječmene během ontogenese. [Photosynthetic capacity of leaves of spring barley during ontogeny.] - Sb. vys. Školy zem. Praze, Ser. A *1976* : 53 - 59, 1976. [In Czech, ab : R.]

35113 - NOVAK, V.A., IVANKINA, N.G. : Zavisimost' membrannogo potentsiala fotosinteziruyushchikh rastitel'nykh kletok ot kisloroda. [Oxygen dependence of the membrane potential of photosynthesizing plant cells.] - Tsitologiya *20* : 896 - 902, 1978. [In R, ab : E.]

35114 - NOVAK-HOFER, I., SIEGENTHALER, P.-A. : Chemical cross-linking of neighboring thylakoid membrane polypeptides. - Plant Physiol. *62* : 368 - 372, 1978.

35115 - NOVIKAVA, A.A. : Intênsiŭnasts' fotasintêzu seyantsaŭ khvoĭnykh parod pry roznaŭ pratsyaglastsi asvyatlennya. [Photosynthetic rate of seedlings of conifers under different photoperiod.] - Vestsi Akad. Navuk belarus. SSR, Ser. biyal. Navuk *1978* (1) : 17 - 20, 138, 1978. [In Belorus., ab : E, R.]

35116 - NOVIKOV, E.P. : Pokhodnyĭ fotointegrator FAR. [Field photointegrator PhAR.] - Fiziol. Biokhim. kul't. Rast. *10* : 211 - 214, 1978. [In R, ab : E.]

35117 - NOWAKOWSKI, W. : Action de l'acide gibbérellique (GA₃) sur la photosynthèse, la respiration et la transpiration des limbes de seigle (*Secale cereale* L.). - Biol. Plant. *20* : 142 - 145, 1978.

35118 - NOY-MEIR, I. : Grazing and production in seasonal pastures: Analysis of a simple model. - J. appl. Ecol. *15* : 809 - 835, 1978.

35119 - NULTSCH, W., BENEDETTI, P.A. : Microspectrometric measurements of the light induced chromatophore movements in a single cell of the brown alga *Dictyota dichotoma*. - Z. Pflanzenphysiol. *87* : 173 - 180, 1978.

35120 - NULTSCH, W., HÄDER, M. : Photoakkumulation bei *Halobacterium halobium*. - Ber. deut. bot. Ges. *91* : 441 - 453, 1978.

35121 - NUTBEAM, A.R., DUFFUS, C.M. : Oxygen exchange in the pericarp green layer of immature cereal grains. - Plant Physiol. *62* : 360 - 362, 1978.

35122 - NYMAN, L.P., DENGLER, N.G. : Cell enlargement during leaf development in *Catharanthus roseus*. - Can. J. Bot. *56* : 592 - 605, 1978.

\*35123 - OBATA, Y., OMORI, M., KATO, M., TAKEO, T., SAIJO, R. : [Relationship between
         the quality and chlorophyll content of black tea.] - Kaseigaku Zasshi 28 :
         292 - 294, 1977. [In Jap., ab : E.]

35124 - OCHIAI, H., SHIBATA, H., MATSUO, T., HASHINOKUCHI, K., INAMURA, I. : Immobi-
        lization of chloroplast photosystems with polyvinyl alcohols. - Agr. biol.
        Chem. 42 : 683 - 685, 1978.

35125 - OCHIAI, H., SHIBATA, H., MATSUO, T., HASHINOKUCHI, K., YUKAWA, M. : [Immo-
        bilization of chloroplast photosystems within polyacrylamide gel formed by
        the redox polymerization.] - J. agr. chem. Soc. Jap. 52 : 31 - 36, 1978. [In
        Jap., ab : E.]

35126 - OCHIAI, H., TANABE, H., NAKAMURA, H. : [Circadian rhythms in photosynthe-
        tic activities of a limno Chlorella and a hyphalmyro Chlorella.] - Shimane
        Daigaku Nogakubu Kenkyu Hokoku [Bull. Fac. Agr., Shimane Univ.] 12 : 117 -
        120, 1978. [In Jap., ab : E.]

35127 - O'CONNORS, H.B.,Jr., WURSTER, C.F., POWERS, C.D., BIGGS, D.C., ROWLAND, R.G. :
        Polychlorinated biphenyls may alter marine trophic pathways by reducing phy-
        toplankton size and production. - Science 201 : 737 - 739, 1978. [Chl.]

35128 - ODINTSOVA, M.S., SAMSONOVA, I.A., TURISCHEVA, M.S., BÖTTCHER, F. : Geneti-
        sche Kontrolle der Plastidendifferenzierung. 1. Die Wirkung der Plastom-
        mutation Pl-alb 1 auf die Ultrastruktur der Plastiden in den Kotyledonen von
        Lycopersicon esculentum. - Biol. Zentralbl. 97 : 69 - 82, 1978.

35129 - OECHEL, W.C., SVEINBJÖRNSSON, B. : Primary production processes in arctic
        bryophytes at Barrow, Alaska. - In : TIESZEN, L.L. (ed.) : Vegetation and
        Production Ecology of an Alaskan Arctic Tundra. Pp. 269 - 298. Springer-Ver-
        lag, New York - Heidelberg - Berlin 1978.

35130 - OELZE, J. : Proteins exposed at the surface of chromatophores of Rhodospi-
        rillum rubrum. The orientation of isolated chromatophores. - Biochim. bio-
        phys. Acta 509: 450 - 461, 1978.

35131 - OELZE, J., FAKOUSSA, R.M., HUDEWENTZ, J. : On the significance of electron
        transport systems for growth of Rhodospirillum rubrum. - Arch. Microbiol.
        118 : 127 - 132, 1978.

35132 - OELZE-KAROW, H., KASEMIR, H., MOHR, H. : Control of chlorophyll b formation
        by phytochrome and a threshold level of chlorophyllide a. - In : AKOYUNOGLOU,
        G., ARGYROUDI-AKOYUNOGLOU, J.H. (ed.) : Chloroplast Development. Pp. 787 -
        792. Elsevier/North-Holland Biomedical Press, Amsterdam - New York - Oxford
        1978.

35133 - OELZE-KAROW, H., MOHR, H. : Control of chlorophyll b biosynthesis by phyto-
        chrome. - Photochem. Photobiol. 27 : 189 - 193, 1978.

35134 - OELZE-KAROW, H., MOHR, H. : Phytochrome control of the development of pho-
        tophosphorylation. - Photochem. Photobiol. 27 : 255 - 258, 1978.

35135 - OESTERHELT, D., HARTMANN, L., MICHEL, H., WAGNER, G. : Light-driven proton
        translocation and energy conservation by Halobacteria. - In : SCHÄFER, G.,
        KLINGENBERG, M. (ed.) : Energy Conservation in Biological Membranes. Pp.
        140 - 151. Springer-Verlag, Berlin - Heidelberg - New York 1978.

35136 - OETTMEIER, W., REIMER, S., LINK, K. : Quantitative structure-activity rela-
        tionship of substituted benzoquinones as inhibitors of photosynthetic elec-
        tron transport. - Z. Naturforsch. 33 C : 695 - 703, 1978.

35137 - OGAWA, M., TSUTSUI, Y., KONISHI, M. : Effects of illumination on absorption
        peak shifts in spectra of intact etiolated cotyledons of Pharbitis nil. II.
        Effects of leaf age on protochlorophyllide regeneration and the Shibata
        shift. - Plant Cell Physiol. 19 : 127 - 132, 1978.

35138 - OGAWA, T., AIBA, S. : $CO_2$ assimilation and growth of a blue-green alga,
        Spirulina platensis, in continuous culture. - J. appl. Chem. Biotechnol.
        28 : 515 - 521, 1978.

35139 - OGAWA, T., ISHIKAWA, H., SHIMADA, K., SHIBATA, K. : Synergistic action of
red and blue light and action spectra for malate formation in guard cells
of *Vicia faba* L. - Planta *142* : 61 - 65, 1978.

*35140 - OGAWA, T., SHIBATA, K. : Two phases of $CO_2$ absorption on leaves. - In :
MITSUI, A., MIYACHI, S., SAN PIETRO, A., TAMURA, S. (ed.) : Biological Solar
Energy Conversion. Pp. 183 - 195. Academic Press, New York - San Francisco -
London 1977.

35141 - OGREN, W.L. : Increasing carbon fixation by crop plants. - In : HALL, D.O.,
COOMBS, J., GOODWIN, T.W. (ed.) : Proceedings of the Fourth International
Congress on Photosynthesis. Pp. 721 - 733. Biochem. Soc., London 1978.

35142 - OGREN, W.L., HUNT, L.D. : Comparative biochemistry of ribulose bisphosphate
carboxylase in higher plants. - In : SIEGELMAN, H.W., HIND, G. (ed.) : Photosynthetic Carbon Assimilation. Pp. 127 - 138. Plenum Press, New York -
London 1978.

35143 - OHAD, I., BAR-NUN, S., CAHEN, D., GERSHONI, J.M., GUREVITZ, M., LAVINTMAN, N.
: Step-wise synthesis and assembly of photosynthetic membranes in algae. -
In : HALL, D.O., COOMBS, J., GOODWIN, T.W. (ed.) : Proceedings of the Fourth
International Congress on Photosynthesis. Pp. 517 - 526. Biochem. Soc.,
London 1978.

35144 - OH-HAMA, T., HASE, E. : Blue light effect on chlorophyll formation in *Chlorella protothecoides*. - Photochem. Photobiol. *27* : 199 - 202, 1978.

35145 - OH-HAMA, T., SENGER, H. : Spectral effectiveness in chlorophyll and 5-ami-
nolevulinic acid formation during regreening of glucose-bleached cells of
*Chlorella protothecoides*. - Plant Cell Physiol. *19* : 1295 - 1299, 1978.

35146 - OHKI, K. : Zinc concentration in soybean as related to growth, photosynthesis, and carbonic anhydrase activity. - Crop Sci. *18* : 79 - 82, 1978.

35147 - OHKI, K., FUJITA, Y. : Photocontrol of phycoerythrin formation in the blue-
-green alga *Tolypothrix tenuis* growing in the dark. - Plant Cell Physiol.
*19* : 7 - 15, 1978.

35148 - OHMORI, M., HATTORI, A. : Transient change in the ATP pool of *Anabaena cylindrica*. - Arch. Microbiol. *117* : 17 - 20, 1978.

35149 - OHYAMA, K. : [Photosynthesis and reproduction of the mulberry (*Morus bom-
bycus*).] - Nippon Sanshigaku Zasshi [J. sericult. Sci. Jap.] *47* : 91 - 100,
1978. [In Jap.]

35150 - OIKAWA, T. : Wind characteristics of the model canopy demonstrated with
wind-blow computer experiments. - In : MONSI, M., SAEKI, T. (ed.) : Ecophysiology of Photosynthetic Productivity. JIBP Synthesis Vol. 19. Pp. 159 -
167. University of Tokyo Press, Tokyo 1978.

35151 - OIKAWA, T. : Canopy photosynthesis of the plant population simulated on
the basis of light and $CO_2$ conditions. - In : MONSI, M., SAEKI, T. (ed.) :
Ecophysiology of Photosynthetic Productivity. JIBP Synthesis Vol. 19. Pp.
167 - 183. University of Tokyo Press, Tokyo 1978.

35152 - OKABE, K., SCHMID, G.H. : Properties of the tobacco aurea mutant Su/su var.
aurea. On photorespiration and on the structure and function relationship
in chloroplasts. - In : AKOYUNOGLOU, G., ARGYROUDI-AKOYUNOGLOU, J.H. (ed.) :
Chloroplast Development. Pp. 501 - 506. Elsevier/North-Holland Biomedical
Press, Amsterdam - New York - Oxford 1978.

35153 - OKADA, M., INOUE, M., IKEDA, T. : Circadian rhythm in photosynthesis of the
green alga *Bryopsis maxima*. - Plant Cell Physiol. *19* : 197 - 202, 1978.

35154 - OKAFO, O.A., HANOVER, J.W. : Comparative photosynthesis and respiration
of trembling and bigtooth aspens in relation to growth and development. -
Forest Sci. *24* : 103 - 109, 1978.

35155 - OKANENKO, A.S., MANUIL'SKIĬ, V.D., IVANISHCHEVA, S.Yu. : Issledovanie roli
kaliya v regulyatornykh funktsiyakh fotosinteticheskogo apparata rasteniĭ
sakharnoĭ svekly. [Study of potassium role in regulatory functions of photosynthetic apparatus of sugar beet plants.] - Fiziol. Biokhim. kul't.
Rast. *10* : 146 - 150, 1978. [In R, ab : E.]

*35156 - OKU, T. : [Determination of chlorophyll in coniferous leaves.]-J. jap. Fo-
rest. Soc. *57* : 239 - 242, 1975. [In Jap.]

35157 - OKU, T., INOUE, Y., SANADA, M., MATSUSHITA, K., TOMITA, G. : Development
of photosynthetic activities in dark-grown spruce seedlings. - Plant Cell
Physiol. *19* : 1 - 6, 1978.

35158 - ŌKUBO, T., KAWANABE, S. : Maximum crop growth rate, photosynthesis and
chlorophyll-index in some pasture-plants. - In : MONSI, M., SAEKI, T. (ed.):
Ecophysiology of Photosynthetic Productivity. JIBP Synthesis Vol. 19. Pp.
194 - 211. University of Tokyo Press, Tokyo 1978.

35159 - OLECH, K., BLAMOWSKI, Z.K. : Formowanie powierzchni asymilacyjnej i biomasy
przez rośliny buraków cukrowych. [Formation of the assimilation area and
biomass by sugar beet plants.] - Acta agrobot. *31* : 21 - 32, 1978. [Growth
analysis ; in Pol., ab : E.]

35160 - OLESEN, P. : Structure of chloroplast membranes as revealed by natural and
experimental fixation with tannic acid : particles in and on the thylakoid
membrane. - Biochem. Physiol. Pflanz. *172* : 319 - 342, 1978.

35161 - OLIKER, M., POLJAKOFF-MAYBER, A., MAYER, A.M. : Changes in weight, nitrogen
accumulation, respiration and photosynthesis during growth and development
of seeds and pods of *Phaseolus vulgaris*. - Amer. J. Bot. *65* : 366 - 371, 1978.

35162 - OLIVER, D.J. : Decrease in the inhibition of net photosynthesis by $O_2$ in
the presence of glyoxylate. - Plant Physiol. *61* (Suppl.) : 7, 1978.

35163 - OLIVER, D.J. : Inhibition of photorespiration and increase of net photosynthesis in isolated maize bundle sheath cells treated with glutamate or
aspartate. - Plant Physiol. *62* : 690 - 692, 1978.

35164 - OLIVER, D.J. : Effect of glyoxylate on the sensitivity of net photosynthesis to oxygen (the Warburg effect) in tobacco. - Plant Physiol. *62* : 938 -
940, 1978.

*B35165 - OLOVYANNIKOVA, I.N. : Vliyanie Lesnykh Kolkov na Solonchakovye Solontsy.
[Influence of Forest Rings on Saline Soils.] - Nauka, Moskva 1976. [Dry-
-matter accumulation ; in R.]

35166 - OLSON, J.M. : Bacteriochlorophyll *a*-proteins from green bacteria. - In :
CLAYTON, R.K., SISTROM, W.R. (ed.) : The Photosynthetic Bacteria. Pp. 161 -
178. Plenum Press, New York - London 1978.

35167 - OLSON, J.M. : Precambrian evolution of photosynthetic and respiratory organisms. - In : HECHT, M.K., STEERE, W.C., WALLACE, B. (ed.) : Evolutionary
Biology. Vol. 11. Pp. 1 - 37. Plenum Press, New York 1978.

35168 - ONDOK, J.P. : Radiation climate in fishpond littoral plant communities. -
In : DYKYJOVÁ, D., KVĚT, J. (ed.) : Pond Littoral Ecosystems. Pp. 113 - 125.
Springer-Verlag, Berlin - Heidelberg - New York 1978.

35169 - ONDOK, J.P. : Estimation of seasonal growth of underground biomass. - In :
DYKYJOVÁ, D., KVĚT, J. (ed.) : Pond Littoral Ecosystems. Pp. 193 - 197.
Springer-Verlag, Berlin - Heidelberg - New York 1978.

35170 - ONDOK, J.P. : Estimation of net photosynthetic efficiency from growth analytical data. - In : DYKYJOVÁ, D., KVĚT, J. (ed.) : Pond Littoral Ecosystems.
Pp. 221 - 227. Springer-Verlag, Berlin - Heidelberg - New York 1978.

35171 - ONDOK, J.P., GLOSER, J. : Net photosynthesis and dark respiration in a stand
of *Phragmites communis* TRIN. calculated by means of a model I. Description
of the model. - Photosynthetica *12* : 328 - 336, 1978.

35172 - ONDOK, J.P., GLOSER, J. : Net photosynthesis and dark respiration in a stand
of *Phragmites communis* TRIN. calculated by means of a model II. Results. -
Photosynthetica *12* : 337 - 343, 1978.

35173 - ONDOK, J.P., GLOSER, J. : Modeling of photosynthetic production in littoral
helophyte stands. - In : DYKYJOVÁ, D., KVĚT, J. (ed.) : Pond Littoral
Ecosystems. Pp. 234 - 245. Springer-Verlag, Berlin - Heidelberg - New York
1978.

35174 - **ONDOK, J.P., KVĚT, J.** : Selection of sampling areas in assessment of production. - In : DYKYJOVÁ, D., KVĚT, J. (ed.) : Pond Littoral Ecosystems. Pp. 163 - 174. Springer-Verlag, Berlin - Heidelberg - New York 1978.

35175 - **O'NEAL, S.W.** : The relationship of photosynthesis and dark respiration to growth in *Caulerpa paspaloides* (*Caulerpaceae*). - J. Phycol. *14* (Suppl.) : 38, 1978.

35176 - **ONG, C.K., COLVILL, K.E., MARSHALL, C.** : Assimilation of $^{14}CO_2$ by the inflorescence of *Poa annua* L. and *Lolium perenne* L. - Ann. Bot. *42* : 855 - 862, 1978.

35177 - **ONO, T.-A., MURATA, N.** : Photosynthetic electron transport and phosphorylation reactions in thylakoid membranes prepared from the blue-green alga *Anacystis nidulans*. - Biochim. biophys. Acta *502* : 477 - 485, 1978.

35178 - **OOHUSA, T., ARAKI, S., SAKURA, T., SAITOH, M.** : [Diurnal variations of the photosynthetic pigments, total nitrogen and total nitrogen/total carbohydrate ratio of cultivated *Porphyra* thali and their relationships to the quality of dried Nori.] - Jap. J. Phycol. *26* : 185 - 187, 1978. [In Jap., ab : E.]

35179 - **OOHUSA, T., ARAKI, S., SAKURAI, T., SAITOH, M., KIRITA, M., YAMASHITA, T.** : [The diurnal variations in the cell size, in the physiological activity and in the contents of some cellular components of *Porphyra yezoensis* f. *narawaensis* in cultivation ground.] - Nippon Suisan Gakkaishi [Bull. jap. Soc. sci. Fisheries] *44* : 299 - 303, 1978. [In Jap., ab : E.]

35180 - **OPRITOV, V.A.** : Rasprostranyayushcheesya vozbuzhdenie i transport assimilyatov vo floeme. [Spreading excitation and phloem translocation of photosynthates.] - Fiziol. Rast. *25* : 1042 - 1054, 1978. [In R, ab : E.]

35181 - **ÖQUIST, G., MÅRTENSSON, O., MARTIN, B., MALMBERG, G.** : Seasonal effects on chlorophyll-protein complexes isolated from *Pinus silvestris*. - Physiol. Plant. *44* : 187 - 192, 1978.

35182 - **OREN, A., PADAN, E.** : Induction of anaerobic, photoautotrophic growth in the cyanobacterium *Oscillatoria limnetica*. - J. Bacteriol. *133* : 558 - 563, 1978. [Ps.]

35183 - **ORITANI, T.** : Effect of nitrogen supply and light intensity on leaf area growth, nitrogen metabolism and photosynthetic capacity in the leaves of two rice varieties. - Toyama-Kenritsu Gijutsu Tanki Daigaku Kenkyu Hokoku *11* : 69 - 76, 1978.

35184 - **ORITANI, T.** : Studies on nitrogen metabolism in crop plants. XV. The effects of light intensity and TIBA on the photosynthetic capacity and nitrogen metabolism in the two contrasting varieties of soybean plants. - Jap. J. Crop Sci. *47* : 124 - 132, 1978.

35185 - **ORMOS, P., DANCSHÁZY, Zs., KARVALY, B.** : Mechanism of generation and regulation of photopotential by bacteriorhodopsin in bimolecular lipid membrane. The quenching effect of blue light. - Biochim. biophys. Acta *503* : 304 - 315, 1978.

35186 - **ORSENIGO, M., PUPILLO, P.** : L'organizacione del carbonio nelle piante ad alta produttività. [Organization of carbon in highly productive plants.] - G. bot. ital. *112* : 251 - 269, 1978. [In Ital.]

35187 - **ORT, D.R.** : Different sensitivities of chloroplasts to uncouplers when ATP formation is induced by continuous illumination, by brief illumination, by pre-illumination, or by acid-base transitions. - Europe. J. Biochem. *85* : 479 - 485, 1978.

35188 - **ORT, D.R., DILLEY, R.A.** : The role of proton gradients in initiating photophosphorylation and in slowing electron transport. - In : HALL, D.O., COOMBS, J., GOODWIN, T.W. (ed.) : Proceedings of the Fourth International Congress on Photosynthesis. Pp. 581 - 590. Biochem. Soc., London 1978.

35189 - **ORT, D.R., PARSON, W.W.** : Flash-induced volume changes of bacteriorhodopsin-containing membrane fragments and their relationship to proton movements and absorbance transients . - J. biol. Chem. *253* : 6158 - 6164, 1978.

35190 - ORTIZ, W., PRICE, C.A. : Physiological conditioning of protein synthesis
       by isolated chloroplasts of *Euglena gracilis*. - Plant Physiol. *61* (Suppl.) :
       104, 1978.

35191 - OSHCHEPKOV, V.P., DYURIAN, I., VOROB'EVA, L.M., CHERNYAD'EV, I.I., ABROS'-
       KINA, L.S. : Sravnitel'noe issledovanie fotoobrazovaniya $H_2$ i $O_2$ mutantami
       zelenykh vodoroslei. [A comparison of $H_2$ and $O_2$ photoproduction by the mu-
       tants of green algae.] - Fiziol. Rast. *25* : 821 - 828, 1978. [In R, ab : E.]

35192 - OSMOND, C.B. : Crassulacean acid metabolism : A curiosity in context. -
       Annu. Rev. Plant Physiol. *29* : 379 - 414, 1978.

35193 - OSNITSKAYA, L.K., CHUDINA, V.I. : Fotosinteziruyushchie bakterii iz ozera
       Vanda (Antarktida). [Photosynthetic bacteria from the Vanda Lake in the
       Antarctic.] - Mikrobiologiya *47* : 131 - 137, 1978. [In R, ab : E.]

35194 - OSONUBI, O., DAVIES, W.J. : Solute accumulation in leaves and roots of
       woody plants subjected to water stress. - Oecologia *32* : 323 - 332, 1978.
       [Resistances.]

*35195 - OSWALD, W.J., BENEMANN, J.R. : A critical analysis of bioconversion with
       microalgae. - In : MITSUI, A., MIYACHI, S., SAN PIETRO, A., TAMURA, S.
       (ed.) : Biological Solar Energy Conversion. Pp. 379 - 396. Academic Press,
       New York - San Francisco - London 1977.

35196 - OUDOT, C. : Intérêt du concept NO (oxygène corrigé de la respiration) pour
       l'étude de l'écosystème de l'upwelling équatorial. - Cah. O.R.S.T.O.M., Sér.
       Océanogr. *16* : 191 - 199, 1978. [Ps.]

35197 - OUTLAW, W., KENNEDY, J. : Anion synthesis in guard cells. - Plant Physiol.
       *61* (Suppl.) : 86, 1978. [Ps.]

35198 - OUTLAW, W.H. Jr., KENNEDY, J. : Enzymic and substrate basis for anaplero-
       tic step in guard cells. - Plant Physiol. *62* : 648 - 652, 1978. [PEPC.]

35199 - OVCHINNIKOV,Yu.A., ABDULAEV, N.G., FEIGINA, M.Yu., KISELEV, A.V., LOBANOV,
       N.A., NAZIMOV, I.V. : K voprosu o pervichnoi strukture bakteriorhodopsina.
       [Primary structure of bacteriorhodopsin.] - Bioorgan. Khim. *4* : 979 - 981,
       1978. [In R , ab : E.]

35200 - OVERDIECK, D. : $CO_2$-Gaswechsel und Transpiration von Sonnen- und Schatten-
       blättern bei unterschiedlichen Strahlungsqualitäten. - Ber. deut. bot. Ges.
       *91* : 633 - 644, 1978.

*35201 - OVSYANNIKOV, A.S. : Fotosintet0icheskaya deyatel'nost' yabloni v zavisimosti
       ot razmeshcheniya list'ev v krone, vozrasta derev'ev i tipa plodonosheniya.
       [Photosynthetic activity of apple tree in dependence on leaf position in
       the crown, tree age and fruiting type.] - Sb. nauch. Rabot vsesoyuz. nauch-
       no-issled. Inst. Sadovod. (Michurinsk) *23* : 70 - 74, 1976. [In R.]

35202 - OVSYANNIKOV, A.S., ANDREEVA, A.N. : Izuchenie zavisimosti mezhdu morfofizio-
       logicheskimi priznakami i urozhainost'yu zemlyaniki v agrotsenoze. [Relation-
       ship between morphophysiological characters and productivity of strawberry
       in agrocoenosis.] - Sb. nauch. Tr. vsesoyuz. nauchno-issled. Inst. Sado-
       vodstva (Michurinsk) *27* (Sovershenstvovanie Sortimenta i Agrotekhniches-
       kikh Priemov v Sadovodstve) : 1 - 4, 1978. [Ps; in R.]

35203 - OVSYANNIKOV, A.S., OVSYANNIKOVA, R.S. : Fotosinteticheskaya produktivnost'
       i potentsial'naya urozhainost' sortov grushi v usloviyach TsChO. [Photo-
       synthetic productivity and potential yielding capacity of pear cultivars
       in conditions of TsChO.] - Sb. nauch. Tr. vsesoyuz. nauchno-issled. Inst.
       Sadovodstva (Michurinsk) *27* (Sovershenstvovanie Sortimenta i Agrotekhni-
       cheskikh Priemov v Sadovodstve) :,1 - 6, 1978. [In R.]

35204 - OWENS, T.G., RIPER, D.M., FALKOWSKI, P.G. : Studies of delta-aminolevulinic
       acid dehydrase from *Skeletonema costatum*, a marine plankton diatom. - Plant
       Physiol. *62* : 516 - 521, 1978.

35205 - OWUSU, J.K., ADOMAKO, D., HUTCHEON, W.V. : Seasonal changes in total free
       sugar content of field cocoa plants. - Physiol. Plant. *44* : 43 - 47, 1978.
       [Photosynthates.]

35206 - **PAAU, A.S., ORO, J., COWLES, J.R.** : Application of flow microfluorometry to the study of algal cells and isolated chloroplasts. - J. exp. Bot. *29* : 1011 - 1020, 1978.

35207 - **PACKER, L., CULLINGFORD, W.** : Stoichiometry of $H_2$ production by an *in vitro* chloroplast, ferredoxin, hydrogenase reconstituted system. - Z. Naturforsch. *33 C* : 113 - 115, 1978.

35208 - **PACKER, L., KONISHI, T.** : Chemical modification of bacteriorhodopsin : an approach to the mechanism of proton translocation. - In : **CAPLAN, S.R., GINZBURG, M.** (ed.) : Energetics and Structure of Halophilic Microorganisms. Pp. 143 - 163. Elsevier/North-Holland Biomedical Press, Amsterdam - New York 1978.

35209 - **PADAN, E., SCHULDINER, S.** : Energy transduction in the photosynthetic membranes of the cyanobacterium (blue-green alga) *Plectonema boryanum*. - J. biol. Chem. *253* : 3281 - 3286, 1978.

35210 - **PADHI, B., CHAKRABARTI, N.K., NAYAK, S.K.** : Effect of blast infection on the photosynthetic activity of rice seedlings. - Biol. Plant. *20* : 418 - 420, 1978.

35211 - **PADMANABHAN, U., GREEN, B.R.** : The kinetic complexity of *Acetabularia* chloroplast DNA. - Biochim. biophys. Acta *521* : 67 - 73, 1978.

35212 - **PAECH, C., McCURRY, S.D., PIERCE, J., TOLBERT, N.E.** : Active site of ribulose 1,5-bisphosphate carboxylase/oxygenase. - In : **SIEGELMAN, H.W., HIND, G.** (ed.) : Photosynthetic Carbon Assimilation. Pp. 227 - 243. Plenum Press, New York - London 1978.

35213 - **PAECH, C., McCURRY, S.D., PIERCE, J., TOLBERT, N.E.** : Inhibition of RuBP carboxylase/oxygenase by RuBP epimerization and degradation products. - In : **SIEGELMAN, H.W., HIND, G.** (ed.) : Photosynthetic Carbon Assimilation. Pp. 422 - 423. Plenum Press, New York - London 1978.

35214 - **PAECH, C., TOLBERT, N.E.** : Active site studies of ribulose-1,5-bisphosphate carboxylase/oxygenase with pyridoxal 5'—phosphate. - J. biol. Chem. *253* : 7864 - 7873, 1978.

35215 - **PAECH, C., TOLBERT, N.E.** : Functional groups of ribulose-1,5-bisphosphate carboxylase oxygenase from spinach. - Plant Physiol. *61* (Suppl.) : 98, 1978.

35216 - **PAERL, H.W.** : Effectiveness of various counting methods in detecting viable phytoplankton. - N. Z. J. mar. Freshw. Res. *12* : 67 - 72, 1978.

35217 - **PAERL, H.W.** : Nannoplankton *vs* netplankton photosynthetic and heterotrophic activities in Fijian waters of the South Pacific Ocean. - Bull. roy. Soc. New Zealand *17* (Lau-Tonga 1977) : 211 - 221, 1978.

35218 - **PAILLOTIN, G.** : Dynamics and structural aspects of energy transport in green plants. - Spectrosc. Lett. *11* : 513 - 523, 1978. [Ps.]

35219 - **PAILLOTIN, G.** : Excitation energy transport in photosynthesis. - In : **FIALA, J.** (ed.) : Third International Seminar on Energy Transfer in Condensed Matter. Pp. 40 - 51. Univerzita Karlova, Praha 1978.

35220 - **PAILLOTIN, G.** : Organization of the photosynthetic pigments and transfer of excitation energy. - In : **HALL, D.O., COOMBS, J., GOODWIN, T.W.** (ed.) : Proceedings of the Fourth International Congress on Photosynthesis. Pp. 33 - 44. Biochem. Soc., London 1978.

35221 - **PALIWAL, K.V., MALIWAL, G.L., NANAWATI, G.C.** : Effect of bicarbonate-rich irrigation waters on the yield and nutrient uptake in cotton (*Gossypium* spp.) and linseed (*Linum usitatissimum*). - Ann. Arid Zone *17* : 164 - 174, 1978.

35222 - **PALLAS, J.E. Jr.** : Photosynthesis,Γ and leaf diffusional resistances of selected peanut cultivars. - Plant Physiol. *61* (Suppl.) : 9, 1978.

35223 - **PALTA, J.P., LI, P.H.** : Cell membrane properties in relation to freezing injury. - In : **LI, P.H., SAKAI, A.** (ed.) : Plant Cold Hardiness and Freezing Stress. Mechanisms and Crop Implications. Pp. 93 - 115. Academic Press, New York - San Francisco - London 1978. [Chloroplast.]

35224 - PANCHENKO, T.M. : O fotosinteze kedrovogo stlanika *Pinus pumila* (PALL.)
REGEL v estestvennykh usloviyakh. [Photosynthesis of dwarf-pine elfin wood
*Pinus pumila* (PALL.) REGEL under natural conditions.]- Bot. Zh. *63* : 736 -
744, 1978. [In R.]

35225 - PANDEY, R.K., SAXENA, M.C., SINGH, V.B., SINGH, B.K. : Assimilation and
utilization of $^{14}C$ assimilates in chickpeas. - J. nucl. agr. Biol. *7* : 56 -
60, 1978.

35226 - PANDEY, R.K., SINGH, B.K. : Genotype variation in photosynthetic rate and
transport and utilization of $C^{14}$ photosynthate in chickpea. - In : Inter-
-Disciplinary Symposium on Photosynthesis and Productivity. Abstract of
Papers. Pp. 49 - 50. Indian nat. Sci. Acad., New Delhi 1978.

35227 - PANT, A. : Productivity of oceans. - In : Inter-Disciplinary Symposium on
Photosynthesis and Productivity. Abstract of Papers. Pp. 28 - 31. Indian
nat. Sci. Acad., New Delhi 1978.

35228 - PAPADOPOULOS, G.K., HSIAO, T.L., CASSIM, J.Y. : Determination of the reti-
nal/protein molar ratios for the purple membranes of *Halobacterium halobium*
and *Halobacterium cutirubrum*. - Biochem. biophys. Res. Commun. *81* : 127 -
132, 1978.

35229 - PAPADOPOULOS, G.K., MUCCIO, D.D., HSIAO, T.L., CASSIM, J.Y. : Comparative
studies on the fine structure of purple membrane from *Halobacterium cutirubrum* and *Halobacterium halobium*. - J. Membrane Biol. *43* : 277 - 294, 1978.

35230 - PAPAGEORGIOU, G.C., ISAAKIDOU, J. : The organizational state of thylakoid
membranes as a determinant of the biochemical and biophysical properties
of glutaraldehyde-fixed chloroplasts. - Z. Pflanzenphysiol. *89* : 449 - 452,
1978.

35231 - PARADIES, H.H., ZIMMERMANN, J., SCHMIDT, U.D. : The conformation of chloroplast coupling factor 1 from spinach in solution. - J. biol. Chem. *253* :
8972 - 8979, 1978.

35232 - PARIDA, R.K., KAR, M., MISHRA, D. : Enhancement of senescence in excised
rice leaves by hydrogen peroxide. - Can. J. Bot. *56* : 2937 - 2941, 1978.
[Chl.]

35233 - PARKER, J. : Seasonal variations in photosynthesis in black oak twigs. -
Photosynthetica *12* : 423 - 427, 1978.

35234 - PARKHURST, D.F. : The adaptive significance of stomatal occurrence on one
or both surfaces of leaves. - J. Ecol. *66* : 367 - 383, 1978. [Resistances.]

35235 - PARKINSON, K.J., LEGG, B.J. : Calibration of infra-red analysers for carbon
dioxide. - Photosynthetica *12* : 65 - 67, 1978.

35236 - PARKS, L.C., NIEDERMAN, R.A. : Membranes of *Rhodopseudomonas sphaeroides*.
V. Identification of bacteriochlorophyll *a*-depleted cytoplasmic membrane
in phototrophically grown cells. - Biochim. biophys. Acta *511* : 70 - 82,
1978.

35237 - PARSHINA, O.V., USHAKOVA, S.A. : Temnoustoĭchivost' pigmentov pshenitsy
v zavisimosti ot intensivnosti i spektral'nogo sostava sveta, predshestvu-
yushchego temnote. [Tolerance of pigments in wheat to darkness as affected
by the intensity and spectral composition of light preceding the darkness.]
- Fiziol. Rast. *25* : 1122 - 1128, 1978. [In R, ab : E.]

35238 - PARSON, W.W. : Role of the reaction center in photosynthesis. - In :
CLAYTON, R.K., SISTROM, W.R. (ed.) : The Photosynthetic Bacteria. Pp. 317 -
322. Plenum Press, New York - London 1978.

35239 - PARSON, W.W. : Quinones as secondary electron acceptors. - In : CLAYTON,
R.K., SISTROM, W.R. (ed.) : The Photosynthetic Bacteria. Pp. 455 - 469.
Plenum Press, New York - London 1978.

35240 - PARSON, W.W. : Thermodynamics of the primary reactions of photosynthesis. -
Photochem. Photobiol. *28* : 389 - 393, 1978.

35241 - **PARTHIER, B.** : Licht-induzierte Transformation von Proplastiden zu Chloroplasten in Algen und höheren Pflanzen. - In : **NOVER, L., LUCKNER, M., PARTHIER, B.** (ed.) : Zelldifferenzierung. Molekulare Grundlagen und Probleme. Pp. 210 - 259. VEB G.Fischer Verlag, Jena 1978.

35242 - **PARTHIER, B., MUELLER-URI, F., KRAUSPE, R.** : The aminoacyl-tRNA synthetases of *Euglena* chloroplasts. - In : **AKOYUNOGLOU, G., ARGYROUDI-AKOYUNOGLOU, J.H.** (ed.) : Chloroplast Development. Pp. 687 - 693. Elsevier/North-Holland Biomedical Press, Amsterdam - New York - Oxford 1978.

35243 - **PARTON, W.J., SINGH, J.S., COLEMAN, D.C.** : A model of production and turnover of roots in shortgrass prairie. - J. appl. Ecol. *15* : 515 - 542, 1978. [Ps.]

35244 - **PÂRVU, C., ENE, E.** : Contributions to the investigation of macrophytic and phytoplanctonic primary productivity from peat-sphagnicol marsh Manta (Romania) in 1976. - Arch. Hydrobiol. Suppl. *52* (2/3) : 229 - 240, 1978.

35245 - **PASHCHENKO, V.N., PODOL'NYĬ, V.Z.** : Razvitie fotosinteticheskoĭ funktsii semyadol'nykh list'ev v svyazi s generativnym razvitiem prorostkov grechikhi. [Development of the photosynthetic function of cotyledonous leaves in connection with the generative growth of buckwheat seedlings.] - Fiziol. Rast. *25* : 525 - 530, 1978. [In R, ab : E.]

35246 - **PASHCHENKO, V.Z., KONONENKO, A.A., RUBIN, A.B., RUBIN, L.B.** : Pikosekundnaya fluorometriya bakteriĭ *Rhodopseudomonas sphaeroides*, shtamm 1760-1. [Picosecond fluorometry of bacterium *Rhodopseudomonas sphaeroides*, strain 1760-1.] - Biofizika *23* : 833 - 838, 1978. [In R, ab : E.]

35247 - **PASSERA, C.** : Effetto della carenza di zolfo sulla fotosintesi e fotorespirazione di *Hordeum vulgare* L., cv. "Astrix". [Effects of sulfur deficiency on photosynthesis and photorespiration of *Hordeum vulgare* L., cv. Astrix.] - Riv. Agron. *12* (3) : 113 - 118, 1978. [In Ital., ab : E.]

35248 - **PASSERA, C., ALBUZIO, A.** : Benzylaminopurine induced changes in photosynthesis and photorespiration of barley plants. - Biol. Plant. *20* : 274 - 280, 1978.

35249 - **PASSERA, C., ALBUZIO, A.** : Effect of salinity on photosynthesis and photorespiration of two wheat species (*Triticum durum* cv. PEPF 2122 and *Triticum aestivum* cv. Marzotto). - Can. J. Bot. *56* : 121 - 126, 1978.

32550 - **PASSERA, C., ALI, M.M.** : Composizione e possibilità di utilizzazione della pianta marina *Cymodocea ciliata*. [Composition and possible use of the sea plant *Cymodocea ciliata*.] - Riv. Agr. subtrop. trop. *72* : 109 - 116, 1978. [Chl, Car; in Ital., ab : E.]

35251 - **PASSERA, C., ALI, M.M.** : Variazione della composizione chimica durante la maturazione del mango. [Variation in chemical composition during ripening of mango fruits.] - Riv. Agr. subtrop. trop. *72* : 117 - 124, 1978. [Chl, Car; in Ital., ab : E.]

35252 - **PASSERA, C., MAGGIONI, A.** : Adenosine triphosphatase (ATPase) activities in *Chlorella vulgaris* chloroplasts. - Agrochimica *22* (2) : 69 - 74, 1978.

35253 - **PASTERNAK, C., SHINITZKY, M.** : Polarized photochromism in the purple membrane from *Halobacterium halobium*. - In : **CAPLAN, S.R., GINZBURG, M.** (ed.) : Energetics and Structure of Halophilic Microorganisms. Pp. 309 - 314. Elsevier/North-Holland Biomedical Press, Amsterdam - New York 1978.

35254 - **PASTERNAK, K., KASZA, H.** : Stosunki chemiczne i produkcja pierwotna fitoplanktonu w podgrzanej wodzie zbiornika rybnickiego. Chemical relations and primary production of phytoplankton in the warmed water of the reservoir Rybnik. - Acta hydrobiol. *20* : 305 - 322, 1978.

35355 - **PATE, J.S., HERRIDGE, D.F.** : Partitioning and utilization of net photosynthate in a nodulated annual legume. - J. exp. Bot. *29* : 401 - 412, 1978.

35256 - **PATE, J.S., HOCKING, P.J.** : Phloem and xylem transport in the supply of minerals to a developing legume (*Lupinus albus* L.) fruit. - Ann. Bot. *42* : 911 - 921, 1978. [Photosynthates.]

35257 - PATRA, H.K., KAR, M., MISHRA, D. : Catalase activity in leaves and cotyledons during plant development and senescence. - Biochem. Physiol. Pflanz.
*172* : 385 - 390, 1978. [Chl.]

35358 - PATTEN, D.T. : Productivity and production efficiency of an upper Sonoran
Desert ephemeral community. - Amer. J. Bot. *65* : 891 - 895, 1978.

35259 - PATTERSON, B.D., PAULL, R., SMILLIE, R.M. : Chilling resistance in *Lycopersicon hirsutum* HUMB & BONPL., a wild tomato with a wide altitudinal distribution. - Aust. J. Plant Physiol. *5* : 609 - 617, 1978. [Chl.]

35260 - PATTERSON, D.T. : Effects of chilling on photosynthesis and stomatal resistance in cotton and two related weeds. - Plant Physiol. *61* (Suppl.) : 4,
1978.

35261 - PATTERSON, D.T., DUKE, S.O., HOAGLAND, R.E. : Effects of irradiance during
growth on adaptive photosynthetic characteristics of velvetleaf and cotton. -
Plant Physiol. *61* : 402 - 405, 1978.

35262 - PATTERSON, D.T., MEYER, C.T., FLINT, E.P., QUIMBY, P.C. Jr. : Mathematical
analysis of the growth of itchgrass (*Rottoboellia exaltata* L.) in 36 combinations of day and night temperature. - Plant Physiol. *61* (Suppl.) : 6,
1978. [Ps.]

35263 - PATTERSON, D.T., MEYER, C.R., QUIMBY, P.C. Jr. : Effects of irradiance on
relative growth rates, net assimilation rates, and leaf area partitioning
in cotton and three associated weeds. - Plant Physiol. *62* : 14 - 17, 1978.

35264 - PATTERSON, G.M.L., COHEN, W.S., HARRIS, D.O. : Inhibition of photosynthetic
electron transport by a toxic substance isolated from the alga *Pandorina
morum*. - Plant Physiol. *61* (Suppl.) : 102, 1978.

35265 - PAUL, J.S., BASSHAM, J.A. : Effects of sulfite on metabolism in isolated
mesophyll cells from *Papaver somniferum*. - Plant Physiol. *62* : 210 - 214,
1978.

35266 - PAUL, J.S., CORNWELL, K.L., BASSHAM, J.A. : Effects of ammonia on carbon
metabolism in photosynthesizing isolated mesophyll cells from *Papaver somniferum* L. - Planta *142* : 49 - 54, 1978.

35267 - PAUL, J.S., CORNWELL, K.L., BASSHAM, J.A. : Effects of ammonia on carbon
metabolism in photosynthesizing leaf-free mesophyll cells from *Papaver somniferum*. - Plant Physiol. *61* (Suppl.) : 38, 1978.

35268 - PAUL, N.K. : Genetic architecture of yield and components of yield in mustard (*Brassica juncea* (L.) CZERN & COSS.). - Theor. appl. Genet. *53* : 233 -
237, 1978.

35269 - PAVLOVA, I.E., MATORIN, D.N., VENEDIKTOV, P.S. : Issledovanie zamedlennoĭ
fluorestsentsii list'ev drevesnykh rasteniĭ, vyrashchennykh v raznykh usloviyakh osveshchennosti. [Delayed fluorescence of leaves of trees grown under
different illuminance.] - Fiziol. Rast. *25* : 97 - 105, 1978. [In R, ab : E.]

35270 - PAWAR, R.B., HEGDE, B.A. : Effect of silicon foliar application on growth,
tillering and photosynthetic efficiency in *Oryza sativa* var. Satya. -
Indian J. Plant Physiol. *21* : 12 - 16, 1978.

35271 - PAWLAK, M., NALBORCZYK, E. : Equipment for simultaneous measurement of
$^{14}CO_2$, $^{12}CO_2$ and water vapour in plant gas exchange studies. - Newslett.
Appl. nucl. Meth. Biol. Agr. *1978* (10 - 11) : 1 - 5, 1978.

35272 - PEAKE, D.C.I., HENZELL, E.F., STIRK, G.B. : Simulation of herbage production and soil water use by Biloela buffel grass in small plot experiments
at Narayen. - Trop. Agron. tech. Memorandum *12* : 1 - 26, 1978.

*35273 - PEARCY, R.W. : Temperature responses of growth and photosynthetic $CO_2$ ex-
change rates in coastal and desert races of *Atriplex lentiformis*. - Oecologia *26* : 245 - 255, 1976.

35274 - PEARLSTEIN, R.M., HEMENGER, R.P. : Bacteriochlorophyll electronic transition moment directions in bacteriochlorophyll *a*-protein. - Proc. nat. Acad.
Sci. USA *75* : 4920 - 4924, 1978.

*35275 - **PEARMAN, G.I.** : A correction for the effect of drying of air samples and
its significance to the interpretation of atmospheric $CO_2$ measurements. -
Tellus *27* : 311 - 317, 1975.

*35276 - **PEARMAN, G.I.** : Measurement of atmospheric composition at the Australian
baseline atmospheric monitoring station. - In : Analytical Techniques in the
Determination of Air Pollutants. Proceedings of Clean Air Society Symposium.
Pp. 16 - 22. Melbourne University, Melbourne 1977. [$CO_2$.]

*35277 - **PEARMAN, G.I.** : The carbon dioxide-climate problem : Recent developments. -
Clean Air *1977* (May) : 21 - 26, 1977. [$\delta^{13}C$.]

*35278 - **PEARMAN, G.I., WEAVER, H.L., TANNER, C.B.** : Boundary layer heat transfer
coefficients under field conditions. - Agr. Meteorol. *10* : 83 - 92, 1972.

35379 - **PEARMAN, I., THOMAS, S.M., THORNE, G.N.** : Effects of nitrogen fertilizer on
the distribution of photosynthate during grain growth of spring wheat. -
Ann. Bot. *42* : 91 - 99, 1978.

35280 - **PEARSON, J.A., THOMAS, K., THOMAS, H.** : Nucleic acids from leaves of a yellowing and a non-yellowing variety of *Festuca pratensis* HUDS. - Planta *144* :
85 - 87, 1978. [Chl.]

35281 - **PEARSON, L.C., BRAMMER, E.** : Rate of photosynthesis and respiration in different lichen tissues by the Cartesian diver technique. - Amer. J. Bot. *65* :
276 - 281, 1978.

35282 - **PEDERSEN, J.A.** : Naturally occurring quinols and quinones studies as semi-
quinones by electron spin resonance. - Phytochemistry *17* : 775 - 778, 1978.

35283 - **PEET, M.M., KRAMER, P.J.** : Effect of leaf and pod excision on photosynthetic rates, stomatal resistances and seed weights in *Phaseolus vulgaris* L. -
Plant Physiol. *61* (Suppl.) : 87, 1978.

35284 - **PEISER, G.D., YANG, S.F.** : Chlorophyll destruction in the presence of bisulfite and linoleic acid hydroperoxide. - Phytochemistry *17* : 79 - 84,
1978.

35285 - **PEISKER, M.** : A comment on the effects of carbon dioxide, oxygen, and temperature on photosynthetic quantum yield in $C_3$ plants. - Acta Physiol. Plant.
*1* : 23 - 26, 1978.

35286 - **PEISKER, M.** : Der Einfluß von Sauerstoff auf die $CO_2$-Kompensationskonzen-
tration von $C_3$- und $C_4$-Pflanzen und von Intermediärformen. - Kulturpflanze
*26* : 81 - 98, 1978.

35287 - **PELEVIN, V.N.** : Otsenka kontsentratsiĭ vzvesi i khlorofilla v more po izme-
ryaemomu s vertoleta spektru vykhodyashchego izlucheniya. [Estimation of
suspended matter and chlorophyll concentrations in the sea from the spectrum
of upwelling radiation measured from a helicopter.]-Okeanologiya *18* : 428 -
434, 1978. [In R, ab : E.]

35288 - **PELEVIN, V.N., RUTKOVSKAYA, V.A.** : Ob oslablenii fotosinteticheski aktivnoĭ
solnechnoĭ radiatsii v vodakh Tikhogo okeana.[Attenuation of photosynthetically active solar radiation in the Pacific Ocean waters.] - Okeanologiya
*18* : 619 - 626, 1978. [In R, ab : E.]

*35289 - **PELLEGRINI, M.** : Présence d'une mitochondrie unique et de proplastes isolés
chez l'*Euglena gracilis* Z, en culture synchrone hétérotrophe. - Compt.
rend. Acad. Sci. Paris, Sér. D *283* : 911 - 913, 1976.

*35290 - **PELLEGRINI, M., PELLEGRINI, L.** : Continuité mitochondriale et discontinuité
plastidale chez l'*Euglena gracilis* Z. - Compt. rend. Acad. Sci. Paris,
Sér. D *282* : 357 - 360, 1976.

35291 - **PELLEGRINO, F., YU, W., ALFANO, R.R.** : Fluorescence kinetics of spinach
chloroplasts measured with a picosecond optical Kerr gate. - Photochem.
Photobiol. *28* : 1007 - 1012, 1978.

35292 - **PELLIN, M.J., WRAIGHT, C.A., KAUFMANN, K.J.** : Modulation of the primary
electron transfer rate in photosynthetic reaction centers by reduction of
a secondary acceptor. - Biophys. J. *24* : 361 - 369, 1978.

35293 - **PENTECOST, A.** : Calcification and photosynthesis in *Corallina officinalis* L. using the $^{14}CO_2$ method. - Br. phycol. J. *13* : 383 - 390, 1978.

35294 - **PEOPLES, M.B., DALLING, M.J.** : Degradation of ribulose-1,5-bisphosphate carboxylase by proteolytic enzymes from crude extracts of wheat leaves. - Planta *138* : 153 - 160, 1978.

35295 - **PEOPLES, T.R., KOCH, D.W.** : Physiological response of three alfalfa cultivars to one chilling night. - Crop Sci. *18* : 255 - 258, 1978. [Ps.]

35296 - **PEOPLES, T.R., KOCH, D.W., SMITH, S.C.** : Relationship between chloroplast membrane fatty acid composition and photosynthetic response to a chilling temperature in four alfalfa cultivars. - Plant Physiol. *61* : 472 - 473, 1978.

35297 - **PEREIRA, J.F., SPLITTSTOESSER, W.E., OGREN, W.L.** : Carbon dioxide assimilation by cassava (*Manihot esculenta* CRANTZ) leaves. - Plant Physiol. *61* (Suppl.) : 100, 1978.

35298 - **PERES EĬRIS, M., PUBIENES, M.A.** : Balans produktsii i destruktsii organicheskogo veshchestva v vodokhranilishche S'erra-del'-Rozario (Kuba). [Balance of production and destruction of organic matter in the reservoir Sierra-del-Rosario (Cuba).] - Gidrobiol. Zh. *14* (1) : 34 - 39, 1978. [In R, ab : E.]

35299 - **PERES EĬRIS, M., PUBIENES, M.A., ROMANENKO, V.I., KUDRYAVTSEV, V.M.** : Osobennosti mikrobiologicheskikh protsessov produktsii i destruktsii organicheskogo veshchestva v evtrofnom vodokhranilishche Ol'gin na Kube. [Peculiarities of microbial processes of production and destruction of organic matter in eutrophic water reservoir Holguin in Cuba.] - Biol. vnutr. Vod *37* : 10 - 16, 1978. [Ps; In R.]

35300 - **PERIASAMY, N., LINSCHITZ, H., CLOSS, G.L., BOXER, S.G.** : Photoprocesses in covalently linked pyrochlorophyllide dimer : Triplet state formation and opening and closing of hydroxylic linkages. - Proc. nat. Acad. Sci. USA *75* : 2563 - 2566, 1978.

35301 - **PERKINS, D.F., JONES, V., MILLAR, R.O., NEEP, P.** : Primary production, mineral nutrients and litter decomposition in the grassland ecosystem. - In : HEAL, O.W., PERKINS, D.F. (ed.) : Production Ecology of British Moors and Montane Grasslands. Pp. 304 - 331. Springer-Verlag, Berlin - Heidelberg - New York 1978.

35302 - **PERL, M.** : Phosphoenol-pyruvate-carboxylase activity in cotton and *Sorghum* seeds and its relation to seedling development. - Planta *139* : 239 - 243, 1978.

35303 - **PERRIER, A.** : L'évapotranspiration et la photosynthèse d'une culture en fonction de ses propriétés physiques. - Zeszyty problem. Postępów Nauk roln. *203*: 485 - 498, 1978.

35304 - **PERSANOV, V.M., GOGOTOV, I.N.** : Gidrogenaznaya aktivnost' kletok *Chlorella vulgaris*. [Activity of hydrogenase in the cells of *Chlorella vulgaris*.] - Mikrobiologiya *47* : 212 - 216, 1978. [In R, ab : E.]

35305 - **PERSANOV, V.M., GOGOTOV, I.N.** : Vydelenie molekulyarnogo vodoroda kletkami khlorelly v temnote. [Molecular hydrogen evolution by *Chlorella* cells in darkness.] - Fiziol. Rast. *25* : 1139 - 1143, 1978. [In R, ab : E.]

*35306 - **PESCHEK, G.A.** : Biologische Wasserstoffgewinnung. - In : GETOFF, N. (ed.) : Wasserstoff als Energieüberträger. Herstellung, Lagerung, Transport. Pp. 291 - 322. Springer-Verlag, Wien - New York 1977. [Ps.]

35307 - **PESCHEK, G.A.** : Reduced sulfur and nitrogen compounds and molecular hydrogen as electron donors for anaerobic $CO_2$ photoreduction in *Anacystis nidulans*. - Arch. Microbiol. *119* : 313 - 322, 1978.

35308 - **PESCHEK, G.A., SCHMETTERER, F.** : Reversible photooxidative loss of pigments and of intracytoplasmic membranes in the blue-green alga *Anacystis nidulans*. - FEMS Microbiol. Lett. *3* : 295 - 297, 1978.

35309 - PÉTERFI, S., NAGY-TÓTH, F., BARNA, A., ŞTIRBAN, M., BERCEA, V. : Some metabolic characteristics of *Scenedesmus acutus* cultivated in media prepared
from waste waters. - Rev. roum. Biol., Sér. Biol. vég. *23* : 45 - 53, 1978.
[Chl.]

35310 - PETERS, J.A., HELL, K.G., HANDRO, W. : Effects of kinetin and gamma radiation on the growth and chlorophyll synthesis of isolated radish cotyledons.
- Plant Cell Physiol. *19* : 1483 - 1487, 1978.

35311 - PETERS, K., AVOURIS, P., RENTZEPIS, P.M. : Picosecond dynamics of primary
electron-transfer processes in bacterial photosynthesis. - Biophys. J. *23* :
207 - 217, 1978.

35312 - PETERS, R., PETERS, J. : Light-dark-adaptation of bacteriorhodopsin in
brown membrane. - In : CAPLAN, S.R., GINZBURG, M. (ed.) : Energetics and
Structure of Halophilic Microorganisms. Pp. 315 - 321. Elsevier/North-Hol-
land Biomedical Press, Amsterdam - New York 1978.

35313 - PETERSON, B.J. : Radiocarbon uptake : Its relation to net particulate carbon
production. - Limnol. Oceanogr. *23* : 179 - 184, 1978.

35314 - PETERSON, D.L., BAZZAZ, F.A. : Life cycle characteristics of *Aster pilosus*
in early successional habitats. - Ecology *59* : 1005 - 1013, 1978. [Ps.]

35315 - PETERSON, R.B., BURRIS, R.H. : Hydrogen metabolism in isolated heterocysts
of *Anabaena* 7120. - Arch. Microbiol. *116* : 125 - 132, 1978.

35316 - PETKE, J.D., MAGGIORA, G.M., SHIPMAN, L.L., CHRISTOFFERSEN, R.E. : Stereo-
electronic properties of photosynthetic and related systems. *Ab initio*
configuration interaction calculations on the ground and lower excited
singlet and triplet states of magnesium porphine and porphine. - J. mol.
Spectroscopy *71* : 64 - 84, 1978.

35317 - PETKE, J.D., MAGGIORA, G.M., SHIPMAN, L.L., CHRISTOFFERSEN, R.E. : Stereo-
electronic properties of photosynthetic and related systems. *Ab initio*
configuration interaction calculations of the ground and lower excited
singlet and triplet states of magnesium chlorin and chlorin. - J. mol. Spectroscopy *73* : 311 - 331, 1978.

*35318 - PETKOVA, R.A. : On the question about a characteristics of a natural redox
system in chloroplast suspension. - Dokl. bolg. Akad. Nauk *27* : 1589 - 1591,
1974.

35319 - PETRIYA, V. : Vliyanie khromistykh soleĭ iz osadochnykh vod na fiziologicheskie protsessy u vodorosleĭ *Chlorella vulgaris*. [Effect of chromic salts
from precipitation water on physiological processes in the alga *Chlorella
vulgaris*.] - Rev. roum. Biol., Sér. Biol. vég. *23* : 55 - 57, 1978. [Ps;
in R, ab : F.]

35320 - PETROV, V.E., LOSKUTNIKOV, A.I., SEĬFULLINA, N.Kh. : Énergetika assimiliruyushcheĭ kletki i usloviya vodoobespecheniya. [Energy balance of assimilating cell and conditions of water supply.] - In : Vodnyĭ Rezhim Rasteniĭ
v Svyazi s Raznymi Ékologicheskimi Usloviyami. Pp. 244 - 252. Izdatel'stvo
Kazanskogo Universiteta, Kazan' 1978. [In R.]

35321 - PETROVIČ, N. : Uticaj nikla na težinu suve materije i zastupljenost pigme-
nata hloroplasta u mladim biljkama pšenice, suncokreta, boranije i kukuruza. [Effect of nickel on dry matter weight and chloroplast pigments content
in young plants of wheat, sunflower, bean and maize.] - Arhiv poljopr.
Nauke *31* : 51 - 60, 1978. [In Croat., ab : E.]

35322 - PETTIGREW, G.W., BARTSCH, R.G., MEYER, T.E., KAMEN, M.D. : Redox potentials
of the photosynthetic bacterial cytochromes $c_2$ and the structural bases for
variability. - Biochim. biophys. Acta *503* : 509 - 523, 1978.

35323 - PFENNIG, N. : General physiology and ecology of photosynthetic bacteria. -
In : CLAYTON, R.K., SISTROM, W.R. (ed.) : The Photosynthetic Bcteria. Pp.
3 - 18. Plenum Press, New York - London 1978.

35324 - PHAM THI, A.T. : Influence du déficit hydrique sur le métabolisme du glycolate chez le Cottonnier. - Physiol. vég. *16* : 301 - 317, 1978. [Photo-
respiration.]

35325 - **PHAM THI, A.T., VIEIRA da SILVA, J.** : Sur la reprise au cours de la rehydra-
tation, du point de compensation de $CO_2$, de la photorespiration et de l'ultrastructure, chez deux espèces de Cotonnier. - Plant Sci. Lett. *11* : 121 -
136, 1978.

*35326 - **PHILLIPS, R.D., JENNINGS, D.H.** : Succulence, cations and organic acids in
leaves of *Kalanchoe daigremontiana* grown in long and short days in soil and
water culture. - New Phytol. *77* : 599 - 611, 1976. [CAM.]

35327 - **PHUNG NHU HUNG, S., HOULIER, B., MOYSE, A.** : Light-induced EPR signal I
and signal II in wheat etioplasts greened under intermittent light without
or with subsequent continuous light. - Plant Sci. Lett. *12* : 107 - 117,
1978.

35328 - **PICCIONI, R.G., MAUZERALL, D.C.** : Calcium and photosynthetic oxygen evolution in cyanobacteria. - Biochim. biophys. Acta *504* : 384 - 397, 1978.

35329 - **PICCIONI, R.G., MAUZERALL, D.C.** : A high-potential redox component located
within cyanobacterial Photosystem II. - Biochim. biophys. Acta *504* : 398 -
405, 1978.

35330 - **PICKARD, W.F., MINCHIN, P.E.H., TROUGHTON, J.H.** : Real time studies of carbon 11 translocation in moonflower 1. The effects of cold blocks. - J. exp.
Bot. *29* : 993 - 1001, 1978. [Photosynthates.]

35331 - **PICKARD, W.F., MINCHIN, P.E.H., TROUGHTON, J.H.** : Real time studies of
carbon-11 translocation in moonflower II. The effects of metabolic and photosynthetic activity and of water stress. - J. exp. Bot. *29* : 1003 - 1009,
1978.

35332 - **PICKARD, W.F., MINCHIN, P.E.H., TROUGHTON, J.H.** : Transient inhibition of
translocation in *Ipomoea alba* L. by small temperature reductions. - Aust.
J. Plant Physiol. *5* : 127 - 130, 1978.

35333 - **PIERSON, B.K., CASTENHOLZ, R.W.** : Photosynthetic apparatus and cell membranes of the green bacteria. - In : **CLAYTON, R.K., SISTROM, W.R.** (ed.) : The
Photosynthetic Bacteria. Pp. 179 - 197. Plenum Press, New York - London
1978.

35334 - **PILARSKI, J.** : Influence of changes in light intensity on photorespiration
in *Nuphar luteum* (L.) SM. leaves. - Bull. Acad. pol. Sci., Sér. Sci. biol.
*26* : 201 - 206, 1978.

35335 - **PILL, W.G., LAMBETH, V.N., HINCKLEY, T.M.** : Effects of nitrogen form and
level on ion concentrations, water stress, and blossom-end rot incidence
in tomato. - J. amer. Soc. hort. Sci. *103* : 265 - 268, 1978. [Resistances.]

35336 - **PIRT, S.J., WALACH, M.** : Biomass yields of *Chlorella* from iron (Yx/Fe)
in iron-limited batch cultures. Effect of specific growth rate. - Arch.
Microbiol. *116* : 293 - 296, 1978. [Ps.]

35337 - **PISICĂ-DONOSE, A.** : Influența temperaturii asupra intensității fotosintezei la grîul de toamnă Bezostaia 1 și secara Danae, în condițiile din
nord-estul Moldovei. [Effect of temperature on photosynthetic rate in the
Bezostaia 1 winter wheat and Danae rye in north-eastern Moldavia.] - Stud.
Cercet. Biol., Ser. Biol. veg. *30* : 165 - 170, 1978. [In Roum., ab : E.]

35338 - **PITELKA, L.F.** : Variation in caloric values of annual and perennial lupines
(*Lupinus* : *Leguminosae*). - Amer. Midland Naturalist *99* : 454 - 462, 1978.

*35339 - **PLANCHON, C.** : Heterosis, combining ability and photosynthesis in bread
wheat (*Triticum aestivum* L.). - Eucarpia *7* : 95 - 103, 1976.

35340 - **PLANCHON, C., VIGNES, D.** : Etude de la transpiration et de la régulation
stomatique de deux variétés de soja (*Glycine max*. L. Merrill) ; conséquen-
ces relatives à la création de types à exigences en eau réduites. - Ann.
Amélior. Plant. *28* : 149 - 155, 1978. [Stomatal resistance.]

35341 - **PLATT, S.G., BASSHAM, J.A.** : Photosynthesis and increased production of
protein. - In : **FRIEDMAN, M.** (ed.) : Nutritional Improvement of Food and
Feed Proteins. Pp. 195 - 247. Plenum Press, New York - London 1978.

35342 - **PLATT, S.G., HENRIQUES, F., RAND, L.** : Photosynthesis in virus infected *Tolmiea menziesii.* - Plant Physiol. *61* (Suppl.) : 9, 1978.

35343 - **PLUMLEY, F.G.** : Atrazine effects on salt marsh edaphic algae (diatoms). - J. Phycol. *14* (Suppl.) : 20, 1978. [Ps.]

35344 - **PLUMMER, G.L.** : Variations in heat emitted during summertime from a grassland. - Ann. Bot. *42* : 1327 - 1332, 1978. [Ps.]

35345 - **POCKER, Y., MIKSCH, R.R.** : Plant carbonic anhydrase. Properties and bicarbonate dehydration kinetics. - Biochemistry *17* : 1119 - 1125, 1978.

35346 - **PODOLÁK, M.** : Vplyv závlahy, hnojenia, hustoty porastu a hybridov na úrodu silážnej kukurice. [Influence of irrigation, fertilizer doses, stand density and hybrids on yield of silage maize.] - Rostl. Výroba (Praha) *24* : 1163 - 1171, 1978. [In Slovak, ab : E, G, R.]

35347 - **POLAK, J., HAFFNER, G.D.** : Oxygen depletion of Hamilton Harbour. - Water Res. *12* : 205 - 215, 1978. [Ps.]

35348 - **POLESCU-IONĂŞESCU, L.** : Rythmicité circadienne et hebdomadaire de quelques processus physiologiques chez les algues *Chlorella coelastroides.* - Rev. roum. Biol., Sér. Biol. vég. *23* : 59 - 63, 1978. [Chl, Car.]

*35349 - **POLONSKIĬ, V.I.** : $CO_2$ - gazoobmen tsenoza pshenitsy pri vysokikh intensivnostyakh FAR v svetokul'ture. [$CO_2$ gas exchange of wheat coenoses at high PhAR intensities in light culture.] - In : Intensivnaya Svetokul'tura Rasteniĭ. Pp. 39 - 47. Inst. Fiz. Sib. Otd. Akad. Nauk SSSR, Krasnoyarsk 1977. [In R.]

35350 - **POLONSKIĬ, V.I., LISOVSKIĬ, G.M.** : Anatomo-morfologicheskaya kharakteristika rasteniĭ pshenitsy pri vysokikh intensivnostyakh fotosinteticheski aktivnoĭ radiatsii (FAR) v svetokul'ture. [Anatomical and morphological description of wheat plants under high intensities of photosynthetically active radiation (PAR) in light culture.] - Bot. Zh. *63* : 263 - 269, 1978. [In R.]

*35351 - **POLONSKIĬ, V.I., LISOVSKIĬ, G.M., TRUBACHEV, I.N.** : Optimizatsiya intensivnosti FAR v techenie vegetatsionnogo perioda dlya tsenoza pshenitsy. [Optimization of PhAR intensity during the vegetation period for wheat coenoses.] - In : Intensivnaya Svetokul'tura Rasteniĭ. Pp. 14 - 34. Inst. Fiz. Sib. Otd. Akad. Nauk SSSR, Krasnoyarsk 1977. [In R.]

*35352 - **POLONSKIĬ, V.I., ZINENKO, G.K., GRIBOVSKAYA, I.V.** : Dinamika vynosa mineral'nykh ĕlementov pshenitseĭ v usloviyakh intensivnogo kul'tivirovaniya. [Dynamics of use of mineral elements by wheat plants under conditions of intensive cultivation.] - In : Intensivnaya Svetokul'tura Rasteniĭ. Pp. 48 - 58. Inst. Fiz. Sib. Otd. Akad. Nauk SSSR, Krasnoyarsk 1977. [In R.]

35353 - **PONGRATZ, P., BECK, E.** : Diurnal oscillation of amylolytic activity in spinach chloroplasts. - Plant Physiol. *62* : 687 - 689, 1978.

35354 - **POOLE, D.K., MILLER, P.C.** : Water related characteristics of some evergreen sclerophyll shrubs in central Chile. - Oecol. Plant *13* : 289 - 299, 1978. [Resistances.]

*35355 - **POOLE, D.K., MILLER, P.C.** : Water relations of selected species of chaparral and coastal sage communities. - Ecology *56* : 1118 - 1128, 1975. [Resistances.]

35356 - **POOLE, R.J.** : Energy coupling for membrane transport. - Annu. Rev. Plant Physiol. *29* : 437 - 460, 1978. [Ps.]

35357 - **POPDIMITROVA, N.** : Razpredelenie na bionaprezhenieto po d"lzhinata na internodalni kletki ot *Nitella* i *Nitellopsis obtusa* pri zat"mnyavane i osvetlyavane. [Distribution of the biopotential along the length of internodal cells in *Nitella* and *Nitellopsis obtusa* at darkening and illuminating.] - Rasteniev. Nauki *15* (4) : 35 - 40, 1978. [Chloroplast; in Bulg., ab : E, R.]

35358 - **POPOV, V.K., POPOVA, N.M.** : Dinamika soderzhaniya karotinoidov v khvoe sosny v chistykh i smeshannykh kul'turakh. [Dynamics of the content of carotenoids in pine needles in pure and mixed cultivations.] - Izv. vyssh. uchebn. Zavedeniĭ, Lesn. Zh. *21* (6) : 22 - 26, 1978. [In R.]

*35359 - **POPOVICI, Gh.** : The influence of blue and red light on the formation of certain free amino acids in leaves exposed in the atmosphere with $^{14}CO_2$. - Rev. roum. Biol., Sér. Bot. *17* : 203 - 208, 1972.

35360 - **POPPE, W.L.** : Time series analysis and forecasting of phytoplankton dynamics and selected limnological parameters. - J. Phycol. *14* (Suppl.) : 24, 1978. [Chl.]

35361 - **PORAT, N., BEN-HAYYIM, G., FRIEDBERG, I.** : Localization of oxidized 3,3'-diaminobenzidine deposits in chloroplasts. - Protoplasma *93* : 397 - 403, 1978.

35362 - **PORAT, N., BEN-HAYYIM, G., FRIEDBERG, I.** : The site of ferricyanide photoreduction in pea chloroplasts pretreated by silicomolybdic acid. - FEBS Lett. *92* : 9 - 11, 1978.

*35363 - **POROKHNYA, A.D., VOLKOVA, N.P.** : Izmenenie pigmentnogo sostava list'ev v usloviyakh razlichnoĭ gustoty stoyaniya risa. [Changes in leaf pigment composition under various rice crop densities.] - Byul. nauch.-tekh. Inform. vsesoyuz. nauch.-issled. Inst. Risa *15* : 37 - 40, 1975. [In R, ab : E.]

*35364 - **POROKHNYA, A.D., VOLKOVA, N.P.** : K voprosu vliyaniya azota na pigmentnyĭ sostav list'ev risa. [Effect of nitrogen on pigment composition of rice leaves.] - Byul. nauch.-tekh. Inform. vsesoyuz. nauch.-issled. Inst. Risa *16* : 27 - 32, 1975. [In R, ab : E.]

35365 - **PORRA, R.J., GRIMME, L.H.** : Tetrapyrrole biosynthesis in algae and higher plants: a discussion of the importance of the 5-aminolaevulinate synthase and the dioxovalerate transaminase pathways in the biosynthesis of chlorophyll. - Int. J. Biochem. *9* : 883 - 886, 1978.

35366 - **PORTER, G.** : The Bakerian lecture, 1977. *In vitro* models for photosynthesis. - Proc. roy. Soc. London, Ser. A *362* : 281 - 303, 1978.

35367 - **PORTER, G., TREDWELL, C.J., SEARLE, G.F.W., BARBER, J.** : Picosecond time-resolved energy transfer in *Porphyridium cruentum* . Part I. In the intact alga. - Biochim. biophys. Acta *501* : 232 - 245, 1978.

35368 - **POSKUTA, J., KOCHAŃSKA, K.** : The effect of potassium glycidate on the rates of $CO_2$-exchange and photosynthetic products of bean leaves. - Z. Pflanzenphysiol. *89* : 393 - 400, 1978.

35369 - **POSPÍŠILOVÁ, J., SOLÁROVÁ, J.** : Epidermal diffusive conductance as related to leaf water potential and photon flux density. - Biol. Plant. *20* : 453 - 457, 1978.

35370 - **POSPÍŠILOVÁ, J., TICHÁ, I., ČATSKÝ, J., SOLÁROVÁ, J.** : Carbon dioxide exchange in primary bean leaves as affected by water stress. - Biol. Plant. *20* : 368 - 372, 1978.

35371 - **POTTER, J.R., BREEN, P.J.** : High rates of photosynthesis in the presence of high leaf starch contents. - Plant Physiol. *61* (Suppl.) : 8, 1978.

35372 - **POULSEN, C.** : Isolation and characterization of cyanogen bromide fragments from the large subunit of ribulose bisphosphate carboxylase from barley. - In : **SIEGELMAN, H.W., HIND, G.** (ed.) : Photosynthetic Carbon Assimilation. P. 423. Plenum Press, New York - London 1978.

35373 - **POWELL, C.E., RYLE, G.J.A.** : Effect of nitrogen deficiency on photosynthesis and the partitioning of $^{14}C$-labelled leaf assimilate in unshaded and partially shaded plants of *Lolium temulentum*. - Ann. appl. Biol. *90* : 241 - 248, 1978.

35374 - **POWLES, S.B., OSMOND, C.B.** : Inhibition of the capacity and efficiency of photosynthesis in bean leaflets illuminated in a $CO_2$-free atmosphere at low oxygen : A possible role for photorespiration. - Aust. J. Plant Physiol. *5* : 619 - 629, 1978.

35375 - **POYARKOVA, N.M., GRISHINA, G.S., VIĬL, Yu.A., VOSKRESENSKAYA, N.P.** : Obrazovanie glikolata v izolirovannykh khloroplastakh shpinata na krasnom i sinem svetu. [Glycolate production in isolated spinach chloroplasts under red and blue light.] - Fiziol. Rast. *25* : 1115 - 1121, 1978. [In R, ab : E.]

35376 - **PRADEL, J., LAVERGNE, J., MOYA, I.** : Formation and development of photosynthetic units in repigmenting *Rhodopseudomonas sphaeroides* wild type and "Phofil" mutant strain. - Biochim. biophys. Acta *502* : 169 - 182, 1978.

35377 - **PRAGA-KUBALSKA, B., GEJ, B.** : Asymilacja $^{14}CO_2$ i przebieg wzrostu pszenicy jarej odm. Negradowicka w warunkach intensywnego żywienia azotem lub fosforem. [Assimilation of $^{14}CO_2$ and growth dynamics of spring wheat cv. Negradowicka under conditions of intensive nitrogen or phosphorus fertilization.] - Acta agrobot. *31* : 107 - 115, 1978. [In Pol., ab : E.]

35378 - **PREISS, J., LEVI, C.** : Regulation of α1,4-glucan metabolism in photosynthetic systems. - In : **HALL, D.O., COOMBS, J., GOODWIN, T.W.** (ed.) : Proceedings of the Fourth International Congress on Photosynthesis. Pp. 457 - 468. Biochem. Soc., London 1978.

35379 - **PREŤOVÁ, A.** : Embryogenesis of the flax : Changes in the contents of pigments and fats *in vitro* and *in situ*. - Biológia (Bratislava) *33* : 29 - 34, 1978.

35380 - **PRÉZELIN, B.B., ALBERTE, R.S.** : Photosynthetic characteristics and organization of chlorophyll in marine dinoflagellates. - Proc. nat. Acad. Sci. USA *75* : 1801 - 1804, 1978.

35381 - **PRÉZELIN, B.B., SWEENEY, B.M.** : Photoadaptation of photosynthesis in *Gonyaulax polyedra*. - Mar. Biol. *48* : 27 - 35, 1978.

*35382 - **PRICE, C.A.** : Protein synthesis by spinach and *Euglena* chloroplasts isolated in gradients of silica sols. - In : **BOGORAD, L., WEIL, J.H.** (ed.) : Acides Nucléiques et Synthèse des Protéines chez les Végétaux. Pp. 473 - 480. CNRS, Paris 1977.

35383 - **PRICE, C.A., ORTIZ, W., GAYNOR, J.J.** : Regulation of protein synthesis in isolated chloroplasts of *Euglena gracilis*. - In : **AKOYUNOGLOU, G., ARGYROUDI-AKOYUNOGLOU, J.H.** (ed.) : Chloroplast Development. Pp. 257 - 266. Elsevier/North-Holland Biomedical Press, Amsterdam - New York - Oxford 1978.

*35384 - **PRICE, D.N., WAIN, R.L.** : Studies on plant growth-regulating substances. XLI. Structure-activity relationships and metabolism of a group of nitrophenols capable of inhibiting chloroplast development. - Ann. appl. Biol. *83* : 115 - 124, 1976. [Chloroplast.]

35385 - **PRIESTLEY, D.A., WOOLHOUSE, H.W.** : Characterisation of the chloroplast envelope of *Phaseolus vulgaris* L. - Plant Physiol. *61* (Suppl.) : 114, 1978.

35386 - **PRINCE, R.C.** : The reaction center and associated cytochromes of *Thiocapsa pfennigii* : their thermodynamic and spectroscopic properties, and their possible location within the photosynthetic membrane. - Biochim. biophys. Acta *501* : 195 - 207, 1978.

35387 - **PRINCE, R.C., DUTTON, P.L.** : Protonation and the reducing potential of the primary electron acceptor. - In : **CLAYTON, R.K., SISTROM, W.R.** (ed.) : The Photosynthetic Bacteria. Pp. 439 - 453. Plenum Press, New York - London 1978.

35388 - **PRINCE, R.C., DUTTON, P.L., CLAYTON, B.J., CLAYTON, R.K.** : EPR properties of the reaction center of *Rhodopseudomonas gelatinosa* *in situ* and in a detergent-solubilized form. - Biochim. biophys. Acta *502* : 354 - 358, 1978.

*35389 - **PRISHCHEPA, A.G.** : Otzyvchivost' na udobreniya razlichnykh sortov ozimoĭ pshenitsy pri oroshenii. [Reaction on nutrition of various cultivars of winter wheat under irrigation.] - In : Pitanie Rasteniĭ i Primenenie Udobreniĭ. Pp. 17 - 23, 99 - 100. Kishinev. sel'skokhoz. Inst. im. M.V. Frunze, Kishinev 1977. [Chl; in R.]

35390 - **PRISTAVU, N., WEGMANN, K.** : The action of various nitrogen sources on the photosynthesis of *Dunaliella*. - Rev. roum. Biol., Sér. Biol. vég. *23* : 81 - 88, 1978.

35391 - **PROCHASKA, L.J., DILLEY, R.A.** : Chloroplast membrane conformational changes measured by chemical modification. - Arch. Biochem. Biophys. *187* : 61 - 71, 1978.

35392 - **PROCHASKA, L.J., DILLEY, R.A.** : Site specific interaction of protons libera-
ted from Photosystem II oxidation with a hydrophobic membrane component of
the chloroplast membrane. - Biochem. biophys. Res. Commun. *83* : 664 - 672,
1978.

35393 - **PROCHASKA, L.J., DILLEY, R.A.** : Site specific interaction of protons libe-
rated from Photosystem II oxidation with a hydrophobic membrane component
of the chloroplast membrane. - In : DUTTON, P.L., LEIGH, J.S., SCARPA, A.
(ed.) : Frontiers of Biological Energetics : Electrons to Tissues. Vol. 1.
Pp. 265 - 274. Academic Press, New York - San Francisco - London 1978.

35394 - **PROCHÁZKA, S.** : To the problem of assimilates redistribution in winter wheat
during the period of grain formation. - Acta Univ. Agr. Fac. agron. *26* (2) :
19 - 25, 1978.

35395 - **PROCTOR, J.T.A.** : Apple photosynthesis : Microclimate of the tree and or-
chard. - HortScience *13* :641 - 643, 1978.

35396 - **PRONINA, N.B.** : Vliyanie 2,4-D na sintez i gidroliz ATF v khloroplastakh
yachmenya i gorokha v svyazi s usloviyami fosfornogo pitaniya. [Effect of
2,4-D on synthesis and hydrolysis of ATP in barley and pea chloroplasts in
connection with phosphorus nutrition.]- Fiziol. Biokhim. kul't. Rast. *10* :
584 - 592, 669, 1978. [In R, ab : E.]

35397 - **PROSKURYAKOV, I.I., PROKHORENKO, I.P., VOZNYAK, V.M., EROKHIN, Yu.E.** :
Izuchenie metodom EPR okislitel'no-vosstanovitel'nogo titrovaniya khromatoforov *Chromatium minutissimum*. [Redox titration of *Chromatium minutissimum*
chromatophores using ESR detection.] - Biofizika *23* : 916 - 918, 1978.
[In R, ab : E.]

35398 - **PROTSENKO, D.F., ZHIVKOVA, T.D.** : Dinamika pigmentov i prochnost' svyazi
khlorofilla s belkom gibridov kukuruzy. [Dynamics of pigments and strength
of chlorophyll bond with protein in maize hybrids.] - Fiziol. Biokhim.
kul't. Rast. *10* : 608 - 612, 670, 1978. [In R, ab : E.]

35399 - **PUCACCO, L.R., CARTER, N.W.** : An improved $pCO_2$ microelectrode. - Anal. Biochem. *90* : 427 - 434, 1978.

35400 - **PUISEUX-DAO, S., HOURSIANGOU-NEUBRUN, D., DUBACQ, J.P.** : Heterogeneity of
the plastid population and cellular streaming in *Acetabularia mediterranea*.
- In : AKOYUNOGLOU, G., ARGYROUDI-AKOYUNOGLOU, J.H. (ed.) : Chloroplast Development. Pp. 345 - 352. Elsevier/North-Holland Biomedical Press, Amsterdam - New York - Oxford 1978.

35401 - **PULICH, W.M. Jr.** : Photocontrol of boron metabolism in sea grasses. - Science *200* : 319 - 320, 1978. [Ps.]

35402 - **PURCZELD, P., CHON, C.J., PORTIS, A.R. Jr., HELDT, H.W., HEBER, U.** : The
mechanism of the control of carbon fixation by the pH in the chloroplast
stroma. Studies with nitrite-mediated proton transfer across the envelope.
- Biochim. biophys. Acta *501* : 488 - 498, 1978.

35403 - **PUROHIT, A.N.** : Productivity patterns at different altitudes. - In : Inter-
-Disciplinary Symposium on Photosynthesis and Productivity. Abstract of Papers. Pp. 38 - 39. Indian nat. Sci. Acad., New Delhi 1978.

*35404 - **PUROHIT, A.N., TREGUNNA, E.B.** : Effect of DCMU and DNP on tobacco mosaic
virus lesion development on *Nicotiana glutinosa*. - Indian J. Plant Physiol.
*19* : 66 - 70, 1976.

*35405 - **PURUSHOTHAMAN, D., KESAVAN, R., MARIMUTHU, T., OBLISAMI, G.** : Chlorophyll
stability index (CSI) of certain algae. - Curr. Sci. *43* : 159 - 161, 1974.

35406 - **P'YANKOV, V.I.** : Vliyanie temperatury i kontsentratsii uglekislogo gaza na
fotosintez arkticheskikh rasteniĭ o. Vrangelya. [Influence of temperature
and concentration of carbon dioxide on photosynthesis of arctic plants of
Vrangel Island.] - In : Mezostruktura i Funktsional'naya Aktivnost' Fotosinteticheskogo Apparata. Pp. 137 - 144. Ural'skiĭ Gosudarstvennyĭ Universitet, Sverdlovsk 1978. [In R.]

*35407 - **PYRINA, I.L., GETSEN, M.V., VAĬNSHTEĬN, M.B.** : Pervichnaya produktsiya fitoplanktona ozer Kharbeĭskoĭ sistemy Bol'shezemel'skoĭ tundry. [Primary
production of phytoplankton in lakes of Kharbeĭskaya system of the Bol'-
shezemel'skaya tundra.] - In : Produktivnost' Ozer Vostochnoĭ Chasti Bol'-
shezemel'skoĭ Tundry. Pp. 63 - 76. Nauka, Leningrad 1976. [In R.]

35408 - **PYRINA, I.L., SIGAREVA, L.E.** : Soderzhanie pigmentov fitoplanktona v Ivan'-
kovskom vodokhranilishche v 1973 - 1974 gg. [Pigment content of phytoplankton in Ivan'kovskoe reservoir in 1973 - 1974.] - Tr. Inst. Biol. vnutr. Vod
Akad. Nauk SSSR *40* (Biologiya Nizshikh Organizmov) : 3 - 17, 1978. [In R.]

35409 - **PYT'EVA, N.F., CHAMOROVSKIĬ, S.K., RUBIN, A.B.** : Sootnoshenie tsiklicheskogo i netsiklicheskogo élektronnogo transporta v khromatoforakh *Rhodospirillum rubrum*. I. Éksperimental'noe issledovanie kineticheskikh kharakteristik
temnovogo vosstanovleniya *P870* pri razlichnykh kontsentratsiyakh donora
v srede. [Relationship of cyclic and non-cyclic electron transport in chromatophores of *Rhodospirillum rubrum*. I. Experimental study of the kinetics
of *P870* dark recovery at different concentrations of donors.] - Biofizika
*23* : 48 - 52, 1978. [In R, ab : E.]

35410 - **QUAIL, P.H., BRIGGS, W.R.** : Irradiation-enhanced phytochrome pelletability.
Requirement for phosphorylative energy *in vivo*. -- Plant Physiol. *62* : 773 -
778, 1978.

35411 - **QUANDT, L., PFENNIG, N., GOTTSCHALK, G.** : Evidence for the key position of
pyruvate synthase in the assimilation of $CO_2$ by *Chlorobium*.-FEMS Microbiol.
Lett. *3* : 227 - 230, 1978.

35412 - **QUEDADO, R.M., FRIEND, D.J.** : Participation of photosynthesis in floral induction of long day plant *Anagallis arvensis* L. - Plant Physiol. *62* : 802 -
806, 1978.

35413 - **QUIMBY, P.C. Jr., POTTER, J.R., DUKE, S.O.** : Photosystem II and hypoxic qui-
escence in alligatorweed. - Physiol. Plant. *44* : 246 - 250, 1978.

35414 - **QUINTANILHA, A.T., MEHLHORN, R.J.** : pH gradients across thylakoid membranes
measured with a spin-labeled amine. - FEBS Lett. *91* : 104 - 108, 1978.

35415 - **QUINTANILHA, A.T., PACKER, L.** : Outer surface potential changes due to energization of the chloroplast thylakoid membrane. - Arch. Biochem. Biophys.
*190* : 206 - 209, 1978.

35416 - **RABEY, G.G., BATE, G.C.** :  The effect of a period of darkness on the translocation of $^{14}C$-labelled assimilates from leaves subtending five- to ten-
-day old cotton bolls (*Gossypium hirsutum* L.). - Rhod. J. agr. Res. *16* : .
61 - 71, 1978.

35417 - **RACHKOVSKAYA, M.M.** : Fiziologo-biokhimicheskie puti adaptatsii rasteniĭ
k usloviyam promyshlennogo zagryazneniya atmosfery. [Physiological and biochemical ways of adaptation of plants to industrial atmosphere pollution.] -
In : Mezhvuzovskiĭ Sbornik Nauchnykh Trudov. Pp. 150 - 155. Kemerovo 1978.
[Chl, Car; in R.]

35418 - **RADMER, R., KOK, B., OLLINGER, O.** : Kinetics and apparent $K_m$ of oxygen
cycle under conditions of limiting carbon dioxide fixation. - Plant Physiol.
*61* : 915 - 917, 1978.

*35419 - **RADUNZ, A.** : Binding of antibodies onto the thylakoid membrane II. Distribution of lipids and proteins at the outer surface of the thylakoid membrane. - Z. Naturforsch. *32 C* : 597 - 599, 1977.

35420 - **RADUNZ, A.** : Binding of antibodies onto the thylakoid membrane III. Proteins in the outer surface of the thylakoid membrane. - Z. Naturforsch.
*33 C* : 731 - 734, 1978.

35421 - **RADUNZ, A.** : Binding of antibodies onto the thylakoid membrane IV. Phospha-
tides and xanthophylls in the outer surface of the thylakoid membrane. -
Z. Naturforsch. *33 C* : 941 - 947, 1978.

35422 - **RAFFERTY, C.N., CLAYTON, R.K.** : Properties of reaction centers of *Rhodopseudomonas sphaeroides* in dried gelatin films. Linear dichroism and low temperature spectra. - Biochim. biophys. Acta *502* : 51 - 60, 1978.

35423 - **RAGGI, V.** : Fotorespirazione, respirazione, fotosintesi e loro influenza sul punto di compensazione per la $CO_2$ in piante di Fagiolo affette da leggeri attacchi di "ruggine". [Photorespiration, respiration, photosynthesis and their correlation with the $CO_2$ compensation point in French bean leaves mildly infected by rust.] - Phytopathol. medit. *17* : 105 - 109, 1978. [In Ital., ab : E.]

35424 - **RAGGI, V.** : The $CO_2$ compensation point, photosynthesis and respiration in rust infected bean leaves. - Physiol. Plant Pathol. *13* : 135 - 139, 1978.

35425 - **RAGHAVENDRA, A.S.** : Plant species intermediate between $C_3$ and $C_4$ photosynthesis and their significance. - In : Inter-Disciplinary Symposium on Photosynthesis and Productivity. Abstract of Papers. Pp. 21 - 23. Indian nat. Sci. Acad., New Delhi 1978.

*35426 - **RAGHAVENDRA, A.S., DAS, V.S.R.** : Distribution of the $C_4$ dicarboxylic acid pathway of photosynthesis in local monocotyledonous plants and its taxonomic significance. - New Phytol. *76* : 301 - 305, 1976.

35427 - **RAGHAVENDRA, A.S., DAS, V.S.R.** : Photochemical characteristics of mesophyll and bundle sheath chloroplasts from $C_4$ plants. - Physiol. Plant. *43* : 107 - 113, 1978.

35428 - **RAGHAVENDRA, A.S., DAS, V.S.R.** : Carbon fixation pattern in chloroplasts isolated from $C_3$- and $C_4$-plants. - Photosynthetica *12* : 166 - 177, 1978.

35429 - **RAGHAVENDRA, A.S., DAS, V.S.R.** : The occurrence of $C_4$-photosynthesis : A supplementary list of $C_4$ plants reported during late 1974 - mid 1977. - Photosynthetica *12* : 200 - 208, 1978.

35430 - **RAGHAVENDRA, A.S., DAS, V.S.R.** : Photosynthetic carbon metabolism in leaves of $C_4$- and $C_3$-plants : A detailed comparative study. - Z. Pflanzenphysiol. *87* : 297 - 311, 1978.

35431 - **RAGHAVENDRA, A.S., DAS, V.S.R.** : Comparative studies on $C_4$ and $C_3$ photosynthetic systems : Enzyme levels in the leaves and their distribution in mesophyll and bundle sheath cells. - Z. Pflanzenphysiol. *87* : 379 - 393, 1978.

35432 - **RAGHAVENDRA, A.S., DAS, V.S.R.** : Photochemical activities of chloroplasts isolated from plants with the $C_4$-pathway of photosynthesis and from plants with the Calvin cycle. - Z. Pflanzenphysiol. *88* : 1 - 11, 1978.

35433 - **RAGHAVENDRA, A.S., DAS, V.S.R.** : ($Na^+$ - $K^+$)-stimulated ATPase in leaves of $C_4$ plants : possible involvement in active transport of $C_4$ acids. - J. exp. Bot. *28* : 39 - 47, 1978.

35434 - **RAGHAVENDRA, A.S., DAS, V.S.R.** : Development of photochemical activities in mesophyll and bundle sheath chloroplasts of $C_4$ and $C_3$ plants during seedling growth. - Plant Sci. Lett. *12* : 355 - 360, 1978.

35435 - **RAGHI-ATRI, F.** : Zur Arbeitsmethodik der Chlorophyll *a*-Bestimmung in Phytoplankton. - Gesundheits-Ingenieur *99* : 380 - 381, 1978.

35436 - **RAI, H.** : Distribution of carbon, chlorophyll-*a* and pheo-pigments in the black water lake ecosystem of Central Amazon Region. - Arch. Hydrobiol. *82* : 74 - 87, 1978.

*35437 - **RAI, R.S.V., MURTHY, K.S.** : Effect of submergence on some physiological changes in rice seedlings. - Ind. J. exp. Biol. *14* : 369 - 370, 1976. [Chl.]

35438 - **RAISON, J.K., BERRY, J.A.** : The physical properties of membrane lipids in relation to the adaptation of higher plants and algae to contrasting thermal regimes. - Carnegie Inst. Year Book *77* : 276 - 282, 1978.

35439 - **RAJENDRAN, A., SUMITRA-VIJAYARAGHAVAN, WAFAR, M.V.M.** : Effect of some metal ions on the photosynthesis of microplankton & nannoplankton. - Indian J. mar. Sci. *7* : 99 - 102, 1978.

35440 - **RAKHMANKULOV, S.A.** : Izuchenie fotosinteticheskogo apparata v pokoleniyakh gibridov khlopchatnika. [Photosynthetic apparatus in generations of cotton hybrids.] - Fiziol. Rast. *25* : 536 - 540, 1978. [In R, ab : E.]

35441 - **RAKITINA, Z.G.** : Narushenie normal'nogo gazoobmena rasteniĭ - faktor, prepyatsvuyushchiĭ zashchitnomu deĭstviyu sakharov i glitserina. [Disturbed normal gas exchange of plants - a factor interfering with the protective action of sugars and glycerol.] - Fiziol. Rast. *25* : 584 - 591, 1978. [In R, ab : E.]

35442 - **RAMACHANDRA REDDY, A., DAS, V.S.R.** : The decarboxylating systems in fourteen taxa exhibiting CAM pathway. - Z. Pflanzenphysiol. *86* : 141 - 146, 1978.

35443 - **RAMADOSS, C.S., PISTORIUS, E.K., AXELROD, B.** : Coupled oxidation of carotene by lipoxygenase requires two isoenzymes. - Arch. Biochem. Biophys. *190* : 549 - 552, 1978.

35444 - **RAMAKRISHNA, J., BHAGWAT, A.S., SANE, P.V.** : Studies on enzymes of C-4 pathway : Part V - Comparative studies of $RuP_2$ carboxylase/oxygenase from maize & spinach. - Indian J. exp. Biol. *16* : 51 - 53, 1978.

35445 - **RAMAKRISHNAN, P.S.** : Observations on productivity of tropical forests ecosystems. - In : Inter-Disciplinary Symposium on Photosynthesis and Productivity. Abstract of Papers. Pp. 46 - 47. Indian nat. Sci. Acad., New Delhi 1978. [Biomass production.]

35446 - **RAMALINGAM, R.S.** : Induced chlorophyll chimeras and breeding behaviour in chilies. - Curr. Sci. *47* : 381 -    , 1978.

35447 - **RAMASWAMY, N.K., NAIR, P.M.** : Specific requirement for copper ions in the reversal of inhibition of photosynthesis in tris-washed potato tuber chloroplasts. - Plant Sci. Lett. *13* : 383 - 388, 1978.

*35448 - **RAMSAY, W.E., McCLOUD, D.E.** : Growth analysis and yield components of holley wheat in a dry year. - Proc. Soil Crop Sci. Soc. Florida *37* : 114 - 117, 1977.

35449 - **RAMUS, J.** : Seaweed anatomy and photosynthetic performance : The ecological significance of light guides, heterogenous absorption and multiple scatter. - J. Phycol. *14* : 352 - 362, 1978.

35450 - **RAND, R.H.** : A theoretical analysis of $CO_2$ absorption in sun versus shade leaves. - J. biomechanic. Eng. *100* : 20 - 24, 1978.

35451 - **RAND, R.H., COOKE, J.R.** : Fluid dynamics of phloem flow : An axisymmetric model. - Trans. ASAE *21* : 898 - 900, 906, 1978. [Photosynthates.]

35452 - **RANDAL, J., MIDDENDORF, H.D., CRESPI, H.L., TAYLOR, A.D.** : Dynamics of protein hydration by quasi-elastic neutron scattering. - Nature *276* : 636 - 638, 1978. [Biliproteins.]

35453 - **RANGER, J.** : Recherches sur les biomasses comparées de deux plantations de Pin laricio de Corse avec ou sans fertilisation. - Ann. Sci. forest. *35* : 93 - 115, 1978.

35454 - **RAO, A.N., TRIVEDI, R.C., DUBEY, P.S.** : Primary production and photosynthetic pigment concentration of ten maize cultivars. - Photosynthetica *12* : 62 - 64, 1978.

35455 - **RAO, D.V.M., MAHALAKSHMI, B.K., ALI, S.M.** : Studies on the relative contribution of various photosynthetic plant parts to the grain development at various moisture levels in *Pennisetum typhoides* (BURM.) S. & H. - Mysore J. agr. Sci. *12* : 363 - 367, 1978.

35456 - **RAO, K.K., GOGOTOV, I.N., HALL, D.O.** : Hydrogen evolution by chloroplast--hydrogenase systems : improvements and additional observations. - Biochimie *60* : 291 - 296, 1978.

35457 - **RAO, P.V., KEISTER, D.L.** : Energy-linked reactions in photosynthetic bacteria. X. Solubilization of the membrane-bound energy-linked inorganic pyrophosphatase of *Rhodospirillum rubrum*. - Biochem. biophys. Res. Commun. *84* : 465 - 473, 1978.

35458 - **RAPER, C.D., THOMAS, J.F.** : Photoperiodic alteration of dry matter partitioning and seed yield in soybeans. - Crop Sci. *18* : 654 - 656, 1978.

35459 - **RAPER, C.D. Jr., PEEDIN, G.F.** : Photosynthetic rate during steady-state
growth as influenced by carbon-dioxide concentration. - Bot. Gaz. *139* : 147
- 149, 1978.

35460 - **RASCHKE, K., HANEBUTH, W.F., FARQUHAR, G.D.** : Relationship between stomatal
conductance and light intensity in leaves of *Zea mays* L. derived from experiments using the mesophyll as shade. - Planta *139* : 73 - 77, 1978. [Ps.]

35461 - **RASCHKE, K., SCHNABL, H.** : Availability of chloride affects the balance
between potassium chloride and potassium malate in guard cells of *Vicia faba* L. - Plant Physiol. *62* : 84 - 87, 1978. [Ps.]

35462 - **RASSASHKO, I.F.** : Zimnyaya produktivnost' vod Amurskogo zaliva Yaponskogo
morya. [Winter productivity of water of the Amur Bay in the Japan Sea.] -
Nauch. Soobshch. Inst. Biol. Morya, dal'nevostoch. nauch. Tsentr Akad. Nauk
SSSR *3* (Biologicheskie Issledovaniya Dal'nevostochnykh Moreĭ) : 81 - 84,
1978. [Chl; in R.]

35463 - **RASTORFER, J.R.** : Composition and bryomass of the moss layers of two wet-
-tundra meadow communities near Barrow, Alaska. - In : TIESZEN, L.L. (ed.) :
Vegetation and Production Ecology of an Alaskan Arctic Tundra. Pp. 169 -
183. Springer-Verlag, New York - Heidelberg - Berlin 1978.

35464 - **RATAJCZAK, L., RATAJCZAK, W., MASZNER, B., PAWŁOWSKA, B.** : The effect of
hormone tratment of the seeds on growth and development of some *Leguminosae*. II. Genus differences in chlorophyll content changes under the effect
of gibberellic acid. - Bull. Soc. Amis Sci. Lett. Poznań, Ser. *D-18* : 35 -
38, 1978.

35465 - **RATHNAM, C.K.M.** : $C_4$ photosynthesis : the path of carbon in bundle sheath
cells. - Sci. Prog. (Oxford) *65*  : 409 - 435, 1978.

35466 - **RATHNAM, C.K.M.** : Heat inactivation of leaf phosphoenolpyruvate carboxylase : Protection by aspartate and malate in $C_4$ plants. - Planta *141* : 289 -
295, 1978.

35467 - **RATHNAM, C.K.M.** : Mechanism of $C_4$-acid decarboxylation in bundle sheath
cells of $C_4$ dicarboxylic acid cycle plants. - What's new Plant Physiol. *9*
(2) : 1 - 4, 1978.

35468 - **RATHNAM, C.K.M.** : Metabolic regulation of carbon flux during $C_4$ photosynthesis. 1. Evidence for parallel $CO_2$ fixation by mesophyll and bundle
sheath cells *in situ*. - Z. Pflanzenphysiol. *87* : 65 - 84, 1978.

35469 - **RATHNAM, C.K.M.** : Heat inactivation of leaf PEP carboxylase : Protection
by aspartate and malate in $C_4$ plants. - Plant Physiol. *61* (Suppl.) :109 ,
1978.

35470 - **RATHNAM, C.K.M.** : Malate and dihydroxyacetone phosphate-dependent nitrate
reduction in spinach leaf protoplasts. - Plant Physiol. *62* : 220 - 223,
1978. [Ps.]

35471 - **RATHNAM, C.K.M., CHOLLET, R.** : $CO_2$ donation by malate and aspartate reduces
photorespiration in *Panicum milioides*, a $C_3$-$C_4$ intermediate species. -
Biochem. biophys. Res. Commun. *85* : 801 - 808, 1978.

35472 - **RATHNAM, C.K.M., CHOLLET, R.** : PEP carboxylase reduces photorespiration in
*Panicum milioides*. - In : SIEGELMAN, H.W., HIND, G. (ed.) : Photosynthetic
Carbon Assimilation. P. 424. Plenum Press, New York - London 1978.

35473 - **RATHORE, V.S., SHROTRI, C.K., SRIVASTAVA, J.P.** : Photosynthesis in relation
to zinc nutrition. - In : Inter-Disciplinary Symposium on Photosynthesis
and Productivity. Abstract of Papers. Pp. 53 - 54. Indian nat. Sci. Acad.,
New Delhi 1978.

35474 - **RATUSHNYAK, Yu.M., VASIL'EVA, I.M., ISHMUKHAMETOVA, N.N.** : Nekotorye storony vodno-solevogo obmena khloroplastov ozimoĭ pshenitsy v usloviyakh zakalivaniya. [Some features of water and salt exchange in chloroplasts of winter wheat during hardening.] - In : Vodnyĭ Rezhim Rasteniĭ v Svyazi s Raznymi Ekologicheskimi Usloviyami. Pp. 331 - 337. Izdatel'stvo Kazanskogo Universiteta, Kazan' 1978. [In R.]

35475 - RAVEN, J.A. : Photosynthesis in cells and tissues. - In : HALL, D.O.,
COOMBS, J., GOODWIN, T.W. (ed.) : Proceedings of the Fourth International
Congress on Photosynthesis. Pp. 147 - 155. Biochem. Soc., London 1978.

35476 - RAWSON, H.M., TURNER, N.C., BEGG, J.E. : Agronomic and physiological responses of soybean and sorghum crops to water deficits. IV. Photosynthesis,
transpiration and water use efficiency of leaves. - Aust. J. Plant Physiol.
5 : 195 - 209, 1978.

35477 - RAY, T.B., PETERS, G.A., MAYNE, B.C., TOIA, R.E. Jr. : Levels of chlorophyll,
protein and enzymes of ammonia assimilation in the Azolla-Anabaena azollae
symbiotic association. - Plant Physiol. 61 (Suppl.) : 2, 1978.

35478 - RAY, T.B., PETERS, G.A., TOIA, R.E., MAYNE, B.C. : Azolla-Anabaena relationship. VII. Distribution of ammonia-assimilating-enzymes, protein, and
chlorophyll between host and symbiont. - Plant Physiol. 62 : 463 - 467,
1978.

35479 - REARDON, E.M., BARTOLF, M., ORTIZ, W., SANTORO, D., ZIELINSKI, R., PRICE,
C.A. : Isolation of chloroplasts active in protein synthesis from spinach
and Euglena. - In : AKOYUNOGLOU, G., ARGYROUDI-AKOYUNOGLOU, J.H. (ed.) :
Chloroplast Development. Pp. 277 - 282. Elsevier/North-Holland Biomedical
Press, Amsterdam - New York - Oxford 1978.

35480 - REBEIZ, C.A., SMITH, B.B., MATTHEIS, J.R., COHEN, C.E., McCARTHY, S.A. :
Chlorophyll biosynthesis : the reactions between protoporphyrin IX and
phototransformable protochlorophyll in higher plants. - In : AKOYUNOGLOU, G.,
ARGYROUDI-AKOYUNOGLOU, J.H. (ed.) : Chloroplast Development. Pp. 59 - 76.
Elsevier/North-Holland Biomedical Press, Amsterdam - New York - Oxford
1978.

35481 - REDGWELL, R.J., BIELESKI,R.L. : Sorbitol-1-phosphate and sorbitol-6-phos-
phate in apricot leaves. - Phytochemistry 17 : 407 - 409, 1978.

35482 - REDLINGER, T.E. : Photo-induced oxygen uptake in developing chloroplasts. -
In : AKOYUNOGLOU, G., ARGYROUDI-AKOYUNOGLOU, J.H. (ed.) : Chloroplast Development. Pp. 495 - 499. Elsevier/North-Holland Biomedical Press, Amsterdam - New York - Oxford 1978.

35483 - REDLINGER, T.E., McDANIEL, R.G. : Comparison of light-dependent oxygen
uptake, protochlorophyll(ide)-650 photoconversion, and chlorophyll disappearance in wheat etioplasts. - Plant Physiol. 61 : 1006 - 1009, 1978.

35484 - REDMANN, R.E. : Seasonal dynamics of carbon dioxide exchange in a mixed
grassland ecosystem. - Can. J. Bot. 56 : 1999 - 2005, 1978.

35485 - REED, M.L., GRAHAM, D. : Effect of inhibitors of protein synthesis on carbonic anhydrase in Chlorella pyrenoidosa. - Plant Physiol. 61 (Suppl.) :
97, 1978.

35486 - REED, T., HESS, B., DOSTER, W. : Photoselection and aggregation in purple
membrane of Halobacterium halobium. - Biochim. biophys. Acta 502 : 188 -
197, 1978.

35487 - REEVES, S.G., HALL, D.O. : Photophosphorylation in chloroplasts. - Biochim.
biophys. Acta 463 : 275 - 297, 1978.

35488 - REHM, S. : Angewandte Botanik in den Tropen. - Angew. Bot. 52 : 89 - 96,
1978. [Ps production.]

35489 - REIBACH, P.H., WONG, W.W., BENEDICT, C.R. : The enzymatic fractionation of
$H^{12}CO_3^- - H^{13}CO_3^-$ by phosphoenolpyruvate carboxylase. - Plant Physiol. 61
(Suppl.) : 110, 1978.

35490 - REICHENBÄCHER, D., BÖRNER, T., RICHTER, J. : Untersuchungen am Fraktion-I-
-Protein der Gerste mit Hilfe quantitativer Immunelektrophoresen. - Biochem.
Physiol. Pflanz. 172 : 53 - 60, 1978.

35491 - REICOSKY, D.A., HANOVER, J.W. : Physiological effects of surface waxes. 1.
Light reflectance for glaucous and nonglaucous Picea pungens. - Plant Physiol. 62 : 101 - 104, 1978.

35492 - REICOSKY, D.C., LAMBERT, J.R. : Field measured and simulated corn leaf water potential. - Soil Sci. Soc. Amer. J. *42* : 221 - 228, 1978. [Stomatal resistance.]

35493 - REID, F.M.H., STEWART, E., EPPLEY, R.W., GOODMAN, D. : Spatial distribution of phytoplankton species in chlorophyll maximum layers off southern California. - Limnol. Oceanogr. *23* : 219 - 226, 1978.

35494 - REIMANN, K. : Sauerstoffproduktion des Phytoplanktons in Abhängigkeit von der Temperatur des Wassers. - Z. Wasser-Abwasser-Forsch. *11* (5) : 151 - 154, 1978.

35495 - REIMER, S., SELMAN, B.R. : Tentoxin-induced energy-independent adenine nucleotide exchange and ATPase activity with chloroplast coupling factor 1. - J. biol. Chem. *253* : 7249 - 7255, 1978.

35496 - REINMAN, S., MARKWELL, J., THORNBER, J.P. : Improved fractionation of chlorophyll-protein complexes. - Plant Physiol. *61* (Suppl.) : 103, 1978.

35497 - REISFELD, A., GRESSEL, J., JAKOB, K.M., EDELMAN, M. : Characterization of the 32,000 dalton membrane protein-I. Early synthesis during photoinduced plastid development of *Spirodela*. - Photochem. Photobiol. *27* : 161 - 165, 1978.

35498 - REISFELD, A., JAKOB, K.M., EDELMAN, M. : Characterization of the 32000 dalton chloroplast membrane protein - II. The molecular weight of chloroplast messenger RNAs translating the precursor to P-32000 and full size RUDP carboxylase large subunit. - In : AKOYUNOGLOU, G., ARGYROUDI-AKOYUNOGLOU, J.H. (ed.) : Chloroplast Development. Pp. 669 - 674. Elsevier/North-Holland Biomedical Press, Amsterdam - New York - Oxford 1978.

35499 - REISFELD, A., JAKOB, K.M., EDELMAN, M. : Molecular weight of chloroplast messenger RNA translating full-size RuDP carboxylase large subunit. - Plant Physiol. *61* (Suppl.) : 104, 1978.

35500 - REJMÁNKOVÁ, E. : Growth, production and nutrient uptake of duckweeds in fishponds and in experimental cultures. - In : DYKYJOVÁ, D., KVĚT, J. (ed.): Pond Littoral Ecosystems. Pp. 278 - 291. Springer-Verlag, Berlin - Heidelberg - New York 1978.

35501 - REMSEN, C.C. : Comparative subcellular architecture of photosynthetic bacteria. - In : CLAYTON, R.K., SISTROM, W.R. (ed.) : The Photosynthetic Bacteria. Pp. 31 - 60. Plenum Press, New York - London 1978.

35502 - REMY, R., HOARAU, J. : New forms of chlorophyll-protein complexes from thylakoids of different photosynthesizing organisms. - In : AKOYUNOGLOU, G., ARGYROUDI-AKOYUNOGLOU, J.H. (ed.) : Chloroplast Development. Pp. 235 - 240. Elsevier/North-Holland Biomedical Press, Amsterdam - New York - Oxford 1978.

35503 - RENGER, G., ECKERT, H.-J., BUCHWALD, H.E. : On the detection of a new rapid recovery kinetics of photo-oxidized chlorophyll-$a_{11}$ in isolated chloroplasts under repetitive flash illumination. - FEBS Lett. *90* : 10 - 14, 1978.

35504 - REPKA, J., JUREKOVÁ, Z. : Changes in endogenous gibberellins in plant organs producing and utilizing photosynthates. - Biol. Plant. *20* : 25 - 33, 1978.

35505 - REPKA, J., RIMÁR, J., LORENČÍK, L. : Vplyv pôdneho typu na produkčné procesy a produkciu porastov pol'nych plodin. [The effect of the soil type on processes of production and on the production of field plant stands.] - Rost. Výroba (Praha) *24* : 1235 - 1245, 1978. [Growth analysis; in Slovak, ab : E, G, R.]

35506 - REVELANTE, N., GILMARTIN, M. : Characteristics of the microplankton and nanoplankton communities of an Australian coastal plain estuary. - Aust. J. mar. Freshwater Res. *29* : 9 - 18, 1978. [Chl.]

35507 - RHOADS, A., BRENNAN, E. : The effect of ozone on chloroplast lamellae and isolated mesophyll cells of sensitive and resistant tobacco selections. - Phytopathology *68* : 883 - 886, 1978.

35508 - RHODES, I., STERN, W.R. : Competition for light. - In : WILSON, J.R. (ed.) :
Plant Relations in Pastures. Pp. 175 - 189. CSIRO, Melbourne 1978. [Ps.]

35509 - RHODES, P.R., KUNG, S.D., MARSHO, T.V. : Regulation of RuBP carboxylase-oxygenase activity by selected combinations of large and small subunits. -
Plant Physiol. 61 (Suppl.) : 108, 1978.

35510 - RICE, S.C., PON, N.G. : Direct spectrophotometric observation of ribulose-
-1,5-bisphosphate carboxylase activity. - Anal. Biochem. 87 : 39 - 48, 1978.

35511 - RICH, D.H., BHATNAGAR, P.K., JASENSKY, R.D., STEELE, J.A., UCHYTIL, T.F.,
DURBIN, R.D. : Two conformations of the cyclic tetrapeptide, [DMeAla$^1$]-
-tentoxin have different biological activities. - Bioorg. Chem. 7 : 207 -
214, 1978. [Ps.]

35512 - RICHARDS, R.A. : Variation between and within species of rapeseed (*Brassica
campestris* and *B.napus*) in response to drought stress. III. Physiological
and physicochemical characters. - Aust. J. agr. Res. 29 : 491 - 501, 1978.
[Chl.]

35513 - RICHARDS, R.A., THURLING, N. : Variation between and within species of rapeseed (*Brassica campestris* and *B.napus*) in response to drought stress. I
Sensivity at different stages of development. - Aust. J. agr. Res. 29 :
469 - 477, 1978. [Dry-matter accumulation.]

35514 - RICHARDS, R.A., THURLING, N. : Variation between and within species of rapeseed (*Brassica campestris* and *B.napus*) in response to drought stress. II
Growth and development under natural drought stresses. - Aust. J. agr. Res.
29 : 479 - 490, 1978. [Growth analysis.]

35515 - RICHTER, G., DIRKS, W. : Blue-light induced development of chloroplasts in
isolated seedling roots. Preferential synthesis of chloroplast ribosomal
RNA species. - Photochem. Photobiol. 27 : 155 - 160, 1978.

35516 - RICKARD, L.H., LANDRUM, H.L., HAWKRIDGE, F.M. : A mediated electrochemical
redox study of soluble spinach ferredoxin using optically coupled methods.
- Bioelectrochem. Bioenerg. 5 : 686 - 696, 1978.

35517 - RICKMAN, R.W., ALLMARAS,R.R., RAMIG, R.E. : Root-sink descriptions of water
supply to dryland wheat. - Agron. J. 70 : 723 - 728, 1978. [Growth analysis.]

35518 - RIDLEY, S.M. : Photosynthesis and food - A symposium report. - In :
HALL, D.O., COOMBS, J., GOODWIN, T.W. (ed.) : Proceedings of the Fourth
International Congress on Photosynthesis. Pp. 745 - 751. Biochem. Soc.,
London 1978.

35519 - RIDLEY, S.M., RIDLEY, J. : Locating the cellular site of action of a caro-
tenoid inhibitor during chloroplast development. - In : AKOYUNOGLOU, G.,
ARGYROUDI-AKOYUNOGLOU, J.H. (ed.) : Chloroplast Development. Pp. 309 - 314.
Elsevier/North-Holland Biomedical Press, Amsterdam - New York - Oxford 1978.

35520 - RIEMANN, B. : Quantitative and qualitative determinations of chlorophylls
and phaeopigments in Lake Mossø. - Verh. int. Verein. Limnol. 20 : 674 -
677, 1978.

35521 - RIEMANN, B. : Absorption coefficients for chlorophylls *a* and *b* in methanol
and a comment on interference of chlorophyll *b* in determinations of chlorophyll *a*. - Vatten 3 : 187 - 194, 1978.

35522 - RIEMANN, B. : Carotenoid interference in the spectrophotometric determination of chlorophyll degradation products from natural populations of phytoplankton. - Limnol. Oceanogr. 23 : 1059 - 1066, 1978.

35523 - RIEMANN, B. : Differentiation between heterotrophic and photosynthetic
plankton by size fractionation, glucose uptake, ATP and chlorophyll content. - Oikos 31 : 358 - 367, 1978.

*35524 - RIES, E., GAUSS, V. : D-glucose as an exogenous substrate of blue light
enhanced respiration in *Chlorella*. - Z. Pflanzenphysiol. 82 : 261 - 273,
1977. [Photosynthates.]

35525 - **RIGHTMIRE, C.T.** : Seasonal variation in $P_{CO_2}$ and $^{13}C$ content of soil atmosphere. - Water Resour. Res. *14* : 691 - 692, 1978. [ $\delta^{13}C$ in aquifers.]

35526 - **RIJGERSBERG, C.P., AMESZ, J.** : Changes in light absorbance and chlorophyll fluorescence in spinach chloroplasts between 5 and 80 K. - Biochim.biophys. Acta *502* : 152 - 160, 1978.

35527 - **RILLING, G., STEFFAN, H.** : Versuche über die $CO_2$-Fixierung und den Assimilatimport durch Blattgallen der Reblaus(*Dyctylosphaera vitifolii* SHIMER) an *Vitis rupestris* 187 G. - Angew. Bot. *52* : 343 - 354, 1978.

35528 - **ŘIMSA, V., JÍLKOVÁ, M., PETRÁK, Z.** : Vztahy mezi výkonností, plochou listů a obsahem chlorofylu u cukrovky infikované uměle virem žloutenky řepy. [Relations between performance, leaf area, and chlorophyll content in sugar beet artificially infected with beet yellows virus.] - Rost. Výroba (Praha) *24* : 661 - 669, 1978. [In Czech, ab : E, G, R.]

35529 - **RIPLEY, E.A., SAUGIER, B.** :  Biophysics of a natural grassland : evaporation. - J. appl. Ecol. *15* : 459 - 479, 1978. [Growth analysis, resistances.]

35530 - **RIZNICHENKO, G.J., PYTEVA, N.F., KRENDELEVA, T.E., KONONENKO, A.A., RUBIN, A.B.** : Light-induced electron transport and coupled processes in digitonin--fractionated pea subchloroplast particles enriched in photosystem I. II. Theoretical study. - Plant Sci. Lett. *11* : 19 - 25, 1978.

35531 - **RIZNICHENKO, G.Yu., CHAMOROVSKIĬ, S.K., PYT'EVA, N.F., RUBIN, A.B.** : Model of rate control of electron transport in purple bacteria by redox potential of the medium. - Photosynthetica *12* : 361 - 368, 1978.

35532 - **RIZNICHENKO, G.Yu., SHINKAREV, V.P., PYT'EVA, N.F., VENEDIKTOV, P.S., RUBIN, A.B.** : Matematicheskoe opisanie ėlektron-transportnykh protsessov s uchetom prirody vzaimodeĭstviya perenoschikov. [Matematical description of the electron transport processes taking into account the nature of the interaction between the carriers.] - Stud. biophys. *70* : 15 - 29, 1978. [In R, ab : E.]

35533 - **RIZNYK, R.Z., EDENS, J.I., LIBBY, R.C.** : Production of epibenthic diatoms in a southern California impounded estuary. - J. Phycol. *14* : 273 - 279, 1978. [Chl.]

35534 - **ROARK, B., QUISENBERRY, J.E.** : Relationships among seedling growth, transpiration rate, and yield under semiarid conditions. - Plant Physiol. *61* (Suppl.) : 80, 1978.

35535 - **ROBERTS, G.R., KEYS, A.J.** : The mechanism of photosynthesis in the tea plant (*Camellia sinensis* L.). - J. exp. Bot. *29* : 1403 - 1407, 1978.

35536 - **ROBERTS, J.** : The use of 'tree cutting' technique in the study of the water relations of Norway spruce, *Picea abies* (L.) KARST. - J. exp. Bot. *29* : 465 - 471, 1978. [Stomatal resistance.]

35537 - **ROBERTS, S.W., KNOERR, K.R.** : *In situ* estimates of variable plant resistance to water flux in *Ilex opaca* AIT. - Plant Physiol. *61* : 311 - 313, 1978. [Stomatal resistance.]

35538 - **ROBERTSON, R.N.** : Charge separation, proton pumps and the hydrophobic region of bilayer membranes. - In : DEAMER, D.W. (ed.) : Light Transducing Membranes : Structure, Function and Evolution. Pp. 215 - 231. Academic Press, New York - San Francisco - London 1978. [Ps.]

35539 - **ROBINSON, H.H., YOCUM, C.F.** : Low potential catalysts of photosystem I cyclic photophosphorylation. - Plant Physiol. *61* (Suppl.) : 76, 1978.

35540 - **ROBINSON, J.M., SMITH, M.G., GIBBS, M.** : Broken chloroplasts in an "intact" chloroplast preparation as a causal agent of the Warburg effect. - Plant Physiol. *61* (Suppl.) : 100, 1978.

35541 - **ROBITAILLE, H.A.** : Dry matter accumulation patterns in indeterminate *Phaseolus vulgaris* L. cultivars. - Crop Sci. *18* : 740 - 743, 1978.

35542 - **ROBSON, M.J., DEACON, M.J.** : Nitrogen deficiency in small closed communities of S24 ryegrass. II. Changes in the weight and chemical composition of single leaves during their growth and death. - Ann. Bot. *42* : 1199 - 1213, 1978.

35543 - **ROBSON, M.J., PARSONS, A.J.** : Nitrogen deficiency in small closed communities of S24 ryegrass. I. Photosynthesis, respiration, dry matter production
and partition. - Ann. Bot. *42* : 1185 - 1197, 1978.

35544 - **ROCHAIX, J.-D., MALNOE, P.** : Gene localization on the chloroplast DNA of
*Chlamydomonas reinhardii*. - In : AKOYUNOGLOU, G., ARGYROUDI-AKOYUNOGLOU,
J.H. (ed.) : Chloroplast Development. Pp. 581 - 586. Elsevier/North-Holland
Biomedical Press, Amsterdam - New York - Oxford 1978.

35545 - **RODDICK, J.G.** : Effect of $\alpha$-tomatine on the integrity and biochemical activities of isolated plant cell organelles. - J. exp. Bot. *29* : 1371 - 1381,
1978. [Ps.]

35546 - **RODGERS, J.H. Jr., DICKSON, K.L., CAIRNS, J. Jr.** : A chamber for *in situ*
evaluation of periphyton productivity in lotic systems. - Arch. Hydrobiol.
*84* : 389 - 398, 1978. [Ps.]

35547 - **RODIONOV, V.S., NYUPPIEVA, K.A., KHOLOPTSEVA, N.P., MARKOVA, L.V., DROZDOV,
S.N.** : Rasshcheplenie fosfolipidov v list'yakh nekotorykh vidov kartofelya
v zavisimosti ot intensivnosti zamorozkov. [Decomposition of phospholipids
in leaves of some potato species dependent on frost intensity.] - Fiziol.
Rast. *24* : 845 - 853, 1977. [Ps; in R, ab : E.]

35548 - **RODIONOVA, M.A., KHOLODENKO, N.Ya., MAKAROV, A.D.** : Lokalizatsiya adenilat-
kinazy v organellakh kletki i substrukturakh khloroplastov list'ev shpinata.
[Localization of adenylate kinase in cellular organelles and chloroplast
substructures of spinach leaves.] - Fiziol. Rast. *25* : 731 - 734, 1978.
[In R, ab : E.]

35549 - **RODRIGUES PEREIRA, A.S.** : Effects of leaf removal on yield components in
sunflower. - Neth. J. agr. Sci. *26* : 133 - 144, 1978.

35550 - **RODSKJER, N.** : Net and solar radiation over bare soil, short grass, winter
wheat and barley. - Swed. J. agr. Res. *8* : 195 - 201, 1978.

35551 - **ROGERS, P.J., MORRIS, C.A.** : Regulation of bacteriorhodopsin synthesis by
growth rate in continuous cultures of *Halobacterium halobium*. - Arch. Microbiol. *119* : 323 - 325, 1978.

35552 - **ROMAN, M.R.** : Tidal resuspension in Buzzards Bay, Massachusetts. II. Seasonal changes in the size distribution of chlorophyll, particle concentration,
carbon and nitrogen in resuspended particulate matter. - Estuar. coast. mar.
Sci. *6* : 47 - 53, 1978.

35553 - **ROMAN, M.R., TENORE, K.R.** : Tidal resuspension in Buzzards Bay, Massachusetts. I. Seasonal changes in the resuspension of organic carbon and chlorophyll *a*. - Estuar. coast. mar. Sci. *6* : 37 - 46, 1978.

35554 - **ROMANOV, V.I., FEDULOVA, N.G., SHRAMKO, V.I., MOLCHANOV, M.I., KRETOVICH,
V.L.** : Metabolizm poli-$\beta$-oksimaslyanoĭ kisloty v bakteroidakh kluben'kov
lyupina i ego svyaz' s protsessami azotfiksatsii i fotosinteza. [Metabolism
of poly-$\beta$-hydroxybutyric acid in lupine root nodule bacteroids and its relation to nitrogen fixation and photosynthesis.] - Fiziol. Rast. *25* : 726 -
730, 1978. [In R, ab : E.]

35555 - **ROMANOVSKAYA, O.I., IRBE, I.K., BLUMBERGA, L.A., VOĬTSEKHOVICH, Z.V.** :
Deĭstvie gerbitsidov na sostoyanie i aktivnost' membran izolirovannykh
khloroplastov. [Effect of herbicides on the condition and activity of membranes of isolated chloroplasts.] - In : Fiziologo-biokhimicheskie Issledovaniya Rasteniĭ. Pp. 49 - 56, 156. Zinatne, Riga 1978. [In R.]

35556 - **ROMANYUK, V.A.** : Metody registratsii svetoindutsirovannykh izmeneniĭ pogloshcheniya i lyuminestsentsii tallomov morskikh vodorosleĭ. [Methods for
recording of light-induced changes in absorption and luminescence of marine
algae.] - Sb. Rabot Akad. Nauk SSSR, dal'nevost. nauch. Tsentr, Inst. Biol.
Morya (Vladivostok) *11* (Ėkologicheskie Aspekty Fotosinteza Morskikh Makrovodorosleĭ) : 64 - 72, 184, 1978. [In R, ab : E.]

35557 - **ROMANYUK, V.A., LAPSHINA, A.A.** : Svetoindutsirovannye izmeneniya poslesvecheniya zelenoĭ vodorosli *Ulva fenestrata*. [Light-induced changes in delayed-light emission of the green alga *Ulva fenestrata*.] - Sb. Rabot Akad. Nauk
SSSR, dal'nevost. nauch. Tsentr, Inst. Biol. Morya (Vladivostok) *11* (Ėkologicheskie Aspekty Fotosinteza Morskikh Makrovodoroslei) : 112 - 120, 185 -
186, 1978. [In R, ab : E.]

*35558 - **RONSAL', G.A., LAGODA, P.P.** : Vliyanie mineral'nykh udobreniĭ i navoza na
urozhaĭ i kachestvo kukuruzy v usloviyakh orosheniya. [Effect of mineral
nutrition and fertilization on yield and maize quality under irrigation.] -
In : Pitanie Rasteniĭ i Primenenie Udobreniĭ. Pp. 28 - 33, 102. Kishinev.
sel'skokhoz. Inst. Im. M.V. Frunze, Kishinev 1977. [Growth analysis; in R.]

35559 - **ROOK, D.A., CORSON, M.J.** : Temperature and irradiance and the total daily
photosynthetic production of the crown of a *Pinus radiata* tree. - Oecologia
*36* : 371 - 382, 1978.

35560 - **ROOS, E.E., SOWA, S., BURTON, G.W.** : Accelerated aging studies of normal
and segregating chlorophyl deficient isolines of pearl millet. - Crop Sci.
*18* : 231 - 233, 1978.

35561 - **ROSA, R.N.** : A energia solar. Seu aproveitamento por conversão fotossin-
tética. [Solar energy. Its use for photosynthetic conversion.] - Lab. Fis.
Engenh. nucl. LFEN-B-N° 20, U.C.N., Sacavém, Portugal 1977. [In Port.,
ab : E.]

35562 - **ROSEN, B.H.** : Chlorophyll "*a*" fluorescence as a measure of cell growth in
a unialgal bioassay. - J. Phycol. *14* (Suppl.) : 20, 1978.

35563 - **ROSEN, P.M., MUSSELLMAN, R.C., KENDER, W.J.** : $O_3$ injury in 'Ives' and
'Delaware' grapevines : relation to stomatal behavior. - HortScience *13* :
377, 1978. [Stomatal resistance.]

35564 - **ROSENBERG, N.J.** : Possible impacts of climatic change and fluctuations on
crop production. - In : **GODBY, E.A., OTTERMAN, J.** (ed.) : COSPAR : The Contribution of Space Observations to Global Food Information Systems. Vol. 2.
Pp. 125 - 136. Pergamon Press, Oxford - New York 1978. [Ps.]

35565 - **ROSENHECK, K., BRITH-LINDNER,M., LINDNER,P., ZAKARIA, A., CAPLAN, S.R.** :
Proteolysis and flash photolysis of bacteriorhodopsin in purple membrane
fragments. - Biophys. Struct. Mechanism *4* : 301 - 313, 1978.

35566 - **ROSHCHINA, V.V., AKULOVA, E.A.** : Deĭstvie floridzina na fotosinteticheskiĭ
ėlektronnyĭ transport. [Effect of phloridzin on photosynthetic electron
transport.] - Biokhimiya *43* : 899 - 903, 1978. [In R, ab : E.]

35567 - **ROSOWSKI, J.R., LEE, K.W.** : *Cryptoglena pigra* : a euglenoid with one chloroplast. - J. Phycol. *14* : 160 - 166, 1978.

35568 - **ROTTENBERG, H.** : The proton electrochemical potential and active transport
in bacterial cells. - In : **AZZONE, G.F., AVRON, M., METCALFE, J.C., QUAGLI-
ARIELLO, E., SILIPRANDI, N.** (ed.) : The Proton and Calcium Pumps. Pp. 125 -
135. Elsevier/North-Holland Biomedical Press, Amsterdam - Oxford - New York
1978. [Ps, Chl.]

35569 - **ROY, H., COSTA, K.A., ADARI, H.** : Free subunits of ribulose-1,5-bisphospha-
te carboxylase in pea leaves. - Plant Sci. Lett. *11* : 159 - 168, 1978.

35570 - **ROY, H., VALERI, A., POPE, D.H., RUECKERT, L., COSTA, K.A.** : Small subunit
contacts in ribulose-1,5-bisphosphate carboxylase.-Biochemistry *17* : 665 -
668, 1978.

35571 - **ROZONOVA, L.N., KALASHNIKOV, Yu.E., ZAKRZHEVSKIĬ, D.A., SHUBIN, L.N.** :
O deĭstvii atmosfery molekulyarnogo vodoroda na vydelenie kisloroda trades-
kantsieĭ pri nedostatke margantsa. [The effect of the atmosphere of molecular hydrogen on oxygen evolution by spidewort in the conditions of manganе-
se deficiency.] - Fiziol. Rast. *25* : 836 - 842, 1978. [In R, ab : E.]

35572 - **RUBIN, A.B.** : Picosecond fluorescence and electron transfer in primary photosynthetic processes. - Photochem. Photobiol. *28* : 1021 - 1028, 1978.

35573 - **RUBIN, A.B., DEVAULT, D.** : The effects of uncoupler on the rates of cytochrome oxidation and reduction in the photosynthetic bacterium, *Chromatium*. Evidence for a possible cytochrome switching . - Biochim. biophys. Acta *501* : 440 - 448, 1978.

*35574 - **RUBIN, B.A.** : O nekotorykh fiziologicheskikh aspektakh problemy produktivnosti rasteniĭ. [Some physiological aspects of the problem of plant productivity.] - Sel'skokhoz. Biol. *12* : 165 - 175, 1977. [In R, ab : E.]

*B35575 - **RUBIN, B.A., GAVRILENKO, V.F.** : Biokhimiya i Fiziologiya Fotosinteza. [Biochemistry and Physiology of Photosynthesis.] - Mosk. Univ., Moskva 1977. [In R.]

35576 - **RUBIN, L.B., RUBIN, A.B.** : Picosecond fluorometry in primary events of photosynthesis. - Biophys. J. *24* : 84 - 92, 1978.

35577 - **RUD', G.Ya., TANAS'EV, V.K., BARBAROSH, M.N., PURIS, M.F.** : Formirovanie listovogo apparata i soderzhanie khlorofilla v list'yakh yabloni sorta Dzhonatan v zavisimosti ot podvoya, ploshchadi pitaniya i doz predplantazhnogo udobreniya. [Formation of leaf apparatus and chlorophyll content of Jonathan cultivar apple tree leaves in relation to rootstock, area of feeding, and dose of presowing fertilizer.] - Trudy kishinev. sel'skokhoz. Inst. im. M.V. Frunze *1978* (Voprosy Intensifikatsii Plodovodstva) : 50 - 53, 1978. [In R.]

35578 - **RUD', G.Ya., TANAS'EV, V.K., CHIMPOESH, G.P.** : Formirovanie listovogo apparata yabloni i ego optiko-fiziologicheskie svoĭstva pod vliyaniem podvoya i doz predplantazhnogo udobreniya. [Formation of leaf apparatus of apple trees and its optical and physiological properties in relation to rootstock and dose of presowing fertilizer.] - Trudy kishinev. sel'skokhoz. Inst. Im. M.V. Frunze *1978* (Voprosy Intensifikatsii Plodovodstva) : 53 - 59, 1978. [In R.]

35579 - **RUDENKO, T.I., MAKAROV, A.D., BORISYUK, G.N.** : Primenenie statisticheskogo analiza dlya issledovaniya konformatsionnykh kolebaniĭ khloroplastov. [Application of statistical analysis for investigation of conformational oscillations of chloroplasts.] - Biofizika *23* : 827 - 832, 1978. [In R, ab : E.]

35580 - **RUDENKO, T.I., MAKAROV, A.D., BORISYUK, G.N.** : Primenenie avtokorrelyatsionnogo analiza dlya issledovaniya konformatsionnykh kolebaniĭ khloroplastov. [Application of autocorrelation analysis for investigation of conformational changes of chloroplasts.] - Biofizika *23* : 1034 - 1040, 1978. [In R, ab : E.]

35581 - **RUDENKO, T.I., MAKAROV, A.D., BUDNITSKIĬ, A.A.** : K voprosu o konformatsionnykh kolebaniyakh khloroplastov. [Conformational changes in chloroplasts.] - Biofizika *23* : 92 - 98, 1978. [In R, ab : E.]

*35582 - **RÜDIGER, W., HEDDEN, P., KÖST, H.-P., CHAPMAN, D.J.** : Esterification of chlorophyllide by geranylgeranyl pyrophosphate in cell-free system from maize shoots. - Biochem. biophys. Res. Commun. *74* : 1268 - 1272, 1977.

*35583 - **RUDOĬ, A.B., VEZITSKIĬ, A.Yu., SHLYK, A.A.** : Obratimaya reaktsiya prevrashcheniya khlorofilla *b* v khlorofill *a* v ètiolirovannykh list'yakh, infil'trirovannykh èkzogennym khlorofillidom *b*. [Inverse reaction of the transformation of chlorophyll *b* into chlorophyll *a* in etiolated leaves infiltrated by exogenous chlorophyllide *b*.] - Dokl. Akad. Nauk SSSR *234* : 974 - 977, 1977. [In R.]

35584 - **RUDY, K.C.** : A review of phytoplankton studies in Lake Livingston,Texas. - Texas J. Sci. *30* : 273 - 282, 1978. [Ps, Chl.]

35585 - **RÜFFER, U., NULTSCH, W., PFAU, J.** : Untersuchungen zur lichtinduzierten Chromatophorenverlagerung bei *Fucus vesiculosus*. - Helgoländer wiss. Meeresunters. *31* : 333 - 346, 1978.

*35586 - **RUGGIU, D., SARACENI, C.** : Popolamento fitoplanctonico e produzione primaria. [Phytoplankton population and primary production.] - In : BARBANTI, L., BONACINA, C., CALDERONI, A., CAROLLO, A., DE BERNARDI, R., GUILIZZONI, P., NOCENTINI, A.M., RUGGIU, D., SARACENI, C., TONOLLI, L. : Indagini Ecologi-

che sul Lago d'Endine. Pp. 151 - 182. Ed. Ist. Ital. Idrobiologia, Verbania
Pallanza 1974. [In Ital.]

*35587 - RUGGIU, D., SARACENI, C. : Fitoplancton, clorofilla e produzione primaria
nel Lago Maggiore durante gli anni 1972 - 1973. [Phytoplankton, chlorophyll
and primary production of Lago Maggiore in years 1972 - 1973.] - Mem. Ist.
ital. Idrobiol. "Dott. Marco De Marchi" *34* : 57 - 78, 1977. [In Ital.,
ab : E.]

35588 - RUPERT, C.S. : Uniform terminology for radiations. - Photochem. Photobiol.
*28* : 1, 1978.

35589 - RUPERT, C.S., LATARJET, R. : Toward a nomenclature and dosimetric scheme
applicable to all radiations. - Photochem. Photobiol. *28* : 3 - 5, 1978.

35590 - RUPPEL, H.G., KESSELMEIER, J., LÜTZ, C. : Biochemical and cytological observations on chloroplast development. 5. Reaggregations of prolamellar body
tubules without protein participation. - Z. Pflanzenphysiol. *90* : 101 - 110,
1978.

35591 - RURAINSKI, H.J., MADER, G. : The effect of $Mg^{2+}$ on the reduction of NADP by
an artificial electron donor. - Z. Naturforsch. *33C* : 664 - 666, 1978.

*35592 - RUSINOVA, N.G., SEITOVA, T.A., DOMAN, N.G., VAKLINOVA, S.G. : Fotosinteti-
cheskiĭ metabolizm ugleroda list'ev volosnetsa sitnikovogo v zavisimosti ot
istochnika azotnogo pitaniya. [Photosynthetic carbon metabolism in *Elymus*
leaves in relation to the nitrogen source.] - Dokl. bolg. Akad. Nauk *30* :
1761 - 1764, 1977. [In R.]

35593 - RUSINOVA, N.G., SEITOVA, T.A., DOMAN, N.G., VAKLINOVA, S.G. : Aktivnost'
fermentov vosstanovitel'nogo pentozofosfatnogo tsikla i fosfopiruvatkarbo-
ksilazy 4.I.I.3I list'ev volosnetsa sitnikovogo v zavisimosti ot istochni-
kov azotnogo pitaniya. [Activity of enzymes of reductive pentose phosphate
cycle and phosphoenolpyruvate carboxylase 4.1.1.31 from *Elymus junceus* leaves in dependence on nitrogen nutrition source.] - Dokl. bolg. Akad. Nauk *31* :
93 - 96, 1978. [In R.]

35594 - RUSSELL,G., GRACE, J. : The effect of wind on grasses . IV. Some influences
of drought or wind on *Lolium perenne*. - J. exp. Bot. *28* : 245 - 255, 1978.
[Ps.]

35595 - RUSSELL,G., GRACE, J. :   The effect of wind on grasses. V. Leaf extension,
diffusive conductance, and photosynthesis in the wind tunnel. - J. exp. Bot.
*29* : 1249 - 1258, 1978.

35596 - RUSSELL, G.K., DRAFFAN, A.G., SCHMIDT, G.W., LYMAN, H. : Light-induced enzyme formation in a chlorophyll-less mutant of *Euglena gracilis*. - Plant Physiol. *62* : 678 - 682, 1978.

35597 - RUSSO, A.R. : Some ecological observations on a permanent pond in southern
England: primary production and planktonic seasonal succession. - Hydrobiologia *60* : 33 - 48, 1978.

35598 - RUTHERFORD, M.C. : Primary production ecology in Southern Africa. - In :
WERGER, M.J.A. (ed.) : Biogeography and Ecology of Southern Africa. Pp.
623 -   . Dr.W. Junk b. v. Publ., The Hague 1978.

35599 - RUTMAN, G.I., SAAKOV, V.S. : K metodike registratsii proizvodnykh spektrov
v fotobiologicheskikh issledovaniyakh. [Technique for the registration of
derivative spectrum in photobiological studies.] - Tr. prikl. Bot., Genet.
Selektsii *61* (3) : 140 - 143, 1978. [In R, ab : E.]

35600 - RUUGE, Ė.K., TIKHONOV, A.N. : Issledovanie ėlektronnogo transporta v fotosinteticheskikh sistemakh vysshikh rasteniĭ metodom ĖPR. VI. Kinetika fotoindutsirovannykh redoks-prevrashcheniĭ P700 v rezhime nepreryvnogo i impul'-
snogo osveshcheniya razlichnoĭ dlitel'nosti. [ESR study of electron transport in photosynthetic systems of higher plants. VI. Kinetics of P700 redox-
-transients induced by continuous light and flashes with different duration.]
- Biofizika *23* : 839  - 844, 1978. [In R, ab : E.]

*35601 - **RYAN, C.A.** : The regulation by carbon dioxide of protein synthesis in tomato leaves. - Biochem. biophys. Res. Commun. *77* : 1004 - 1008, 1977.

35602 - **RYBERG, H., VIRGIN, H.I.** : Red light (phytochrome) - induced acceleration of chloroplast differentiation. - In : **AKOYUNOGLOU, G., ARGYROUDI-AKOYUNO-GLOU, J.H.** (ed.) : Chloroplast Development. Pp. 793 - 800. Elsevier/North--Holland Biomedical Press, Amsterdam - New York - Oxford 1978.

35603 - **RYBKINA, G.V.** : K voprosu ob avtonomnosti vodoobmena khloroplastov. [Question of independence of chloroplast water regime.] - In : Vodnyĭ Rezhim Rasteniĭ v Svyazi s Raznymi Ékologicheskimi Usloviyami. Pp. 309 - 317. Izdatel'stvo Kazanskogo Universiteta, Kazan' 1978. [In R.]

35604 - **RYBKINA, G.V.** : Nekotorye rezul'taty sravnitel'nogo izucheniya *in vivo* vodoobmena kletok i khloroplastov. [Some results of a comparative study *in vivo* of cells and chloroplasts water exchange.] - Fiziol. Biokhim. kul't. Rast. *10* : 54 - 59, 1978. [In R, ab : E.]

35605 - **RYBKINA, G.V., GUSEV, N.A.** : O vodoobmene khloroplastov *in vivo* v usloviyakh zasukhi. [Study *in vivo* of water exchange in chloroplasts under the conditions of drought.] - Sel'skokhoz. Biol. *13* : 224 - 229, 1978. [In R, ab : E.]

35606 - **RYBOVÁ, R., JANÁČEK, K., SLAVÍKOVÁ, M.** : Electrical and pH transients in the alga *Hydrodictyon reticulatum*. - In : Échanges Ioniques Transmembranaires chez les Végétaux. Coll. CNRS No. 258. Edit. CNRS, Paris 1978.

35607 - **RYCHNOVSKÁ, M.** : Water relations. Water balance, transpiration, and water turnover in selected reedswamp communities. - In : **DYKYJOVÁ, D., KVĚT, J.** (ed.) : Pond Littoral Ecosystems. Pp. 246 - 256. Springer-Verlag, Berlin - Heidelberg - New York 1978. [Growth analysis.]

35608 - **RYCHNOVSKÁ, M.** : Energy flow and relevant processes in a meadow ecosystem. - In : **SEN, D.N., BANSAL, R.P.** (ed.) : Environmental Physiology and Ecology of Plants. Pp. 315 - 322. Bishen Singh Mehendra Pal Singh, Dehra Dun 1978.

*35609 - **RYCZKOWSKI, M., SZEWCZYK, E.** : Changes of the chlorophyll concentration and photosynthesis in the developing embryo (mono- and dicotyledonous plants). - In : **MALIK, C.P.** (ed.) : Advances in Plant Reproductive Physiology. Pp. 222 - 229. Kalyani Publ., New Delhi 1977.

35610 - **RYHINER, A.H., MATSUDA, M.** : Effect of plant density and water supply on wheat production. - Neth. J. agr. Sci. *26* : 200 - 209, 1978. [Resistances.]

35611 - **RYLE, G.J.A., POWELL, C.E., GORDON, A.J.** : Effect of source of nitrogen on the growth of Fiskeby soya bean : The carbon economy of whole plants. - Ann. Bot. *42* : 637 - 648, 1978. [Ps.]

35612 - **SAAKOV, V.A., BARANOW, A.A., SHIRYAJEVA, G.A., HOFFMANN, P.** : Pigmentphysiologische Untersuchungen mit Hilfe der Derivativspektrophotometrie. - Stud. biophys. *70* : 129 - 142, 1978. [Chl.]

35613 - **SAAKOV, V.S., BARANOV, A.A., HOFFMANN, P.** : Derivativ-spektroskopische Charakteristik des pigmentphysiologischen Zustandes des Photosyntheseapparates unter besonderer Berücksichtigung der Temperatur. - Stud. biophys. *70* : 163 - 173, 1978.

35614 - **SABATER, B., RODRIGUEZ, M.T.** : Control of chlorophyll degradation in detached leaves of barley and oat through effect of kinetin on chlorophyllase levels. - Physiol. Plant. *43* : 274 - 276, 1978.

35615 - **SADEWASSER, D.A., DILLEY, R.A.** : A dual requirement for plastoquinone in chloroplast electron transport. - Biochim. biophys. Acta *501* : 208 - 216, 1978.

35616 - **SADOVOĬ, A.F., SHALIN, Yu.P.** : Vliyanie temperatury i obluchennosti na fotosintez rasteniĭ pshenitsy. [Effect of temperature and irradiance on photosynthesis of wheat plants.] - Fiziol. Biokhim. kul't. Rast. *10* : 26 - 29, 110, 1978. [In R, ab : E.]

35617 - **SAGROMSKY, H.** : Ein Beitrag zur physiologischen Charakterisierung von drei Chlorophyll b-freien Mutanten von *Hordeum vulgare*. - Biochem. Physiol. Pflanz. *173* : 146 - 159, 1978.

35618 - **SAGROMSKY, H.** : Nicht-zyklische Photophosphorylierung in Chloroplasten von drei Chlorophyll b-freien Mutanten von *Hordeum vulgare*. - Biochem. Physiol. Pflanz. *173* : 541 - 543, 1978.

35619 - **SAIJO, R., TAKEO, T.** : [Changes in some metabolic abilities in tea shoots during aging and by sun-shade treatment.] - Chagyo Gijutsu Kenkyu *54* : 37 - 43, 1978. [Chl; in Jap., ab : E.]

35620 - **SAITO, T.** : [Studies on fruiting of muskmelons - II Effect of fertilizer kinds, method of fertilizer application and manure on plant growth and quality.] - Bull. Coll. Agr. vet. Med. Nihon Univ. *35* : 97 - 110, 1978. [In Jap., ab : E.]

35621 - **SAITO, T., ISO, N., MIZUNO, H., KITAMURA, I.** : Solution properties of phycocyanin. V. Studies of the self-association reaction of phycocyanin in a pH 5.4 solution. - Bull. chem. Soc. Jap. *51* : 3471 - 3474, 1978.

*35622 - **SAKAI, S.** : Recent studies and problems of photosynthesis of tea plant. - Jap. agr. Res. quart. *9* : 101 - 106, 1975.

35623 - **SALAMA, F.M.** : Svetovye krivye i kinetika vydeleniya kisloroda list'yami pshenitsy v usloviyakh vodnogo stressa. [The light curves and kinetics of $O_2$ evolution from the wheat leaves under water deficiency.] - Dokl. Akad. Nauk AzSSR *34* : 61 - 65, 1978. [In R, ab : E.]

35624 - **SALATENKO, V.N., YAĬLO, A.L.** : Povyshenie intensivnosti fotosinteza i produktivnosti kleshcheviny pod vliyaniem orosheniya i udobreniĭ. [Increase in photosynthetic rate and productivity of castor bean plants as affected by irrigation and fertilizers.] - Sel'skokhoz. Biol. *13* : 142 - 145, 1978. [In R.]

35625 - **SALEMA, R., BRANDÃO, I.** : Development of microtubules in chloroplasts of two halophytes forced to follow Crassulacean acid metabolism. - J. Ultrastruct. Res. *62* : 132 - 136, 1978.

35626 - **SALISBURY, J.L., FLOYD, G.L.** : Molecular, enzymatic and ultrastructure characterization of the pyrenoid of the scaly green monad *Micromonas squamata*. - J. Phycol. *14* : 362 - 368, 1978.

35627 - **SALUJA, A.K., McFADDEN, B.A.** : Inhibition of ribulose bisphosphate carboxylase/oxygenase by sedoheptulose-1,7-bisphosphate. - FEBS Lett. *96* : 361 - 363, 1978.

35628 - **SALVADOR, G.F.** : Δ-aminolevulinic acid synthesis during greening of *Euglena gracilis*. - In : **AKOYUNOGLOU, G., ARGYROUDI-AKOYUNOGLOU, J.H.** (ed.) : Chloroplast Development. Pp. 161 - 165. Elsevier/North-Holland Biomedical Press, Amsterdam - New York - Oxford 1978.

35629 - **SALVADOR, G.F.** : δ-aminolevulinic acid synthesis from γ-δ-dioxovaleric acid by cellular preparations of *Euglena gracilis*. - Plant Sci. Lett. *13* : 351 - 355, 1978. [Chl.]

35630 - **SALVADOR, G.F.** : La synthèse d'acide δ-aminolévulinique par des chloroplastes isolés d' *Euglena gracilis*. - Compt. rend. Acad. Sci. Paris, Sér. D *286* : 49 - 52, 1978.

35631 - **SAMEJIMA, M., MIYACHI, S.** : Photosynthetic and light-enhanced dark fixation of $^{14}CO_2$ from the ambient atmosphere and $^{14}C$-bicarbonate infiltrated through vascular bundles in maize leaves. - Plant Cell Physiol. *19* : 907 - 916 , 1978.

35632 - **SAMMONS, D.J., PETERS, D.B., HYMOWITZ, T.** : Screening soybeans for drought resistance. I. Growth chamber procedure. - Crop Sci. *18* : 1050 - 1055, 1978. [Ps.]

35633 - **SAMSONOVA, I.A., BÉTTKHER, F.** : Issledovanie mutabil'nosti plastoma. Soob-
shchenie VI. Vliyanie akridinovykh soedineniĭ na chastotu obratnykh mutatsiĭ
u plastomnogo mutanta Pl-alb 1 tomata. [Mutability of plastom. VI. Effect of
acridine compounds on the frequency of reverse mutations in the plastom mutant Pl-alb 1 of tomato.] - Genetika *14* : 1928 - 1934, 1978. [Chi; in R,
ab : E.]

35634 - **SAMSUDDIN, Z., IMPENS, I.** : Water vapour and carbon dioxide diffusion resistances of four *Hevea brasiliensis* clonal seedlings. - Exp. Agr. *14* : 173 -
177, 1978.

35635 - **SAMUELSSON, G., ÖQUIST, G., HALLDAL, P.** : The variable chlorophyll *a* fluorescence as a measure of photosynthetic capacity in algae. - Mitt. int. Ver.
theor. angew. Limnol. *21* : 207 - 215, 1978.

35636 - **SAMUILOV, F.D.** : Fosfornoe pitanie, énergeticheskiĭ obmen i ustoĭchivost'
rasteniĭ k neblagopriyatnym usloviyam sredy. [Phosphorus nutrition, energy
balance and plant resistance to unfavourable environmental conditions.] -
In : Vodnyĭ Rezhim Rasteniĭ v Svyazi s Raznymi Ékologicheskimi Usloviyami.
Pp. 217 - 225. Izdatel'stvo Kazanskogo Universiteta, Kazan' 1978. [Ps; in R.]

35637 - **SAMUILOV, F.D.** : Vliyanie fosfornogo pitaniya na énergeticheskiĭ obmen i
ustoĭchivost' rasteniĭ k neblagopriyatnym usloviyam sredy. [Effect of phosphorus nutrition on the energy metabolism and plant resistance to unfavourable environmental conditions.] - Izv. Akad. Nauk SSSR, Ser. biol. *1978* (6) :
828 - 839, 1978. [Ps; in R, ab : E.]

35638 - **SAMUILOV, F.D., SHVALEVA, L.S., BEZUGLOV, V.K.** : Vodnyĭ rezhim list'ev i énergeticheskiĭ obmen khloroplastov kukuruzy pri narushenii fosfornogo pitaniya.
[Leaf water relations and energy exchange in chloroplasts of maize under
damaged phosphorus nutrition.] - In : Vodnyĭ Rezhim Rasteniĭ v Svyazi s Raznymi Ékologicheskimi Usloviyami. Pp. 265 - 269. Izdatel'stvo Kazanskogo Universiteta, Kazan' 1978. [In R.]

B35639 - **SANADI, D.R., VERNON, L.P.** (ed.) : Current Topics in Bioenergetics. Vol.7.
Photosynthesis: Part A. - Academic Press, New York - San Francisco - London
1978.

B35640 - **SANADI, D.R., VERNON, L.P.** (ed.) : Current Topics in Bioenergetics. Vol.8.
Photosynthesis: Part B. - Academic Press, New York - San Francisco - London
1978.

35641 - **SANADZE, G.A., BLĚK, K.K., TEVZADZE, I.T., TARKHNISHVILI, G.M.** : Izmenenie
otnosheniya $C^{13}O_2/C^{12}O_2$ pri fotosinteze rasteniyami $C_3$, $C_4$. [A change in the
$^{13}CO_2/^{12}CO_2$ ratio during photosynthesis by $C_3$ and $C_4$ plants.] - Fiziol. Rast.
*25* : 171 - 172, 1978. [In R.]

35642 - **SANDERS, J.K.M., WATERTON, J.C., DENNISS, I.S.** : Spin-lattice relaxation,
nuclear Overhauser enhancements, and long range coupling in chlorophylls
and metalloporphyrins. - J. chem. Soc., Perkin Trans. I *1978* : 1150 - 1157,
1978.

35643 - **SAND-JENSEN, K.** : Metabolic adaptation and vertical zonation of *Littorella
uniflora* (L.)ASCHERS. and *Isoetes lacustris* L. - Aquat. Bot. *4* : 1 - 10,
1978. [Ps.]

35644 - **SANE, P.V.** : Chloroplast structure and function. - In : Inter-Disciplinary
Symposium on Photosynthesis and Productivity. Abstract of Papers. Pp. 13 -
14. Indian nat. Sci. Acad., New Delhi 1978.

35645 - **SANKHLA, N.** : Photosynthetic adaptations for water deficiency. - In : Inter-
-Disciplinary Symposium on Photosynthesis and Productivity. Abstract of Papers. Pp. 43 - 45. Indian nat. Sci. Acad., New Delhi 1978.

35646 - **SANO, C., ITO, O., YONEYAMA, T., KUMAZAWA, K.** : Incorporation of $^{14}CO_2$ and
$^{15}NH_4$ into amino acids of the two subunits of fraction 1 protein in spinach
leaves. - Soil Sci. Plant Nutr. *24* : 581 - 586, 1978.

35647 - **SANO, C., YONEYAMA, T., KUMAZAWA, K.** : Incorporation of $^{15}N$ into subcellular
fraction and soluble proteins in rice seedlings. - Soil Sci. Plant Nutr. *24* :
503 - 513, 1978. [RuBPC.]

35648 - **SANTARIUS, K.A.** : Biochemical basis of frost resistance in higher plants. - Acta Hort. *81* (Winter Hardiness in Woody Perennials) : 9 - 21, 1978. [Chloroplast.]

*35649 - **SAPHON, S., CROFTS, A.R.** : Protolytic reactions in photosystem II: a new model for the release of protons accompanying the photooxidation of water. - Z. Naturforsch. *32C* : 617 - 626, 1977.

*35650 - **SAPHON, S., CROFTS, A.R.** : The $H^+/e$ ratio in chloroplasts is 2. Possible errors in its determination. - Z. Naturforsch. *32C* : 810 - 816, 1977.

35651 - **SAPHON, S., GRÄBER, P.** : External proton uptake, internal proton release and internal pH changes in chromatophores from *Rps. sphaeroides* following single turnover flashes. - Z. Naturforsch. *33C* : 715 - 722, 1978.

35652 - **SAPOZHNIKOV, D.I., MASLOVA, T.G., POPOVA, O.F., POPOVA, I.A., KOROLEVA, O.Ya.:** Metod fiksatsii i khraneniya list'ev dlya kolichestvennogo opredeleniya pigmentov plastid. [The method of fixation and keeping leaves for quantitative estimation of plastid pigments.] - Bot. Zh. *63* : 1586 - 1592, 1978. [In R.]

35653 - **SAPOZHNIKOV, D.I., POPOVA, I.A., POPOVA, O.F., MASLOVA, T.G., KOROLEVA, O.Ya.:** Deĭstvie perekisi vodoroda na reaktsiyu ėpoksidatsii violaksantinovogo tsikla, zatormozhennuyu deĭstviem povyshennoĭ temperatury. [The effect of hydrogen peroxide on the reaction of epoxidation of the violaxanthin cycle inhibited by the action of increased temperature.] - Fiziol. Rast. *25* : 356 - 360, 1978. [In R, ab : E.]

35654 - **SARACENI, C., RUGGIU, D., NAKANISHI, M.** : Phytoplankton dynamics, chlorophyll *a* and phaeophytin in Lago di Mergozzo (Northern Italy). - Mem. Ist. ital. Idrobiol. "Dott. Marco de Marchi" *36* : 215 - 237, 1978.

35655 - **SARANTOGLOU, V., IMBAULT, P., WEIL, J.-H.** : Partial purification and properties of *Euglena* cytoplasmic and chloroplastic valyl-tRNA synthetases. - In : **AKOYUNOGLOU, G., ARGYROUDI-AKOYUNOGLOU, J.H.** (ed.) : Chloroplast Development. Pp. 695 - 700. Elsevier/North-Holland Biomedical Press, Amsterdam - New York - Oxford 1978.

35656 - **SARIĆ, M., STANKOVIĆ, Ž., KRSTIĆ, B.** : Usvajanje i metabolizam $^{14}CO_2$ u zavisnosti od kvaliteta svetlosti u kukuruzu, pasulju i suncokretu. [Intake and metabolism of $^{14}CO_2$ depending on the quality of light in maize, beans and sunflower plants.] - Arh. poljopr. Nauke *31* : 3 - 11, 1978. [In Croat., ab : E.]

35657 - **SÁRVÁRI, É., HALÁSZ, G., TÖRÖK, Sz., LÁNG, F.** : Light-induced fluorescence decay during the greening of normal and lincomycin-treated maize leaves. - Planta *141* : 135 - 139, 1978.

35658 - **SÁRVÁRI, É., KERESZTES, Á., HALÁSZ, G., FRIDVALSZKY, L.** : Effect of lincomycin on the dimorphic chloroplasts of *Zea mays* L. leaves. - Cytobios *22* : 17 - 24, 1978.

*35659 - **SASAHARA, T.** : Genetic variations in cell and tissue forms in relation to plant growth II. Total cell surface area in the palisade parenchyma and total cell surface area. Total nitrogen content ratio in relation to photosynthetic activity in *Brassica*. - Jap. J. Breed. *21* : 61 - 68, 1971.

*35660 - **SASAHARA, T., TSUNODA, S.** : Genetic variations in cell and tissue forms in relation to plant growth. I. Relationship between growth rates of the *Brassica* species and cell sizes of the shoot apex and palisade parenchyma. - Jap. J. Breed *21* : 1 - 8, 1971. [Growth analysis.]

35661 - **SATAROVA, N.A.** : Regulyatsiya nekotorykh fiziologicheskikh i metabolicheskikh protsessov u rasteniĭ v svyazi s adaptatsieĭ k zasukhe. [Regulation of some physiological and metabolic processes in plants as related to their adaptation to drought.] - In : Problemy Zasukhoustoĭchivosti Rasteniĭ. Pp. 20 - 59. Nauka, Moskva 1978. [Chloroplast; in R.]

35662 - **SATO, H., NAKASEKO, K., GOTOH, K.** : Physio-ecological studies on prolificacy in maize II. Differences in dry matter accumulation between prolific and single-ear type hybrid. - Jap. J. Crop Sci. *47* : 206 - 211, 1978.

35663 - **SATO, N., MURATA, N.** : Preparation of chlorophyll $a$, chlorophyll $b$ and bacteriochlorophyll $a$ by means of column chromatography with diethylaminoethyl-
cellulose. - Biochim. biophys. Acta *501* : 103 - 111, 1978.

35664 - **SATO, T., KAWAI, M., FUKUYAMA, T.** : [Studies on matter production of taro plant
(*Colocasia esculenta* SCHOTT). I. Changes with growth in photosynthetic rate
of single leaf.] - Jap. J. Crop Sci. *47* : 425 - 430, 1978. [In Jap., ab : E.]

35665 - **SATOH, K., BUTLER, W.L.** : Low temperature spectral properties of subchloroplast fractions purified from spinach. - Plant Physiol. *61* : 373 - 379,
1978.

35666 - **SATOH, K., BUTLER, W.L.** : Competition between the 735 nm fluorescence and
the photochemistry of Photosystem I in chloroplasts at low temperature. -
Biochim. biophys. Acta *502* : 103 - 110, 1978.

35667 - **SATTERLUND, D.R., MEANS, J.E.** : Estimating solar radiation under variable
cloud conditions. - Forest Sci. *24* : 363 - 373, 1978.

35668 - **SAUER, K.** : Photosynthetic membranes. - Accounts chem. Res. *11* : 257 - 264,
1978.

35669 - **SAUER, K., BREWINGTON, G.T.** : Fluorescence lifetimes of chloroplasts, sub-
chloroplast particles and *Chlorella* using single photon counting. - In :
**HALL, D.O., COOMBS, J., GOODWIN, T.W.** (ed.) : Proceedings of the Fourth
International Congress on Photosynthesis. Pp. 409 - 421. Biochem. Soc.,
London 1978.

35670 - **SAUER, K., MATHIS, P., ACKER, S., VAN BEST, J.A.** : Electron acceptors
associated with $P$-700 in Triton solubilized Photosystem I particles from
spinach chloroplasts. - Biochim. biophys. Acta *503* : 120 - 134, 1978.

*35671 - **SAVCHENKO, G.E.** : Vliyanie khloramfenikola na nakoplenie i prevrashchenie
protokhlorofillida v zelenykh list'yakh yachmenya. [Effect of chloramphenicol on the accumulation and transformation of protochlorophyllide in barley
leaves.] - In : Biologiya i Nauchno-Tekhnicheskiĭ Progress. Pp. 60 - 64.
Akad. Nauk SSSR, Pushchino 1974. [In R.]

35672 - **SAVIDGE, G.** : Variations in the progress of $^{14}C$ uptake as a source of error
in estimates of primary production. - Mar. Biol. *49* : 295 - 301, 1978.

35673 - **SAWADA, S.** : On midday depression of photosynthesis in wheat seedlings. -
Jap. J. Crop Sci. *47* : 18 - 24, 1978.

35674 - **SAWADA, S., IWAKI, H.** : Photosynthetic features of some grassland plants
of Japan.-In : **MONSI, M., SAEKI, T.** (ed.) : Ecophysiology of Photosynthetic
Productivity. JIBP Synthesis. Vol. 19. Pp. 11 - 18. Univ. of Tokyo Press,
Tokyo 1978.

35675 - **SAWADA, S., YAMADA, M.** : Characterization of the castor bean fruit as a
photosynthetic organ. - Jap. J. Crop Sci. *47* : 602 - 608, 1978.

35676 - **SAWHNEY, S.K., NAIK, M.S., NICHOLAS, D.J.D.** : Regulation of NADH supply
for nitrate reduction in green plants *via* photosynthesis and mitochondrial
respiration. - Biochem. biophys. Res. Commun. *81* : 1209 - 1216, 1978.

35677 - **SAXENA, O.P.** : Seed size in relation to growth, metabolism and yield in
mung. - In : Inter-Disciplinary Symposium on Photosynthesis and Productivity.
Abstract of Papers. Pp. 58 - 59. Indian nat. Sci. Acad., New Delhi 1978.
[Production.]

35678 - **SAYRE, R.T., KENNEDY, R.A.** : The relationship between $C_4$ activity and photorespiration in *Mollugo verticillata*. - Plant Physiol. *61* (Suppl.) : 37,
1978.

35679 - **SCAWEN, M.D., RAMSHAW, J.A.M., BROWN, R.H., BOUTLER, D.** : The amino acid
sequences of plastocyanin from *Mercurialis perennis* and *Capsella bursa-
-pastoris*. - Phytochemistry *17* : 901 - 905, 1978.

35680 - **SCHAEFER, H., BECKER, H.** : Untersuchungen über die Chloroseresistenz von
Rebenneuzüchtungen - Ergebnisse aus dem Zeller Prüfgarten. - Weinberg Keller
*25* : 204 - 225, 1978.

35681 - **SCHÄFER, G., ONUR, G., EDELMAN, K., BICKEL-SANDKÖTTER, S., STROTMANN, H.** :
Energy transfer inhibition in photosynthesis by 3'-aryl-$N_3$-ADP, an ADP ana-
log. - FEBS Lett. *87* : 318 - 322, 1978.

35682 - **SCHÄFER, G., ONUR, G., STROTMANN, H.** : Energy transfer inhibition in photophosphorylation and oxidative phosphorylation by 3'- esters of ADP. - In :
**SCHÄFER, G., KLINGENBERG, M.** (ed.) : Energy Conservation in Biological Membranes. Pp. 220 - 227. Springer-Verlag, Berlin - Heidelberg - New York 1978.

35683 - **SCHEER, H., KATZ, J.J.** : Peripheral metal complexes: Chlorophyll "isomers"
with magnesium bound to the ring E β-keto ester system. - J. amer. chem.
Soc. *100* : 561 - 571, 1978.

*35684 - **SCHEER, H., KUFER, W.** : Studies on plant bile pigments, IV: Conformational
studies on C-phycocyanin from *Spirulina platensis*. - Z. Naturforsch. *32C* :
513 - 519, 1977.

*35685 - **SCHENK, H.E.A., HANF, J.** : Thioacylamide und thioacylharnstoffderivate als
Inhibitoren des Photosystem II. - Z. Naturforsch. *32C* : 880 - 883, 1977.

35686 - **SCHIDLOWSKI, M.** : A model for the evolution of photosynthetic oxygen (exten-
ded abstract). - Pageoph *116* : 234 - 238, 1978.

35687 - **SCHIEDER, O.** : Production and uses of metabolic and chlorophyll deficient
mutants. - In : **THORPE, T.A.** (ed.) : Frontiers of Plant Tissue Culture 1978.
Pp. 393 - 401. Int. Assoc. Plant Tissue Culture, Calgary 1978.

35688 - **SCHIEWER, U., ERDMANN, N., KUHNKE, K.-H.** : NaCl-Wirkung auf die Photosyntheseintensität von Blaualgen. Die Wirkung unterschiedlicher NaCl-Konzen-
trationen auf die Photosyntheseintensität der Blaualgen *Microcystis firma*
und *Synechococcus aquatilis*. - Biochem. Physiol. Pflanz. *172* : 351 - 368,
1978.

35689 - **SCHIFF, J.A.** : Photocontrol of chloroplast development in *Euglena*. - In :
**AKOYUNOGLOU, G., ARGYROUDI-AKOYUNOGLOU, J.H.** (ed.) : Chloroplast Develop-
ment. Pp. 747 - 767. Elsevier/North-Holland Biomedical Press, Amsterdam -
New York - Oxford 1978.

35690 - **SCHINDLER, D.W .** : Factors regulating phytoplankton production and standing
crop in the world's freshwaters. - Limnol. Oceanogr. *23* : 478 - 486, 1978.

35691 - **SCHLESINGER, W.H.** : Community structure, dynamics and nutrient cycling in
the Okefenokee cypress swamp-forest. - Ecol. Monogr. *48* : 43 - 65, 1978.
[Primary production.]

35692 - **SCHLOSS, J.V., NORTON, I.L., STRINGER, C.D., HARTMAN, F.C.** : Inactivation
of ribulosebisphosphate carboxylase by modification of arginyl residues
with phenylglyoxal. - Biochemistry *17* : 5626 - 5631, 1978.

35693 - **SCHLOSS, J.V., PHARES, E.F., LONG, M.V., STRINGER, C.D., HARTMAN, F.C.** :
Increased levels of ribulose bisphosphate carboxylase in *Rhodospirillum
rubrum*. - In : **SIEGELMAN, H.W., HIND, G.** (ed.) : Photosynthetic Carbon Assimilation. Pp. 424 - 425. Plenum Press, New York - London 1978.

35694 - **SCHLOSS, J.V., STRINGER, C.D., HARTMAN, F.C.** : Identification of essential
lysyl and cysteinyl residues in spinach ribulosebisphosphate carboxylase/
/oxygenase modified by the affinity label *N*-bromoacetylethanolamine phosphate. - J. biol. Chem. *253* : 5707 - 5711, 1978.

35695 - **SCHMERDER, B., RABENSTEIN, F., BORRISS, H.** : Steuerung des Ergrünungsprozes-
ses in den Kotyledonen nachgereifter *Agrostemma*-Embryonen durch Phytohormone.
- Biochem. Physiol. Pflanz. *173* : 97 - 113, 1978.

*35696 - **SCHMID, G.H., LIST, H., RADUNZ, A.** : Inhibition of photosystem II-reactions
in blue-green algae by the antisera to lutein and neoxanthin. - Z. Naturforsch. *32C* : 118 - 124, 1977.

*35697 - SCHMID, G.H., MENKE, W., KOENIG, F., RADUNZ, A. : Inhibition of electron transport on the oxygen-evolving side of photosystem II by an antiserum to a polypeptide isolated from the thylakoid membrane. - In : MITSUI, A., MIYA-CHI, S., SAN PIETRO, A., TAMURA, S. (ed.) : Biological Solar Energy Conversion. Pp. 129 - 141. Academic Press, New York - San Francisco - London 1977.

35698 - SCHMID, G.H., MENKE, W., RADUNZ, A., KOENIG, F. : Polypeptides of the thylakoid membrane and their functional characterization. - Z. Naturforsch. 33C : 723 - 730, 1978.

35699 - SCHMID, G.H., RADUNZ, A., KOENIG, F., MENKE, W. : Characterization of the oxygen-evolving side of photosystem II by antisera to polypeptides and carotenoids. - In : METZNER, H. (ed.) : Photosynthetic Oxygen Evolution. Pp. 91 - 104. Academic Press, London - New York - San Francisco 1978.

*35700 - SCHMID, G.H., RADUNZ, A., MENKE, W. : Localization and function of cytochrome $f$ in the thylakoid membrane. - Z. Naturforsch. 32C : 271 - 280, 1977.

35701 - SCHMIDT, H.-L., WINKLER, F.J., LATZKO, E., WIRTH, E. : $^{13}C$-kinetic isotope effects in photosynthetic carboxylation reactions and $\delta^{13}C$-values of plant material. - Isr. J. Chem. 17 : 223 - 224, 1978.

35702 - SCHMIDT, K. : Biosynthesis of carotenoids. - In : CLAYTON, R.K., SISTROM, W.R. (ed.) : The Photosynthetic Bacteria. Pp. 729 - 750. Plenum Press, New York - London 1978.

35703 - SCHMITT, J.M. : Oenothera plastome mutants lacking Fraction 1 protein. - In : AKOYUNOGLOU, G., ARGYROUDI-AKOYUNOGLOU, J.H. (ed.) : Chloroplast Development. Pp. 739 - 743. Elsevier/North-Holland Biomedical Press, Amsterdam - New York - Oxford 1978.

35704 - SCHMITT, M.R., KU, M.S.B., EDWARDS, G.E. : Quantitative determination of RuBP carboxylase-oxygenase in leaves of several $C_3$ and $C_4$ plants by immunochemical assay. - Plant Physiol. 61 (Suppl.) : 108, 1978.

35705 - SCHNEIDER, E., SCHWULÉRA, U., MÜLLER, H.W., DOSE, K. : Solubilization of an oligomycin-sensitive ATPase complex from Rhodospirillum rubrum chromatophores and its inhibition by various antibiotics. - FEBS Lett. 87 : 257 - 260, 1978.

35706 - SCHNEIDER, H.A.W., BOGORAD, L. : On the regulation of phycobiliprotein accumulation in relation to chlorophyll and ALA formation in Cyanidium caldarium. - In : AKOYUNOGLOU, G., ARGYROUDI-AKOYUNOGLOU, J.H. (ed.) : Chloroplast Development. Pp. 823 - 826. Elsevier/North-Holland Biomedical Press, Amsterdam - New York - Oxford 1978.

35707 - SCHNEITER, A.A. : Non-destructive leaf area estimation in sunflower. - Agron. J. 70 : 141 - 142, 1978.

35708 - SCHNETTER, M.-L. : Der Einfluß von Außenfaktoren auf die Struktur des Blattes von Avicennia germinans (L.) L. unter natürlichen Bedingungen. - Beitr. Biol. Pflanzen 54 : 13 - 28, 1978.

35709 - SCHOCH, P.G., SIBI, M. : Action du rayonnement sur l'indice stomatique in situ et en culture in vitro de feuilles du Vigna sinensis L. - Compt. rend. Acad. Sci. Paris, Sér. D 287 : 1285 - 1287, 1978.

35710 - SCHOCH, S. : The esterification of chlorophyllide a in greening bean leaves. - Z. Naturforsch. 33 C : 712 - 714, 1978.

35711 - SCHOCH, S., SCHÄFER, W. : Tetrahydrogeranylgeraniol,a precursor of phytol in the biosynthesis of chlorophyll a - localization of the double bonds. - Z. Naturforsch. 33 C : 408 - 412, 1978.

35712 - SCHOLEFIELD, P.B., NEALES, T.F., MAY, P. : Carbon balance of the Sultana vine (Vitis vinifera L.) and the effects of autumn defoliation by harvest--pruning. - Aust. J. Plant Physiol. 5 : 561 - 570, 1978.

35713 - SCHONBECK, M., NORTON, T.A. : Factors controlling the upper limits of fucoid algae on the shore. - J. exp. mar. Biol. Ecol. 31 : 303 - 313, 1978. [Ps.]

35714 - SCHÖNBOHM, E. : Bewegungen. Lichtorientierte Chloroplastenbewegungen. -
        Progr. Bot. *40* : 185 - 196, 1978.

*35715 - SCHÖNBORN, W., PROFT, G. : Periphyton und Sauerstoffhaushalt der mittleren
         Saale. - Limnologica *10* : 171 - 176, 1976. [Ps.]

*35716 - SCHOPF, R. : The degree of $CF_1$ release and the reconstituting capacity of
         the depleted membranes. - Z. Naturforsch. *32 C* : 600 - 604, 1977.

*35717 - SCHOPF, R., HARNISCHFEGER, G. : Studies on the retention of $CF_1$ with or
         without induced ATPase activity by pyrophosphate treated thylakoids and its
         relation to the regeneration of photophosphorylation. - Z. Naturforsch.
         *32 C* : 398 - 404, 1977.

35718 - SCHRECKENBACH, T., WALCKHOFF, B., OESTERHELT, D. : Specificity of the reti-
        nal binding site of bacteriorhodopsin: chemical and stereochemical require-
        ments for the binding of retinol and retinal. - Biochemistry *17* : 5353 -
        5359, 1978.

35719 - SCHREIBER, U., ARMOND, P.A. : Heat-induced changes of chlorophyll fluores-
        cence in isolated chloroplasts and related heat-damage at the pigment level.
        - Biochim. biophys. Acta *502* : 138 - 151, 1978.

35720 - SCHREIBER, U., VIDAVER, W., RUNECKLES, V.C., ROSEN, P. : Chlorophyll fluo-
        rescence assay for ozone injury in intact plants. - Plant Physiol. *61* : 80 -
        84, 1978.

35721 - SCHROEDER, H.J.,Jr., WEBB, W.L. : Carbon flow in plants: a three-compartment
        model. - Photosynthetica *12* : 406 - 411, 1978.

35722 - SCHULTEN, K. : An isomerization model for the photocycle of bacteriorhodo-
        psin. - In : CAPLAN, S.R., GINZBURG, M. (ed.) : Energetics and Structure of
        Halophilic Microorganisms. Pp. 331 - 334. Elsevier/North-Holland Biomedical
        Press, Amsterdam - New York 1978.

35723 - SCHULTEN, K., TAVAN, P. : A mechanism for the light-driven proton pump of
        *Halobacterium halobium*. - Nature *272* : 85 - 86, 1978.

35724 - SCHULTEN, K., WELLER, A. : Exploring fast electron transfer processes by mag-
        netic fields. - Biophys. J. *24* : 295 - 305, 1978.

35725 - SCHUMACHER, A., DREWS, G. : The formation of bacteriochlorophyll-protein
        complexes of the photosynthetic apparatus of *Rhodopseudomonas capsulata* du-
        ring early stages of development. - Biochim. biophys. Acta *501* : 183 - 194,
        1978.

35726 - SCHÜRMANN, P., WOLOSIUK, R.A. : Studies on the regulatory properties of
        chloroplast fructose-1,6-bisphosphatase. - Biochim. biophys. Acta *522* :
        130 - 138, 1978.

35727 - SCHUSTER, H., KOHLER, A., KREEB, K. : Experimentelle Untersuchungen zur
        Zinkbelastung von submersen Makrophyten. - Verh. Ges. Ökol. (7.Jahresver-
        sammlung, Kiel 1977) : 261 - 271, 1978. [Ps.]

35728 - SCHUURMANS, J.J., CASEY, R.P., KRAAYENHOF, R. : Transmembrane electrical
        potential formation in spinach chloroplasts. Investigation using a rapidly-
        -responding extrinsic probe. - FEBS Lett. *94* : 405 - 409, 1978.

35729 - SCHWARTZ, R.M., DAYHOFF, M.O. : Origins of prokaryotes, eukaryotes, mito-
        chondria, and chloroplasts. - Science *199* : 395 - 403, 1978.

35730 - SCHWARZ, Zs., KÖSSEL, H., HOBOM, G., GROSS, J. : Analysis of RNA polymerase
        binding sites on the rDNA of maize chloroplasts. - In : AKOYUNOGLOU, G.,
        ARGYROUDI-AKOYUNOGLOU, J.H. (ed.) : Chloroplast Development. Pp. 701 - 708.
        Elsevier/North-Holland Biomedical Press, Amsterdam - New York - Oxford
        1978.

35731 - SCHWELITZ, F.D., CISNEROS, P.L., JAGIELO, J.A. : The effect of glucose on
        the biochemical and ultrastructural characteristics of developing *Euglena*
        chloroplasts. - J. Protozool. *25* : 398 - 403, 1978.

*35732 - SCHWENN, J.-D., DEPKA, B. : Assimilatory sulfate reduction by chloroplasts:
The regulatory influence of adenosine-mono- and adenosine-diphosphate. - Z.
Naturforsch. *32 C* : 792 - 797, 1977.

35733 - SCOTT, B.D. : Nutrient cycling and primary production in Port Hacking, New
South Wales. - Aust. J. mar. Freshwater Res. *29* : 803 - 815, 1978.

35734 - SCOTT, B.D. : Phytoplankton distribution and light attenuation in Port Hacking estuary. - Aust. J. mar. Freshwater Res. *29* : 31 - 44, 1978. [Chl.]

35735 - SEAPY, R.R., LITTLER, M.M. : The distribution, abundance, community structure, and primary productivity of macroorganisms from two central California
rocky intertidal habitats. - Pacific Sci. *32* : 293 - 314, 1978.

*35736 - SEARLE, G.F.W. : A chloroplast photosystem 2 reaction resistant to salicyl-
aldoxime. - Z. Naturforsch. *32 C* : 968 - 972, 1977.

35737 - SEARLE, G.F.W., BARBER, J. : The involvement of the electrical double layer
in the quenching of 9-aminoacridine fluorescence by negatively charged surfaces. - Biochim. biophys. Acta *502* : 309 - 320, 1978. [Chloroplast.]

35738 - SEARLE, G.F.W., BARBER, J., PORTER, G., TREDWELL, C.J. : Picosecond time-
-resolved energy transfer in *Porphyridium cruentum*. Part II. In the isolated
light harvesting complex (phycobilisomes). - Biochim. biophys. Acta *501* :
246 - 256, 1978.

35739 - SEARLE, G.F.W., WESSELS, J.S.C. : Role of β-carotene in the reaction centres
of photosystems I and II of spinach chloroplasts prepared in non-polar solvents. - Biochim. biophys. Acta *504* : 84 - 99, 1978.

35740 - SEELEY, E.J. : Apple tree photosynthesis: An introduction. - HortScience
*13* : 640 - 641, 1978.

35741 - SEELY, G.R. : Photochemistry of chlorophyll in solution: Modeling photosystem II. - In : SANADI, D.R., VERNON, L.P. (ed.) : Current Topics in Bioenergetics. Vol.7. Pp. 3 - 37. Academic Press, New York - San Francisco - London
1978.

35742 - SEELY, G.R. : The energetics of electron-transfer reactions of chlorophyll
and other compounds. - Photochem. Photobiol. *27* : 639 - 654, 1978.

35743 - SEGAL, M.G., SYKES, A.G. : Kinetic studies on 1:1 electron-transfer reactions
involving blue copper proteins. I. Evidence for an unreactive form of the reduced protein (pH<5) and for protein-complex association in reactions of
parsley (and spinach) plastocyanin. - J. amer. chem. Soc. *100* : 4585 - 4592,
1978.

35744 - SEGOVIA, A.J., BROWN, R.H. : Relationship of phloem size to leaf size and
position. - Crop Sci. *18* : 90 - 93, 1978.

35745 - SEIBERT, M. : Picosecond events and their measurement. - In : SANADI, D.R.,
VERNON, L.P. (ed.) : Current Topics in Bioenergetics. Vol.7. Pp. 39 - 73.
Academic Press, New York - San Francisco - London 1978. [Chl, Car.]

35746 - SEILHEIMER, A.V. : Chlorophyll and lipid changes on germination in the non-
-green spores of *Thelypteris dentata*. - Amer. Fern J. *68* : 67 - 70, 1978.

35747 - SEITZ, K. : Zur Wirkung des ATP im Mechanismus der lichtinduzierten Chloro-
plastenbewegung. - Ber. deut. bot. Ges. *91* : 455 - 458, 1978.

35748 - SEKIYA, J., KAJIWARA, T., HATANAKA, A. : Effects of inhibitors on the enzyme system producing $C_6$-aldehydes from $C_{18}$-unsaturated fatty acids in chloroplasts of Japanese silver (*Farfugium japonicum*) leaves. - Plant Cell Physiol.
*19* : 553 - 559, 1978.

*35749 - SELGA, M.P., STRAUYAĬS, Yu.Yu. : Reaktsiya ul'trastruktury khloroplastov
na snizhenie intensivnosti osveshcheniya. [Response of chloroplast ultrastructure to a decrease in illuminance.] - In : Adaptatsiya Fiziologo-Biokhimicheskikh Sistem Rasteniĭ k Peremene Osveshcheniya. Part 1. Pp. 44 - 55.
Zinatne, Riga 1977. [In R.]

35750 - **SELGA, M.P., STRAUYAÏS, Yu.Yu.** : Adaptatsionnye izmeneniya v organizatsii
khloroplastov pri smene intensivnosti sveta v usloviyakh raznoĭ obespechen-
nosti mineral'nym pitaniem. [Adaptational changes in chloroplast organization
at the shift of illumination under various mineral nutrition.] - In : Fizio-
go-biokhimicheskie Issledovaniya Rasteniĭ. Pp. 17 - 27, 155. Zinatne, Riga
1978. [In R.]

35751 - **SELGA, M.P., STRAUYAÏS, Yu.Yu.** : O nekotorykh submikroskopicheskikh osobennostyakh otkladyvaniya i ottoka assimilyatsionnogo krakhmala. [Some submic-
roscopic characteristics of deposition and outflow of assimilation starch.]
- Latv. PSR Zināt. Akad. Vēstis *1978* (3) : 126 - 130, 1978. [In R.]

35752 - **SELMAN, B.R., DURBIN, R.D.** : Evidence for a catalytic function of the coupling factor 1 protein reconstituted with chloroplast thylakoid membranes. -
Biochim. biophys. Acta *502* : 29 - 37, 1978.

35753 - **SELMAN, B.R., SMITH, D.D., VOEGELI, K.K., JOHNSON, G., DILLEY, R.A.** : Chloroplast membrane sidedness. Location of plastocyanin determined by chemical mo-
difiers. - In : HALL, D.O., COOMBS, J., GOODWIN, T.W. (ed.) : Proceedings
of the Fourth International Congress on Photosynthesis. Pp. 793 - 798. Biochem. Soc., London 1978.

35754 - **SELSTAM, E.** : Photodecomposition of monogalactosyl diglyceride mediated by
chlorophyll. - Physiol. Plant. *44* : 26 - 30, 1978.

35755 - **SEMENENKO, V.E.** : Molekulyarno-biologicheskie aspekty ėndogennoĭ regulyatsii
fotosinteza. [Molecular-biological aspects of endogenous regulation of photosynthesis.] - Fiziol. Rast. *25* : 903 - 921, 1978. [In R, ab : E.]

35756 - **SEMEN'KOVA, E.A., ZVALINSKIĬ, V.I.** : Spektral'no-kineticheskoe izuchenie
dolgozhivushchikh komponentov poslesvecheniya morskikh vodorosleĭ. [The
spectral-kinetic study of long-lived components of the sea-weed delayed
light emission.] - Sb. Rabot Akad. Nauk SSSR, dal'nevost. nauch. Tsentr,
Inst. Biol. Morya (Vladivostok) *11* (Ėkologicheskie Aspekty Fotosinteza Morskikh Makrovodorosleĭ) : 73 - 82, 184, 1978. [In R, ab : E.]

35757 - **SEMICHAEVSKIĬ, V.D.** : Regulyatsiya sostoyaniya i svoĭstv khlorofilla v pigment-belkovykh kompleksakh. [Control of state and properties of chlorophylls
in pigment-protein complexes.] - In : Biologiya i Nauchno-Tekhnicheskiĭ
Progress. Pp. 64 - 67. Akad. Nauk SSSR, Pushchino 1974. [In R.]

35758 - **SEMYANOVICH, N.D.** : Ab vydzyalenni rėaktsyĭnykh tsėntraŭ z khramataforaŭ
nyasernykh fotasintėzuyuchykh baktėryĭ. [Isolation of reaction centres from
chromatophores of non-sulphur photosynthetic bacteria.] - Vestsi Akad. Navuk belarus.SSR, Ser. biyal. Navuk *1978* (4) : 28 - 32, 139, 1978. [In Belorus., ab : E, R.]

35759 - **SEN, S.P., DAS, B.K., SEN GUPTA, D.** : Photosynthesis and crop productivity :
utilization of photosynthates. - In : Inter-Disciplinary Symposium on Photosynthesis and Productivity. Abstract of Papers. Pp. 39 - 40. Indian nat.
Sci. Acad., New Delhi 1978.

35760 - **SENCHENKOVA, E.M.** : Raboty M.S. Tsveta po izucheniyu fotosinteticheskogo apparata rasteniĭ i sozdaniyu metoda khromatografii. [M.S. Tsvet's work on
photosynthetic apparatus in plants and the creation of chromatographic method.] - Istoriko-biol. Issledovaniya *6* : 28 - 53, 1978. [In R, ab : E.]

35761 - **SENFT, W.H.** : Dependence of light-saturated rates of algal photosynthesis
on intracellular concentrations of phosphorus. - Limnol. Oceanogr. *23* :
709 - 718, 1978.

35762 - **SENGER, H., FLEISCHHACKER, P.** : Adaptation of the photosynthetic apparatus of *Scenedesmus obliquus* to strong and weak light conditions. I. Differences in pigments, photosynthetic capacity, quantum yield and dark reactions. - Physiol. Plant. *43* : 35 - 42, 1978.

35763 - **SENGER, H., STRAβBERGER, G.** : Development of the photosystems in greening
algae. - In : AKOYUNOGLOU, G., ARGYROUDI-AKOYUNOGLOU, J.H. (ed.) : Chloroplast Development. Pp. 367 - 378. Elsevier/North-Holland Biomedical Press,
Amsterdam - New York - Oxford 1978.

35764 - SENSER, M., BECK, E. : Photochemically active chloroplasts from spruce (*Picea abies* (L.) KARST.). - Photosynthetica *12* : 323 - 327, 1978.

35765 - SERVAITES, J.C., OGREN, W.L. : Oxygen inhibition of photosynthesis and stimulation of photorespiration in soybean leaf cells. - Plant Physiol. *61* : 62 - 67, 1978.

35766 - SERVAITES, J.C., SCHRADER, L.E., EDWARDS, G.E. : Glycolate synthesis in a $C_3$ , $C_4$ and intermediate photosynthetic plant type. - Plant Cell Physiol. *19* : 1399 - 1405, 1978.

35767 - ŠESTÁK, Z. : Ontogenetic effects in oxygen evolution. - In : METZNER, H. (ed.) : Photosynthetic Oxygen Evolution. Pp. 489 - 493. Academic Press, London - New York - San Francisco 1978.

35768 - ŠESTÁK, Z. : Photosynthetic characteristics during ontogenesis of leaves. 3. Carotenoids. - Photosynthetica *12* : 89 - 109, 1978.

*35769 - ŠESTÁK, Z., ČATSKÝ, J. : Bibliography of reviews and methods of photosynthesis - 39, 40, 41. - Photosynthetica *11* : 93 - 105, 220 - 233, 343 - 351, 1977.

35770 - ŠESTÁK, Z., ČATSKÝ, J. : Bibliography of reviews and methods of photosynthesis - 42, 43. - Photosynthetica *12* : 209 - 237, 452 - 479, 1978.

*B35771 - ŠESTÁK, Z., ČATSKÝ, J. (ed.) : Photosynthesis Bibliography. Vol. 3-1972. - Dr. W. Junk b. v. - Publ., The Hague 1977.

B35772 - ŠESTÁK, Z., ČATSKÝ, J. (ed.) : Photosynthesis Bibliography. Vol. 4-1973. - Dr. W. Junk b. v. - Publ., The Hague 1978.

35773 - ŠESTÁK, Z., SOLÁROVÁ, J., ZIMA, J., VÁCLAVÍK, J. : Effect of growth irradiance on photosynthesis and transpiration in *Phaseolus vulgaris* L. - Biol. Plant. *20* : 234 - 238, 1978.

35774 - ŠESTÁK, Z., ZIMA, J., WILHELMOVÁ, N. : Ontogenetic changes in the internal limitations to bean-leaf photosynthesis. 4. Effect of pH of the isolation and/or reaction medium on the activities of photosystems 1 and 2. - Photosynthetica *12* : 1 - 6, 1978.

35775 - SETTER, T.L., SCHRADER, L.E., BINGHAM, E.T. : Carbon dioxide exchange rates, transpiration, and leaf characters in genetically equivalent ploidy levels of alfalfa. - Crop Sci. *18* : 327 - 332, 1978.

35776 - SHABEL'SKAYA, É.F. : Sravnitel'naya temnoustoĭchivost' plastidnogo apparata filogeneticheski drevnikh i molodykh poryadkov tsvetkovykh rasteniĭ. [Comparative dark resistance of the plastid apparatus of phylogenetically old and young families of flowering plants.] - In : Voprosy Estestvoznaniya. Pp. 28 - 32. Minsk. gos. ped. Inst. Im. A.M. Gor'kogo, Minsk 1978. [In R.]

35777 - SHABEL'SKAYA, É.F., GVARDIYAN, V.N. : Vliyanie prodolzhitel'nogo polnogo zatemneniya na soderzhanie fotolabil'noĭ formy khlorofilla v list'yakh rasteniĭ. [Effect of prolonged full darkening on the content of photolabile form of chlorophyll in leaves of plants.] - Vestsi Akad. Navuk belarus.SSR, Ser. biyal. Navuk *1978* (1) : 115 - 116, 1978. [In R.]

35778 - SHABEL'SKAYA, É.F., KHOMENKO, V.A. : K voprosu o zashchitnoĭ roli katalazy v protsesse temnovoĭ degradatsii khlorofilla. [Protective role of catalase in the process of dark chlorophyll degradation.] - In : Voprosy Estestvoznaniya. Pp. 32 - 36. Minsk. gos. ped. Inst. Im. A.M. Gor'kogo, Minsk 1978. [In R.]

*35779 - SHADRIKOV, O.A., DOTLOV, M.M. : K voprosu o primenenii modulyatora sveta ML-3 v ustanovkakh dlya issledovaniya fotoprovodimosti rastvora khlorofilla. [Use of light modulator ML-3 in devices for measuring photoconductivity of chlorophyll solutions.] - In : SHAKHOV, A.A. (ed.) : Problemy Fotoénergetiki Rasteniĭ. Vol. 3. Fiziko-Tekhnicheskie Voprosy. Pp. 103 - 106. Shtiintsa, Kishinev 1975. [In R.]

35780 - SHAFFER, P.W., LOCKAU, W., WOLK, C.P. : Isolations of mutants of the cyanobacterium, *Anabaena variabilis*, impaired in photoautotrophy. - Arch. Microbiol. *117* : 215 - 219, 1978.

35781 - **SHANNON, M.C., FRANCOIS, L.E.** : Salt tolerance of three muskmelon cultivars. - J. amer. Soc. hort. Sci. *103* : 127 - 130, 1978. [Primary production.]

35782 - **SHAPIRO, S.L., CAMPILLO, A.J., LEWIS, A., PERREAULT, G.J., SPOONHOWER, J.P., CLAYTON, R.K., STOECKENIUS, W.** : Picosecond and steady state, variable intensity and variable temperature emission spectroscopy of bacteriorhodopsin. - Biophys. J. *23* : 383 - 393, 1978.

35783 - **SHARKEY, T., RASCHKE, K.** : Photosynthesis not inhibited by phaseic acid. - Plant Physiol. *61* (Suppl.) : 50, 1978.

35784 - **SHARMA, D.P., FEREE, D.C., HARTMAN, F.O.** : Influence of pesticides on photosynthesis in apple (*Malus domestica* BORK). - Pesticides *12* (7) : 16 - 19, 1978.

35785 - **SHARMA, R., KUMAR, S., NANDA, K.K.** : The effect of gibberellic acid and guanosine monophosphates on extension growth, leaf production and flowering of *Impatiens balsamina*. - Physiol. Plant. *44* : 359 - 364, 1978.

35786 - **SHARMAN, I.M.** : Isoprene units : their role in structure and function of terpenes, carotenoids and other fat soluble vitamins. - In : **CAMA, H.R., SASTRY, P.S.** (ed.) : Vitamin and Carrier Functions of Polyprenoids. World Review of Nutrition and Dietetics, Vol. 31. Pp. 10 - 15. S.Karger, Basel 1978.

*35787 - **SHARUPICH, V.P., MARKOV, I.E.** : Radiatsionnyĭ rezhim i opticheskie svoĭstva kul'tury ogurtsov. [Radiation regime and optical properties of cucumber cultures.] - In : Intesivnaya Svetokul'tura Rasteniĭ. Pp. 158 - 164. Inst. Fiz. Sib. Otd. Akad. Nauk SSSR, Krasnoyarsk 1977. [In R.]

35788 - **SHATILOV, I.S.** : Maksimal'noe akkumulirovanie solnechnoĭ ėnergii kul'turnymi rasteniyami - vazhneĭshaya zadacha sovremennogo zemledeliya. [Maximal accumulation of solar energy by agricultural plants - the most important problem of agriculture of today.] - In : Problemy Zemledeliya. Pp. 12 - 21. Kolos, Moskva 1978. [Ps; in R.]

35789 - **SHATILOV, I.S., NAZARYAN, G.Kh.** : Intensivnost' fotosinteza i dykhaniya kartofelya pri ponizhennoĭ vlagoobespechennosti. [Photosynthetic and respiration rates of potato at reduced water availability.] - Sel'skokhoz. Biol. *13* : 934 - 935, 957, 1978. [In R.]

35790 - **SHATILOV, I.S., SHAROV, A.F.** : Dinamika assimiliruyushcheĭ poverkhnosti, intensivnost' i produktivnost' fotosinteza i formirovanie urozhaya ozimoĭ pshenitsy. [Dynamics of assimilating surface, photosynthetic rate and productivity and yield formation in winter wheat.] - Izv. TSKhA *1978* (1) : 23 - 35, 1978. [In R.]

35791 - **SHATILOV, I.S., SHAROV, A.F.** : Fotosinteticheskiĭ potentsial, intensivnost' fotosinteza i rol' otdel'nykh organov rasteniĭ v formirovanii biologicheskogo urozhaya ozimoĭ pshenitsy na raznykh agrofonakh. [Photosynthetic potential, photosynthetic rate and the role of separate plant organs in the formation of the photosynthetic surface of winter wheat grown under different soil fertility conditions.] - Sel'skokhoz. Biol. *13* : 36 - 43, 1978. [In R, ab : E.]

35792 - **SHAVER, G.R.** : Leaf angle and light absorptance of *Arctostaphylos* species (*Ericaceae*) along environmental gradients. - Madroño *25* : 133 - 138, 1978.

35793 - **SHEATH, R.G., COLE, K.** : Salinity adaptations of a recent migrant into the great lakes *Bangia atropurpurea* (*Rhodophyta*). - J. Phycol. *14* (Suppl.) : 23, 1978. [Ps.]

35794 - **SHEEN, S.J.** : Chemical modification of RuBP carboxylase activity in tobacco chlorophyll genotypes. - Plant Physiol. *61* (Suppl.) : 98, 1978.

35795 - **SHEFFER, K.M., WATSCHKE, T.L., DUICH, J.M.** : Effect of mowing height on leaf angle, leaf number, and tiller density of 62 Kentucky bluegrasses. - Agron. J. *70* : 686 - 689, 1978.

35796 - **SHEĬTANOV, Kh., MANOLOVA, N.** : V"rkhu khronoamperometrichniya metod za sledene intenzivnostta na fotosintezata *in vivo*. [Chronoamperometric method for following the rate of photosynthesis *in vivo*.] - Fiziol. Rast. (Sofia) *4* (1) : 24 - 32, 1978. [In Bulg., ab : E, R.]

35797 - SHELDON, R.B., BOYLEN, C.W. : An underwater survey method for estimating submerged macrophyte population density and biomass. - Aquat. Bot. *4* : 65 - 72, 1978.

35798 - SHELDON, R.W., SUTCLIFFE, W.H.Jr. : Generation times of 3h for Sargasso Sea microplankton determined by ATP analysis. - Limnol. Oceanogr. *23* : 1051 - 1055, 1978.

35799 - SHELEF, G., MORAINE, R., BERNER, T., LEVI, A., ORON, G. : Solar energy conversion via algal wastewater tratment and protein production. - In : HALL, D.O., COOMBS, J., GOODWIN, T.W. (ed.) : Proceedings of the Fourth International Congress on Photosynthesis. Pp. 657 - 675. Biochem. Soc., London 1978.

35800 - SHEPARD, D.V., MOORE, K.G. : Concanavalin A-mediated agglutination of plant plastids. - Planta *138* : 35 - 39, 1978.

35801 - SHEPHERD, W.D., KAPLAN, S. : Changes in *Rhodopseudomonas sphaeroides* isoaccepting phenylalanyl-tRNA species during transitions from chemoheterotrophic to photoheterotrophic growth. - Arch. Microbiol. *116* : 161 - 167, 1978.

35802 - SHERIDAN, R.P. : Toxicity of bisulfite to photosynthesis and respiration. - J. Phycol. *14* : 279 - 281, 1978.

35803 - SHERMA, J., LATTA, M. : Reversed-phase thin-layer chromatography of chloroplast pigments on chemically bonded $C_{18}$ plates. - J. Chromatogr. *154* : 73 - 75, 1978.

35804 - SHERMAN, L.A. : Differences in photosynthesis-associated properties of the bluegreen alga *Synechococcus cedrorum* grown at 30 and 40 C. - J. Phycol.' *14* : 427 - 433, 1978.

35805 - SHERMAN, W.V., CAPLAN, S.R. : Influence of membrane lipids on the photochemistry of bacteriorhodopsin in the purple membrane of *Halobacterium halobium*. - Biochim. biophys. Acta *502* : 222 - 231, 1978.

35806 - SHEVCHUK, S.N., GAPONENKO, V.I., SHLYK, A.A. : Issledovanie prirody vliyaniya fitokhroma na zelenenie êtiolirovannykh list'ev kukuruzy. [Nature of phytochrome effect on greening of etiolated maize leaves.] - Dokl. Akad. Nauk belorus.SSR *22* :1115 - 1118, 1150 - 1151, 1978. [In R, ab : E.]

35807 - SHIBA, Y. : Seasonal-changes in rates of photosynthesis and respiration of three poplar clones in relation to leaf age. - In : MONSI, M., SAEKI, T. (ed.) : Ecophysiology of Photosynthetic Productivity. JIBP Synthesis Vol. 19. Pp. 67 - 72. University of Tokyo Press, Tokyo 1978.

35808 - SHIDEI, T. : Studies on productivity of terrestrial communities. - In : TAMIYA, H. (ed.) : Summary Report on the Contribution of the Japanese National Committee for IBP, 1964 - 1974. JIBP Synthesis Vol. 20. Pp. 23 - 56. University of Tokyo Press, Tokyo 1978.

35809 - SHIEH, J., MILLER, G.W., PSENAK, M. : Properties of S-adenosyl-L-methionine- -magnesium-protoporphyrin IX methyltransferase from barley. - Plant Cell Physiol. *19* : 1051 - 1059, 1978.

35810 - SHIMADA, S., SHIMOKAWA, K. : [Ethylene-activated chlorophyllase in Satsuma mandarin (*Citrus unshiu* MARC.) fruits.] - J. agr. chem. Soc. Jap. *52* : 489 - 491, 1978. [In Jap., ab : E.]

35811 - SHIMAZAKI, K., TAKAMIYA, K., NISHIMURA, M. : Studies on electron transfer systems in the marine diatom *Phaeodactylum tricornutum*. I. Isolation and characterization of cytochromes. - J. Biochem. (Tokyo) *83* : 1631 - 1638, 1978.

35812 - SHIMAZAKI, K., TAKAMIYA, K., NISHIMURA, M. : Studies on electron transfer systems in the marine diatom *Phaeodactylum tricornutum*. II. Identification and determination of quinones, cytochromes, and flavins. - J. Biochem. (Tokyo) *83* : 1639 - 1642, 1978.

35813 - **SHIMOKAWA, K., SAKANOSHITA, A., HORIBA, K.** : Ethylene-induced changes of chloroplast structure in Satsuma mandarin (*Citrus unshiu* MARC.). - Plant Cell Physiol. *19* : 229 - 236, 1978.

35814 - **SHIMOKAWA, K., SHIMADA, S., YAEO, K.** : Ethylene-enhanced chlorophyllase activity during degreening of *Citrus unshiu* MARC. - Sci. Hort. *8* : 129 - 135, 1978.

35815 - **SHIMSHI, D., KAFKAFI, U.** : The effect of supplemental irrigation and nitrogen fertilization on wheat (*Triticum aestivum* L.). - Irrig. Sci. *1* : 27 - 38, 1978.

35816 - **SHIN, M., OSHINO, R.** : Ferredoxin-Sepharose 4B as a tool for the purification of ferredoxin-NADP$^+$ reductase. - J. Biochem. (Tokyo) *83* : 357 - 361, 1978.

*35817 - **SHINAR, R., DRUCKMANN, S., OTTOLENGHI, M., KORENSTEIN, R.** : Electric field effects in bacteriorhodopsin.- Biophys. J. *19* : 1 - 5, 1977.

35818 - **SHIOI, Y., TAMAI, H., SASA, T.** : Effects of copper on photosynthetic electron transport systems in spinach chloroplasts. - Plant Cell Physiol. *19* : 203 - 209, 1978.

35819 - **SHIOI, Y., TAMAI, H., SASA, T.** : Inhibition of Photosystem II in the green alga *Ankistrodesmus falcatus* by copper. - Physiol. Plant. *44* : 434 - 438, 1978.

*35820 - **SHIPMAN, L.L.** : Antenna chlorophyll *a* and *P*700. Exciton transitions in chlorophyll *a* arrays. - J. phys. Chem. *81* : 2180 - 2184, 1977.

*35821 - **SHIPMAN, L.L.** : Oscillator and dipole strengths for chlorophyll and related molecules. - Photochem. Photobiol. *26* : 287 - 292, 1977.

35822 - **SHIRAIWA, Y., MIYACHI, S.** : Form of inorganic carbon utilized for photosynthesis across the chloroplast membrane. - FEBS Lett. *95* : 207 - 210, 1978.

35823 - **SHIRAKI, M., YOSHIURA, M., IRIYAMA, K.** : Rapid and easy separation of chlorophylls, their derivatives, and plant yellow pigments by thin-layer chromatography. - Chem. Lett. *1978* : 103 - 104, 1978.

35824 - **SHIROYA, M.** : Translocation of organic substances in sunflower II. Role of sugar phosphate in translocation of sugars. - Plant Cell Physiol. *19* : 1363 - 1369, 1978.

35825 - **SHIRYAEV, A.I., REINGARD, T.A., OSTROVSKAYA, L.K.** : Architectonics of structural elements of photosynthetic apparatus. - In : 9$^{th}$ Int. Congr. Electron Microsc. Pp. 420 - 421. Toronto 1978.

*B35826 - **SHISHKANU, G.V.** : Fotosintez Yabloni. [Photosynthesis of Apple Trees.] - Shtiintsa, Kishinev 1973. [In R.]

35827 - **SHIVE, J.B. Jr., BROWN, K.W.** : Quaking and gas exchange in leaves of cottonwood (*Populus deltoides*, MARSH.). - Plant Physiol. *61* : 331 - 333, 1978.

35828 - **SHKLYAEV, Yu.N.** : Regulyatsiya vodnogo rezhima i fotosinteza u bobovykh kul'tur. [Regulation of water regime and photosynthesis in legumes.] - In : Vodnyĭ Rezhim Rasteniĭ v Svyazi s Raznymi Ėkologicheskimi Usloviyami. Pp. 141 - 146. Izdatel'stvo Kazanskogo Universiteta, Kazan' 1978. [In R.]

35829 - **SHKROB, A.M., RODIONOV, A.V.** : Shchelochnaya denaturatsiya bakteriorodopsina v purpurnykh membranakh. [Alkali-induced denaturation of bacteriorhodopsin in purple membranes.] - Bioorg. Khim. *4* : 360 - 368, 1978. [In R, ab : E.]

35830 - **SHKROB, A.M., RODIONOV, A.V.** : Mnozhestvennost' form relaksiruyushchikh molekul bakteriorodopsina. [The state multiplicity in relaxing molecules of the bacteriorhodopsin.] - Bioorg. Khim. *4* : 500 - 513, 1978. [In R, ab : E.]

35831 - **SHKROB, A.M., RODIONOV, A.V., OVCHINNIKOV, Yu.A.** : Obratimyĭ fotoindutsirovannyĭ gidroliz al'dimina retinalya v solyubilizirovannom bakteriorodopsine. [Reversible photoinduced hydrolysis of retinal aldimine in solubilized bacteriorhodopsin.] - Bioorg. Khim. *4* : 354 - 359, 1978. [In R, ab : E.]

35832 - **SHKUROPATOV, A.Ya., KURBANOV, K.B., STOLOVITSKIĬ, Yu.M., EVSTIGNEEV, V.B.** :
Fotokhimicheskie i fotoêlektronnye svoĭstva komponentov fotosinteticheskogo
apparata. III. Termostimulirovannaya provodimost' v lamellyarnoĭ sisteme
khlorofilla *a* - n-khloranil. [Photochemical and photoelectron properties of
photosynthetic apparatus components. III. Thermostimulated conductivity in
the lamellar system chlorophyll *a* - n-chloranil.] - Biofizika *23* : 16 - 19,
1978. [In R, ab : E.]

*35833 - **SHKUROPATOV, A.Ya., MEL'NIKOV, V.I., STOLOVITSKIĬ, Yu.M.** : Fotoêlektronnye
svoĭstva khlorofill-bel'kovykh sloev. [Photoelectric properties of chlorophyll-protein layers.] - In : Biologiya i Nauchno-Tekhnicheskiĭ Progress. Pp.
42 - 45. Pushchino 1974. [In R.]

*35834 - **SHKUROPATOV, A.Ya., VANKEVICH, M.M.** : Fotoêlektrokhimicheskiĭ êffekt v plen-
kakh Mg-ftalotsianina pri impul'snom osveshchenii. [Photoelectric effect in
films of Mg-phthalocyanine under impulse irradiation.] - In : Biologiya i
Nauchno-Tekhnicheskiĭ Progress. Pp. 45 - 47. Pushchino 1974. [In R.]

35835 - **SHLYK, A.A., FRADKIN, L.I., RUDOI, A.B., PRUDNIKOVA, I.V., SAVCHENKO, G.E.** :
Group mechanism of formation of pigment assembly in centers of chlorophyll
biosynthesis. - In : **AKOYUNOGLOU, G., ARGYROUDI-AKOYUNOGLOU, J.H.** (ed.) :
Chloroplast Development. Pp. 119 - 130. Elsevier/North-Holland Biomedical
Press, Amsterdam - New York - Oxford 1978.

*35836 - **SHLYK, A.A., MIKHAĬLOVA, S.A., MANANKINA, E.E.** : Izmenenie sostoyaniya khlo-
rofillovogo fonda v tsikle razvitiya kletok khlorelly. [Variation of the
chlorophyll fund in the cycle of *Chlorella* cell development.] - Dokl. Akad.
Nauk SSSR *236* : 1513 - 1516, 1977. [In R.]

35837 - **SHOAF, W.T.** : Rapid method for the separation of chlorophylls *a* and *b* by
high-pressure liquid chromatography. - J. Chromatogr. *152* : 247 - 249, 1978.

35838 - **SHOSHAN, V., SHAVIT, N., CHIPMAN, D.M.** : Kinetics of nucleotide binding to
chloroplast coupling factor ($CF_1$). - Biochim. biophys. Acta *504* : 108 - 122,
1978.

*35839 - **SHUBIN, L.N., STOLOVITSKIĬ, Yu.M., EVSTIGNEEV, V.B.** : O vliyanii nekotorykh
fizicheskikh i fiziko-khimicheskikh vozdeĭstviĭ na poslesvechenie khloroplastov. [Effect of some physical and physico-chemical factors on delayed
light emission of chloroplasts.] - In : Itogi Issledovaniya Mekhanizma Fotosinteza. Pp. 53 - 60. Pushchino 1974. [In R.]

35840 - **SHUKANAŬ, A.S., KAKHNOVICH, L.V., KARPUK, V.V., LEMYAZA, M.A.** : Uplyŭ gryba
*Peronospora schleidenii* UNGER. na kol'kasts' pigmentaŭ i fotakhimichnuyu
aktyŭnasts' khlaraplastaŭ taybuli rêpchataŭ. [Effect of *Peronospora schleidenii* UNGER. on pigment amount and photochemical activity of onion chloroplasts.] - Vestsi Akad. Navuk belarus.SSR, Ser. biyal. Navuk *1978* (3) : 53 -
57, 140, 1978. [In Belorus., ab : E, R.]

35841 - **SHULENBERGER, E.** : The deep chlorophyll maximum and mesoscale environmental
heterogeneity in the western half of the North Pacific central gyre. -
Deep-Sea Res. *25* : 1193 - 1208, 1978.

35842 - **SHUL'GIN, I.A., MUREĬ, I.A., NICHIPOROVICH, A.A.** : O strukturno-funktsional'noĭ organizatsii lista kak tselostnoĭ fotosinteziruyushcheĭ sistemy.
[Structural-functional organization of the leaf as an integral photosynthetic system.] - Fiziol. Rast. *25* : 76 - 84, 1978. [In R, ab : E.]

35843 - **SHUMILOV, N.G., PAKHOMOVA, G.I.** : Vliyanie kinetina na fotokhimicheskuyu
aktivnost' khloroplastov pri razlichnom vodnom rezhime list'ev. [Effect of
kinetin on photochemical activity of chloroplasts under different leaf water regime.] - In : Vodnyĭ Rezhim Rasteniĭ v Svyazi s Raznymi Êkologicheski-
mi Usloviyami. Pp. 274 - 278. Izdatel'stvo Kazanskogo Universiteta, Kazan'
1978. [In R.]

*35844 - **SHUTILOVA, N.I., ZHIGAL'TSOVA, Z.V., KUTYURIN, V.M.** : Sravnitel'noe izuchenie svetoindutsiruemykh okislitel'no-vosstanovitel'nykh svoĭstv pigment-bel-
kovolipidnykh kompleksov fotosistemy I i II, izolirovannykh iz khloroplastov.
[Comparative study of light-induced redox properties of pigment-lipoprotein

complexes of photosystems I and II isolated from chloroplasts.] - In : Itogi
Issledovaniya Mekhanizma Fotosinteza. Pp. 130 - 141. Pushchino 1974. [In R.]

35845 - SHUTTLEWORTH, W.J. : A simplified one-dimensional theoretical description
of the vegetation-atmosphere interaction. - Boundary-Layer Meteorol. *14* :
3 - 27, 1978. [Resistances.]

*35846 - SHUVALOV, V.A., ASADOV, A.A., KRAKHMALEVA, I.N. : Linear dichroism of light-
-induced absorbance changes of reaction centers of *Rhodospirillum rubrum*. -
FEBS Lett. *76* : 240 - 245, 1977.

35847 - SHUVALOV, V.A., KLEVANIK, A.V., SHARKOV, A.V., MATVEETZ, Ju.A., KRUKOV, P.G.:
Picosecond detection of BChl-800 as an intermediate electron carrier between selectively-excited $P_{870}$ and bacteriopheophytin in *Rhodospirillum rubrum* reaction centers. - FEBS Lett. *91* : 135 - 139, 1978.

35848 - SICHER, R.C., JENSEN, R.G. : Regulation of $CO_2$ fixation by ribulose 1,5-diP
levels in isolated spinach chloroplasts. - Plant Physiol. *61* (Suppl.) : 109,
1978.

*35849 - SID'KO, F.Ya., LISOVSKII, G.M., SARYCHEV, G.S., TIKHOMIROV, A.A., ZOLOTUKHIN,
I.G., PRIKUPETS, L.B. : Deĭstvie sveta razlichnoĭ intensivnosti i spektral'-
nogo sostava na produktsionnye protsessy tsenozov redisa. [Influence of
light of different intensity and spectral composition on the production processes of radish coenoses.] - In : Intensivnaya Svetokul'tura Rasteniĭ.
Pp. 3 - 14. Inst. Fiz. Sib. Otd. Akad. Nauk SSSR, Krasnoyarsk 1977. [In R.]

*35850 - SID'KO, F.Ya., MARKOV, I.E., SHARUPICH, V.P. : Raschet oblucheniya rasteniĭ
v sooruzheniyakh iskusstvennogo klimata. [Estimation of plant irradiation
in artificial climate.] - In : Intensivnaya Svetokul'tura Rasteniĭ. Pp. 148 -
158. Inst. Fiz. Sib. Otd. Akad. Nauk SSSR, Krasnoyarsk 1977. [In R.]

35851 - SIEBERTZ, H.P., HEINZ, E., BERGMANN, L. : Acyl lipids in photosynthetically
active tissue cultures of tobacco. - Plant Sci. Lett. *12* : 119 - 126, 1978.

35852 - SIEBURTH, J.McN. : Bacterioplankton : nature, biomass, activity and relationships to the protist plankton. - J. Phycol. *14* (Suppl.) : 31, 1978.
[Chl.]

35853 - SIEFERMANN-HARMS, D. : The accumulation of neutral red in illuminated thylakoids. - Biochim. biophys. Acta *504* : 265 - 277, 1978.

35854 - SIEFERMANN-HARMS, D., JOYARD, J., DOUCE, R. : Light-induced changes of the
carotenoid levels in chloroplast envelopes. - Plant Physiol. *61* : 530 - 533,
1978.

35855 - SIEFERT, E., IRGENS, R.L., PFENNIG, N. : Phototrophic purple and green bacteria in a sewage treatment plant. - Appl. environ. Microbiol. *35* : 38 - 44,
1978.

35856 - SIEFERT, E., PFENNIG, N. : Hydrogen metabolism and nitrogen fixation in
wild type and Nif⁻ mutants of *Rhodopseudomonas acidophila*. - Biochimie *60* :
261 - 265, 1978.

*35857 - SIEVERS, G., HYNNINEN, P.H. : Thin-layer chromatography of chlorophylls and
their derivatives on cellulose layer. - J. Chromatogr. *134* : 359 - 364,
1977.

35858 - SIGFRIDSSON, B. : Factors affecting ethanol-induced luminescence in dark-
-treated chloroplasts. - Physiol. Plant. *44* : 256 - 260, 1978.

35859 - SIGRIST-NELSON, K., SIGRIST, H., AZZI, A. : Characterization of the dicy-
clohexylcarbodiimide-binding protein isolated from chloroplast membranes. -
Europe. J. Biochem. *92* : 9 - 14, 1978.

35860 - SIKES, C.S. : Calcification and cation sorption of *Cladophora glomerata*
(*Chlorophyta*). - J. Phycol. *14* : 325 - 329, 1978. [Carbonic anhydrase.]

35861 - SILAEVA, A.M. : Rol' svetovogo faktora v organizatsii struktury khloroplastov. [Role of light factor in organization of chloroplast structure.] -
Fiziol. Biokhim. kul't. Rast. *10* : 563 - 572, 1978. [In R, ab : E.]

B35862 - **SILAEVA, A.M.** : Struktura Khloroplastov i Faktory Sredy. [Chloroplast Structure and Environmental Factors. ]- Naukova Dumka, Kiev 1978. [In R.]

*35863 - **SILBERSTEIN, B.R., EPEL, B.L., MALKIN, S., GROMET-ELHANAN, Z.** : The effect of electron donors and acceptors on light-induced absorbance changes and photophosphorylation in *Rhodospirillum rubrum* chromatophores. - Europe. J. Biochem. *80* : 135 - 141, 1977.

35864 - **SILINA, A.V., RASSASHKO, I.F.** : Modelirovanie sezonnoĭ dinamiki pervichnoĭ produktsii v Amurskom zalive Yaponskogo morya. [Modelling of seasonal dynamics of the primary production of phytoplankton in the Amur Bay of the Japan Sea.] - Nauch. Soobshch. Inst. Biol. Morya, dal'nevostoch. nauch. Tsentr Akad. Nauk SSSR *3* (Biologicheskie Issledovaniya Dal'nevostochnykh Moreĭ) : 85 - 88, 1978. [In R.]

35865 - **SILVERT, W., PLATT, T.** : Energy flux in the pelagic ecosystem : A time-dependent equation. - Limnol. Oceanogr. *23* : 813 - 816, 1978.

35866 - **SILVIUS, J.E., KREMER, D.F., LEE, D.R.** : Carbon assimilation and translocation in developing soybean leaves. - Plant Physiol. *61* (Suppl.) : 100, 1978.

35867 - **SILVIUS, J.E., KREMER, D.F., LEE, D.R.** : Carbon assimilation and translocation in soybean leaves at different stages of development. - Plant Physiol. *62* : 54 - 58, 1978.

35868 - **SIMIONESCU, C.I., MORA, R., SIMIONESCU, B.C.** : Porphyrins and the evolution of biosystems. - Bioelectrochem. Bioenerg. *5* : 1 - 17, 1978.

35869 - **SIMMONDS, P.G.** : Direct determination of ambient carbon dioxide and nitrous oxide with a high-temperature $^{63}$Ni electron-capture detector. - J. Chromatogr. *166* : 593 - 598, 1978.

35870 - **ŠIMON, J.** : Analýza produkce biomasy silážní kukuřice na lehkých půdách s řízeným vláhovým režimem. [Analysis of silage maize biomass production on light soils with a controlled moisture regime.] - Rost. Výroba (Praha) *24* : 641 - 650, 1978. [Growth analysis; in Czech, ab : E, G, R.]

35871 - **ŠIMON, J.** : Různá organizace porostu zavlažované cukrovky na lehké půdě. [Various organization of the irrigated sugar beet stand on light soil.] - Rost. Výroba (Praha) *24* : 1183 - 1192, 1978. [Growth analysis; in Czech, ab : E, G, R.]

35872 - **SIMONSEN, J.F., HARREMOËS, P.** : Oxygen and pH fluctuations in rivers. - Water Res. *12* : 477 - 489, 1978.

35873 - **SIMPSON, D., HØYER-HANSEN, G., CHUA, N.-H., WETTSTEIN, D. von** : The use of gene mutants in barley to correlate thylakoid polypeptide composition with the structure of the photosynthetic membrane. - In : **HALL, D.O., COOMBS, J., GOODWIN, T.W.** (ed.) : Proceedings of the Fourth International Congress on Photosynthesis. Pp. 537 - 548. Biochem. Soc., London 1978.

35874 - **SIMPSON, D., MØLLER, B.L., HØYER-HANSEN, G.** : Freeze-fracture structure and polypeptide composition of thylakoids of wild-type and mutant barley plastids. - In : **AKOYUNOGLOU, G., ARGYROUDI-AKOYUNOGLOU, J.H.** (ed.) : Chloroplast Development. Pp. 507 - 512. Elsevier/North-Holland Biomedical Press, Amsterdam - New York - Oxford 1978.

35875 - **SIMPSON, D.J.** : Freeze-fracture studies on barley plastid membranes. I. Wild-type etioplast. - Carlsberg Res. Commun. *43* : 145 - 170, 1978.

35876 - **SIMPSON, D.J.** : Freeze-fracture studies on barley plastid membranes. II. Wild-type chloroplast. - Carslberg Res. Commun. *43* : 365 - 389, 1978.

35877 - **SIMPSON, E.** : Biochemical and genetic studies of the synthesis and degradation of RuBP carboxylase. - In : **SIEGELMAN, H.W., HIND, G.** (ed.) : Photosynthetic Carbon Assimilation. Pp. 113 - 125. Plenum Press, New York - London 1978.

35878 - **SINCLAIR, J., SARAI, A.** : Variations in photosynthetic electron-transport pathways in *Chlorella vulgaris*. - Biochem. Soc. Trans. *6* : 904 - 906, 1978.

35879 - SINCLAIR, J., SARAI, A., ARNASON, T., GARLAND, S. : Recent studies on pho-
        tosystem II with the modulated oxygen electrodes. - In : METZNER, H. (ed.) :
        Photosynthetic Oxygen Evolution. Pp. 295 - 320. Academic Press, London -
        New York - San Francisco 1978.

35880 - SINCLAIR, M. : Summer phytoplankton variability in the lower St. Lawrence
        estuary. - J. Fish. Res. Board Can. 35 : 1171 - 1185, 1978. [Chl.]

*35881 - SINENSKY, M. : Specific deficit in the synthesis of 6-sulfoquinovsyl digly-
        ceride in *Chlorella pyrenoidosa*. - J. Bacteriol. 129 : 516 - 524, 1977.
        [Ps.]

35882 - SINGH, B.P. : Effect of shade on growth of spring barley. - Fyton 36 : 53 -
        59, 1978.

*35883 - SINGH, D., SINGH, H.P., SINGH, P. : Pre-harvest forecasting of wheat yield.
        - Indian J. agr. Sci. 46 : 445 - 450, 1976.

35884 - SINGH, D.P. : Relation of soil moisture and air conditioning irrigation to
        plant water balance, growth characteristics and nutrients uptake in rye
        and wheat. - Biol. Plant. 20 : 161 - 166, 1978. [Growth analysis.]

35885 - SINGH, J.S. : Photosynthesis and productivity in grasslands. - In : Inter-
        -Disciplinary Symposium on Photosynthesis and Productivity. Abstract of
        Papers. Pp. 35 - 38. Indian nat. Sci. Acad., New Delhi 1978.

35886 - SINGH, M.K., TSUNODA, S. : Photosynthetic and transpirational response of
        a cultivated and a wild species of *Triticum* to soil moisture and air humidi-
        ty. - Photosynthetica 12 : 280 - 283, 1978.

35887 - SINGH, M.P., KALIA, C.S. : LSD induced chlorophyll mutations in barley. -
        Experientia 34 : 1437 - 1438, 1978.

35888 - SINGH, S.P., RAO, D.V.M. : Contribution of different photosynthetic sites
        to grain yield in *Pennisetum typhoides*. - Indian J. Plant Physiol. 21 :
        287 - 291, 1978.

35889 - SINGHAL, G.S. : Techniques in photosynthesis. - In : Inter-Disciplinary
        Symposium on Photosynthesis and Productivity. Abstract of Papers. Pp. 4 - 9.
        Indian nat. Sci. Acad., New Delhi 1978.

35890 - SINHA, B.D., KUMAR, H.D. : Growth and pigments of nitrosoguanidine treated
        cells of blue-green alga *Anacystis nidulans* in different nitrogen sources.
        - Biochem. Physiol. Pflanz. 173 : 82 - 85, 1978.

35891 - SINHA, S.K. : The futility of the controversy "whether 'source' or 'sink'
        limits the crop yield. - In : Inter-Disciplinary Symposium on Photosynthe-
        sis and Productivity. Abstract of Papers. Pp. 60 - 62. Indian nat. Sci.
        Acad., New Delhi 1978.

*35892 - SINHA, S.K., BALASUBRAMANIAN, V., KHANNA-CHOPRA, R., SHANTHAKUMARI, P. :
        Growth analysis & photosynthetic systems in relation to hybrid vigour in
        maize *Zea mays* L. - Indian J. exp. Biol. 14 : 459 - 462, 1976.

*35893 - SINHA, S.K., KHANNA, R. : Physiological, biochemical, and genetic basis of
        heterosis. - Adv. Agron. 27 : 123 - 174, 1975. [Ps, Chl.]

35894 - SIRECI, J.E., PLOTNER, A., BARR, R., CRANE, F.L. : Selective thiol inhibi-
        tion of ferricyanide reduction in Photosystem II of spinach chloroplasts. -
        Biochem. biophys. Res. Commun. 85 : 976 - 982, 1978.

35895 - SIROHI, G.S., SHRIVASTAVA, A.K. : Carbon dioxide compensation concentration
        and its relationship to photorespiration and net carbon exchange — a review.
        - Indian J. Plant Physiol. 21 : 70 - 89, 1978.

35896 - SIROIS, D.L., FREDRICK, S.W. : Phytoplankton and primary production in the
        Lower Hudson River estuary. - Estuar. coastal marine Sci. 7 : 413 - 423,
        1978.

35897 - SIRONVAL, C., JOUY, M., MICHEL, J.M. : Early photoactivity in illuminated,
        etiolated bean leaves. - In : METZNER, H. (ed.) : Photosynthetic Oxygen
        Evolution. Pp. 495 - 509. Academic Press, London - New York - San Francisco
        1978.

35898 - SISLER, E.C., WYLIE, P.A. : Effect of some α-olefin derivatives in relation
to  the ethylene response. - Tobacco *180* (17) : 60 - 63, 1978.  Tobacco Sci.
*22* : 102 - 105, 1978. [Chl.]

35899 - SISTROM, W.R. : Control of antenna pigment components. - In : CLAYTON, R.K.,
SISTROM, W.R. (ed.) : The Photosynthetic Bacteria. Pp. 841 - 848. Plenum
Press, New York - London 1978.

35900 - SISTROM, W.R. : Phototaxis and chemotaxis. - In : CLAYTON, R.K., SISTROM,
W.R. (ed.) : The Photosynthetic Bacteria. Pp. 899 - 905. Plenum Press,
New York - London 1978. [Ps.]

35901 - SISTROM, W.R. : Lists of mutant strains. - In : CLAYTON, R.K., SISTROM, W.R.
(ed.) : The Photosynthetic Bacteria. Pp. 927 - 934. Plenum Press, New York -
London 1978.

35902 - SIVAKUMAR, M.V.K. : Prediction of leaf area index in soya bean (*Glycine
max* (L.) MERRILL).- Ann. Bot. *42* : 251 - 253, 1978.

35903 - SIVAKUMAR, M.V.K., SHAW, R.H. : Methods of growth analysis in field-grown
soya beans (*Glycine max* (L.) MERRILL). - Ann. Bot. *42* : 213 - 222, 1978.

35904 - SIVAKUMAR, M.V.K., SHAW, R.H. : Leaf response to water deficits in soybeans.
- Physiol. Plant. *42* : 134 - 138, 1978.

35905 - SIVAKUMAR, M.V.K., SHAW, R.H. : Relative evaluation of water stress indicators for soybeans. - Agron. J. *70* : 619 - 623, 1978. [Stomatal conductance.]

35906 - SIVAKUMARAN, S., HALL, M.A. : Effects of age and water stress on endogenous
levels of plant growth regulators in *Euphorbia lathyrus* L. - J. exp. Bot.
*28* : 195 - 205, 1978. [Chl.]

35907 - SKIDMORE, E.L., HAGEN, L.J. : Sheltering 3-dwarf with taller 2-dwarf grain
sorghum. - Fyton *36* : 7 - 14, 1978. [Stomatal resistance.]

35908 - SKJOLDAL, H.R., LÄNNERGREN, C. : The spring phytoplankton bloom in Lindåspollene, a land-locked Norwegian fjord. II. Biomass and activity of net
and nanoplankton. - Mar. Biol. *47* : 313 - 323, 1978.

35909 - SLATYER, R.O. : Altitudinal variation in the photosynthetic characteristics
of snow gum, *Eucalyptus pauciflora* SIEB. *ex* SPRENG. VII Relationship between gradients of field temperature and photosynthetic temperature optima
in the Snowy Mountains area. - Aust. J. Bot. *26* : 111 - 121, 1978.

35910 - SLATYER, R.O., FERRAR, P.J. : Photosynthetic characteristics of tree-line
populations of the Australian snow gum, *Eucalyptus pauciflora*. - Photosynthetica *12* : 137 - 144, 1978.

35911 - SLAWYK, G., COLLOS, Y., MINAS, M., GRALL, J.-R. : On the relationship between carbon-to-nitrogen composition ratios of the particulate matter and
growth rate of marine phytoplankton from the northwest African upwelling
area. - J. exp. mar. Biol. Ecol. *33* : 119 - 131, 1978.

35912 - SLEMNEV, N.N. : Vliyanie poliva na rasteniya pustynno-stepnoĭ zony MNR.
[Effect of irrigation on plants in desert-steppe area of Mongolia.] - In :
Geografiya i Dinamika Rastitel'nogo i Zhivotnogo Mira MNR. Pp. 55 - 59.
Nauka, Moskva 1978. [Ps; in R.]

35913 - SLIFKIN, M.A., GARTY, H., CAPLAN, S.R. : Modulation-excitation methods in
the study of bacteriorhodopsin. - In : CAPLAN, S.R., GINZBURG, M. (ed.) :
Energetics and Structure of Halophilic Microorganisms. Pp. 165 - 184.
Elsevier/North-Holland Biomedical Press, Amsterdam - New York 1978.

35914 - SLINGER, L.A., BIRD, G.W. : Ontogeny of *Daucus carota* infected with *Meloidogyne hapla*. - J. Nematol. *10* : 188 - 194, 1978. [Growth analysis.]

35915 - SLOVACEK, R., HIND, G. : Antimycin and uncoupler effects on P518 and cytochrome *f* in intact chloroplasts. - Plant Physiol. *61* (Suppl.) : 76, 1978.

*35916 - SLOVACEK, R.E., HANNAN, P.J. : *In vivo* fluorescence determinations of phytoplankton chlorophyll *a*. - Limnol. Oceanogr. *22* : 919 - 925, 1977.

35917 - SLOVACEK, R.E., HIND, G. : Flash spectroscopic studies of cyclic electron
flow in intact chloroplasts. - Biochem. biophys. Res. Commun. *84* : 901 -
906, 1978.

35918 - SLOVACEK, R.E., MILLS, J.D., HIND, G. : The function of cyclic electron
transport in photosynthesis. - FEBS Lett. *87* : 73 - 76, 1978.

35919 - SLUKA, Z.A. : O soderzhanii khlorofilla u mkhov v proizvodnykh tipakh lesa.
[Content of chlorophyll in mosses in production forests.] - Vestn. mosk.
Univ., Ser. Biol. *1978* (1) : 23 - 27, 1978. [In R, ab : E.]

35920 - SMALL, E., DESJARDINS, R.L. : Comparative gas exchange physiology in the
*Daucus carota* complex. - Can. J. Bot. *56* : 1739 - 1743, 1978.

35921 - SMILLIE, R.M., CRITCHLEY, C., BAIN, J.M., NOTT, R. : Effect of growth temperature on chloroplast structure and activity in barley. - Plant Physiol.
*62* : 191 - 196, 1978.

35922 - SMILLIE, R.M., HENNINGSEN, K.W., BAIN, J.M., CRITCHLEY, C., FESTER, T.,
WETTSTEIN, D. von : Mutants of barley heat-sensitive for chloroplast development. - Carlsberg Res. Commun. *43* : 351 - 364, 1978.

35923 - SMITH, B.B., REBEIZ, C.A. : Lifetime of Mg-photoporphyrin chelatase. - Plant
Physiol. *61* (Suppl.) : 84, 1978.

35924 - SMITH, B.N., NAMBUDIRI, E.M.V., HEBBERT, N.P., TIDWELL, W.D. : Evidence for
a fossil $C_4$ grass from the Pliocene Ricardo Formation near Mohave; California. - Plant Physiol. *61* (Suppl.) : 37, 1978.

35925 - SMITH, C., BOWN, A.W. : Regulation of $H^+$ and malate levels in *Avena sativa*
coleoptile tissue. - Plant Physiol. *61* (Suppl.) : 97, 1978.

35926 - SMITH, F.A., RAVEN, J.A. : The evolution of $H^+$ transport and its role in
photosynthetic energy transduction. - In : DEAMER, D.W. (ed.) : Light Transducing Membranes. Structure, Function, Evolution. Pp. 233 - 251. Academic
Press, New York - San Francisco - London 1978.

35927 - SMITH, L., PINDER, P.B. : Oxygen-linked electron transport and energy conservation. - In : CLAYTON, R.K., SISTROM, W.R. (ed.) : The Photosynthetic
Bacteria. Pp. 641 - 654. Plenum Press, New York - London 1978. [Ps and respiration in Ps bacteria.]

35928 - SMITH, R.A.H., FORREST, G.I. : Field estimates of primary production. -
In : HEAL, O.W., PERKINS, D.F. (ed.) : Production Ecology of British Moors
and Montane Grasslands. Pp. 17 - 37. Springer-Verlag, Berlin - Heidelberg -
New York 1978.

35929 - SMITH, R.C., BAKER, K.S. : The remote sensing of chlorophyll. - In :
GODBY, E.A., OTTERMAN, J. (ed.) : COSPAR : The Contribution of Space Observations to Global Food Information Systems. Vol. 2. Pp. 161 - 172. Pergamon Press, Oxford - New York 1978.

35930 - SMITH, W.K. : Temperatures of desert plants : another perspective on the
adaptability of leaf size. - Science *201* : 614 - 616, 1978. [Resistances.]

35931 - SMITH, W.K., NOBEL, P.S. : Influence of irradiation, soil water potential,
and leaf temperature on leaf morphology of a desert broadleaf, *Encelia farinosa* GRAY (*Compositae*). - Amer. J. Bot. *65* : 429 - 432, 1978. [Resistan-
ces.]

35932 - SMOLOV, A.P., POLEVAYA, V.S., SHUMILINA, T.N. : Vliyanie sakharozy i sveta
na rost i dykhanie kul'tury tkani ruty pri perekhode ot geterotrofnogo k
fotogeterotrofnomu sposobu pitaniya. [Effect of sucrose and light on the
growth and respiration of rue tissue culture on passing from heterotrophic
to photoheterotrophic nutrition.] - Fiziol. Rast. *25* : 531 - 535, 1978.
[In R, ab : E.]

35933 - SNYDER, F.W., CARLSON, G.E. : Photosynthate partitioning in sugarbeet. -
Crop Sci. *18* : 657 - 661, 1978.

35934 - SOE, G., NISHI, N., KAKUNO, T., YAMASHITA, J. : Reversible conversion from
$Ca^{2+}$-ATPase activity to $Mg^{2+}$- and $Mn^{2+}$-ATPase activities of coupling factor
purified from acetone powder of *Rhodospirillum rubrum* chromatophores. - J.

Biochem. (Tokyo) *84* : 805 - 814, 1978.

35935 - SOFROVÁ, D., WILHELM, J., NAUŠ, J., LEBLOVÁ, S. : Effect of Tris and analogous hydroxycompounds on Photosystem II of blue-green algae. - Photosynthetica *12* : 391 - 398, 1978.

35936 - SOIKKELI, S. : Seasonal changes in mesophyll ultrastructure of needles of Norway spruce (*Picea abies*). - Can. J. Bot. *56* : 1932 - 1940, 1978. [Chloroplast.]

35937 - SOJKA, G.A. : Metabolism of nonaromatic organic compounds. - In : CLAYTON, R.K., SISTROM, W.R. (ed.) : The Photosynthetic Bacteria. Pp. 707 - 718. Plenum Press, New York - London 1978. [Ps.]

35938 - SOLANSKY, S., HEYLAND, K.-U. : Verteilung von Fahnenblattassimilaten in Monokulturweizen. - Z. Acker-Pflanzenbau *147* : 171 - 180, 1978.

35939 - SOLÁROVÁ, J., POSPÍŠILOVÁ, J., SLAVÍK, B. : Možnosti využití chemických látek omezujících výdej vodní páry u kulturních rostlin. [Possibility to use some materials decreasing transpiration rate in agricultural plants.] - Stud. Inform. ÚVTIZ (Praha), rostl. Výroba *1978* (4) : 1 - 75, 1978. [Ps; in Czech, ab : E, R.]

35940 - SOLDATINI, G.F., ZIEGLER, I., ZIEGLER, H. : Sulfite : preferential sulfur source and modifier of $CO_2$ fixation in *Chlorella vulgaris*. - Planta *143* : 225 - 231, 1978.

*35941 - SOLOV'EVA, A.A. : Dinamika chislennosti fitoplanktona i soderzhaniya khlofilla *a* v gube Dal'nezelenetskoĭ (Barentsovo more). [Dynamics of the phytoplankton number and chlorophyll *a* content in Dal'nezelenetsian inlet (the Barents Sea).] - Gidrobiol. Zh. *11* (4) : 26 - 31, 1975. [In R, ab : E.]

*35942 - SOLOV'EVA, A.A., GALKINA, V.N., GARKAVAYA, G.P. : Eksperimental'noe izuchenie vliyaniya rastvorennogo organicheskogo veshchestva metabolitov midiĭ na prirodnoe soobshchestvo fitoplanktona Belogo morya. [Experimental study of the influence produced by dissolved organic matter of mussel metabolites on the natural phytoplankton community of the White Sea.] - Okeanologiya *17* : 449 - 458, 1977. [Ps,Chl; in R, ab : E.]

35943 - SONNEVELD, C., VOOGT, S.J. : Effects of saline irrigation water on glasshouse cucumbers. - Plant Soil *49* : 595 - 606, 1978. [Chlorosis.]

35944 - SØRENSEN, L., HOLMEN, AA.T., HALLDAL, P. : Excitation of System I and System II in photosynthesis as a possible control mechanism of glycolate metabolism in the blue-green alga *Anacystis nidulans*. - Physiol. Plant. *42* : 425 - 427, 1978.

35945 - SOROKIN, E.M. : Decay of luminescence of chlorophyll *a* light-collecting molecules in photosystem II. - Photosynthetica *12* : 250 - 273, 1978.

35946 - SOROKIN, Yu.I., KONOVALOVA, G.V. : Issledovanie podlednogo "tsveteniya" fitoplanktona v Amurskom zalive Yaponskogo morya. [Under-ice phytoplankton bloom in the Amur Bay of the Japan Sea.] - Nauch. Soobshch. Inst. Biol. Morya, dal'nevostoch. nauch. Tsentr Akad. nauk SSSR *3* (Biologicheskie Issledovaniya Dal'nevostochnykh Moreĭ) : 89 - 94, 1978. [Ps; in R.]

35947 - SOUZA MACHADO, V., ARNTZEN, C.J., BANDEEN, J.D., STEPHENSON, G.R. : Comparative triazine effects upon system II photochemistry in chloroplast of two common lambsquarters (*Chenopodium album*) biotypes. - Weed Sci. *26* : 318 - 322, 1978.

35948 - SOUZA MACHADO, V., BANDEEN, J.D., STEPHENSON, G.R., LAVIGNE, P. : Uniparental inheritance of chloroplast atrazine tolerance in *Brassica campestris*. - Can. J. Plant Sci. *58* : 977 - 981, 1978.

35949 - SPAKHOVA, A.S., RYAZANTSEVA, L.A. : Povrezhdaemost' nekotorykh drevesnykh rasteniĭ sernistym gazom. Damage of some trees by sulphur dioxide. - Fiziol. Rast. *25* : 407 - 409, 1978. [In R.]

35950 - SPALDING, M.H., EDWARDS, G.E. : Photosynthesis in enzymatically isolated leaf cells from the CAM plant *Sedum telephium* L. - Planta *141* : 59 - 63, 1978.

35951 - **SPERLING, W., RAFFERTY, C.N., KOHL, K.-D., DENCHER, N.A.** : Effect of light
on the isomer composition of bacteriorhodopsin in the purple membrane of
*Halobacterium halobium*. - In : **CAPLAN, S.R., GINZBURG, M.** (ed.) : Energetics and Structure of Halophilic Microorganisms. Pp. 323 - 330. Elsevier/
North-Holland Biomedical Press, Amsterdam - New York 1978.

39552 - **SPIERTZ, J.H.J.** : Grain production and assimilate utilization of wheat in
relation to cultivar characteristics, climatic factors and nitrogen supply.
- Agr. Res. Rep. (Wageningen) *881* : 1 - 35, 1978.

35953 - **SPIERTZ, J.H.J., ELLEN, J.** : Effects of nitrogen on crop development and
grain growth of winter wheat in relation to assimilation and utilization of
assimilates and nutrients. - Neth. J. agr. Sci. *26* : 210 - 231, 1978.

35954 - **SPIERTZ, J.H.J., VAN DE HAAR, H.** : Differences in grain growth, crop photosynthesis and distribution of assimilates between a semi-dwarf and a standard cultivar of winter wheat. - Neth. J. agr. Sci. *26* : 233 - 249, 1978.

35955 - **SPILLER, H., ERNST, A., KERFIN, W., BÖGER, P.** : Increase and stabilization
of photoproduction of hydrogen in *Nostoc muscorum* by photosynthetic electron-
transport inhibitors. - Z. Naturforsch. *33 C* : 541 - 547, 1978.

35956 - **SPILLER, S., TERRY, N.** : Iron deficiency reduces photochemical capacity
*via* an effect on the number but not size of photosynthetic units. - Plant
Physiol. *61* (Suppl.) : 87, 1978.

35957 - **SPIRESKU, I.** : Vozdeĭstvie gerbitsidov Karagarda i Étazina na rost, fotosin-
tez i dykhanie vodorosli *Chlorella*. [Effect of herbicides Caragard and
Etazine on growth, photosynthesis and respiration of the alga *Chlorella*.] -
Rev. roum. Biol., Sér. Biol. vég. *23* : 75 - 80, 1978. [In R, ab : E.]

35958 - **SPOONER, J., DAILEY, L., WARE, G., VINES, M.** : Determination of irrigation
and fertilizer practices for jade plant (*Crassula argentea* (L.) THUNB.). -
J. amer. Soc. hort. Sci. *103* : 306 - 308, 1978. [Chi.]

*35959 - **SPREY, B., GLIEM, G., JÁNOSSY, A.G.S.** : Changes in the iron and phosphorus
content of stroma inclusions during etioplast-chloroplast development in
*Nicotiana*. - Z. Naturforsch. *32 C* : 138 - 139, 1977.

35960 - **SPREY, B., GLIEM, G., JÁNOSSY, A.G.S.** : Iron and phosphorus containing inclusions in chloroplasts of *Nicotiana clevelandii* x *Nicotiana glutinosa* II.
Development of etioplasts to chloroplasts in cotyledons. - Z. Pflanzenphysiol. *88* : 69 - 82, 1978.

35961 - **SPREY, B., LAETSCH, W.M.** : Structural studies of peripheral reticulum in
$C_4$ plant chloroplasts of *Portulaca oleracea* L. - Z. Pflanzenphysiol. *87* :
37 - 53, 1978.

35962 - **SPYROPOULOS, C.G., MAVROMMATIS, M.** : Effect of water stress on pigment formation in *Quercus* species. - J. exp. Bot. *29* : 473 - 477, 1978.

35963 - **SQUIRE, G.R.** : Stomatal behaviour of tea (*Camellia sinensis*) in relation
to environment. - J. appl. Ecol. *15* : 287 - 301, 1978.

35964 - **SREDOJEVIĆ, S., MIRŽINSKI-STEFANOVIĆ, L.** : Uticaj herbicida Gesaprim-a
multy i kombinacije Gesaprima 500 T sa Lassom na zastupljenost karotinoida
u nekim linijama i hibridima kukuruza. [Effects of herbicide Gesaprim multy
and combination Gesaprim 500 T with Lasso on the content of carotenoids in
some maize inbreds and hybrids.] - Arhiv poljopr. Nauke *31* : 143 - 150,
1978. [In Croat., ab : E.]

*35965 - **SREENIVASAN, A.** : Limnology studies on Parambikulam-Aliyar project II Limnology and fisheries of Tirumoorthy Reservoir (Tamilnadu), India. - Arch.
Hydrobiol. *80* : 70 - 84, 1977.

35966 - **SRINIVASA RAO, N.K., SINGH, S.P.** : Contribution of stem sugars and different photosynthetic plant parts to grain development in sorghum. - In :
Inter-Disciplinary Symposium on Photosynthesis and Productivity. Abstract
of Papers. Pp. 56 - 57. Indian nat. Sci. Acad., New Delhi 1978.

35967 - **SSYMANK, V.** : Studies on chloroplast DNA of *Chlamydomonas*. - In : **AKOYUNO-
GLOU, G., ARGYROUDI-AKOYUNOGLOU, J.H.** (ed.) : Chloroplast Development.

Pp. 587 - 592. Elsevier/North-Holland Biomedical Press, Amsterdam - New York - Oxford 1978.

35968 - STAEHELIN, L.A., GOLECKI, J.R., FULLER, R.C., DREWS, G. : Visualization of the supramolecular architecture of chlorosomes (*Chlorobium* type vesicles) in freeze-factured cells of *Chloroflexus aurantiacus*. - Arch. Microbiol. *119* : 269 - 277, 1978.

35969 - STAKHOV, L.F., ZOLOTAREVA, E.K., MAKAROV, A.D. : Ob uchastii prirodnykh pteridinov v svetoindutsirovannom élektronnom transporte i vosstanovlenii kisloroda v khloroplastakh. [Participation of natural pteridines in light-induced electron transport and oxygen reduction in chloroplasts.] - In : KARPILOV, Yu.S., ROMANOVA, A.K. (ed.) : Mekhanizm Fotodykhaniya i Ego Osobennosti u Rasteniĭ Razlichnykh Tipov. Pp. 6 - 17. Pushchino 1978. [In R.]

35970 - STAMP, P. : Der Chlorophyllgehalt, die PEP- und RuDP-carboxylase-Aktivitäten während der Blattentwicklung einer ergrünenden Chlorophyllmutante und einer normalen Linie von *Zea mays* L. - Z. Pflanzenphysiol. *86* : 395 - 404, 1978.

35971 - STANEV, V., ANGELOV, M. : Vliyanie na temperaturata v zonata na korenovata sistema v"rkhu nyakoi pokazateli na fotosintetichnata deĭnost na domatite. [Effect of temperature in the root zone on some indices of tomato photosynthetic activity.] - Fiziol. Rast. (Sofia) *4* (1) : 33 - 41, 1978. [In Bulg., ab : E, R.]

35972 - STANEV, V., POPOV, G. : Vliyanie na azotnoto i fosfornoto khranene v"rkhu ottoka na asimilatite pri mladi sl"nchogledovi rasteniya. [Influence of nitrogen and phosphorus nutrition on the outflow of assimilates in young sunflower plants.] - Fiziol. Rast. (Sofia) *4* (1) : 10 - 18, 1978. [In Bulg., ab : E, R.]

35973 - STANEV, V., TSVETKOVA, S. : Izmeneniya na nyakoi pokazateli na fotosintetichnata deĭnost na sl"nchogleda v zavisimost ot azotnoto, fosfornoto i kallevoto gladuvane. [Changes in some characteristics of sunflower photosynthesis in relation to nitrogen, phosphorus and potassium stress.] - Fiziol. Rast. (Sofia) *4* (4) : 3 - 11, 1978. [In Bulg., ab : E, R.]

*35974 - STANIER, R.Y. : The utilization of organic substrates by cyanobacteria. - Biochem. Soc. Trans. *3* (3) : 352 - 357, 1975. [Ps.]

35975 - STANKOVIĆ, Ž. : Spektralni sastav svetlosti i fotosinteze. [Spectral composition of light and photosynthesis.] - Savremena Poljoprĭvreda *26* (1-2) : 63 - 78, 1978. [In Croat., ab : E.]

35976 - STANKOVIĆ, Ž.S. : Starch formation in isolated pea chloroplasts. - Plant Sci. Lett. *12* : 371 - 377, 1978.

35977 - STANLEY, C.D., SHAW, R.H. : The relationship of evapotranspiration to open-pan evaporation throughout the growth cycle of soybeans. - Iowa State J. Res. *53* : 129 - 136, 1978. [Growth analysis.]

35978 - STANLEY, P.E. : Quantitation of picomole amounts of NADH, NADPH, and FMN using bacterial luciferase. - In : COLOWICK, S.P., KAPLAN, N.O. (ed.) : Methods in Enzymology. Vol. 57. Pp. 215 - 222. Academic Press, New York 1978.

35979 - STARCK, Z. : Dystrybucja asymilatóv jako jeden z czynników determinujących plon rolniczy. [Photosynthate distribution as one of factors determining agricultural yield.] - Postępy Nauk roln. *1978* (1) : 17 - 34, 1978. [In Pol.]

35980 - STARCK, Z., KARWOWSKA, R. : Effect of salt-stresses on the hormonal regulation of growth, photosynthesis and distribution of $^{14}C$-assimilates in bean plants. - Acta Soc. Bot. Pol. *47* : 245 - 267, 1978.

35981 - STARK, B.C., URIBE, E.G. : The relation of conformational change to phosphorylation and ATPase activities of spinach chloroplasts. - Plant Physiol. *61* (Suppl.) : 77, 1978.

35982 - STEELE, J.A., DURBIN, R.D., UCHYTIL, T.F., RICH, D.H. : Tentoxin. An uncompetitive inhibitor of lettuce chloroplast coupling factor 1. - Biochim. biophys. Acta *501* : 72 - 82, 1978.

35983 - STEELE, J.A., UCHYTIL, T.F., DURBIN, R.D. : The stimulation of coupling
factor 1 ATPase by tentoxin. - Biochim. biophys. Acta *504* : 136 - 141, 1978.

35984 - STEELE, J.A., UCHYTIL, T.F., DURBIN, R.D. : A coupled enzyme assay for $CF_1$
ATPase. - Photosynthetica *12* : 68 - 69, 1978.

35985 - STEELE, J.A., UCHYTIL, T.F., DURBIN, R.D., BHATNAGAR, P.K., RICH, D.H. :
The stimulation of coupling factor 1 ATPase by tentoxin and its analogs. -
Biochem. biophys. Res. Commun. *84* : 215 - 218, 1978.

35986 - STEFANIS, E. DE, GALTERIO, G., NOVARO MANMANA, P., SGRULLETTA, D., STEFA-
NINI, R. : Phosphoenolpyruvate, ribulose 1,5-diphosphate carboxylase and
glutamate dehydrogenase activities in seedlings of maize inbreds and their
hybrids. - Maydica *23* : 63 - 74, 1978.

35987 - STEINBACK, K.E., BURKE, J.J., ARNTZEN, C.J. : The effects of trypsin on cation induced changes in chloroplast structure and function. - Plant Physiol.
*61* (Suppl.) : 75, 1978.

35988 - STEINBACK, K.E., BURKE, J.J., MULLET, J.E., ARNTZEN, C.J. : The role of the
light-harvesting complex in cation-mediated grana formation. - In :
AKOYUNOGLOU, G., ARGYROUDI-AKOYUNOGLOU, J.H. (ed.) : Chloroplast Development. Pp. 389 - 400. Elsevier/North-Holland Biomedical Press, Amsterdam -
New York - Oxford 1978.

35989 - STEINBISS, H.-H. : Physiologische und cytomorphologische Veränderungen
im Blatt von *Vicia faba* L. nach Kurzzeitbehandlung mit Mikrowellen. - Protoplasma *94* : 155 - 166, 1978. [Ps, chloroplast.]

35990 - STEINBISS, H.-H. : Eine Methode zur Lokalisation $^{14}$C-markierter Assimilate
im Blatt von *Vicia faba* L. mit Hilfe der elektronenmikroskopischen Autora-
diographie. - Protoplasma *94* : 281 - 297, 1978.

35991 - STEINMETZ, A., MUBUMBILA, M., KELLER, M., BURKARD, G., WEIL, J.H. : Mapping
of tRNA genes on the circular DNA molecule of *Spinacia oleracea* chloroplasts. - In : AKOYUNOGLOU, G., ARGYROUDI-AKOYUNOGLOU, J.H. (ed.) : Chloroplast Development. Pp. 573 - 580. Elsevier/North-Holland Biomedical Press,
Amsterdam - New York - Oxford 1978.

35992 - STELLA, A.M., BATLLE, A.M. del C. : Porphyrin biosynthesis - immobilized
enzymes and ligands. VIII. Studies on the purification of δ-aminolaevulinate dehydratase from *Euglena gracilis*. - Plant Sci. Lett. *11* : 87 - 92, 1978.

35993 - STEMLER, A. : A dynamic binding of $HCO_3^-$ to chloroplast reaction centers in
the light. - Carnegie Inst. Year Book *77* : 302 - 305, 1978.

35994 - STEMLER, A. : Photosystem II activity depends on membrane-bound bicarbonate.
- In : METZNER, H. (ed.) : Photosynthetic Oxygen Evolution. Pp. 283 - 293.
Academic Press, London - New York - San Francisco 1978.

35995 - STENGEL, E., RECKERMANN, J. : Methodische Vorarbeiten zur Messung der pho-
tosynthetischen Sauerstoffproduktion in offenen Algengroßkulturen. - Arch.
Hydrobiol. *82* : 263 - 294, 1978.

35996 - STEPANOVA, A.M., FEDOSEENKO, A.A., YARMUKHAMEDOVA, F.M., DUDICH, G.K. :
Issledovanie zavisimosti vosstanovleniya nitratnogo azota v list'yakh kukuruzy i makhorki na svetu ot fotosinteza i dykhaniya. [Dependence of nitra-
te-nitrogen reduction in maize and makhorka leaves in the light on photosynthesis and respiration.] - Fiziol. Biokhim. kul't. Rast. *10* : 38 - 43,
111, 1978. [In R, ab : E.]

35997 - STEPAN-SARKISSIAN, G., FOWLER, M.W. : Changes in the levels of metabolites
of the pathways of carbohydrate metabolism during the induction of nitrate
assimilation in pea roots. - Biochem. Physiol. Pflanz. *172* : 1 - 13, 1978.

35998 - STEPHENSON, R.A., WILSON, G.L. : Patterns of assimilate distribution in
soybeans at maturity. III. The contribution of assimilate to pods near the
apex in determinate types. - Aust. J. agr. Res. *29* : 1 - 8, 1978.

35999 - STEPONKUS, P.L., WIEST, S.C. : Plasma membrane alterations following cold
acclimation and freezing. - In : LI, P.H., SAKAI, A. (ed.) : Plant Cold

Hardiness and Freezing Stress. Mechanisms and Crop Implications. Pp. 75 -
91. Academic Press, New York - San Francisco - London 1978. [Chloroplast.]

36000 - STEVENS, A.F., ACOCK, B. : Automated $CO_2$ injection system with analogue
and digital outputs for measuring photosynthesis in crop enclosures. -
Agr. Meteorol. *19* : 113 - 120, 1978.

*36001 - STEVENS, S.E. Jr., FOX, J.L. : Simple equipment for the growth of photosyn-
thetic bacteria. - J. appl. Bacteriol. *42* : 275 - 278, 1977.

36002 - STEWART, W.D.P., ROWELL, P., CODD, G.A., APTE, S.K. : $N_2$ fixation and pho-
tosynthesis in photosynthetic prokaryotes. - In : HALL, D.O., COOMBS, J.,
GOODWIN, T.W. (ed.) : Proceedings of the Fourth International Congress on
Photosynthesis. Pp. 133 - 146. Biochem. Soc., London 1978.

36003 - STILLWELL, W., TIEN, H.T. : Oxygen evolution from broken thylakoids fused
with liposomes. - Biochem. biophys. Res. Commun. *81* : 212 - 216, 1978.

36004 - ŞTIRBAN, M., BERCEA, V., SPÎRCHEZ, C., DUMITRU, G. : Acţiunea radiaţiilor
ionizante şi a neutronilor termici asupra conţinutului în pigmenţi asimilato-
ri şi proteine la plantulele de porumb şi soia. [Effects of ionizing radiations and thermic neutrons on the contents of assimilatory pigments and
proteins in maize and soybean seedlings.] - Stud. Univ. Babes-Bolyai, Biol.
*1978* (2) : 3 - 10, 1978. [In Roum., ab : E.]

36005 - STITT, M., BULPIN, P.V., AP REES, T. : Pathway of starch breakdown in photosynthetic tissues of *Pisum sativum*. - Biochim. biophys. Acta *544* : 200 -
214, 1978.

36006 - STOECKENIUS, W. : Bacteriorhodopsin. - In : CLAYTON, R.K., SISTROM, W.R.
(ed.) : The Photosynthetic Bacteria. Pp. 571 - 592. Plenum Press, New York -
London 1978.

36007 - STOECKENIUS, W. : Bioenergetic mechanisms in *Halobacteria*. - In : CAPLAN,
S.R., GINZBURG, M. (ed.) : Energetics and Structure of Halophilic Microorga-
nisms. Pp. 185 - 200. Elsevier/North-Holland Biomedical Press, Amsterdam -
New York 1978.

*36008 - STOECKENIUS, W., LOZIER, R.H., NIEDERBERGER, W. : Photoreactions of bacterio-
rhodopsin. - Biophysics Struct. Mech. *3* : 65 - 68, 1977.

*36009 - STOEV, K., IVANCHEV, V. : Opyt dlya ustanovlenia vzaimosvyazi mezhdu plod-
nymi i besplodnymi pobegami v pitanii vinogradnoĭ lozy. [Relation between
fertile and infertile shoots in the nutrition of grape-vine.] - In :
Fiziologiya Vinogradnoĭ Lozy (Simposium). Pp. 415 - 422. Varna 1971. [Pho-
tosynthates; in R, ab : E.]

36010 - STOJANOVIĆ, J. : Prilog proučavanju fotosintetske produktívnosti nekih
sorta pšenice. [Photosynthetic characteristics of some cultivars of winter
wheat.] - Arhiv poljopr. Nauke *31* : 149 - 155, 1978. [In Croat., ab : E.]

36011 - STONE, J.F. : Evapotranspiration control on agricultural lands. - Soil Crop
Sci. Soc. Florida Proc. *37* : 1 - 11, 1978. [Stomatal resistance.]

36012 - STONER, W.A., MILLER, P.C., MILLER, P.M. : A test of a model of irradiance
within vegetation canopies at northern latitudes. - Arctic alpine Res. *10* :
761 - 767, 1978.

36013 - STONER, W.A., MILLER, P.C., OECHEL, W.C. : Simulation of the effect of the
tundra vascular plant canopy on the productivity of four moss species. -
In : TIESZEN, L.L. (ed.) : Vegetation and Production Ecology of an Alaskan
Arctic Tundra. Pp. 371 - 387. Springer-Verlag, New York - Heidelberg -
Berlin 1978.

36014 - STONER, W.A., MILLER, P.C., TIESZEN, L.L. : A model of plant growth and
phosphorus allocation for *Dupontia fisheri* in coastal, wet-meadow tundra. -
In : TIESZEN, L.L. (ed.) : Vegetation and Production Ecology of an Alaskan
Arctic Tundra. Pp. 559 - 576. Springer-Verlag, New York - Heidelberg -
Berlin 1978.

36015 - STORBART, A.K., HENDRY, G.A.F. : Chlorophyll formation and glycine metabolism in laevulinic acid treated barley leaves. - Phytochemistry *17* : 993 -
994, 1978.

36016 - **STOUT, D.G., KANNANGARA, T., SIMPSON, G.M.** : Drought resistance of *Sorghum bicolor*. 2. Water stress effects on growth. - Can. J. Plant Sci. *58* : 225 - 233, 1978.

36017 - **STOUT, D.G., SIMPSON, G.M.** : Drought resistance of *Sorghum bicolor*. 1. Drought avoidance mechanisms related to leaf water status. - Can. J. Plant Sci. *58* : 213 - 224, 1978. [Stomatal resistance.]

36018 - **STOUT, D.G., STEPONKUS, P.L., COTTS, R.M.** : Plasmalemma alteration during cold acclimation of *Hedera helix* bark. - Can. J. Bot. *56* : 196 - 205, 1978. [$CO_2$ effect on plasmalemma permeability.]

36019 - **STOUT, R.G., CLELAND, R.E.** : Effects of fusicoccin on the activity of a key pH-stat enzyme, PEP-carboxylase. - Planta *139* : 43 - 45, 1978.

36020 - **STOWE, L.G., TERRI, J.A.** : The geographic distribution of $C_4$ species of the *Dicotyledonae* in relation to climate. - Amer. Naturalist *112* : 609 - 623, 1978.

36021 - **STOYANOV, I., CHICHEV, P., TUNGAROV, G.** : Vliyanie na magneziya i nyakoi mikroelementi v"rkhu fotosintezata i vklynchvaneto na $^{14}C$ v nyakoi grupi s"edineniya v listata na v"zstanovyavani tsarevichni rasteniya, prekarali magnezievo gladuvane. [Effect of magnesium and some trace elements on the rate of photosynthesis and $^{14}C$ incorporation in some groups of compounds in leaves of recovering maize plants following magnesium starvation.] - Fiziol. Rast. (Sofia) *4* (1) : 84 - 91, 1978. [In Bulg., ab : E, R.]

36022 - **STRANSKY, H.** : Die quantitative Bestimmung von Chloroplastenpigmenten im picomol-Bereich mit Hilfe einer isochratischen HPLC-Methode. - Z. Natur-forsch. *33 C* : 836 - 840, 1978.

36023 - **STRASSER, R.J.** : The grouping model of plant photosynthesis. - In : **AKOYU-NOGLOU, G., ARGYROUDI-AKOYUNOGLOU, J.H.** (ed.) : Chloroplast Development. Pp. 513 - 524. Elsevier/North-Holland Biomedical Press, Amsterdam - New York - Oxford 1978.

36024 - **STRASSER, R.J., BUTLER, W.L.** : Energy coupling in the photosynthetic appa-ratus during development. - In : **HALL, D.O., COOMBS, J., GOODWIN, T.W.** (ed.) : Proceedings of the Fourth International Congress on Photosynthesis. Pp. 527 - 536. Biochem. Soc., London 1978.

36025 - **STREKAS, T., KRASNA, A.I.** : Nature of the iron sulfur core and stability of *Chromatium* hydrogenase. - In : **SCHLEGEL, H.G., SCHNEIDER, K.** (ed.) : Hydrogenases : Their Catalytic Activity, Structure and Function. Pp. 141 - 150. E.Goltze KG, Göttingen 1978.

36026 - **STRINGER, C.D., HARTMAN, F.C.** : Sequences of two active-site peptides from spinach ribulosebisphosphate carboxylase/oxygenase. - Biochem. biophys. Res. Commun. *80* : 1043 - 1048, 1978.

36027 - **STRUGNELL, R.G., PIGOTT, C.D.** : Biomass, shoot-production and grazing of two grasslands in the Rwenzori National Park, Uganda. - J. Ecol. *66* : 73 - 96, 1978.

36028 - **STUART, T.J., STANFORD, J.A.** : A case of thermal pollution limited primary productivity in a southwestern U.S.A. reservoir. - Hydrobiologia *58* : 199 - 211, 1978.

36029 - **STUMMANN, B.M.** : Tetrapyrrol protein complexes from wild type barley and barley mutants affecting chlorophyll synthesis. - Biochem. Physiol. Pflan-zen *173* : 249 - 269, 1978.

36030 - **STUMMANN, B.M.** : The capacity of various detergents to solubilize and sta-bilize protochlorophyll(ide) holochrome. - Physiol. Plant. *43* : 173 - 176, 1978.

36031 - **STUTZ, E., JENNI, B., KNOPF, U.C., GRAF, L.** : Mapping of genes on *Euglena gracilis* chloroplast DNA. - In : **AKOYUNOGLOU, G., ARGYROUDI-AKOYUNOGLOU, J.H.** (ed.) : Chloroplast Development. Pp. 609 - 618. Elsevier/North-Holland Biomedical Press, Amsterdam - New York - Oxford 1978.

36032 - SUD'ÏNA, O.G., DOVBYSH, K.P., GOLOD, M.G. : Zminy stanu i aktyvnosti khlo-
rofilazy pry porushenni struktury khloroplastiv. [Changes in the chlorophyllase state and activity with disturbance in chloroplasts structure.] - Ukr.
bot. Zh. *35* : 646 - 651, 672, 1978. [In Ukr., ab : E, R.]

36033 - SUGIYAMA, K.- I., MURATA, N. : Analyses of absorption and fluorescence
spectra of water-soluble chlorophyll proteins, pigment system II particles
and chlorophyll $a$ in diethylether solution by the curve-fitting method. -
Biochim. biophys. Acta *503* : 107 - 119, 1978.

36034 - SUGIYAMA, T. : [Light, temperature, and enzymes in $C_4$-photosynthesis.] -
J. agr. chem. Soc. Jap. *52* : R159 - R166, 1978. [In Jap.]

36035 - SUGIYAMA, Y., MUKOHATA, Y. : Energy transfer inhibition induced by modification of membrane-bound chloroplast coupling factor 1 by pyridoxal phosphate.
- FEBS Lett. *85* : 211 - 214, 1978.

36036 - SUMMERFIELD, R.J., MINCHIN, F.R., STEWART, K.A., NDUNGURU, B.J. : Growth,
reproductive development and yield of effectively nodulated cowpea plants
in contrasting aerial environments. - Ann. appl. Biol. *90* : 277 - 291, 1978.
[Growth analysis.]

36037 - SUMMERS, C.F. : Production in montane dwarf shrub communities. - In :
HEAL, O.W., PERKINS, D.F. (ed.) : Production Ecology of British Moors and
Montane Grasslands. Pp. 263 - 276. Springer-Verlag, Berlin - Heidelberg -
New York 1978.

36038 - SUNDQVIST, C. : Red light stimulated accumulation of protochlorophyllide
in dark grown leaves treated with δ-aminolevulinic acid. - Plant Sci. Lett.
*12* : 69 - 76, 1978.

36039 - SUNDQVIST, C. : Factors influencing protochlorophyllide forms in plants. -
In : AKOYUNOGLOU, G., ARGYROUDI-AKOYUNOGLOU, J.H. (ed.) : Chloroplast Deve-
lopment. Pp. 77 - 82. Elsevier/North-Holland Biomedical Press, Amsterdam -
New York - Oxford 1978.

36040 - SUPONEVA, E.P., KISELEV, B.A., EVSTIGNEEV, V.B., EROKHIN, Yu.E. : Polyaro-
graficheskoe issledovanie khlorofilla v vodnykh rastvorakh detergentov i
v sostave khlorofill-belkovogo kompleksa. [Polarographic study of chlorophyll in aqueous solutions of detergents and in chlorophyll-protein complex.]
- Biofizika *23* : 441 - 444, 1978. [In R, ab : E.]

36041 - SÜSS, K.-H., DAMASCHUN, H., DAMASCHUN, G., ZIRWER, D. : Chloroplast coupling
factor $CF_1$ in solution. Small-angle X-ray scattering and circular dichroism
measurements. - FEBS Lett. *87* : 265 - 268, 1978.

36042 - SUZUKI, H., KOBAYASHI, H. : Quantum-chemical calculations on photoreceptor
pigments. - Photochem. Photobiol. *27* : 815 - 818, 1978. [Bacteriorhodopsin.]

36043 - SWADER, J.A., HOWE, C.M. : 1-($m$-$t$-butyacetamidophenyl)-3-methyl-3-methoxy
urea inhibition of photosystem II. - Plant Physiol. *61* (Suppl.) : 75, 1978.

36044 - SWANSON, C.A., HODDINOTT, J. : Effect of light and ontogenetic stage on
sink strength in bean leaves. - Plant Physiol. *62* : 454 - 457, 1978.

*36045 - SWEENEY, B.M. : Pros and cons of the membrane model for circadian rhythms
in the marine algae *Gonyaulax* and *Acetabularia*. - In : DECOURSEY, P.J.
(ed.) : Biological Rhythms in the Marine Environment. Pp. 63 - 76. Univ.
South Carolina Press, Columbia 1976. [Chloroplast.]

36046 - SWENBERG, C.E., GEACINTOV, N.E., BRETON, J. : Laser pulse excitation studies of the fluorescence of chloroplasts. - Photochem. Photobiol. *28* :
999 - 1006, 1978.

36047 - SYAROVA, Z.Ya., RËUTSKAYA, L.M., PADCHUFARAVA, G.M., GINTS, T.A. : Zmyanenne ab'ёmu khlaraplastaŭ u raslin zhyta, zarazhanykh *Fusarium nivale* (FR.)
CES. [Volume changes of chloroplasts of rye infected with *Fusarium nivale*
(FR.) CES.] - Vestsi Akad. Navuk belarus.SSR, Ser. biyal. Navuk *1978* (1) :
54 - 56, 140, 1978. [In Belorus., ab : E, R.]

36048 - **SZALAY, L.** : Problems of excitation transfer in systems of photosynthetic pigments. - In : **FIALA, J.** (ed.) : Third International Seminar on Energy Transfer in Condensed Matter. Pp. 29 - 39. Univerzita Karlova, Praha 1978.

36049 - **SZANIAWSKI, R.K., WIERZBICKI, B.** : Net photosynthetic rate of some coniferous species at diffuse high irradiance. - Photosynthetica *12* : 412 - 417, 1978.

36050 - **SZAREK, S.R., WOODHOUSE, R.M.** : Ecophysiological studies of Sonoran Desert plants. III. The daily course of photosynthesis for *Acacia greggii* and *Cercidium microphyllum*. - Oecologia *35* : 285 - 294, 1978.

36051 - **SZEWCZYK, E.** : Studies in the production of biomass in *Pellia borealis*. I. Characteristics of the material. - Acta biol. cracov., Ser. Bot. *21* : 75 - 84, 1978. [Ps, Chl, Car.]

36052 - **TACHIKAWA, H., FAULKNER, L.R.** : Electrochemical and solid state studies of phthalocyanine thin film electrodes. - J. amer. chem. Soc. *100* : 4379 - 4385, 1978.

36053 - **TÂCU, D., BALTAC, M., TUŞA, C.** : Caloric values in some healthy and *Tilletia*-species-infected wheat varieties. - Rev. roum. Biol., Sér. Biol. vég. *23* : 175 - 177, 1978.

36054 - **TAGEEVA, S.V., POPOV, V.I., ALLAKHVERDOV, B.L.** : Complex electron microscopic study of structural and functional arrangement of chloroplast membranes. - In : IX International Congress on Electron Microscopy. Vol. 2. Pp. 418 - 419. Toronto 1978.

36055 - **TAKAHAMA, U.** : Supression of lipid peroxidation by β-carotene in illuminated chloroplast fragments : Evidence for β-carotene as a quencher of singlet molecular oxygen in chloroplasts. - Plant Cell Physiol. *19* : 1565 - 1569, 1978.

36056 - **TAKAHASHI, E.** : [Environment and plant nutrition [7].] - Nogyo oyobi Engei (Agricult. Horticult.) *53* : 929 - 934, 1978. [Chloroplast; in Jap.]

36057 - **TAKAHASHI, E.** : [Environment and plant nutrition [8].] - Nogyo oyobi Engei (Agricult. Horticult.) *53* : 1062 - 1066, 1978. [Ps; in Jap.]

36058 - **TAKAHASHI, E.** : [Environment and plant nutrition [9].] - Nogyo oyobi Engei (Agricult. Horticult.) *53* : 1178 - 1184, 1978. [Ps, Chl; in Jap.]

36059 - **TAKAHASHI, E.** : [Nutrition of plants and environments.] - Nogyo oyobi Engei (Agricult. Horticult.) *53* : 1295 - 1300, 1978. [Ps; in Jap.]

36060 - **TAKAHASHI, E.** : [Environment and plant nutrition [11].] - Nogyo oyobi Engei (Agricult. Horticult.) *53* : 1427 - 1431, 1978. [Ps; in Jap.]

36061 - **TAKAHASHI, M., BARWELL-CLARKE, J., WHITNEY, F., KOELLER, P.** : Winter condition of marine plankton populations in Saanich Inlet, B.C., Canada. I. Phytoplankton and its surrounding environment. - J. exp. mar. Biol. Ecol. *31* : 283 - 301, 1978. [Chl.]

36062 - **TAKAHASHI, M., GROSS, E.L.** : Use of immobilized light-harvesting chlorophyll *a/b* protein to study the stoichiometry of its self-association. - Biochemistry *17* : 806 - 810, 1978.

36063 - **TAKAICHI, S., MAKINO, K., IWASAKI, A., SUZUKI, K.** : Photochemical reaction systems of photosynthesis in *Phytolacca americana*. III. Two ferredoxins from pokeweed, *Phytolacca americana* ; isolation and characterization. - J. Biochem. (Tokyo) *83* : 1151 - 1158, 1978.

36064 - **TAKANO, Y., TSUNODA, S.** : Curvilinear regression of the leaf photosynthetic rate on leaf nitrogen content among strains of *Oryza* species. - Jap. J. Breed. *21* : 69 - 76, 1971.

36065 - **TAKAOKI, T.** : A simple volumetric apparatus for measuring the rate of gas exchange in microorganisms. - Plant Cell Physiol. *19* : 61 - 70, 1978.

36066 - TAKASU, K., KIMURA, K. : Carbon-dioxide-microclimate in crop fields. - In :
MONSI, M., SAEKI, T. (ed.) : Ecophysiology of Photosynthetic Productivity.
JIBP Synthesis Vol. 19. Pp. 140 - 144. University of Tokyo Press, Tokyo 1978.

36067 - TAKEDA, G. : [Photosynthesis and dry-matter reproduction system in winter
cereals. I. Photosynthetic function.] - Bull. nat. Inst. agr. Sci., Ser. D
*29* : 1 - 65, 1978. [In Jap., ab : E.]

36068 - TAKEDA, G. : [Photosynthesis and dry-matter reproduction system in winter
cereals. II. Model simulation of dry-matter growth.] - Bull. nat.Inst. agr.
Sci., Ser. D *29* : 67 - 112, 1978. [In Jap., ab : E.]

36069 - TAKEDA, T., SUGIMOTO, H., AGATA, W. : [Water and crop production. I. The
relationship between photosynthesis and transpiration in corn leaf.] - Jap.
J. Crop Sci. *47* : 82 - 89, 1978. [In Jap., ab : E.]

36070 - TAKEDA, T., TSUCHIYA, M., AGATA, W. : [Studies on the effects of oxygen concentration on the photosynthesis and the growth of crop plants. IV. The
effect of subambient oxygen concentration during light or darkness on the
growth and the leaf expansion of rice plant and barnyard millet.] - Jap.
J. Crop Sci. *47* : 344 - 353, 1978. [In Jap., ab : E.]

36071 - TALARICO, L., KOSOVEL, V. : Properties and ultrastructure of R-phycoerythrin
from *Gracilaria verrucosa (Gigartinales, Florideae)* (HUDS.) PAPENFUSS. - Photosynthetica *12* : 369 - 347, 1978.

B36072 - TAMIYA, H. (ed.) : Summary Report on the Contribution of the Japanese National Committee for IBP, 1964 - 1974. JIBP Synthesis Vol.20.-University of Tokyo Press, Tokyo 1978.  [Ps.]

36073 - TAMKIVI, R.P., AVARMAA, R.A. : Proyavlenie neodnorodnoĭ struktury i vremen
relaksatsii v skorostyakh zatukhaniya fluorestsentsii khlorofilla. [Mani-
festation of heterogeneous structure and relaxation times in rates of chlorophyll fluorescence quenching.] - Izv. Akad. Nauk SSSR, Ser. fiz. *42* :
568 - 572, 1978. [In R.]

36074 - TAMURA, G., HOSOI, T., AKETAGAWA, J. : Ferredoxin-dependent sulfite reductase
from spinach leaves. - Agr. biol. Chem. (Tokyo) *42* : 2165 - 2167, 1978.

36075 - TAN, C.S., BLACK, T.A., NNYAMAH, J.U. : A simple diffusion model of transpiration applied to a thinned Douglas-fir stand. - Ecology *59* : 1221 - 1229,
1978. [Resistances.]

36076 - TAN, G.-Y., TAN, W.-K., WALTON, P.D. : Effects of temperature and irradiance
on seedling growth of smooth bromegrass. - Crop Sci. *18* : 133 - 136, 1978.
[Growth analysis.]

*36077 - TANAKA, A. : Photosynthesis and respiration in relation to productivity of
crops. - In : MITSUI, A., MIYACHI, S., SAN PIETRO, A., TAMURA, S. (ed.) :
Biological Solar Energy Conversion. Pp. 213 - 229. Academic Press, New York
- San Francisco - London 1977.

36078 - TANAS'EV, V.K., CHIMPOESH, G.P. : Svetovoĭ rezhim i ispol'zovanie ênergii
solnechnoĭ radiatsii nasazhdeniyami yabloni v zavisimosti ot podvoya i ploshchadi pitaniya. [Light regime and the utilization of the energy of solar
radiation by apple-tree stands in dependence on rootstock and nutritional
area.] - Tr. kishinev. sel'skokhoz. Inst. im. M.V. Frunze *1978* (Voprosy
Intensifikatsii Plodovodstva) : 59 - 63, 1978. [In R.]

*36079 - TANIGUCHI, A., KAWAMURA, T. : Primary production in the western tropical
and subtropical Pacific Ocean. - In : SUGAWARA, K. (ed.) : The Kuroshio II.
Proceedings of the Second Symposium on the Results of the Cooperative Study of the Kuroshio and Adjacent Regions. Pp. 159 - 168. Saikon Publ. Co.,
Tokyo 1972.

36080 - TARILA, A.G.I., ORMROD, D.P., ADEDIPE, N.O. : Stomatal responses of cowpea
(*Vigna unguiculata* L.) to light intensity. - Biochem. Physiol. Pflanz. *172* :
541 - 545, 1978.

B36081 - TARUSOV, B.N., VESELOVSKIĬ, V.A. : Sverkhslabye Svecheniya Rasteniĭ i Ikh
Prikladnoe Znachenie. [Ultra-weak Luminescence of Plants and Its Application.]-Izd. Moskovskogo Universiteta, Moskva 1978. [Chl; In R.]

36082 - **TATAKE, V.G.** : Techniques in the study of  photosynthesis. - In : Inter-Disciplinary Symposium on Photosynthesis and Productivity. Abstract of Papers.
Pp. 9 - 10. Indian nat. Sci. Acad., New Delhi 1978.

36083 - **TATKOWSKA, E., KOBYLAŃSKA, D.** : The efect of sodium humate on cultures of
*Spirodela polyrrhiza* (L.) SCHLEIDEN under aseptic conditions. - Ekol. pol.
*26* : 213 - 220, 1978. [Chl.]

36084 - **TATKOWSKA, E., TOPOROWSKA, E.** : The effect of detergents on cultures of
*Spirodela polyrrhiza* (L.) SCHLEIDEN under aseptic conditions. - Ekol. pol.
*26* : 221 - 229, 1978. [Chl.]

*36085 - **TAUBAEV, T.T., BERDYKULOV, Kh.A., SADYKOVA, A.Sh., AKHUNDOV, Kh.** : Vliyanie
pryamogo deĭstviya i posledeĭstviya IKSS na intensivnost' fotosinteza i produktivnost' stsenedesmusa. [The effect of direct and post-action of impulse
concentrated solar radiation on photosynthetic rate and productivity of *Scenedesmus*.] - In : Vodorosli i Griby Sredneĭ Azii. Vol.1. Pp. 83 - 87. Fan,
Tashkent 1974. [In R.]

36086 - **TAYLOR, B.H., FERREE, D.C.** : Influence of summer pruning on photosynthesis,
transpiration, dry weight and leaf area of apple trees. - HortScience *13* :
365, 1978.

36087 - **TAYLOR, G.E.,Jr.** : Plant and leaf resistance to gaseous air pollution stress.
- New Phytol. *80* : 523 - 534, 1978. [Ps, Chl, stomatal resistance.]

*36088 - **TAYLOR, M.K., MARZLOF, G.R.** : Benthic, planktonic, and total photosynthetic
production in the Kansas River. - Trans. Kansas  Acad. Sci. *79* : 56, 1976.

36089 - **TAYLOR, W.D., WILLIAMS, L.R., HERN, S.C., LAMBOU, V.W.** : Comparison of some
new and old indices and measurements of lake trophic state. - J. Phycol. *14*
(Suppl.) : 22, 1978. [Chl.]

36090 - **TAZAKI, T., USHIJIMA, T.** : General discussion of photosynthesis of single
leaves. - In : **MONSI, M., SAEKI, T.** (ed.) : Ecophysiology of Photosynthetic
Productivity. JIBP Synthesis Vol.19. Pp. 72 - 75. University of Tokyo Press,
Tokyo 1978.

36091 - **TAZUKE, S., KITAMURA, N.** : Photofixation of carbon dioxide to formic acid
*in vitro* using water as hydrogen source. - Nature *275* : 301 - 302, 1978.

*36092 - **TEERI, J.A., STOWE, L.G.** : Climatic patterns and the distribution of $C_4$ grasses in North America. - Oecologia *23* : 1 - 12, 1976.

36093 - **TEERI, J.A., STOWE, L.G., MURAWSKI, D.A.** : The climatology of two succulent
plant families : *Cactaceae* and *Crassulaceae*. - Can. J. Bot. *56* : 1750 - 1758,
1978. [Ps.]

36094 - **TELFER, A., BARBER, J.** : Dual action of ionophore A23187 on intact chloroplasts. - Biochim. biophys. Acta *501* : 94 - 102, 1978.

36095 - **TELFER, A., BARBER, J., HEATHCOTE, P., EVANS, M.C.W.** : Variable chlorophyll
*a* fluorescence from *P*-700 enriched Photosystem I particles dependent on the
redox state of the reaction centre. - Biochim. biophys. Acta *504*: 153 - 164,
1978.

36096 - **TEL-OR, E., LUIJK, L.W., PACKER, L.** : Hydrogenase in $N_2$-fixing cyanobacteria.
- Arch. Biochem. Biophys. *185* : 185 - 194, 1978.

*36097 - **TERESHKOVA, G.M.** : Issledovanie ustoĭchivosti khlorelly k temnovomu periodu,
ravnomu "lunnoĭ nochi". [Dark resistance of *Chlorella* to "moon night".] - In :
Intensivnaya Svetokul'tura Rasteniĭ. Pp. 214 - 227. Inst. Fiz. Sib. Otd.
Akad. Nauk SSSR, Krasnoyarsk 1977. [Chl; In R.]

36098 - **TERPSTRA, W.** : Chlorophyllase in *Phaeodactylum tricornutum* photosynthetic
membranes. Extractability, small-scale purification and molecular weight
determination by SDS-gel-electrophoresis. - Physiol. Plant. *44* : 329 - 334,
1978.

36099 - **TETLEY, R.M., BISHOP, N.I.** : Differential inhibition of photosynthesis, nitrogenase activity and $H_2$ uptake by metronidazole in blue-green algae. -
Plant Physiol. *61* (Suppl.) : 2, 1978.

36100 - TETT, P., GALLEGOS, C., KELLY, M.G., HORNBERGER, G.M., COSBY, B.J. : Relationships among substrate, flow, and benthic microalgal pigment density in the
Mechums River, Virginia. - Limnol. Oceanogr. 23 : 785 - 797, 1978.

36101 - TEVINI, M., HERM, K., UHRIG, H., IWANZIK, W. : Acyllipids and plastid development. - In : AKOYUNOGLOU, G., ARGYROUDI-AKOYUNOGLOU, J.H. (ed.) : Chloroplast Development. Pp. 827 - 835. Elsevier/North-Holland Biomedical Press,
Amsterdam - New York - Oxford 1978.

36102 - THAUER, R., SCHIRRMACHER, H., SCHYMANSKI, W., SCHÖNHEIT, P. : A rapid procedure for the purification of ferredoxin from spinach using polyethyleneimine.
- Z. Naturforsch. 33 C : 495 - 497, 1978.

36103 - THAYER, G.W., PARKER, P.L., LaCROIX, M.W., FRY, B. : The stable carbon isotope ratio of some components of an eelgrass, Zostera marina, bed. - Oecologia 35 : 1 - 12, 1978.

36104 - THEIMER, R.R., SCHUSTER, R. : Light-dependent inhibition of germination and
early seedling development of Borago officinalis. - Z. Pflanzenphysiol. 90 :
111 - 118, 1978. [Chl.]

36105 - THIBAULT, P. : A new attempt to study the oxygen evolving of photosynthesis:
determination of transition probabilities of a state $i$. - J. theor. Biol. 73 :
271 - 284, 1978.

36106 - THIBAULT, P. : Essai d'interprétation des séquences d'émission photosynthétique d'oxygène sous éclairs saturants: mise en évidence théorique d'une for-
te probabilité de double transition sous le premier éclair. - Compt. rend.
Acad. Sci. Paris, Sér. D 287 : 725 - 728, 1978.

36107 - THIHATMER, J. : Ermittlung und Quantifizierung von Ertragsfaktoren auf Grund-
wassersanden. - Arch. Acker- Pflanzenbau Bodenk. 22 : 495 - 499, 1978.

36108 - THIJSSE, T.R. : Gas chromatographic measurement of nitrous oxide and carbon
dioxide in air using electron capture detection. - Atmos. Environ. 12 :
2001 - 2003, 1978.

36109 - THIMANN, K.V. : Senescence. - Bot. Mag. (Tokyo) 1978 (Special Issue 1) : 19 -
43, 1978. [Chl.]

36110 - THINH, L.-V. : Photosynthetic lamellae of Prochloron (Prochlorophyta) associated with the ascidian Diplosoma virens (HARTMEYER) in the vicinity of
Townsville. - Aust. J. Bot. 26 : 617 - 620, 1978.

36111 - THOMAS, H. : Enzymes of nitrogen mobilization in detached leaves of Lolium
temulentum during senescence. - Planta 142 : 161 - 169, 1978. [Chl.]

36112 - THOMAS, H., DAVIES, A. : Effect of shading on the regrowth of Lolium perenne
swards in the field. - Ann. Bot. 42 : 705 - 715, 1978. [Ps.]

36113 - THOMAS, R.J., STANTON, D.S., LONGENDORFER, D.H., FARR, M.E. : Physiological
evaluation of the nutritional autonomy of a hornwort sporophyte. - Bot. Gaz.
139 : 306 - 311, 1978. [Ps.]

36114 - THOMAS, S.M., HALL, N.P., MERRETT, M.J. : Ribulose 1,5-bisphosphate carboxy-
lase/oxygenase activity and photorespiration during the ageing of flag leaves of wheat. - J. exp. Bot. 29 : 1161 - 1168, 1978.

36115 - THOMAS, S.M., LONG, S.P. : $C_4$ photosynthesis in Spartina townsendii at low
and high temperatures. - Planta 142 : 171 - 174, 1978.

36116 - THOMAS, S.M., THORNE, G.N., PEARMAN, I. : Effect of nitrogen on growth, yield
and photorespiratory activity in spring wheat. - Ann. Bot. 42 : 827 - 837,
1978.

36117 - THOMAS, W.H., DODSON, A.N., REID, F.M.H. : Diatom productivity compared to
other algae in natural marine phytoplankton assemblages. - J. Phycol. 14 :
250 - 253, 1978. [Ps.]

36118 - THORNBER, J.P., DUTTON, P.L., FAJER, J., FORMAN, A., HOLTEN, D., OLSON, J.M.,
PARSON, W.W., PRINCE, R.C., TIEDE, D.M., WINDSOR, M.W. : Isolated photoche-
mical reaction centers from bacteriochlorophyll b-containing organism. - In :
HALL, D.O., COOMBS, J., GOODWIN, T.W. (ed.) : Proceedings of the Fourth In-

ternational Congress on Photosynthesis. Pp. 55 - 70. Biochem. Soc., London 1978.

36119 - THORNBER, J.P., TROSPER, T.L., STROUSE, C.E. : Bacteriochlorophyll *in vivo* : Relationship of spectral forms to specific membrane components. - In : CLAYTON, R.K., SISTROM, W.R. (ed.) : The Photosynthetic Bacteria. Pp. 133 - 160. Plenum Press, New York - London 1978.

36120 - THORPE, M.R. : Net radiation and transpiration of apple trees in rows. - Agr. Meteorol. *19* : 41 - 57, 1978.

36121 - THORPE, N., BRADY, C.J., MILTHORPE, F.L. : Stomatal metabolism : primary carboxylation and enzyme activities. - Aust. J. Plant Physiol. *5* : 485 - 493, 1978.

*36122 - THROM, G. : Einfluß von Chinonen auf die lichtabhängige und die redoxabhängige Änderung des Membranpotentials bei *Griffithsia setacea* (ELLIS) AG. - Arch. Protistenk. *118* : 1 - 10, 1976.

36123 - THRONDSEN, J. : Productivity and abundance of ultra- and nanoplankton in Oslofjorden. - Sarsia *63* : 273 - 284, 1978.

36124 - THRONEBERRY, G.O., BOOTH, J.A., KASUNIC, D., McKINNEY, H. : Nitrogen and chlorophyll in crop plants exposed to sulfur dioxide. - Bull. N.M. agr. exp. Sta. *659* : 1 - 9, 1978.

*36125 - TIBBITTS, T.W., McFARLANE, J.C., KRIZEK, D.T., BERRY, W.L., HAMMER, P.A., LANGHANS, R.W., LARSON, R.A., ORMROD, D.P. : Radiation environment of growth chambers. - J. amer. Soc. hort. Sci. *101*: 164 - 170, 1976. [Comparison of sensors for measuring PhAR.]

*36126 - TIBBITTS, T.W., READ, M. : Rate of metabolite accumulation into latex of lettuce and proposed association with tipburn injury. - J. amer. Soc. hort. Sci. *101*: 406 - 409, 1976. [Photosynthates.]

36127 - TIBONI, O., PASQUALE, G.DI, CIFERRI, O. : Purification of the elongation factors present in spinach chloroplasts. - Europe. J. Biochem. *92* : 471 - 477, 1978.

36128 - TIBONI, O., PASQUALE, G.DI, CIFERRI, O. : Purification, characterization and site of synthesis of chloroplast elongation factors. - In : AKOYUNOGLOU, G., ARGYROUDI-AKOYUNOGLOU, J.H. (ed.) : Chloroplast Development. Pp. 675 - 678. Elsevier/North-Holland Biomedical Press, Amsterdam - New York - Oxford 1978.

36129 - TIEDE, D.M., LEIGH, J.S., DUTTON, P.L. : Structural organization of the *Chromatium vinosum* reaction center associated *c*-cytochromes. - Biochim. biophys. Acta *503* : 524 - 544, 1978.

36130 - TIEN, H.T. : Bilayer lipid membranes in aqueous media : incorporation of photosynthetic material from broken thylakoids. - In : METZNER, H. (ed.) : Photosynthetic Oxygen Evolution. Pp. 411 - 438. Academic Press, London - New York - San Francisco 1978.

36131 - TIEN, H.T. : Light transduction by pigmented bilayer lipid membranes. - Bioelectrochem. Bioenerg. *5* : 318 - 334, 1978.

36132 - TIESZEN, L.L. : Photosynthesis in the principal Barrow, Alaska, species : A summary of field and laboratory responses. - In : TIESZEN, L.L. (ed.) : Vegetation and Production Ecology of an Alaskan Arctic Tundra. Pp. 241 - 268. Springer-Verlag, New York - Heidelberg - Berlin 1978.

36133 - TIESZEN, L.L. : Summary. - In : TIESZEN, L.L. (ed.) : Vegetation and Production Ecology of an Alaskan Arctic Tundra. Pp. 621 - 646. Springer-Verlag, New York - Heidelberg - Berlin 1978. [Ps.]

B36134 - TIESZEN, L.L. (ed.) : Vegetation and Production Ecology of an Alaskan Arctic Tundra. Ecological Studies Vol. 29. - Springer-Verlag, New York - Heidelberg - Berlin 1978. [Ps, Chl.]

36135 - TIESZEN, L.L., IMBAMBA, S.K. : Gas exchange of finger millet inflorescences. - Crop Sci. *18* : 495 - 498, 1978.

36136 - **TIETEMA, T., VROMAN, J.** : Ecophysiology of the sand sedge, *Carex arenaria* L. I. Growth and dry matter distribution. - Acta bot. neerl. *27* : 161 - 173, 1978. [Photosynthates.]

*36137 - **TIKHOMIROV, A.A.** : Formirovanie struktury i fotosintez tsenozov pri ispol'- zovanii sveta razlichnoĭ intensivnosti v otdel'nykh oblastyakh FAR. [Formation of structure and photosynthesis of coenoses when light of different intensity in individual PhAR parts is used.] - In : Intensivnaya Svetokul'- tura Rasteniĭ. Pp. 58 - 80. Inst. Fiz. sib. Otd. Akad. Nauk SSSR, Krasnoyarsk 1977. [In R.]

*36138 - **TIKHOMIROV, A.A.** : Opticheskie kharakteristiki otdel'nykh fitoělementov i fitotsenozov pri svete razlichnoĭ intensivnosti v otdel'nykh oblastyakh FAR. [Optical characteristics of individual phytoelements and phytocoenoses in light of different intensity in individual parts of PhAR.] - In : Intensivnaya Svetokul'tura Rasteniĭ. Pp. 81 - 100. Inst. Fiz. sib. Otd. Akad. Nauk SSSR, Krasnoyarsk 1977. [In R.]

36139 - **TIKHONOV, A.N., PAVLOVA, I.E.** : Kinetika fotoindutsirovannykh okislitel'no- -vosstanovitel'nykh prevrashcheniĭ $P_{700}$ v list'yakh drevesnykh rasteniĭ, vyrashchennykh pri razlichnoĭ osveshchennosti. [Kinetics of photoinduced oxidative-reductive transformations of $P_{700}$ in the leaves of trees grown under different illuminance.] - Fiziol. Rast. *25* : 477 - 483, 1978. [In R, ab : E.]

36140 - **TIKHONOV, A.N., RUUGE, Ě.K.** : Issledovanie ělektronnogo transporta v foto- sinteticheskikh sistemakh metodom ĚPR. VII. Vliyanie temperatury na protses- sy perenosa ělektrona mezhdu fotosistemami i strukturnoe sostoyanie membra- ny khloroplastov. [ESR study of the electron transport in photosynthetic systems. VII. Effects of temperature on the processes of electron transport between two photosystems and the structural state of chloroplast membrane.] - Mol. Biol. (Moskva) *12* : 1028 - 1036, 1978. [In R, ab : E.]

36141 - **TILZER, M.M., GOLDMAN, C.R.** : Importance of mixing, thermal stratification and light adaptation for phytoplankton productivity in Lake Tahoe (California - Nevada). - Ecology *59* : 810 - 821, 1978. [Ps, Chl.]

36142 - **TIMOFEEV, M.M.** : Vliyanie prometrina i linurona na pigmentnuyu sistemu raste- niĭ kartofelya i red'ki dikoĭ pri razlichnykh usloviyakh pitaniya. [Effect of prometrin and linuron on pigment system of potato and wild radish plants under different nutrition.] - Fiziol. Biokhim. kul't. Rast. *10* : 171 - 176, 1978. [In R, ab : E.]

36143 - **TINDALL, D.R., YOPP, J.H., MILLER, D.M., SCHMID, W.E.** : Physico-chemical parameters governing the growth of *Aphanothece halophytica (Chroococcales)* in hypersaline media. - Phycologia *17* : 179 - 185, 1978. [Production.]

36144 - **TINGEY, D.T., RATSCH, H.C.** : Factors influencing isoprene emissions from live oak. - Plant Physiol. *61* (Suppl.) : 86, 1978. [Photosynthates.]

36145 - **TISCHNER, R., HEISE, K.-P., NELLE, R., LORENZEN, H.** : Changes in pigment content, lipid pattern and ultrastructure of synchronous *Chlorella* after heat and cold shocks. - Planta *139* : 29 - 33, 1978.

36146 - **TISHCHENKO, N.N.** : Ěffect "sverkhsinteza" azotistykh soedineniĭ na svetu v rasteniyakh razlichnykh ěkologicheskikh grupp, vyrashchennykh pri oprede- lennom defitsite azota. [Effect of "oversynthesis" of nitrogen compounds in light in plants of various ecological groups grown at a given nitrogen deficit.] - Tr. petergof. biol. Inst. *27* : 147 - 165, 1978. [Chl, Car; in R, ab : E.]

36147 - **TISHCHENKO, N.N., MAGOMEDOV, I.M.** : Opredelenie izofermentnogo sostava i aktivnosti alanin- i aspartataminotransferaz u vysshikh rasteniĭ. [Determination of isoenzyme composition and activity of aminotransferases in higher plants.] - Tr. priklad. Bot. Genet. Selektsii *61* (3) : 144 - 152, 1978. [Chloroplast; in R, ab : E.]

36148 - **TISHCHENKO, N.N., SOKOLOVA, E.N.** : Aktivnost' nitratreduktazy u rasteniĭ s $C_3$ i $C_4$ tipom fotosinteza, vyrashchennykh pri razlichnykh rezhimakh azotnogo pitaniya. [Activity of nitrate reductase in plants with $C_3$ and $C_4$ types of photosynthesis grown at various nitrogen nutrition.] - Tr. petergof. biol. Inst. *27* : 166 - 180, 1978. [In R, ab : E.]

36149 - **TITLYANOV, É.A.** : Metodicheskie podkhody k ékologo-fiziologicheskomu izucheniyu fotosinteza morskikh prikleplennykh vodorosleĭ. [Some approaches to eco-physiological study of photosynthesis in marine attached algae.] - Sb. Rabot Akad. Nauk SSSR, dal'nevost. nauch. Tsentr, Inst. Biol. Morya (Vladivostok) *11* (Ékologicheskie Aspekty Fotosinteza Morskikh Makrovodorosleĭ) : 5 - 20, 1978. [In R, ab : E.]

36150 - **TITLYANOV, É.A.** : Izuchenie fotosinteza zooksantell rifostroyashchikh korallov v pyatom reĭse NIS "Kallisto". [The 5th tropical trip of the research vessel "Callisto": Studies on photosynthesis of zooxanthellae in reef-building corals.] - Sb. Rabot Akad. Nauk SSSR, dal'nevost. nauch. Tsentr, Inst. Biol. Morya (Vladivostok) *12* [Biologiya Korallovykh Rifov (Fotosintez Zooksantell i Vodorosleĭ-Makrofitov)] : 7 - 14, 1978. [In R, ab : E.]

36151 - **TITLYANOV, É.A.** : Vliyanie sveta na soderzhanie i nativnoe sostoyanie khlorofillov v zelenykh vodoroslyakh tropicheskikh i umerennykh shirot. [Content and ratio of chlorophylls in green algae of tropical and moderate areas.] - Sb. Rabot Akad. Nauk SSSR, dal'nevost. nauch. Tsentr, Inst. Biol. Morya (Vladivostok) *12* [Biologiya Korallovykh Rifov (Fotosintez Zooksantell i Vodorosleĭ-Makrofitov)] : 75 - 82, 1978. [In R, ab : E.]

36152 - **TITLYANOV, É.A., KOLMAKOV, P.V., KOROBEĬNIKOVA, L.S.** : Dnevnye izmeneniya skorosti vidimogo i potentsial'nogo fotosinteza v techenie goda u nekotorykh benticheskikh vodorosleĭ Yaponskogo morya. [Seasonal and daily changes of visible and potential photosynthesis in some benthic algae of the sea of Japan.] - Sb. Rabot Akad. Nauk SSSR, dal'nevost. nauch. Tsentr, Inst. Biol. Morya (Vladivostok) *11* (Ékologicheskie Aspekty Fotosinteza Morskikh Makrovodorosleĭ) : 136 - 149, 186, 1978. [In R, ab : E.]

36153 - **TITLYANOV, E.A., KOLMAKOV, P.V., LEE, B.D., HORVÁTH, I.** : Functional states of the photosynthetic apparatus of the marine green alga *Ulva fenestrata* during the day. - Acta bot. Acad. Sci. hung. *24* : 167 - 177, 1978.

36154 - **TITLYANOV, É.A., LI, B. D.** : Adaptatsiya benticheskikh rasteniĭ k svetu. IV. Izmenenie soderzhaniya i sootnosheniya fotosinteticheskikh pigmentov zelenykh vodorosleĭ pri smene svetovogo rezhima. [Adaptation of benthic plants to light. IV. Variations of photosynthetic pigments contents and ratio in green algae under changing light conditions.] - Biol. Morya *1978* (4) : 36 - 41, 1978. [In R, ab : E.]

36155 - **TITLYANOV, É.A., ZVALINSKIĬ, V.I., LELËTKIN, V.A.** : Nekotorye mekhanizmy adaptatsii zooksantell korallov k svetu. [Some mechanisms of adaptation of corall zooxanthellae to light.] - Dokl. Akad. Nauk SSSR *238* : 1231 - 1234, 1978. [Ps, Car; in R.]

36156 - **TITOV, A.F., DROZDOV, S.N., OLIMPIENKO, G.S.** : Izuchenie korrelyatsii mezhdu khozyaĭstvenno-poleznymi i morfofiziologicheskimi priznakami u ovsyanitsy lugovoĭ. [Correlations between commercially valuable and morpho-physiological properties in the meadow fescue.] - Sel'skokhoz. Biol. *13* : 579 - 582, 1978. [Chl; in R, ab : E.]

36157 - **TITOV, A.F., OLIMPIENKO, G.S., PAVLOVA, N.A.** : O vozmozhnoĭ selektivnoĭ tsennosti temperaturo-chuvstvitel'nykh khlorofil'nykh mutatsiĭ ovsyanitsy lugovoĭ. [Possible selective value of temperature-sensitive chlorophyll mutations in *Festuca pratensis* HUDS.] - Zh. obshch. Biol. *39* : 628 - 632, 1978. [In R, ab : E.]

36158 - **TIȚU, H., SUSHROVA BORȘAN, I.** : Ultrastructure of peroxisomes in cucumber cotyledons (*Cucumis sativus* L.) treated with 60Co gamma irradiations. - Rev. roum. Biol., Sér. Biol. vég. *23* : 115 - 119, 1978.

36159 - **TOBIN, E.** : Light regulation of mRNA coding for the small subunit of RuBP carboxylase. - In : SIEGELMAN, H.W., HIND, G. (ed.) : Photosynthetic Carbon Assimilation. Pp. 425. - 426. Plenum Press, New York - London 1978.

36160 - **TOKUNAGA, F., EBREY, T.** : The blue membrane : The 3-dehydroretinal-based artificial pigment of the purple membrane. - Biochemistry *17* : 1915 - 1922, 1978.

*36161 - TOKUNAGA, F.,GOVINDJEE, R., EBREY, T.G., CROUCH, R. : Synthetic pigment analogues of the purple membrane protein. - Biophys. J. *19* : 191 - 198, 1977.

*36162 - TOLBERT, N.E. : Regulation of products of photosynthesis by photorespiration and reduction of carbon. - In : MITSUI, A., MIYACHI, S., SAN PIETRO, A., TAMURA, S. (ed.) : Biological Solar Energy Conversion. Pp. 243 - 263. Academic Press, New York - San Francisco - London 1977.

36163 - TOLLENAAR, M., DAYNARD, T.B. : Relationship between assimilate source and reproductive sink in maize grown in a short-season environment. - Agron. J. *70* : 219 - 223, 1978.

36164 - TOLLIN, G., RIZZUTO, F. : Effect of chloride on chlorophyll photochemistry in solution : enhancement of cation radical and semiquinone yields. - Photochem. Photobiol. *27* : 487 - 490, 1978.

36165 - TOŁWIŃSKA, M. : Produktywność i dynamika przyrostu biomasy netto zbiorowisk trawiastych łąk w Jaktorowie oraz zmiany w ich składzie florystycznym. [Productivity and dynamics of biomass net increment and exchanges of plant species in the grass communities of permanent meadows in Jaktorow. ]- Acta agrobot. *31* : 121 - 149, 1978. [In Pol., ab : E.]

36166 - TOMBESI, L., FAVOLA, G., MORETTI, R., BIONDI, A., BARONI, R. : Contributo allo studio del bilancio energetico delle piante nell'ambiente climatico di Roma. Nota II.  A contribution to the study of energetic balance of plants in the climatic environment of Rome. Note II.  - Ann. Ist. sperim. Nutr. Piante *8* (2) : 2 - 36, 1978. [PhAR; in E, Ital.]

36167 - TOMLINSON, J.A., WEBB, M.J.W. : Ultrastructural changes in chloroplasts of lettuce infected with beet western yellows virus. - Physiol. Plant Pathol. *12* : 13 - 18, 1978.

36168 - TOMOMATSU, A., ASAHI, T. : Non-synchronous increases in activities of peroxisomal enzymes in etiolated mung bean seedling leaves after illumination. - Plant Cell Physiol. *19* : 183 - 188, 1978. [Chl.]

36169 - TOURNIER, P., ESPINASSE, A., GERSTER, R. : Décarboxylation par $H_2O_2$ de céto-acides en relation avec le métabolisme de la photorespiration. - Compt. rend. Acad. Sci. Paris, Sér. D *287* : 729 - 732, 1978.

36170 - TOWE, K.M. : Early precambrian oxygen : a case against photosynthesis. - Nature *274* : 657 - 661, 1978.

36171 - TOWNE, C.A., BARTELS, P.G., HILTON, J.L. : Interaction of surfactant and herbicide tratments on single cells of leaves. - Weed Sci. *26* : 182 - 188, 1978. [Ps.]

36172 - TRACHTENBERG, S., ZAMSKI, E. : Conduction of ionic solutes and assimilates in the leptom of *Polytrichum juniperinum* WILLD. - J. exp. Bot. *29* : 719 - 727, 1978.

36173 - TREBST, A. : Plastoquinones in photosynthesis. - Phil. Trans. roy. Soc. London, Ser. B *284* : 591 - 599, 1978.

36174 - TREBST, A. : Organization of the photosynthetic electron transport system of chloroplasts in the thylakoid membrane. - In : SCHÄFER, G., KLINGENBERG, M. (ed.) : Energy Conservation in Biological Membranes. Pp. 84 - 95. Springer-Verlag, Berlin - Heidelberg - New York 1978.

36175 - TREBST, A., WIETOSKA, H., DRABER, W., KNOPS, H.J. : The inhibition of photosynthetic electron flow in chloroplasts by the dinitrophenylether of bromo- or iodo-nitrothymol. - Z. Naturforsch. *33 C* : 919 - 927, 1978.

36176 - TREDWELL, C.J., SYNOWIEC, J.A., SEARLE, G.F., PORTER, G., BARBER, J. : Picosecond time resolved fluorescence of chlorophyll *in vivo*. - Photochem. Photobiol. *28* : 1013 - 1020, 1978.

36177 - TREFFRY, T. : Biogenesis of the photochemical apparatus. - Int. Rev. Cytol. *52* : 159 - 196, 1978.

36178 - TREGUBENKO, M.Ya., FILIPPOV, G.L., VISHNEVSKIĬ, N.V. : Osobennosti vodnogo
rezhima i dykhaniya razlichnykh po zasukhoustoĭchivosti gibridov kukuruzy
v usloviyakh nedostatochnogo vodoobespecheniya. [Peculiarities of water regime and respiration of maize hybrids different in drought-resistance under
conditions of unsufficient water supply.] - Fiziol. Biokhim. kul't. Rast.
*10* : 257 - 263, 1978. [In R, ab : E.]

36179 - TRENCH, R.K., POOL, R.R. Jr., LOGAN, M., ENGELLAND, A. : Aspects of the relation between *Cyanophora paradoxa* (KORSCHIKOFF) and its endosymbiotic cyanelles *Cyanocyta korschikoffiana* (HALL & CLAUS). I. Growth, ultrastructure,
photosynthesis and the obligate nature of the association. - Proc. roy.
Soc. London, Ser. B *202* : 423 - 443, 1978.

36180 - TRENCH, R.K., RONZIO, G.S. : Aspects of the relation between *Cyanophora*
*paradoxa* (KORSCHIKOFF) and its endosymbiotic cyanelles *Cyanocyta korschi-*
*koffiana* (HALL & CLAUS). II. The photosynthetic pigments. - Proc. roy. Soc.
London, Ser. B *202* : 445 - 462, 1978.

36181 - TRENCH, R.K., SIEBENS, H.C. : Aspects of the relation between *Cyanophora*
*paradoxa* (KORSCHIKOFF) and its endosymbiotic cyanelles *Cyanocyta korsci-*
*koffiana* (HALL & CLAUS). IV. The effects of rifampicin, chloramphenicol
and cycloheximide on the synthesis of ribosomal ribonucleic acids and chlorophyll. - Proc. roy. Soc. London, Ser. B *202* : 473 - 482, 1978.

*36182 - TREVISAN, R. : Fluttuazione stagionale della densità e della biomassa
fitoplanctonica del Lago Trasimeno (luglio 1976 - agosto 1977). [Seasonal
fluctuations in density and biomass of phytoplankton of Lake Trasimeno
(July 1976 - August 1977).] - Riv. Idrobiol. *16* : 297 - 331, 1977. [In Ital.,
ab : E.]

36183 - TRIBOI-BLONDEL, A.-M. : Effets de différents régimes d'alimentation hydri-
que sur l'activité *in vivo* de la nitrate-réductase dans les feuilles de
Dactyle. - Compt. rend. Acad. Sci. Paris, Sér. D *286* : 1795 - 1798, 1978.
[Ps.]

36184 - TRIMBOLI, D., FAHY, P.C., BAKER, K.F. : Apical chlorosis and leaf spot of
*Tagetes* spp. caused by *Pseudomonas tagetis* HELLMERS. - Aust. J. agr. Res.
*29* : 831 - 839, 1978.

*36185 - TRIPATHY, B.C., MURTY, K.S. : Glycine decarboxylation by excised leaf tis-
sues of rice. - Indian J. exp. Biol. *14* : 714 - 716, 1976. [Ps.]

36186 - TROXLER, R.F., BROWN, A.S., KÖST, H.-P. : Quantitative degradation of ra-
diolabeled phycobiliproteins. Chromic acid degradation of C-phycocyanin. -
Europe.J. Biochem. *87* : 181 - 189, 1978.

36187 - TROXLER, R.F., OFFNER, G.D. : $\delta$-aminolevulinic acid synthesis in a pigment-
less mutant of *Cyanidium caldarium*. - Plant Physiol. *61* (Suppl.) : 84, 1978.

36188 - TRÜPER, H.G. : Sulfur metabolism. - In : CLAYTON, R.K., SISTROM, W.R. (ed.)
: The Photosynthetic Bacteria. Pp. 677 - 690. Plenum Press, New York -
London 1978. [Ps.]

36189 - TRÜPER, H.G., PFENNIG, N. : Taxonomy of the *Rhodospirillales*. - In :
CLAYTON, R.K., SISTROM, W.R. (ed.) : The Photosynthetic Bacteria. Pp. 19 -
27. Plenum Press, New York - London 1978. [Ps.]

36190 - TSEL'NIKER, Yu.L. : Replikatsiya khloroplastov, ee regulyatsiya i znachenie dlya fotosinteza. [Chloroplasts replication, its regulation and significance for photosynthesis.] - In : Mezoatruktura i Funktsional'naya Aktivnost' Fotosinteticheskogo Apparata. Pp. 31 - 45. Ural'skiĭ gosudarstvennyĭ
Universitet, Sverdlovsk 1978. [In R.]

B36191 - TSEL'NIKER, Yu.L. : Fiziologicheskie Osnovy Tenevynoslivosti Drevesnykh
Rasteniĭ. [Physiological Bases of Shade Tolerance of Woody Plants.] - Nauka, Moskva 1978. [Ps; in R.]

36192 - TSENOVA, E.N., FEDINA, I.S. : On the light regulation of NADP-dependent gly-
ceraldehyde-3-phosphate dehydrogenase in greening pea seedlings. - Dokl.
bolg. Akad. Nauk *31* : 241 - 241 - 244, 1978.

36193 - TSIMILLI-MICHAEL, M., AKOYUNOGLOU, G. : The 520 nm light-induced absorbance
         changes in developing bean leaves : Correlation with composition, structure
         and function of the photosynthetic apparatus. - In : AKOYUNOGLOU, G., ARGY-
         ROUDI-AKOYUNOGLOU, J.H. (ed.) : Chloroplast Development. Pp. 525 - 532.
         Elsevier/North-Holland Biomedical Press, Amsterdam - New York - Oxford
         1978.

36194 - TSIVION, Y. : Loading of assimilates and some sugars into the translocation
         system of Cuscuta. - Aust. J. Plant Physiol. 5 : 851 - 857, 1978.

36195 - TSUJI, H., ISA, Y., HATAKEYAMA, I. : Changes in two parameters characteri-
         zing the light-photosynthesis curve of growing bean leaves. - In : MONSI,
         M., SAEKI, T. (ed.) : Ecophysiology of Photosynthetic Productivity. JIBP
         Synthesis, Vol. 19. Pp. 46 - 54. University of Tokyo Press, Tokyo 1978.

36196 - TSUJI, H., NAITO, K., HATAKEYAMA, I. : Effect of benzyladenine on the chan-
         ges in two parameters of the light-photosynthesis curve of bean leaves du-
         ring aging. - In : MONSI, M., SAEKI, T. (ed.) : Ecophysiology of Photosyn-
         thetic Productivity. JIBP Synthesis, Vol. 19. Pp. 55 - 58. University of
         Tokyo Press, Tokyo 1978.

*36197 - TSUKADA, O., KAWAHARA, T. : Mass culture of Chlorella in Asian countries. -
         In : MITSUI, A., MIYACHI, S., SAN PIETRO, A., TAMURA, S. (ed.) : Biological
         Solar Energy Conversion. Pp. 363 - 365. Academic Press, New York - San Fran-
         cisco - London 1977.

36198 - TSUNO, Y., HIRAYAMA, T. : Analysis of light-factor and photosynthesis in
         citrus trees. - In : MONSI, M., SAEKI, T. (ed.) : Ecophysiology of Photo-
         synthetic Productivity. JIBP Synthesis, Vol. 19. Pp. 100 - 111. University
         of Tokyo Press, Tokyo 1978.

36199 - TSUNO, Y., TAKEUCHI, Y., HIRAO, M. : [Photosynthesis, respiration, and re-
         spiration/photosynthesis ratio of melon plants grown under growth cabinet
         conditions.] - Bull. Sand Dune Res. Inst. Tottori Univ. 17 : 11 - 18, 1978.
         [In Jap., ab : E.]

36200 - TULBU, G.V., REMENNIKOV, S.M., KRENDELEVA, T.E. : Izuchenie kinetiki foto-
         indutsirovannykh izmeneniĭ pogloshcheniya v oblasti 520 nm pri vozbuzhdenii
         lazernoĭ vspyshkoĭ v usloviyakh funktsionirovaniya fotosistemy I i fotosis-
         temy II khloroplastov gorokha. [Kinetic of photoinduced changes in absorban-
         ce at 520 nm under laser flash excitation at functioning of photosystems I
         and II in pea chloroplasts.] - Nauch. Dokl. vyssheĭ Shkoly,biol. Nauki 21
         (10) : 39 - 44, 1978. [In R.]

*36201 - TUMIDAJOWICZ, D. : Biomass dynamics and primary production of the herb layer
         in a beech forest Fagetum carpaticum in the Gorce Mts. (Western Carpathians).
         - Bull. Acad. pol. Sci., Sér. Sci. biol. Cl. II. 24 : 341 - 348, 1976.

36202 - TURCOTTE, E.L., FEASTER, C.V. : Inheritance of three genes for plant color
         in American Pima cotton. - Crop Sci. 18 : 149 - 150, 1978.

36203 - TURITZIN, S.N. : Canopy structure and potential light competition in two
         adjacent annual plant communities. - Ecology 59 : 161 - 167, 1978.

36204 - TURNER, C.H.C., EVANS, L.V. : Translocation of photoassimilated $^{14}C$ in the
         red alga Polysiphonia lanosa. - Brit.phycol. J. 13 : 51 - 55, 1978.

36205 - TURNER, N.C., BEGG, J.E. : Responses of pasture plants to water deficits. -
         In : WILSON, J.R. (ed.) : Plant Relations in Pastures. Pp. 50 - 66. C.S.I.R.
         O., Melbourne 1978. [Ps.]

36206 - TURNER, N.C., BEGG, J.E., LORRAINE TONNET, M. : Osmotic adjustment of sorg-
         hum and sunflower crops in response to water deficits and its influence on
         the water potential at which stomata close. - Aust. J. Plant Physiol. 5 :
         597 - 608, 1978.

36207 - TURNER, N.C., BEGG, J.E., RAWSON, H.M., ENGLISH, S.D., HEARN, A.B. : Agro-
         nomic and physiological responses of soybean and sorghum crops to water de-
         ficits. III. Components of leaf water potential, leaf conductance, $^{14}CO_2$
         photosynthesis, and adaptation to water deficits. - Aust. J. Plant Physiol.
         5 : 179 - 194, 1978.

36208 - **TURNER, R.E.** : Variability of the reverse-flow concentration technique of measuring plankton respiration. - Estuaries *1* : 65 - 68, 1978. [Chl.]

36209 - **TYAGI, V.V.S., AHLUWALIA, A.S.** : Heterocyst formation in the blue-green alga *Anabaena doliolum*.- A study of some aspects of photoregulation. - Ann. Bot. *42* : 1333 - 1342, 1978. [Chl.]

36210 - **TYANKOVA, L.A.** : Dependence of the resistance to cold of biological membranes on the molecular structure of the amino acids. - Dokl. bolg. Akad. Nauk *31* : 1465 - 1468, 1978. [Ps.]

36211 - **TYNISSOO, V., TAMKIVI, R.** : Sravnenie metodov ochistki khlorofilla po intensivnosti svecheniya postoronnikh primeseĭ. [Comparison of the methods of chlorophyll purification on the basis of the emission intensity of contaminating impurities.] - Izv. Akad. Nauk ēst.SSR, Khim. *1978* (4) : 219 - 224, 1978. [In R, ab : E, Estonian.]

36212 - **UCHIJIMA, Z., UDAGAWA, T.** : Carbon dioxide environment and $CO_2$-transfer above and within crop canopies - Measurements and simulation. - In : **MONSI, M., SAEKI, T.** (ed.) : Ecophysiology of Photosynthetic Productivity. JIBP Synthesis, Vol. 19. Pp. 129 - 139. University of Tokyo Press, Tokyo 1978.

36213 - **UDOVENKO, G.V., GONCHAROVA, Ė.A., KORZH, B.V., DOBREN'KOVA, L.G.** : Vliyanie plodov na fotosinteticheskiĭ apparat rasteniĭ pri normal'nykh i neblagopriyatnykh usloviyakh. [Effect of fruits on photosynthetic apparatus of plants under normal and unfavourable conditions.] - Fiziol. Biokhim. kul't. Rast. *10* : 184 - 189, 1978. [In R, ab : E.]

26214 - **UENO, T., SASAKI, K.** : Light dependency of the mating process in *Closterium acerosum*. - Plant Cell Physiol. *19* : 245 - 252, 1978. [Ps.]

36215 - **UHEDA, E., KURAISHI, S.** : The relationship between transpiration and chlorophyll synthesis in etiolated squash cotyledons. - Plant Cell Physiol. *19* : 825 - 831, 1978.

36216 - **UHLMANN, D.** : The upper limit of phytoplankton production as a function of nutrient load, temperature, retention time of the water, and euphotic zone depth. - Int. Rev. ges. Hydrobiol. *63* : 353 - 363, 1978.

36217 - **UHRING, J.** : Leaf anatomy of petunia in relation to pollution damage. - J. amer. Soc. hort. Sci. *103* : 23 - 27, 1978. [Chl.]

36218 - **ULLIMAN, J.J., HATCH, C. R.** : Test of an electronic planimeter. - J. Forest. *76* : 346 - 347, 1978.

36219 - **ULLRICH-EBERIUS, C.I., NOVACKY, A., LÜTTGE, U.** : Active hexose uptake in *Lemna gibba* G 1. - Planta *139* : 149 - 153, 1978. [Ps.]

36220 - **ULRICH, E.L., MARKLEY, J.L., KROGMANN, D.W.** : The structure of plastocyanin from NMR spectroscopy. - In : **HALL, D.O., COOMBS, J., GOODWIN, T.W.** (ed.) : Proceedings of the Fourth International Congress on Photosynthesis. Pp. 815 - 819. Biochem. Soc., London 1978.

36221 - **UMPELEV, V.L., BORISOVA, I.S.** : Opredelenie ploshchadi list'ev u ogurtsov. [Determination of leaf area in cucumbers.] - In : Mezostruktura i Funktsional'naya Aktivnost' Fotosinteticheskogo Apparata. Pp. 145 - 147. Ural'skiĭ gosudarstvennyĭ Universitet, Sverdlovsk 1978. [In R.]

36222 - **UPADHYA, M.D.** : Genotypic variability in photosynthesis efficiency and it exploitation through breeding. - In : Inter-Disciplinary Symposium on Photosynthesis and Productivity. Abstract of Papers. P. 47. Indian nat. Sci. Acad., New Delhi 1978.

36223 - **URBANOVICH, T.A., IVANCHANKA, V.M.** : Dzeyanne analagaŭ kafeinu na fasfarylyuyuchuyu aktyŭnasts' khlaraplastaŭ. [Effect of caffeine analogues on the phorphorylating activity of chloroplasts.] - Vestsi Akad. Navuk belarus. SSR, Ser. biyal. Navuk *1978* (1) : 14 - 16, 137, 1978. [In Belorus., ab : E, R.]

36224 - URBANOVICH, T.A., MIKUL'SKAYA, S.A. : Aktyŭnasts' rĕaktsyi fotafasfarylya-
vannya pry roznykh rĕzhymakh vodazabespyachĕnnya raslin. [Activity of the
photophosphorylation reaction under different conditions of water supply
to plants.] - Vestsi Akad. Navuk belarus.SSR, Ser. biyal. Navuk *1978* (2) :
48 - 51, 1978. [In Belorus., ab : E.]

36225 - USHIJIMA, T., TAZAKI, T. : Photosynthetic activity and water metabolism in
some higher plants. - In : MONSI, M., SAEKI, T. (ed.) : Ecophysiology of
Photosynthetic Productivity. JIBP Synthesis, Vol. 19. Pp. 37 - 46. University of Tokyo Press, Tokyo 1978.

36226 - USIK, G.E., BUSHKIĬ, V.D. : Vliyanie usloviĭ vyrashchivaniya na rost i plo-
donoshenie baklazhanov. [Effect of growing conditions on growth and yield
of fruits in egg-plant.] - In : Agrotekhnicheskie Priemy Promyshlennoĭ Tekhnologii v Ovoshchevodstve. Pp. 28 - 30. Min. sel'. Khoz. SSSR, Kishinev
1978. [Ps, Chl; In R.]

36227 - UTSUNOMIYA, H., YAMAGATA, M. : [Scanning electron microscopy of the endosperm of cereal crops. 6. Development of starch granules in the endosperm
cells of non-glutinous rice kernel, *Japonica* type.] - Bull. Fac. Agr. Yama-
guti Univ. *29* : 73 - 87, 1978. [In Jap.]


36228 - VACEK, K. : Model systems and their extrapolation to the photosynthesis. -
In : FIALA, J. (ed.) : Third International Seminar on Energy Transfer in
Condensed Matter. Pp. 11 - 18. Univerzita Karlova, Praha 1978.

36229 - VACEK, K., VAVŘINEC, E., NAUŠ, J. : Polarization spectra of chlorophyll-*a*
in oriented polymer films. - In : FIALA, J. (ed.) : Third International
Seminar on Energy Transfer in Condensed Matter. Pp. 141 - 146. Univerzita
Karlova,Praha 1978.

36230 - VAKLINOVA, S., KUEN, M. : Aktivnost na nitritreduktazata, s"d"rzhanie na
belt"k i dobiv pri bezligulni i ligulni formi tsarevitsa. [Nitrite reductase activity, protein content and yield in liguleless and ligule-possessing
maize forms.] - Fiziol. Rast. (Sofia) *4* (1) : 5 - 9, 1978. [Chl; In Bulg.,
ab : E, R.]

36231 - VALADON, L.R.G., MUMMERY, R.S. : Effects of two triethylamines on carotenogenesis of Turkish lemons and oranges. - Z. Pflanzenphysiol. *90* : 11 - 19,
1978.

36232 - VALADON, L.R.G., MUMMERY, R.S. : Carotenoid synthesis in Turkish lemons and
oranges as influenced by triethylamine derivatives. - Phytochemistry *17* :
818 - 819, 1978.

36233 - VALANNE, N. : The turnover rate of chlorophyll-protein complexes during
prolonged darkness. - In : AKOYUNOGLOU, G., ARGYROUDI-AKOYUNOGLOU, J.H.
(ed.) : Chloroplast Development. Pp. 241 - 244. Elsevier/North-Holland Biomedical Press, Amsterdam - New York - Oxford 1978.

36234 - VALANNE, N., ARO, E.-M., REPO, E. : Changes in photosynthetic capacity and
activity of RuBPC-ase and glycolate oxidase during the early growth of moss
protonemata in continuous and rhythmic light. - Z. Pflanzenphysiol. *88* :
123 - 131, 1978.

36235 - VALCKE, R. : Influence of relative humidity and age on the development of
the photosynthetic apparatus in barley (*Hordeum vulgare* c.v. Union). - In :
AKOYUNOGLOU, G., ARGYROUDI-AKOYUNOGLOU, J.H. (ed.) : Chloroplast Development. Pp. 871 - 874. Elsevier/North-Holland Biomedical Press, Amsterdam -
New York - Oxford 1978.

36236 - VALLEJOS, R.H., LESCANO, W.I.M., LUCERO, H.A. : Involvement of an essential
arginyl residue in the coupling activity of *Rhodospirillum rubrum* chromatophores. - Arch. Biochem. Biophys. *190* : 578 - 584, 1978.

36237 - VAN ARKEL, H. : Leaf area determinations in sorghum and maize by the length-
-width method. - Neth. J. agr. Sci. *26* : 170 - 180, 1978.

36238 - **VAN BESOUW, A., WINTERMANS, J.F.G.M.** : Galactolipid formation in chloroplast envelopes I. Evidence for two mechanisms in galactosylation. - Biochim. biophys. Acta *529* : 44 - 53, 1978.

36239 - **VAN BEST, J.A., MATHIS, P.** : Kinetics of reduction of the oxidized primary electron donor of Photosystem II in spinach chloroplasts and in *Chlorella* cells in the microsecond and nanosecond time ranges following flash excitation. - Biochim biophys. Acta *503* : 178 - 188, 1978.

36240 - **VANCE, W.A., STUMPF, P.K.** : Fat metabolism in higher plants. The elongation of saturated and unsaturated acyl-CoAs by a stromal system from isolated spinach chloroplasts. - Arch. Biochem. Biophys. *190* : 210 - 220, 1978.

36241 - **VANDERHOVEN, C., ZRŸD, J.-P.** : Changes in malate content and in enzymes involved in dark $CO_2$ fixation during growth of *Acer pseudoplatanus* cells in suspension culture. - Physiol. Plant. *43* : 99 - 103, 1978.

36242 - **VAN DER PLOEG, R.R., BEESE, F., STREBEL, O., RENGER, M.** : The water balance of a sugar beet crop : a model and some experimental evidence. - Z. Pflanzenernähr. Bodenk. *141* : 313 - 328, 1978. [Growth analysis.]

36243 - **VAN GINKEL, G.** : Formation of (proteo)lipid vesicles by means of a French pressure cell. - Carnegie Inst. Year Book *77* : 294 - 297, 1978. [Chl.]

36244 - **VAN GINKEL, G., BROWN, J.** : Light-induced proton uptake in liposomes containing photosystem I reaction centers. - Plant Physiol. *61* (Suppl.) : 103, 1978.

36245 - **VAN GINKEL, G., BROWN, J.S.** : Endogenous catalase and superoxide dismutase activities in photosynthetic membranes. - FEBS Lett. *94* : 284 - 286, 1978. [Chl.]

36246 - **VAN GORKOM, H.J., PULLES, M.P.J., ETIENNE, A.-L.** : Fluorescence and absorbance changes in tris-washed chloroplasts. - In : METZNER, H. (ed.) : Photosynthetic Oxygen Evolution. Pp. 135 - 145. Academic Press, London - New York - San Francisco 1978.

36247 - **VAN GRONDELE, R., HOLMES, N.G., RADEMAKER, H., DUYSENS, L.N.M.** : Bacteriochlorophyll fluorescence of purple bacteria at low redox potentials. The relationship between reaction center triplet yield and the emission yield. - Biochim. biophys. Acta *503* : 10 - 25, 1978.

36248 - **VAN LIERE, L., LOOGMAN, J.G., MUR, L.R.** : Measuring light-irradiance in cultures of phototrophic micro-organisms. - FEMS Microbiol. Lett. *3* : 161 - 164, 1978.

36249 - **VANNINI, G.L., FASULO, M.P., BRUNI, A., DALL'OLIO, G.** : Structural and developmental aspects during the greening process of *Euglena gracilis* treated with myomycin. - Protoplasma *96* : 335 - 349, 1978.

36250 - **VAN RENSEN, J.J.S., WONG, D., GOVINDJEE** : Characterization of the inhibition of photosynthetic electron transport in pea chloroplasts by the herbicide 4,6-dinitro-*o*-cresol by comparative studies with 3-(3,4-dichlorophenyl)-1,1-dimethylurea. - Z. Naturforsch. *33 C* : 413 - 420, 1978.

*36251 - **VARFOLOMEEV, S.D., ZAĬTSEV, S.V., BELOGUROVA, N.G., BEREZIN, I.V., NIKITINA, K.A., GUSEV, M.V.** : Issledovanie fotovosstanovleniya ėkzogennykh aktseptorov ėlektrona kletkami sine-zelenykh vodoroslei. Mediatornyĭ mekhanizm ėlektronnogo transporta cherez biomembranu. [Photoreduction of exogenous electron acceptors by blue-green algae. Mediator mechanism of electron transport through a biomembrane.] - Bioorg. Khim. *2* : 1395 - 1403, 1976. [In R, ab : E.]

36252 - **VARSHNEY, K.A., BAIJAL, B.D.** : An analysis of growth of some salt-stressed grasses. - Comp. Physiol. Ecol. *3* : 233 - 236, 1978.

36253 - **VASIL'EV, B.R., LEBSKIĬ, V.K., MIROSLAVOVA, S.A., TISHCHENKO, N.N.** : Osobennosti stroeniya i metabolizma list'ev *Bryophyllum daigremontianum* BERGER v raznykh usloviyakh azotnogo pitaniya. [Characteristics of the structure and metabolism of *Bryophyllum daigremontianum* BERGER leaves under different nitrogen nutrition.] - Tr. petergof. biol. Inst. *27* (Voprosy Ėkologicheskoĭ Anatomii i Fiziologii Rasteniĭ) : 51 - 60, 1978. [Ps; in R.]

36254 - VASIL'EV, B.R., SHMIDT, V.M. : Ob allometricheskom kharaktere svyazi mezhdu parametrami rastushchego lista *Bryophyllum daigremontianum* BERGER. [On the allometric character of the correlation between the parameters of growing leaf of *Bryophyllum daigremontianum* BERGER.] - Bot. Zh. *63* : 1449 - 1456, 1978. [In R.]

36255 - VASSILIOU, G., MÜLLER, F. : Metabolismus von Metoxuron bei unterschiedlich empfindlichen Kultur-Umbelliferen. - Med. Fac. Landbouw. Rijksuniv. Gent *43* : 1181 - 1191, 1978. [Ps.]

*36256 - VATER, J., RENGER, G., STIEHL, H.H., WITT, H.T. : Intermediates and kinetics in the water splitting part of photosynthesis. - Naturwissenschaften *55* : 220 - 221, 1968.

36257 - VAUGHN, K.C., FIELDS, M.B., WILSON, K.G. : Correlation between photosystem deficiency and chloroplast ultrastructure in *Mimulus* and *Hosta*. - Plant Physiol. *61* (Suppl.) : 103, 1978.

36258 - VAUGHN, K.C., WILSON, K.G., STEWART, K.D. : Light-harvesting pigment-protein complex deficiency in *Hosta* (*Liliaceae*). - Planta *143* : 275 - 278, 1978.

*36259 - VAULIN, A.V. : Registratsiya svetovykh kompensatsionnykh punktov yachmenya v polevykh usloviyakh. [Recording of light compensation points of barley in field conditions.] - Sb. Tr. nauch.-issled. Inst. sel'sk. Khoz. tsentr. Raĭ. nechernoz. Zony *36* (Intensifikatsiya Zemledeliya) : 113 - 116, 1975. [In R.]

36260 - VAVŘINEC, E., URBANOVÁ, M. : A possibility of the interpretation of the chlorophyll-$a$ absorption band at 615 nm. - In : FIALA, J.(ed.) : Third International Seminar on Energy Transfer in Condensed Matter. Pp. 147 - 151. Univerzita Karlova, Praha 1978.

36261 - VECHAR, A.S., RASHĚTNIKAŬ, U.M., LEMYAZA, Z.F., MAS'KO, A.A. : Zmyanenne kol'kastsi nukleinavykh kislot u pratsěse prarastannya nasennya i farmiravannya fotasintězuyuchaĭ sistěmy prarostkaŭ. [Changes in the content of nucleic acids during seed germination and formation of the photosynthetic system in seedlings.] - Vestsi Akad. Navuk belarus.SSR, Ser. biyal. Navuk *1978* (5) : 29 - 32, 138, 1978. [In Belorus., ab : E, R.]

36262 - VECHER, A.S., NENADOVICH, R.A., MAS'KO, A.A., RESHETNIKOV, V.N. : Lipidy i plastokhinony khloroplastov kartofelya i rzhi. [Lipids and plastoquinones of potato and rye chloroplasts.]- Fiziol. Biokhim. kul't. Rast. *10* : 269 - 275, 1978. [In R, ab : E.]

36263 - VELTHUYS, B., KOK, B. : Photosynthetic oxygen evolution from hydrogen peroxide. - Biochim. biophys. Acta *502* : 211 - 221, 1978.

36264 - VELTHUYS, B., KOK, B. : Observations on the $O_2$ evolution system. - In : HALL, D.O., COOMBS, J., GOODWIN, T.W. (ed.) : Proceedings of the Fourth International Congress on Photosynthesis. Pp. 397 - 407. Biochem. Soc., London 1978.

36265 - VELTHUYS, B.R. : A third site of proton translocation in green plant photosynthetic electron transport. - Proc. nat. Acad. Sci. USA *75* : 6031 - 6034, 1978.

36266 - VENKATESH, C.S., ARYA, R.S., THAPLIYAL, R.C. : An albina-type natural chlorophyll mutant in *Gmelina arborea* ROXB. - Silvae Genet. *27* : 40 - 41, 1978.

36267 - VENKATESWARLU, B. : Accumulated and current photosynthetic contribution to grain yield in rice (*Oryza sativa* L.) - In : Inter-Disciplinary Symposium on Photosynthesis and Productivity. Abstract of Papers. Pp. 42 - 43. Indian nat. Sci. Acad., New Delhi 1978.

36268 - VENKATESWARLU, S., SINGH, R.M., SINGH, R.B., SINGH, B.D. : Radiosensitivity and frequency of chlorophyll mutations in pigeon pea. - Indian J. Genet. Plant Breed. *38* : 90 - 94, 1978.

36269 - VERASAN, V., PHILLIPS, R.E. : Effects of soil water stress on growth and nutrient accumulation in corn. - Agron. J. *70* : 613 - 618, 1978.

36270 - VERBELEN, J.P., MOEREELS, E., SPRUYT, E., DE GREEF, J.A. : Influence of the
         length of the etiolation period on the photosynthetic rate of bean leaves. -
         Arch. int. Physiol. Biochim. 86 : 894, 1978.

36271 - VERDUIN, J., ILMAVIRTA, V. : A comparison of $^{14}$C-based and ΔpH-based esti-
         mates of phytoplankton production in the brown-water lake Pääjärvi, southern
         Finland. - Ann. bot. fenn. 15 : 27 - 31, 1978.

36272 - VERMEGLIO, A., BRETON, J., PAILLOTIN, G., COGDELL, R. : Orientation of chro-
         mophores in reaction centers of Rhodopseudomonas sphaeroides : a photose-
         lection study. - Biochim. biophys. Acta 501 : 514 - 530, 1978.

36273 - VERWER, W., VERVERGAERT, P.H.J.T., LEUNISSEN-BIJVELT, J., VERKLEIJ, A.J. :
         Particle aggregation in photosynthetic membranes of the blue-green alga
         Anacystis nidulans. - Biochim. biophys. Acta 504 : 231 - 234, 1978.

36274 - VIDAL, J., JACQUOT, J.P., MEMBRE, H., GADAL, P. : Detection and study of
         protein factors involved in dithiothreitol activation of NADP-malate dehy-
         drogenase from a $C_4$ plant. - Plant Sci. Lett. 11 : 305 - 310, 1978.

*36275 - VIDOVIČ, J., POKORNÝ, V. : Významnosť rôznych listových skupín pre tvorbu
         výnosu zrna u kukurice. [The importance of various leaf groups for the for-
         mation of grain yield in maize.] - Rost. Výroba (Praha) 18 : 175 - 186,
         1972. [In Slovak, ab : E, R.]

36276 - VIERKE, G. : Kinetics of deactivation of the charged state formed upon illu-
         mination on the oxidizing site of photosystem II in the presence of DCMU in
         Chlorella after extraction of membrane-bound manganese by $NH_2OH$. - In :
         METZNER, H. (ed.) : Photosynthetic Oxygen Evolution. Pp. 345 - 370. Acade-
         mic Press, London - New York - San Francisco 1978.

*36277 - VIERKE, G., STRUCKMEIER, P. : Binding of copper (II) to proteins of the
         photosynthetic membrane and its correlation with inhibition of electron
         transport in class II chloroplasts of spinach. - Z. Naturforsch. 32 C :
         605 - 610, 1977.

36278 - VIERKE, G., STRUCKMEIER, P. : Inhibition of millisecond luminescence by
         copper(II) in spinach chloroplasts. - Z. Naturforsch. 33 C : 266 - 270,
         1978.

36279 - VIERLING, E., ALBERTE, R.S. : Functional organization and plasticity of the
         photosynthetic unit in Cyanobacteria. - Plant Physiol. 61 (Suppl.) : 87,
         1978.

36280 - VIIL, J., PÄRNIK, T. : On the control of the glycolate pathway by light
         and oxygen. - Z. Pflanzenphysiol. 88 : 219 - 226, 1978.

36281 - VIL'YAMS, M.V., RUMYANTSEVA, V.B. : Svetovoĭ gazoobmen i éffektivnost'
         ispol'zovaniya uglekisloty na postroenie biomassy v poseve rasteniĭ pshe-
         nitsy. [A study of gas exchange in the light and effectivity of $CO_2$ utili-
         zation in biomass formation in wheat planting.] - Fiziol. Rast. 25 : 681 -
         687, 1978. [In R, ab : E.]

36282 - VINKLER, C., AVRON, M., BOYER, P.D. : Initial formation of ATP in photophos-
         phorylation does not require a proton gradient. - FEBS Lett. 96 : 129 - 134,
         1978.

36283 - VINKLER, C., ROSEN, G., BOYER, P.D. : Light-driven ATP formation from $^{32}P_i$
         by chloroplast thylakoids without detectable labeling of ADP, as measured
         by rapid mixing and acid quench techniques. - J. biol. Chem. 253 : 2507 -
         2510, 1978.

36284 - VINTILĂ, R., SORAN, V., CRĂCIUN, C., FABIAN, A. : Efectul cloramfenicolului
         asupra ultrastructurii   cloroplastului de grîu. [Effect of chlorampheni-
         col on the ultrastructure of wheat chloroplast.] - Stud. Cercet. Biol., Sér.
         Biol. vég. 30 : 109 - 112, 1978. [In Roum., ab : E.]

36285 - VIOLLIER, M., BAUSSART, N., LECOMTE, P. : Interpretation de la signature
         spectrale des eaux marines. - In : Proceedings of an International Confe-
         rence on Earth Observation from Space and Management of Planetary Resour-
         ces. Pp. 81 - 88. Toulouse 1978. [Chl.]

36286 - **VIOLLIER, M., DESCHAMPS, P.Y., LECOMTE, P.** : Airborne remote sensing of chlorophyll content under cloudy sky as applied to the tropical waters in the Gulf of Guinea. - Remote Sens. Environ. 7 : 235 - 248, 1978.

36287 - **VIOLLIER, M., LECOMTE, P., BOUGARD, M., RICHARD, A.** : Expérience aéroportée de télédétection (température et couleur de la mer) dans le détroit du Pas-de Calais. - Oceanol. Acta 1 : 265 - 269, 1978. [Chl.]

36288 - **VĪTOLA, Ā.** : Ogļhidrātu fondi un to saistība ar augu produktivitāti un adaptāciju. [Carbohydrate reserves and their link with plant productivity and the process of adaptation to the light factor.] - Latv. PSR Zināt. Akad. Vēstis 1978 (1) : 39 - 46, 1978. [In Latvian.]

*36289 - **VITOLA, A.K.** : Reaktsiya fondov uglevodov v protsesse adaptatsii k snizheniyu intensivnosti osveshcheniya. [Response of carbohydrates resources in the process of adaptation to lower irradiance.] - In : Adaptatsiya Fiziologo-Biokhimicheskikh Sistem Rasteniya k Peremene Osveshcheniya. Vol. I. Pp. 32 - 43. Zinatne, Riga 1977. [In R.]

36290 - **VITOLA, A.K., GROSA, V.F.** : Uroven' uglevodov v list'yakh i tempy protsessa adaptatsii rasteniĭ ogurtsa k snizhennoĭ intensivnosti sveta. [Level of carbohydrates in leaves and the rate of adaptation of cucumber plants to lowered light intensity.] - In : Fiziologo-biokhimicheskie Issledovaniya Rasteniĭ. Pp. 28 - 37, 155 - 156. Zinatne, Riga 1978. [In R.]

36291 - **VLADIMIROVA, M.G.** : Ul'trastrukturnaya organizatsiya kletki *Dunaliella salina* i ee funktsional'nye izmeneniya v zavisimosti ot intensivnosti sveta i temperatury. [Ultrastructural organization of the *Dunaliella salina* cell and its functional changes depending on the light intensity and temperature.] - Fiziol. Rast. 25 : 571 - 576, 1978. [In R, ab : E.]

36292 - **VOELSKOW, H., SCHÖN, G.** : Pyruvate fermentation in light-grown cells of *Rhodospirillum rubrum* during adaptation to anaerobic dark conditions. - Arch. Microbiol. 119 : 129 - 133, 1978. [Chl.]

36293 - **VOGEL, P.** : Untersuchungen über Kürbiskernöl. - Fette, Seifen, Anstrichmittel 80 : 315 - 317, 1978. [Chl.]

36294 - **VOGELMANN, T.C., SCHEIBE, J.** : Action spectra for chromatic adaptation in the blue-green alga *Fremyella diplosiphon*. - Planta 143 : 233 - 239, 1978.

36295 - **VOITURIEZ, B., HERBLAND, A.** : Étude de la production pélagique de la zone équatoriale de l'Atlantique à 4°W I - Relations entre la structure hydrologique et la production primaire. - Cah. ORSTOM,Sér. Océanogr. 15 : 313 - 331, 1977. [Chl.]

36296 - **VOLFOVÁ, A., CHVOJKA, L., FRIEDRICH, A.** : The effect of kinetin and auxin on the chloroplast structure and chlorophyll content in wheat coleoptiles. - Biol. Plant. 20 : 440 - 445, 1978.

B36297 - **VOL'KENSHTEĬN, M.V.** : Obshchaya Biofizika. [General Biophysics.] - Nauka, Moskva 1978. [Ps, pigments; in R.]

*36298 - **VOLKMAR, R.D.** : Primary productivity in relation to chemical parameters in Cheat Lake,West Virginia. - Proc. West Virginia Acad. Sci. 44 : 14 - 22, 1972.

36299 - **VOLKOVA, M.A., MOTKALYUK, O.B.** : Intensivnost' dykhaniya zlakovykh rasteniĭ v kriticheskiĭ k nedostatku vlagi period. [Respiration intensity in cereals during the period critical to moisture deficiency.] - Fiziol. Rast. 25 : 1244 - 1250, 1978. [In R, ab : E.]

36300 - **VOLODARSKIĬ, N.I., BYSTRYKH, E.E., NIKOLAEVA, E.K.** : Fotosinteticheskaya aktivnost' verkhnego lista pshenitsy u sortov razlichnoĭ produktivnosti. [Photosynthetic activity of the upper leaf of wheat in cultivars with different productivity.] - Sel'skokhoz. Biol. 13 : 703 - 710, 1978. [In R, ab : E.]

36301 - **VOLODARSKIĬ, N.I., NIKOLAEVA, E.K.** : Reaktsiya fotovosstanovleniya NADF v svyazi s usloviyami azotnogo pitaniya v ontogeneze dvukh sortov pshenitsy razlichnoĭ produktivnosti. [NADP photoreduction in relation to nitrogen supply during ontogenesis of two wheat varieties of different productivity.]

- Dokl. VASKHNIL *1978* (10) : 1 - 4, 1978. [In R.]

36302 - **VOLYNETS, A.P., PROKHORCHIK, R.A.** : Fenol'nye soedineniya khloroplastov gorokha. [Phenolic compounds in pea chloroplasts.] - Fiziol. Rast. *25* : 778 -
782, 1978. [In R, ab : E.]

36303 - **VONG, N.Q., MURATA, Y.** : The effect of air temperature on the dark respiration and nutrient absorption of $C_3$ and $C_4$ crop species. - Jap. agr. Res.
Quart. *12* : 64 - 68, 1978.

36304 - **VONG, N.Q., MURATA, Y.** : Studies on the physiological characteristics of $C_3$
and $C_4$ crop species II. The effects of air temperature and solar radiation
on the dry matter production of some crops. - Jap. J. Crop Sci. *47* : 90 -
100, 1978.

36305 - **VOROB'EVA, L.M., ABROS'KINA, L.S., KVITKO, K.V., KRASNOVSKIĬ, A.A.** : Lyumi-
nestsentsiya khlorofilla v mutantakh *Scenedesmus obliquus*. [Chlorophyll luminescence in *Scenedesmus obliquus* mutants.] - Fiziol. Rast. *25* : 341 - 349,
1978. [In R, ab : E.]

36306 - **VOROB'EVA, L.M., KRASNOVSKIĬ, A.A.** : Obratimye izmeneniya sostoyaniya pig-
menta pri osveshchenii list'ev. [Reversible changes in the pigment state
during the illumination of leaves.] - Dokl. Akad. Nauk SSSR *236* : 1243 -
1246, 1977. [In R.]

36307 - **VORONKOV, L.A., PEROVA, I.A.** : Fotosinteticheskiĭ apparat rasteniĭ pri pa-
togeneze. [Photosynthetic apparatus of plants in pathogenesis.] - Sel'skokhoz. Biol. *13* : 683 - 694, 1978. [In R, ab : E.]

*36308 - **VORONKOVA, N.M., SEMKIN, B.I.** : Raspredelenie mechenykh assimilyatov u luka
i chesnoka v rannie fazy razvitiya. [Distribution of labelled assimilates
in onion and garlic in early phases of development.] - In : Fiziologicheskie i Biokhimicheskie Issledovaniya Rasteniĭ na Dal'nem Vostoke. Pp. 1 - 7.
Sib. Otd. Akad. Nauk SSSR, Biol.-pochv. Inst., Vladivostok 1970. [In R.]

36309 - **VOSKOBOĬNIKOV, G.M., TITLYANOV, É.A.** : Izuchenie anatomii i ul'trastruktury
krasnoĭ vodorosli *Grateloupia turuturu* iz razlichnykh po osveshchennosti
mest obitaniya. [Anatomy and ultrastructure of the red alga *Grateloupia
turuturu* from habitats of different illumination.] - Sb. Rabot Akad. Nauk
SSSR, dal'nevost. nauch. Tsentr, Inst. Biol. Morya (Vladivostok) *11* (Ėkologicheskie Aspekty Fotosinteza Morskikh Makrovodorosleĭ) : 83 - 87, 184 - 185,
1978. [In R, ab : E.]

36310 - **VOZILOVA, L.D., STVOLOVA, A.P., LUGOVTSOVA, K.A.** : Formirovanie fotokhimicheskikh sistem khloroplastov u prorostkov sosny. [Formation of photochemical systems in the chloroplasts of pine seedlings.] - Fiziol. Rast. *25* :
118 - 122, 1978. [In R, ab : E.]

*36311 - **VOZNYAK, V.M., KIM, V.A., EVSTIGNEEV, V.B.** : Izuchenie metodom ĖPR pervich-
no vosstanovlennykh form khlorofilla i feofitina. [EPR study of primary reduced chlorophyll and pheophytin forms.] - In : Itogi Issledovaniya Mekhanizma Fotosinteza. Pp. 33 - 46. Pushchino 1974. [In R.]

36312 - **VOZVYSHAEVA, L.V., GOL'DFEL'D, M.G., TSAPIN, A.I.** : Vliyanie khelatorov na
reaktsiyu Khilla v khloroplastakh s kremniĭmolibdatom v kachestve aktsepto-
ra ėlektronov. [Effect of chelators on Hill reaction in chloroplasts with
silicomolybdate as electron acceptor.] - Biofizika *23* : 918 - 920, 1978.
[In R, ab : E.]

*36313 - **VRANSKI, V., PETROVA, R., RADENKOV, S.** : Optichen transmisionen planimet'r
za opredelyane ploshchi na golemi lista. [An optical transmission planimeter
for determination of large leaf areas.] - Rast. Nauki *12* (10) : 3 - 8, 1975.
[In Bulg., ab : E, R.]

36314 - **VREDENBERG, W.J., SCHAPENDONK, A.H.C.M.** : Evidence for a light-induced
blue band shift of part of the *P*515 pigment pool in intact chloroplasts. -
FEBS Lett. *91* : 90 - 93, 1978.

36315 - **VRKOČ, F.** : Některé růstové charakteristiky a složky výnosů pšenice a jarního ječmene. [Some growth characteristics and yield components in wheat
and spring barley.] - Rost. Výroba (Praha) *24* : 1277 - 1284, 1978.
[Growth analysis; in Czech, ab E, G, R.]

36316 - VRUBLEVSKAYA, K.G., ZAĬTSEVA, T.A., MANDEL', T.E. : Fotokhimicheskaya aktivnost' khloroplastov pshenitsy v protsesse zeleneniya na svetu razlichnogo
spektral'nogo sostava. [Photochemical activity of chloroplasts during the
greening process in wheat plants in the light of various spectral composition.] - Fiziol. Rast. 25 : 1109 - 1114, 1978. [In R, ab : E.]

36317 - VSEVOLODOV, N.N., CHEKULAEVA, L.N. : Spektral'nye prevrashcheniya fotopro-
duktov v kletkakh Halobacterium halobium. [Spectral transformations of pho-
toproducts in Halobacterium halobium cells.] - Biofizika 23 : 99 - 104, 1978.
[In R, ab : E.]

36318 - VSEVOLODOV, N.N., CHEKULAEVA, L.N. : Spektral'nye prevrashcheniya v purpurnykh membranakh iz Halobacterium halobium : vliyanie perekhoda 560↔570
i sinego sveta na fotokhimicheskie protsessy. [Spectral transformations in
purple membranes of Halobacterium halobium. Effect of the transition 560↔
570 and blue light on photochemical processes.] - Biofizika 23 : 1019 -
1023, 1978. [In R, ab : E.]

36319 - VYAS, L.N., SHRIMAL, R.L., JINDAL, K. : Plant biomass and net production
relations of Anogeissus pendula EDGEW. at deciduous forest near Udaipur
(Rajasthan), India. - Flora 167 : 457 - 465, 1978.


*36320 - WAFAR, M.V.M., QUASIM, S.Z. : Carbon fixation & excretion in symbiotic algae
(Zooxanthellae) in the presence of host homogenates. - Indian J. mar. Sci.
4 : 43 - 46, 1975.

36321 - WAGNER, G. : Halobacterial potassium transport in orange and near UV light.
- In : CAPLAN, S.R., GINZBURG, M. (ed.) : Energetics and Structure of Halo-
philic Microorganisms. Pp. 335 - 340. Elsevier/North-Holland Biomedical
Press, Amsterdam - New York 1978. [Bacteriorhodopsin.]

36322 - WAGNER, G., HARTMANN, R., OESTERHELT, D. : Potassium uniport and ATP synthesis in Halobacterium halobium. - Europe. J. Biochem. 89 : 169 - 179,
1978.

36323 - WAGNER, G., KLEIN, K. : Differential effect of calcium on chloroplast movement in Mougeotia. - Photochem. Photobiol. 27 : 135 - 140, 1978.

36324 - WAGNER, W., FOLLMANN, H., SCHMIDT, A. : Multiple functions of thioredoxins.
- Z. Naturforsch. 33 C : 517 - 520, 1978.

36325 - WAHUA, T.A.T., MILLER, D.A. : Leaf water potentials and light transmission
of intercropped sorghum and soybeans. - Exp. Agr. 14 : 373 - 380, 1978.

36326 - WALCZAK, T. : Application of two independent light beams for measurements
of transmission changes corresponding to chloroplast movements in leaves. -
Acta protozool. 18 : 217 - 218, 1978.

36327 - WALDEN, R., LEAVER, C.J. : Regulation of chloroplast protein synthesis during germination and early development of cucumber (Cucumis sativus). -
In : AKOYUNOGLOU, G., ARGYROUDI-AKOYUNOGLOU, J.H. (ed.) : Chloroplast Development. Pp. 251 - 256. Elsevier/North-Holland Biomedical Press, Amsterdam -
New York - Oxford 1978.

36328 - WALK, R.-A., HOCK, B. : Cell-free synthesis of glyoxysomal malate dehydrogenase. - Biochem. biophys. Res. Commun. 81 : 636 - 643, 1978.

36329 - WALKER, A.J., HO, L.C., BAKER, D.A. : Carbon translocation in the tomato :
pathways of carbon metabolism in the fruit. - Ann. Bot. 42 : 901 - 909,
1978.

*36330 - WALKER, D.A. : Regulatory mechanisms in photosynthetic carbon metabolism. -
Curr. Top. cell. Reg. 11 : 203 - 241, 1976.

36331 - WALKER, D.A., ROBINSON, S.P. : Regulation of photosynthetic carbon assimilation. - In : SIEGELMAN, H.W., HIND, G. (ed.) : Photosynthetic Carbon
Assimilation. Pp. 43 - 59. Plenum Press, New York - London 1978.

36332 - **WALKER, R.B., REED, K.L., SHUMWAY, J.** : Effects of nitrogen supply and shading on the photosynthetic capacity of Douglas fir (*Pseudotsuga menziesii*).
- Plant Physiol. *61* (Suppl.) : 51, 1978.

36333 - **WALLACE, L.L., HARRISON, A.T.** : Carbohydrate mobilization and movement in
alpine plants. - Amer. J. Bot. *65* : 1035 - 1040, 1978.

36334 - **WALLENTINUS, I.** : Productivity studies on Baltic macroalgae. - Bot. mar.
*21* : 365 - 380, 1978.

36335 - **WALLIHAN, E.F., SHARPLESS, R.G., PRINTY, W.L.** : Cumulative toxic effects of
boron, lithium, and sodium in water used for hydroponic production of tomatoes. - J. amer. Soc. hort. Sci. *103* : 14 - 16, 1978. [Chl.]

36336 - **WALLIN, R., SELSET, R., SLETTEN, K.** : Characterization of chromophoric peptides from C-phycocyanin. - Biochem. biophys. Res. Commun. *81* : 1319 - 1328,
1978.

36337 - **WALSBY, A.E.** : The gas vesicles of aquatic prokaryotes. - Symp. Soc. gen.
Microbiol. *28* (Relations between Structure and Function in the Prokaryotic
Cell) : 327 - 358, 1978.

36338 - **WALTER, W.M. Jr., PURCELL, A.E., McCOLLUM, G.K.** : Laboratory preparation of
a protein-xanthophyll concentrate from sweet potato leaves. - J. agr. Food
Chem. *26* : 1222 - 1226, 1978.

36339 - **WANG, L.K., VIELKIND, D., WANG, M.H.** : Mathematical models of dissolved
oxygen concentration in fresh water. - Ecol. Model. *5* : 115 - 123, 1978.

36340 - **WANG, W.** : Photoactivation of the chlorophyll pathway in the absence of chlorophyll synthesis. - Plant Physiol. *61* (Suppl.) : 82, 1978.

36341 - **WANG, W.** : Effect of dim light on the *y-1* mutant of *Chlamydomonas reinhardtii*. - Plant Physiol. *61* : 842 - 846, 1978.

36342 - **WANG, W.-Y.** : Genetic control of chlorophyll biosynthesis in *Chlamydomonas
reinhardtii*. - Int. Rev. Cytol. *8* (Suppl.) : 335 - 354, 1978.

36343 - **WANG BANG-XI, DU YUAN-SHOU, QI MING-QI, WANG BAO-MIN** : [Physiological changes of wheat under the dry-hot-wind condition II. Effect of dry-hot-wind
on the $^{14}CO_2$-assimilation and accumulation of $^{14}C$-assimilates during grain
filling period in wheat.] - Acta bot. sin. *20* : 37 - 43, 1978. [In Chin.,
ab : E.]

36344 - **WANN, M., RAPER, C.D. Jr., LUCAS, H.L. Jr.** : A dynamic model for plant
growth : A simulation of dry matter accumulation for tobacco. - Photosynthetica *12* : 121 - 136, 1978.

36345 - **WAREING, P.F.** : Determination in plant development. - Bot. Mag. (Tokyo)
*1978* (Spec.Issue 1): 3 - 17, 1978. [Production.]

36346 - **WARING, R.H., EMMINGHAM, W.H., GHOLZ, H.L., GRIER, C.C.** : Variation in maximum leaf area of coniferous forests in Oregon and its ecological significance. - Forest Sci. *24* : 131 - 140, 1978. [Ps.]

36347 - **WARRINGTON, I.J., DIXON, T., ROBOTHAM, R.W., ROOK, D.A.** : Lighting systems
in major New Zealand controlled environment facilities. - J. agr. Eng. Res.
*23* : 23 - 36, 1978. [Photosynthetic irradiance.]

36348 - **WARRINGTON, I.J., EDGE, E.A., GREEN, L.M.** : Plant growth under high radiant
energy fluxes. - Ann. Bot. *42* : 1305 - 1313, 1978.

36349 - **WARSHEL, A.** : Charge stabilization mechanism in visual and purple membrane
pigments. - Proc. nat. Acad. Sci. USA *75* : 2558 - 2562, 1978.

36350 - **WARTENBERG, D.E.** : Spectrophotometric equations : An intercalibration technique. - Limnol. Oceanogr. *23* : 566 - 570, 1978. [Chl.]

36351 - **WASHITANI, I., SATO, S.** : Studies on the function of proplastids in the metabolism of *in vitro* cultured tobacco cells V. Primary transamination. -
Plant Cell Physiol. *19* : 43 - 50, 1978.

36352 - WASIELEWSKI, M.R. : Excited and ionic states of dimeric chlorophyll derivatives. Biomimetic modelling of the primary events of photosynthesis. - In :
DUTTON, P.L., LEIGH, J.S., SCARPA, A. (ed.) : Frontiers of Biological Energetics : Electrons to Tissues. Vol. 1. Pp. 63 - 72. Academic Press, New York
- San Francisco - London 1978.

36353 - WASIELEWSKI, M.R., SVEC, W.A., COPE, B.T. : Bis(chlorophyll)cyclophanes.
New models of special pair chlorophyll. - J. amer. chem. Soc. *100* : 1961 -
1962, 1978.

36354 - WASIELEWSKI, M.R., THOMPSON, J.F. : 9-Desoxo-9,10-dehydrochlorophyll *a*,
a new chlorophyll with an effective 20-PI electron macrocycle. - Tetrahedron Lett. *1978* (12) : 1043 - 1046, 1978.

36355 - WASYLEWSKI, Z., BIELAŃSKI, W. Jr., WIĘCKOWSKI, S. : Heterogeneity of the
thylakoid membrane fractions derived from spinach chloroplasts by the action
of *Triton X-100* in low salt medium. I. Isolation, molecular weights and photochemical activities of two chlorophyll-protein complexes. - Photosynthetica *12* : 185 - 192, 1978.

36356 - WATANABE, T., MIYASAKA, T., FUJISHIMA, A., HONDA, K. : Photoelectrocnemical
study on chlorophyll monolayer electrodes. - Chem. Lett. *4* : 443 - 446,
1978.

36357 - WATERTON, J.C., SANDERS, J.K.M. : Hyperfine coupling in chlorophyll radical
cations. A nuclear magnetic resonance approach. - J. amer. chem. Soc. *100* :
4044 - 4049, 1978.

36358 - WATTS, W.R., NEILSON, R.E. : Photosynthesis in Sitka spruce (*Picea sitchensis* (BONG.) CARR.). VIII. Measurements of stomatal conductance and $CO_2$ uptake in controlled environments. - J. appl. Ecol. *15* : 245 - 255, 1978.

36359 - WEARE, N.M. : The photoproduction of $H_2$ and $NH_4^+$ fixed from $N_2$ by a depressed mutant of *Rhodospirillum rubrum*. - Biochim. biophys. Acta *502* : 486 -
494, 1978.

36360 - WEBB, W., SZAREK, S., LAUENROTH, W., KINERSON, R., SMITH, M. : Primary productivity and water use in native forest, grassland, and desert ecosystems.
- Ecology *59* : 1239 - 1247, 1978.

36361 - WEBBER, P.J. : Spatial and temporal variation of the vegetation and its
productivity. - In : TIESZEN, L.L. (ed.) : Vegetation and Production Ecology of an Alaskan Arctic Tundra. Ecological Studies 29. Pp. 37 - 112.
Springer-Verlag, New York - Heidelberg - Berlin 1978.

36362 - WEBER, J.A., TENHUNEN, J.D., YOCUM, C.S., GATES, D.M. : Inhibition of photosynthesis in *Egeria densa* at high concentration of inorganic carbon. -
Plant Physiol. *61* (Suppl.) : 101, 1978.

36363 - WEBSTER, G.D., JACKSON, J.B. : Affinity chromatography of $H^+$-translocating
adenosine triphosphatase isolated by chloroform extraction of *Rhodospirillum rubrum* chromatophores. Modification of binding affinity by divalent cations and activating anions. - Biochim. biophys. Acta *503* : 135 - 154, 1978.

36364 - WEGMANN, K. : Osmotic regulation in *Dunaliella* - facts and questions. -
In : CAPLAN, S.R., GINZBURG, M. (ed.) : Energetics and Structure of Halo-
philic Microorganisms. Pp. 615 - 618. Elsevier/North-Holland Biomedical
Press, Amsterdam - New York 1978. [Ps.]

36365 - WEGMANN, K., PRISTAVU, N. : Influence of cycloserine on the photosynthetic
$^{14}C$ incorporation into amino acids in various algae. - Biochem. Physiol.
Pflanz. *173* : 86 - 90, 1978.

36366 - WEIDNER, M., BURCHARTZ, N. : Inhibition of phosphoenolpyruvate carboxylase
by formulated herbicides and anionic detergents. - Biochem. Physiol. Pflanz.
*173* : 381 - 389, 1978.

36367 - WEINSTEIN, J.D., CASTELFRANCO, P.A. : Mg-protoporphyrin-IX and δ-aminolevulinic acid synthesis from glutamate in isolated greening chloroplasts.
δ-aminolevulinic acid synthesis. - Arch. Biochem. Biophys. *186* : 376 - 382,
1978.

36368 - **WEISS, P.W.** : Reproductive efficiency and growth of *Emex australis* in relation to stress. - Aust. J. Ecol. *3* : 57 - 65, 1978. [Primary production.]

*36369 - **WELCH, E.B., HENDREY, G.R., STOLL, R.K.** : Nutrient supply and the production
and biomass of algae in four Washington lakes. - Oikos *26* : 47 - 54, 1975.
[Chl.]

36370 - **WELLBURN, A.R.** : Red light induced changes of the levels of bound and free
abscisic acid during plastid development. - In : AKOYUNOGLOU, G., ARGYROUDI-
-AKOYUNOGLOU, J.H. (ed.) : Chloroplast Development. Pp. 837 - 841. Elsevier/
North-Holland Biomedical Press, Amsterdam - New York - Oxford 1978.

36371 - **WELLER, S.C., FERREE, D.C.** : Effect of a pinolene-base antitranspirant on
fruit growth, net photosynthesis, transpiration, and shoot growth of "Golden
Delicious" apple trees. - J. amer. Soc. hort. Sci. *103* : 17 - 19, 1978.

36372 - **WENKERT, W., LEMON, E.R., SINCLAIR, T.R.** : Leaf elongation and turgor pressure in field-grown soybean. - Agron. J. *70* : 761 - 764, 1978. [Stomatal resistance.]

36373 - **WERBER, M.M., MEVARECH, M.** : Induction of a dissimilatory reduction pathway of nitrate in *Halobacterium* of the Dead Sea. A possible role for the 2Fe-
-ferredoxin  isolated from this organism. - Arch. Biochem. Biophys. *186* :
60 - 65, 1978.

36374 - **WERBER, M.M., MEVARECH, M.** : Purification and characterization of a highly
acidic 2Fe-ferredoxin from *Halobacterium* of the Dead Sea. - Arch. Biochem.
Biophys. *187* : 447 - 456, 1978.

36375 - **WERBER, M.M., MEVARECH, M., LEICHT, W., EISENBERG, H.** : Structure-function
relationships in proteins and enzymes of *Halobacterium*. - In : CAPLAN, S.R.,
GINZBURG, M. (ed.) : Energetics and Structure of Halophilic Microorganisms.
Pp. 427 - 445. Elsevier/North-Holland Biomedical Press, Amsterdam - New York
1978. [Ferredoxin.]

36376 - **WERNER, H.-J., SCHULTEN, K., WELLER, A.** : Electron transfer and spin exchange contributing to the magnetic field dependence of the primary photochemical reaction of bacterial photosynthesis. - Biochim. biophys. Acta *502* :
255 - 268, 1978.

36377 - **WESSELS, J.S.C., BORCHERT, M.T.** : Polypeptide profiles of chlorophyll·pro-
tein complexes and thylakoid membranes of spinach chloroplasts. - Biochim.
biophys. Acta *503* : 78 - 93, 1978.

36378 - **WETTERAU, J.R., NEWMAN, D.W., JAWORSKI, J.G.** : Quantitative changes of fatty
acids in soybean cotyledons during senescence and regreening. - Phytochemistry *17* : 1265 - 1268, 1978.

36379 - **WETTSTEIN, D. von, POULSEN, C., HOLDER, A.A.** : Ribulose-1,5-bisphosphate
carboxylase as a nuclear and chloroplast marker.  - Theor. appl. Genet.
*53* : 193 - 197, 1978.

36380 - **WEY, C.-L., AHL, P.L., CONE, R.A.** : Bacteriorhodopsin induces a light-scattering change in *Halobacterium halobium*. - J. Cell Biol. *79* : 657 - 662,
1978.

*36381 - **WEYBREW, J.A., HAMANN, H.K.** : Growth measurements of tobacco under field
conditions. - In : Recent Advances in the Chemical Composition of Tobacco
and Tobacco Smoke. Proceedings of American Chemical Society Symposium.
Pp. 164 - 183. R.J.Reynolds Tob. Co., Winston - Salem 1977. [Ps.]

36382 - **WHATLEY, F.R.** : Photophosphorylation and ion  transport. - In : HALL, D.O.,
COOMBS, J., GOODWIN, T.W. (ed.) : Proceedings of the Fourth International
Congress on Photosynthesis. Pp. 611 - 618. Biochem. Soc., London 1978.

36383 - **WHATLEY, J.M.** : A suggested cycle of plastid developmental interrelationships. - New Phytol. *80* : 489 - 502, 1978.

36384 - **WHEELER, C.T.** : Carbon dioxide fixation in the legume root nodule. - Ann.
appl. Biol. *88* : 481 - 484, 1978. [Connections with Ps.]

36385 - **WHERLAND, S., PECHT, I.** : Protein-protein electron transfer. A Marcus theory
analysis of reactions between $c$ type cytochromes and blue copper proteins. -
Biochemistry *17* : 2585 - 2591, 1978. [Plastocyanin.]

36386 - **WHIGHAM, D.F., McCORMICK, J., GOOD, R.E., SIMPSON, R.L.** : Biomass and pri-
mary production in freshwater tidal wetlands of the Middle Atlantic coast. -
In : GOOD, R.E., WHIGHAM, D.F., SIMPSON, R.L. (ed.) : Freshwater Wetlands.
Pp. 3 - 20. Academic Press, New York - London 1978.

36387 - **WHIGHAM, D.F., SIMPSON, R.L.** : The relationship between aboveground and
belowground biomass of freshwater tidal wetland macrophytes. - Aquat. Bot.
*5* : 355 - 364, 1978.

36388 - **WHIGHAM, D.K., MINOR, H.C.** : Agronomic characteristics and environmental
stress. - In : NORMAN, A.G. (ed.) : Soybean Physiology, Agronomy, and Utilization. Pp. 77 - 118. Academic Press, New York - London - San Francisco
1978. [Ps.]

36389 - **WHITE, C.C., CHAIN, R.K., MALKIN, R.** : Duroquinol as an electron donor for
chloroplast electron transfer reactions. - Biochim. biophys. Acta *502* :
127 - 137, 1978.

36390 - **WHITE, E., PAYNE, G.W.** : Chlorophyll production, in response to nutrient
additions, by the algae in Lake Rotorua water. - New Zeal. J. mar. Freshwater Res. *12* : 131 - 138, 1978.

36391 - **WHITE, G.C.** : Estimation of plant biomass from quadrat data using the log-
normal distribution. - J. Range Manage. *31* : 118 - 120, 1978.

36392 - **WHITE, T.L., KNOPP, J.A.** : Conelet abortion and ATP levels in longleaf pine.
- Can. J. Bot. *56* : 680 - 685, 1978.

36393 - **WHITEHEAD, D.** : The estimation of foliage area from sapwood basal area in
Scots pine. - Forestry *51* : 137 - 149, 1978.

36394 - **WHITENBERG, D.C., HARDESTY, W.D.** : Environmental factors affecting growth
and development of the Texas madrone. II. Interaction of light intensity
and water stress. - Texas J. Sci. *30* : 347 - 350, 1978. [Chl.]

36395 - **WHITMAN, W.B., TABITA, F.R.** : Modification of *Rhodospirillum rubrum* ribulose
bisphosphate carboxylase with pyridoxal phosphate. I. Identification of a
lysyl residue at the active site. - Biochemistry *17* : 1282 - 1287, 1978.

36396 - **WHITMAN, W.B., TABITA, F.R.** : Modification of *Rhodospirillum rubrum* ribulose
bisphosphate carboxylase with pyridoxal phosphate. 2. Stoichiometry and kinetics of inactivation. - Biochemistry *17* : 1288 - 1293, 1978.

36397 - **WHITMARSH, J., CRAMER, W.A.** : A pathway for the reduction of cytochrome
$b$-559 by photosystem II in chloroplasts. - Biochim. biophys. Acta *501* :
83 - 93, 1978.

36398 - **WHITTED, B.E., BARR, R., CRANE, F.L.** : The effect of quinolines on electron
transport and proton gradients associated with photophosphorylation in spinach chloroplasts. - Plant Sci. Lett. *11* : 41 - 50, 1978.

36399 - **WHITTEN, W.B., NAIRN, J.A., PEARLSTEIN, R.M.** : Derivative absorption spectroscopy from 5 - 300 K of bacteriochlorophyll $a$-protein from *Prostecochlo-
ris aestuarii*. - Biochim. biophys. Acta *503*: 251 - 262, 1978.

36400 - **WHITTEN, W.B., PEARLSTEIN, R.M., PHARES, E.F., GEACINTOV, N.E.** : Linear dichroism of electric field oriented bacteriochlorophyll $a$-protein from green
photosynthetic bacteria. - Biochim. biophys. Acta *503* : 491 - 498, 1978.

36401 - **WHITTINGHAM, C.P.** : Photosynthesis and productivity. Comments of reporter. -
In : HALL, D.O., COOMBS, J., GOODWIN, T.W. (ed.) : Proceedings of the Fourth
International Congress on Photosynthesis. Pp. 281 - 286. Biochem. Soc.,
London 1978.

36402 - **WIDER DE XIFRA, E.A., STELLA, A.M., BATLLE, A.M. del C.** : Porphyrin biosynthesis - immobilized enzymes and ligands. IX. Studies on δ-aminolaevulinate
synthetase from cultured soybean cells. - Plant Sci. Lett. *11* : 93 - 98,
1978.

36403 - **WIEBELT, J.A., HENDERSON, J.B.** : Theoretical thermal modeling of a leaf with experimental verification. - Agr. Meteorol. *19* : 101 - 111, 1978.

*36404 - **WIĘCKOWSKI, S.** : Czynniki sprzęgające z błon przetwarzających energię. [Coupling factors from energy transducing membranes.] - Zesz. nauk. Uniw. jagielloń. *464* (Pr. Biol. mol. 4) : 131 - 141, 1977. [In Pol., ab : E.]

36405 - **WIĘCKOWSKI, S., DROBA, M.** : Heterogeneity of the thylakoid membrane fractions derived from spinach chloroplasts by the action of *Triton X-100* in low salt medium : II, Isolation of β-carotene by the DEAE cellulose column chromatography. - Plant Sci. Lett. *13* : 397 - 404, 1978.

36406 - **WIEDENROTH, E., POSKUTA, J.** : Photosynthesis, photorespiration, respiration of shoots, and respiration of roots of wheat seedlings as influenced by oxygen concentration. - Z. Pflanzenphysiol. *89* : 217 - 225, 1978.

36407 - **WIEDMAIER, J., KULL, U.** : Die Aufnahme von cAMP und seine Wirkungen auf den Kohlenhydrat- und Lipidstoffwechsel grüner Blätter. - Biochem. Physiol. Pflanz. *172* : 421 - 437, 1978.

36408 - **WIELGOLASKI, F.-E.** : Primary production of alpine communities in Norway estimated by $CO_2$-exchange and harvesting techniques. - In : **HALL, D.O., COOMBS, J., GOODWIN, T.W.** (ed.) : Proceedings of the Fourth International Congress on Photosynthesis. Pp. 245 - 257. Biochem. Soc., London 1978.

36409 - **WILD, A.** : Studies on chloroplast development of a mutant of *Chlorella fusca*. - In : **AKOYUNOGLOU, G., ARGYROUDI-AKOYUNOGLOU, J.H.** (ed.) : Chloroplast Development. Pp. 533 - 538. Elsevier/North-Holland Biomedical Press, Amsterdam - New York - Oxford 1978.

36410 - **WILD, A., HÖHLER, T.** : Die Wirkung unterschiedlicher Lichtintensitäten während der Anzucht auf die $CO_2$-Kompensationslage, die Glykolsäure-Oxidase und Ribulose-biphosphat-Carboxylase-Aktivitäten bei *Sinapis alba*. - Z. Pflanzenphysiol. *87* : 413 - 428, 1978.

36411 - **WILD, A., TROSTMANN, U., KIETZMANN, I., FULDNER, K.-H.** : Development of the photosynthetic apparatus during light-dependent greening of a mutant of *Chlorella fusca*. - Planta *140* : 45 - 52, 1978.

36412 - **WILDMAN, S.G., KWANYUEN, P.** : Fraction I protein and other products from tobacco for food. - In : **SIEGELMAN, H.W., HIND, G.** (ed.) : Photosynthetic Carbon Assimilation. Pp. 1 - 18. Plenum Press, New York - London 1978.

36413 - **WILDNER, G.F.** : The effect of glycidate on the translocation of photoassimilates. - In : **SIEGELMAN, H.W., HIND, G.** (ed.) : Photosynthetic Carbon Assimilation. Pp. 426 - 427. Plenum Press, New York - London 1978.

*36414 - **WILDNER, G.F., HENKEL, J.** : Temperature dependent conformation changes of ribulose-1,5-biphosphate carboxylase studied by the use of 1-anilino-8-naphthalene sulfonate. - Z. Naturforsch. *32 C* : 226 - 228, 1977.

36415 - **WILDNER, G.F., HENKEL, J.** : Differential reactivation of RuBP oxygenase with low carboxylase activity. - In : **SIEGELMAN, H.W., HIND, G.** (ed.) : Photosynthetic Carbon Assimilation. P. 426. Plenum Press, New York - London 1978.

36416 - **WILDNER, G.F., HENKEL, J.** : Differential reactivation of ribulose 1,5-bisphosphate oxygenase with low carboxylase activity by $Mn^{2+}$. - FEBS Lett. *91* : 99 - 103, 1978.

36417 - **WILHELM, W.W., NELSON, C.J.** : Irradiance response of tall fescue genotypes with contrasting levels of photosynthesis and yield. - Crop Sci. *18* : 405 - 408, 1978.

36418 - **WILHELM, W.W., NELSON, C.J.** : Leaf growth, leaf aging, and photosynthetic rate of tall fescue genotypes. - Crop Sci. *18* : 769 - 772, 1978.

36419 - **WILHELM, W.W., NELSON, C.J.** : Growth analysis of tall fescue genotypes differing in yield and leaf photosynthesis. - Crop Sci. *18* : 951 - 954, 1978.

36420 - WILHM, J., COOPER, J., NAMMINGA, H. : Species composition, diversity, biomass and chlorophyll of periphyton in Greasy Creek, Red Cock Creek and the
Arkansas River Oklahoma. - Hydrobiologia *57* : 17 - 23, 1978.

36421 - WILLATT, S.T., TAYLOR, H.M. : Water uptake by soya-bean roots as affected
by their depth and by soil water content. - J. agr. Sci. *90* : 205 - 213,
1978. [Growth analysis.]

36422 - WILLENBRINK, J., SCHUSTER, W.-B. : Localized inhibition of translocation
of $^{14}C$-assimilates in the phloem by valinomycin and other metabolic inhibitors. - Planta *139* : 261 - 265, 1978.

36423 - WILLIAMS, A.M., WILLIAMS, R.R. : Regulation of movement of assimilate into
ovules of *Pisum sativum* cv. Greenfeast : a "remote" effect of the pod. -
Aust. J. Plant Physiol. *5* : 295 - 300, 1978.

36424 - WILLIAMS, G.J. III, KEMP, P.R. : Simultaneous measurement of leaf and root
gas exchange of  short-grass prairie species. - Bot. Gaz. *139* : 150 - 157,
1978.

36425 - WILLIAMS, L.E., KENNEDY, R.A. : Photosynthetic products of mature and se-
nescent *Zea mays* tissue during long term pulse-chase experiments. - Plant
Physiol. *61* (Suppl.) : 110, 1978.

36426 - WILLIAMS, L.E., KENNEDY, R.A. : Photosynthetic carbon metabolism during
leaf ontogeny in *Zea mays* L. : enzyme studies. - Planta *142* : 269 - 274,
1978.

36427 - WILLIAMS, M.E., RUDOLPH, E.D., SCHOFIELD, E.A., PRASHER, D.C. : The role of
lichens in the structure, productivity, and mineral cycling of the wet-coas-
tal Alaskan tundra. - In : TIESZEN, L.L. (ed.) : Vegetation and Production
Ecology of an Alaskan Arctic Tundra. Pp. 185 - 206. Springer-Verlag, New
York - Heidelberg - Berlin 1978.

36428 - WILLIAMS, P.F. : Growth of broad beans infected by *Uromyces viciae-fabae*. -
Ann. appl. Biol. *90* : 329 - 334, 1978.

36429 - WILLIAMS, V.P., GLAZER, A.N. : Structural studies on phycobiliproteins. I.
Bilin-containing peptides of C-phycocyanin. - J. biol. Chem. *253* : 202 -
211, 1978.

36430 - WILLIAMS-SMITH, D.L., HEATHCOTE, P., SIHRA, C.K., EVANS, M.C.W. : Quantitative electron-paramagnetic-resonance measurements of the electron-transfer
components of the Photosystem-I reaction centre. The free-radical signal
I and the bound iron-sulphur centre A. - Biochem. J. *170* : 365 - 371, 1978.

36431 - WILLMER, C.M., DON, R., PARKER, W. : Levels of short-chain fatty acids and
of abscisic acid in water-stressed and non-stressed leaves and their effects
on stomata in epidermal strips and excised leaves. - Planta *139* : 281 - 287,
1978. [Stomatal resistance.]

36432 - WILLMER, C.M., THORPE, N., RUTTER, J.C., MILTHORPE, F.L. : Stomatal metabolism : carbon dioxide fixation in attached and detached epidermis of
*Commelina*. - Aust. J. Plant Physiol. *5* : 767 - 778, 1978.

36433 - WILMAN, D., WRIGHT, P.T. : Dry-matter content, leaf water potential and
digestibility of three grasses in the early of regrowth after defoliation
with and without applied nitrogen. - J. agr. Sci. *91* : 365 - 380, 1978.

36434 - WILSON, D.R., FERNANDEZ, C.J., McCREE, K.J. : $CO_2$ exchange of subterranean
clover in variable light environments. - Crop Sci. *18* : 19 - 22, 1978.

36435 - WILSON, J.M. : Leaf respiration and ATP levels at chilling temperatures. -
New Phytol. *80* : 325 - 334, 1978.

36436 - WILSON, J.R., MANNETJE, L.'t: Senescence, digestibility and carbohydrate
content of buffel grass and green panic leaves in swards. - Aust. J. agr.
Res. *29* : 503 - 516, 1978.

36437 - WILSON, L.G., BRESSAN, R.A., LeCUREUX, L., FILNER, P. : Bisulfite induced
ethylene and ethane production : Dependence on light and photosynthetically
competent tissue. - Plant Physiol. *61* (Suppl.) : 93, 1978.

36438 - WILSON, V.E., HUDSON, L.W. : Inheritance of deleterious factors causing chlorophyll deficiency and seed sterility in lentils. - J. Hered. *69* : 267 - 269, 1978.

36439 - WILTENS, J., SCHREIBER, U., VIDAVER, W. : Chlorophyll fluorescence induction : an indicator of photosynthetic activity in marine algae undergoing desiccation. - Can. J. Bot. *56* : 2787 - 2794, 1978.

36440 - WINKLER, F.J., WIRTH, E., LATZKO, E., SCHMIDT, H.-L., HOPPE, W., WIMMER, P.: Einfluß von Wachstumsbedingungen und Entwicklung auf $\delta^{13}$C-Werte in verschiedenen Organen und Inhaltsstoffen von Weizen, Hafer und Mais. - Z. Pflanzenphysiol. *87* : 255 - 263, 1978.

36441 - WINTER, K. : Short-term fixation of $^{14}$carbon by the submerged aquatic angiosperm *Potamogeton pectinatus*. - J. exp. Bot. *29* : 1169 - 1172, 1978.

36442 - WINTER, K., GREENWAY, H. : Phosphoenolpyruvate carboxylase from *Mesembryanthemum crystallinum* : its isolation and inactivation *in vitro*. - J. exp. Bot. *29* : 539 - 546, 1978.

36443 - WINTER, K., LÜTTGE, U., WINTER, E., TROUGHTON, J.H. : Seasonal shift from $C_3$ photosynthesis to crassulacean acid metabolism in *Mesembryanthemum crystallinum* growing in its natural environment. - Oecologia *34* : 225 - 237, 1978.

36444 - WINTER, K., TROUGHTON, J.H. : Carbon assimilation pathways in *Mesembryanthemum nodiflorum* L. under natural conditions. - Z. Pflanzenphysiol. *88* : 153 - 162, 1978.

36445 - WINTER, K., TROUGHTON, J.H. : Photosynthetic pathways in plants of coastal and inland habitats of Israel and the Sinai. - Flora *167* : 1 - 34, 1978.

36446 - WITHERS, N., VIDAVER, W., LEWIN, R. : Pigment composition, photosynthesis and fine structure of a non-blue-green prokaryotic algal symbiont (*Prochloron* sp.) in a didemnid ascidian from Hawaiian waters. - Phycologia *17* : 167 - 171, 1978.

36447 - WITHERS, N.W., ALBERTE, R.S., LEWIN, R.A., THORNBER, J.P., BRITTON, G., GOODWIN, T.W. : Photosynthetic unit size, carotenoids, and chlorophyll-protein composition of *Prochloron* sp., a prokaryotic green alga. - Proc. nat. Acad. Sci. USA *75* : 2301 - 2305, 1978.

36448 - WITHERS, N.W., HAXO, F.T. : Isolation and characterization of carotenoid--rich lipid globules from *Peridinium foliaceum*. - Plant Physiol. *62* : 36 - 39, 1978.

36449 - WITT, H.T., TIEMANN, R., GRÄBER, P., RENGER, G. : The plastoquinone pool as possible hydrogen pump. - In : SCHÄFER, G., KLINGENBERG, M. (ed.) : Energy Conservation in Biological Membranes. Pp. 109 - 112. Springer-Verlag, Berlin - Heidelberg - New York 1978.

36450 - WITTENBACH, V.A. : Changes in proteolytic activity and the level of RuBPCase in the flag leaf of wheat during grain development and senescence. - Plant Physiol. *61* (Suppl.) : 25, 1978.

36451 - WITTENBACH, V.A. : Breakdown of ribulose bisphosphate carboxylase and change in proteolytic activity during dark-induced senescence of wheat seedlings. - Plant Physiol. *62* : 604 - 608, 1978.

36452 - WITTENBACH, V.A. : Breakdown of RuBP carboxylase during dark-induced senescence. - In : SIEGELMAN, H.W., HIND, G. (ed.) : Photosynthetic Carbon Assimilation. P. 416. Plenum Press, New York - London 1978.

36453 - WITZTUM, A. : Transcellular chloroplast banding patterns in leaves of *Elodea densa* induced by light and DCMU. - Ann. Bot. *42* : 1459 - 1462, 1978.

36454 - WITZTUM, A. : Ultraviolet radiation and sugar-induced chlorosis in detached leaves of *Elodea densa*. - Bot. Gaz. *139* : 295 - 298, 1978.

36455 - WODZINSKI, R.S., LABEDA, D.P., ALEXANDER, M. : Effects of low concentrations of bisulfite-sulfite and nitrite on microorganisms. - Appl. environ. Microbiol. *35* : 718 - 723, 1978. [Ps.]

36456 - **WOJNOWSKA, T.** : Wpływ intensywnego nawożenia użytków zielonych na plonowanie i wartość pokarmową roślin. Cz. VI. Zawartość chlorofilu, karotenu i ksantofili w roślinności intensywnie nawożonego pastwiska. [Effect of intensive fertilization of grasslands on the yield and fodder value of plants. Part VI. Chlorophyll, carotene and xanthophyll content in plants of an intensively fertilized pasture.] - Zesz. probl. Postępów Nauk roln. *210* : 205 - 215, 1978. [In Pol., ab : E, R.]

36457 - **WOLEDGE, J.** : The effect of shading during vegetative and reproductive growth on the photosynthetic capacity of leaves in a grass sward. - Ann. Bot. *42* : 1085 - 1089, 1978.

36458 - **WOLF, D.D.** : Nonstructural carbohydrate and dry matter relationships in alfalfa tap roots. - Crop Sci. *18* : 690 - 692, 1978.

36459 - **WOLFF, B., SCHANTZ, R.** : Thylakoid membrane proteins during chloroplast development in *Euglena gracilis* $\underline{Z}$. - In : **AKOYUNOGLOU, G., ARGYROUDI-AKOYU-NOGLOU, J.H.**(ed.): Chloroplast Development.Pp. 245 - 250. Elsevier/North-Holland Biomedical Press, Amsterdam - New York - Oxford 1978.

*36460 - **WOLFF, C., BÜCHEL, K.H., RENGER, G.** : Studies on the nature of the primary reactions of photosystem II in photosynthesis. II. The modification of the functional integrity of the photochemical active centres of system II by $\alpha$-bromo-$\alpha$-benzylmalodinitril and accompanying effects on chloroplasts. - Z. Naturforsch. *30 C* : 172 - 182, 1975.

36461 - **WOLLGIEHN, R., LERBS, S., MUNSCHE, D.** : Eigenschaften einer Chloroplasten--RNA vom Molekulargewicht 0,5 x $10^6$ aus *Nicotiana rustica*. - Biochem. Physiol. Pflanz. *173* : 60 - 69, 1978.

36462 - **WOLLMAN, F.-A.** : Determination and modification of the redox state of the secondary acceptor of Photosystem II in the dark. - Biochim. biophys. Acta *503* : 263 - 273, 1978.

36463 - **WOLOSIUK, R.A., BUCHANAN, B.B.** : Activation of chloroplast NADP-linked glyceraldehyde-3-phosphate dehydrogenase by the ferredoxin/thioredoxin system. - Plant Physiol. *61* : 669 - 671, 1978.

36464 - **WOLOSIUK, R.A., BUCHANAN, B.B.** : Regulation of chloroplast phosphoribulokinase by the ferredoxin/thioredoxin system. - Arch. Biochem. Biophys. *189* : 97 - 101, 1978.

36465 - **WOLWERTZ, M.-R.** : Two alternative pathways of chlorophyll biosynthesis in *Pinus jeffreyi*. - In : AKOYUNOGLOU, G., ARGYROUDI-AKOYUNOGLOU, J.H. (ed.) : Chloroplast Development. Pp. 111 - 118. Elsevier/North-Holland Biomedical Press, Amsterdam - New York - Oxford 1978.

36466 - **WONG, B.S., MILLER, D.M., YOPP, J.H.** : Proton pulsed NMR study on the cell constituents of *Aphanothece halophytica*, a blue-green alga. - In : AGRIS, P.F., LOEPPKY, R.N., SYKES, B.D. (ed.) : Biomolecular Structure and Function. Pp. 239 - 245. Academic Press, New York - San Francisco - London 1978. [Chloroplast.]

36467 - **WONG, C.S.** : Atmospheric input of carbon dioxide from burning wood. - Science *200* : 197 - 200, 1978.

36468 - **WONG, C.S.** : Carbon dioxide - a global environmental problem into the future. - Mar. Pollut. Bull. *9* (10) : 257 - 264, 1978. [Ps.]

36469 - **WONG, D., GOVINDJEE, JURSINIC, P.** : Analysis of microsecond fluorescence yield and delayed light emission changes after a single flash in pea chloroplasts : effects of mono- and divalent cations. - Photochem. Photobiol. *28* : 963 - 974, 1978.

36470 - **WONG, D., VACEK, K., MERKELO, H., GOVINDJEE** : Excitation energy transfer among chlorophyll *a* molecules in polystyrene : Concentration dependence of quantum yield, polarization and lifetime of fluorescence. - Z. Naturforsch. *33 C* : 863 - 869, 1978.

36471 - **WONG, K.K.** : Protochlorophyllide excretion by a purple sulphur bacterium. - Microbios Lett. *7* : 15 - 18, 1978.

36472 - WONG, S.C., COWAN, I.R., FARQUHAR, G.D. : Leaf conductance in relation to
assimilation in *Eucalyptus pauciflora* SIEB. *ex* SPRENG. Influence of irradiance and partial pressure of carbon dioxide. - Plant Physiol. *62* : 670 -
674, 1978.

36473 - WONG, W.W., BENEDICT, C.R., KOHEL, R.J. : The enzymatic fractionation of
$^{12}CO_2$ - $^{13}CO_2$ by ribulose-1,5-bisphosphate carboxylase. - Plant Physiol.
*61* (Suppl.) : 109, 1978.

36474 - WONG, W.W., SACKETT, W.M. : Fractionation of stable carbon isotopes by marine phytoplankton. - Geochim. cosmochim. Acta *42* : 1809 - 1815, 1978.

36475 - WOO, K.C., BERRY, J.A., TURNER, G.L. : Release and refixation of ammonia
during photorespiration. - Carnegie Inst. Year Book *77* : 240 - 245, 1978.

36476 - WOOD, P.M. : Interchangeable copper and iron proteins in algal photosynthesis. Studies on plastocyanin and cytochrome *c*-552 in *Chlamydomonas*. -
Europe. J. Biochem. *87* : 9 - 19, 1978.

36477 - WOODWELL, G.M. : The carbon dioxide question. - Sci. Amer. *238* (1) : 34 -
43, 1978. [Ps production.]

36478 - WOOLHOUSE, H.W. : Senescence processes in the life cycle of flowering plants.
- BioScience *28* : 25 - 31, 1978. [Ps.]

36479 - WOOLHOUSE, H.W. : Cellular and metabolic aspects of senescence in higher
plants. - In : BEHNKE, J.A., FINCH, C.E., MOMENT, G.B. (ed.) : The Biology
of Aging. Pp. 83 - 99. Plenum Press, New York - London 1978.

36480 - WOOLHOUSE, H.W. : Light-gathering and carbon assimilation processes in photosynthesis; their adaptive modifications and significance for agriculture.
- Endeavour N.S. *2* : 35 - 46, 1978.

36481 - WOŹNY, A. : Niektóre poglądy na budowę i tworzenie się gran. [Grana structure and formation.] - Wiad. bot. *22* : 241 - 248, 1978. [In Pol.]

36482 - WRAIGHT, C.A. : Involvement of the "protein" in redox-coupled protonation
events of the quinone acceptor-complex in bacterial photosynthetic reaction
centers. - In : DUTTON, P.L., LEIGH, J.S., SCARPA, A. (ed.) : Frontiers
of Biological Energetics : Electrons to Tissues. Vol. 1. Pp. 218 - 226.
Academic Press, New York - San Francisco - London 1978.

36483 - WRAIGHT, C.A. : Iron-quinone interactions in the electron acceptor region
of bacterial photosynthetic reaction centers. - FEBS Lett. *93* : 283 - 288,
1978.

36484 - WRAIGHT, C.A., COGDELL, R.J., CHANCE, B. : Ion transport and electrochemi-
cal gradients in photosynthetic bacteria. - In : CLAYTON, R.K., SISTROM,
W.R. (ed.) : The Photosynthetic Bacteria. Pp. 471 - 511. Plenum Press,
New York - London 1978.

36485 - WRAIGHT, C.A., LUEKING, D.R., FRALEY, R.T., KAPLAN, S. : Synthesis of photopigments and electron transport components in synchronous phototrophic
cultures of *Rhodopseudomonas sphaeroides*. - J. biol. Chem. *253* : 465 - 471,
1978.

36486 - WRIGHT, S.W., GRANT, B.R. : Properties of chloroplasts isolated from siphonous algae. Effects of osmotic shock and detergent treatment on intactness.
- Plant Physiol. *61* : 768 - 771, 1978.

36487 - WRISCHER, M. : Ultrastructural localization of diaminobenzidine photooxidation in etiochloroplasts. - Protoplasma *97* : 85 - 92, 1978. [Ps.]

36488 - WRISCHER, M. : Ultrastructural changes in plastids of detached spinach
leaves. - Z. Pflanzenphysiol. *86* : 95 - 106, 1978.

36489 - WURR, D.C.E. : The effects of the date of defoliation of the seed potato
crop and the storage temperature of the seed on subsequent growth. 2. Field
growth. - J. agr. Sci. *91* : 747 - 756, 1978. [Dry-matter accumulation.]

36490 - WYDRZYNSKI, T.J., MARKS, S.B., SCHMIDT, P.G., GOVINDJEE, GUTOWSKY, H.S. :
Nuclear magnetic relaxation by the manganese in aqueous suspensions of chloroplasts. - Biochemistry *17* : 2155 - 2162, 1978.

36491 - **YADYKIN, A.A.** : Opredelenie aktivnosti fotosinteticheskikh fermentov u morskikh zelenykh vodorosleĭ. [Determination of photosynthetic enzymes activity in marine green algae.] - Sb. Rabot Akad. Nauk SSSR, dal'nevost. nauch. Tsentr, Inst. Biol. Morya (Vladivostok) *11* (Ekologicheskie Aspekty Fotosinteza Morskikh Makrovodorosleĭ) : 28 - 37, 183, 1978. [In R, ab : E.]

36492 - **YAGI, T., OCHIAI, H.** : Attempts to produce hydrogen by coupling hydrogenase and chloroplast photosystems. - In : **VEZIROGLU, T.N., SEIFRITZ, W.** (ed.) : Hydrogen Energy System. Vol. 3. Pp. 1293 - 1307. Pergamon Press, Oxford - New York 1978.

*36493 - **YAKOVLEV, B.V., YAKOVLEVA, V.F., ALESHIN, E.P.** : Primenenie pozdnoĭ vnekormovoĭ podkormki fosforom i azotom v sochetanii s 2,4-DA dlya uskoreniya sozrevaniya zerna i povysheniya urozhaya risa. [Utilization of late foliar phosphorus and nitrogen nutrition together with 2,4-DA for accelerating grain ripening and yield increase.] - Byull. nauch.-tekh. Inf. vsesoyuz. nauch.-issled. Inst. Risa *15* : 30 - 36, 1975. [Chl; in R, ab : E.]

36494 - **YAMADA, Y., SATO, F.** : The photoautotrophic culture of chlorophyllous cells. - Plant Cell Physiol. *19* : 691 - 699, 1978.

36495 - **YAMADA, Y., SATO, F., HAGIMORI, M.** : Photoautotropism in green cultured cells. - In : **THORPE, T.A.** (ed.) : Frontiers of Plant Tissue Culture 1978. Pp. 453 - 462. Int. Assoc. Plant Tissue Culture, Calgary 1978. [Ps, Chl.]

36496 - **YAMAGISHI, A., SATOH, K., KATOH, S.** : Fluorescence induction in chloroplasts isolated from the green alga *Bryopsis maxima*. III. A fluorescence transient indicating proton gradient across the thylakoid membrane. - Plant Cell Physiol. *19* : 17 - 25, 1978.

36497 - **YAMAGUCHI, J.** : Respiration and the growth efficiency in relation to crop productivity. - J. Fac. Agr. Hokkaido Univ. *59* : 59 - 129, 1978. [Ps.]

36498 - **YAMAMOTO, H.Y., BANGHAM, A.D.** : Carotenoid organization in membranes. Thermal transition and spectral properties of carotenoid-containing liposomes. - Biochim. biophys. Acta *507* : 119 - 127, 1978.

36399 - **YAMAMOTO, H.Y., HIGASHI, R.M.** : Violaxanthin de-epoxidase. Lipid composition and substrate specificity. - Arch. Biochem. Biophys. *190* : 514 - 522, 1978.

36500 - **YAMAMOTO, Y., MAEDA, K., HAYASHI, K.** : [Studies on transplanting injury in rice plant. I. The effects of root cutting treatments on the early growth of rice seedlings after transplanting.] - Jap. J. Crop Sci. *47* : 31 - 38, 1978. [Growth analysis; in Jap., ab : E.]

36501 - **YAMAMOTO, Y., MAEDA, K., HAYASHI, K.** : [Studies on transplanting injury in rice plant. II. The effects of root cutting treatments on the contents of organic constituents and growth rate of the rice seedlings after transplanting.] - Jap. J. Crop Sci. *47* : 39 - 47, 1978. [In Jap., ab : E.]

36502 - **YAMAMOTO, Y., NISHIMURA, M.** : Discrimination and characterization of photosystem I- and photosystem II-induced 515-nm absorbance change in chloroplasts. I. Dissipation of electric field and related processes. - Plant Cell Physiol. *19* : 785 - 790, 1978.

36503 - **YAMAMOTO, Y., NISHIMURA, M.** : Discrimination and characterization of photosystem I- and photosystem II-induced 515-nm absorbance changes in chloroplasts. II. Separate local fields in thylakoids indicated by differential responses of the 515-nm absorbance change. - Plant Cell Physiol. *19* : 975 - 983, 1978.

36504 - **YAMANAKA, G., GLAZER, A.N., WILLIAMS, R.C.** : Cyanobacterial phycobilisomes. Characterization of the phycobilisomes of *Synechococcus* sp. 6301. - J. biol. Chem. *253* : 8303 - 8310, 1978.

36505 - **YAMANAKA, T., FUKUMORI, Y., WADA, K.** : Cytochrome *c*-554 derived from the blue-green alga *Spirulina platensis*. - Plant Cell Physiol. *19* : 117 - 126, 1978.

36506 - **YAMAOKA, T., SATOH, K., KATOH, S.** : Photosynthetic activities of a thermophilic blue-green alga. - Plant Cell Physiol. *19* : 943 - 954, 1978.

36507 - **YAMAOKA, T., SATOH, K., KATOH, S.** : Preparation of thylakoid membranes active in oxygen evolution at high temperature from a thermophilic blue-green alga. - In : **METZNER, H.** (ed.) : Photosynthetic Oxygen Evolution. Pp. 105 - 115. Academic Press, London - New York - San Francisco 1978.

36508 - **YAMASHITA, T., INOUE, Y., KOBAYASHI, Y., SHIBATA, K.** : Flash reactivation of Tris-inactivated chloroplasts. - Plant Cell Physiol. *19* : 895 - 900, 1978.

36509 - **YAMASHITA, T., KOHDA, H., NANRI, J., TOMITA, G.** : The simultaneous measurement of $O_2$-evolving and $CO_2$-fixing activities in fresh leaves. - J. Fac. Agr. Kyushu Univ. *22* : 107 - 118, 1978.

36510 - **YARISH, C., CASEY, S., EDWARDS, P.** : Acclimation responses to salinity of three estuarine red algae from New Jersey. - J. Phycol. *14* (Suppl.) : 27, 1978.

36511 - **YE JI-YU, TANG CHONG-QIN, WANG MEI-QI, HAN QI** : [The effect of snake venom on intact spinach chloroplasts.] - Chih Wu Hsueh Pao *20* : 114 - 121, 1978. [In Chin., ab : E.]

36512 - **YENTSCH, C.S.** : On the contribution of plant physiology to the study of primary production. - In : **HALL, D.O., COOMBS, J., GOODWIN, T.W.** (ed.) : Proceedings of the Fourth International Congress on Photosynthesis. Pp. 269 - 280. Biochem. Soc., London 1978.

36513 - **YOCH, D.C.** : Nitrogen fixation and hydrogen metabolism by photosynthetic bacteria. - In : **CLAYTON, R.K., SISTROM, W.R.** (ed.) : The Photosynthetic Bacteria. Pp. 657 - 676. Plenum Press, New York - London 1978. [Ps.]

\*36514 - **YOCH, D.C., ARNON, D.I.** : Comparison of two ferredoxins from *Rhodospirillum rubrum* as electron carriers for the native nitrogenase. - J. Bacteriol. *121* : 743 - 745, 1975.

36515 - **YODA, K.** : Estimation of community respiration. - In : **KIRA, T., ONO, Y., HOSOKAWA, T.** (ed.) : Biological Production in a Warm-Temperate Evergreen Oak Forest of Japan. JIBP Synthesis, Vol. 18. Pp. 112 - 131. University of Tokyo Press, Tokyo 1978.

36516 - **YOKOI, Y., KIMURA, M., HOGETSU, K.** : Quantitative relationships between growth and respiration. I. Components of respiratory loss and growth efficiencies of etiolated red bean seedlings. - Bot. Mag. (Tokyo) *91* : 31 - 41, 1978.

36517 - **YOKOYAMA, Z.-I., ABE, A., SHIN, M.** : The roles of carboxyl-terminal amino acid residues in higher plant ferredoxin. - Agr. biol. Chem. *42* : 2169 - 2171, 1978.

36518 - **YOMOSA, S.** : Temperature dependence of the excitation tranfer in photosynthetic systems. - J. phys. Soc. Jap. *45* : 967 - 975, 1978.

36519 - **YORDANOV, I.** : Influence of high temperature on the formation and activity of the photosynthetic apparatus and spectral characteristics of the pigment-protein complexes of *Phaseolus vulgaris* plants. - Photosynthetica *12* : 344 - 348, 1978.

\*36520 - **YORDANOV, I.T., STOYANOV, I.G., CHICHEV, P.N.** : Pigment contents and lamellar chloroplast protein composition of maize plants, grown at magnesium deficiency. - Dokl. bolg. Akad. Nauk *30* : 1625 - 1628, 1977.

36521 - **YOSHIDA, T.** : Effect of stomatal frequency on photosynthesis and its use for breeding in barley. - Bull. Kyushu nat. agr. Exp. Sta. *20* : 129 - 193, 1978. [Stomatal resistance.]

36522 - **YOSHIDA, T.** : On the stomatal frequency in barley IV. Estimating mesophyll resistance taking into account respiration during photosynthesis. - Ikushugaku Zasshi [Jap. J. Breed.] *28* : 13 - 20, 1978.

36523 - YOSHIDA, T. : On the stomatal frequency in barley. V. The effect of stomatal
size on transpiration and photosynthetic rate. - Ikushugaku Zasshi [Jap. J.
Breed.] *28* : 87 - 96, 1978.

36524 - YOSHIDA, T., ONO, T. : Environmental differences in leaf stomatal frequency
of rice. - Jap. J. Crop Sci. *47* : 506 - 514, 1978. [In Jap., ab : E.]

36525 - YOSHIURA, M., IRIYAMA, K., SHIRAKI, M. : High-performance liquid chromatography of chlorophylls and some of their derivatives. - Chem. Lett. *3* :
281 - 282, 1978.

36526 - YOUNIS, H.M., BOYER, J.S., GOVINDJEE : Conformation and activity of chloroplast coupling factor exposed to low leaf water potentials. - Plant Physiol.
*61* (Suppl.) : 77, 1978.

36527 - YUKIMOTO, M., ISHITANI, A., YOSHIDA, K., KOBAYASHI, N. : [Phytotoxicity of
MBCP (4-bromo-2,5-dichlorophenyl methyl phenylphosphonothionate) to Chinese
cabbage seedlings (*Brassica pekinensis* RUPR.).] - Nihon Noyakugaku Kaishi
[J. Pestic. Sci.] *3* : 243 - 247, 1978. [Ps, Chl; in Jap., ab : E.]

*36528 - YULDASHEV, S.Kh., NAZAROV, M., ISMAILOV, G. : Formirovanie fotosinteticbes-
kogo apparata khlopchatnika i nekotorye storony obmena veshchestv v svyazi
s produktivnost'yu. [Formation of the photosynthetic apparatus of cotton
and some sides of metabolism in relation to productivity.] - Tr. vsesoyuz.
nauch.-issled. Inst. Khlopkovodstva *24* (Voprosy Fiziologii i Biokhimii
Khlopchatnika) : 18 - 32, 1973. [In R, ab : Uzb.]

36529 - YURINA, N.P., ODINTSOVA, M.S., MALIGA, P. : An altered chloroplast ribosomal protein in a streptomycin resistant tobacco mutant. - Theor. appl.
Genet. *52* : 125 - 128, 1978.

*36530 - YUSHKOV, V.I. : Vliyanie iskusstvennoĭ defoliatsii na rost i intensivnost'
potentsial'nogo fotosinteza u sosny obyknovennoĭ. [The effect of artificial
defoliation on growth and potential photosynthesis in *Pinus silvestris*.] -
Tr. Inst. Ėkol. Rast. Zhiv. (Sverdlovsk) *100* (Ėkologo-fiziologicheskie Issledovaniya Khvoĭnykh Drevesnykh Vidov na Urale) : 14 - 23, 96, 1976. [In R.]

*36531 - YUSHKOV, V.I. : Nekotorye osobennosti raspredeleniya assimilirovannogo ugle-
roda-14 u sosny obyknovennoĭ. [$^{14}$C-photosynthate distribution in *Pinus silvestris*.] - Tr. Inst. Ėkol. Rast. Zhiv. (Sverdlovsk) *100* (Ėkologo-fiziologicheskie Issledovaniya Khvoĭnykh Drevesnykh Vidov na Urale) : 41 - 53,
97 - 98, 1976. [In R.]

36532 - YUSUPOV, M.I., RAKHMATOV, N.A., IBRAGIMOV, A.P. : Issledovanie poliribosom
khloroplastov khlopchatnika. [Study of polyribosomes of cotton plant chloroplasts.] - Dokl. Akad. Nauk Uz.SSR *1978* (10) : 61 - 63, 1978. [In R.]

36533 - YUZBEKOV, A.K. : Svetoindutsirovannyĭ sintez fermentov $C_4$-puti fotosinteza
v ėtiolirovannykh list'yakh kukuruzy. [Light-induced synthesis of enzymes
of the $C_4$-pathway of photosynthesis in etiolated maize leaves.] - Vestn.
leningr. Univ. *1978* (3) : 111 - 115, 148, 1978. [In R, ab : E.]


36534 - ZABOTINA, L.N. : Sezonnaya dinamika soderzhaniya khlorofilla v list'yakh
ėfemeroidov lesostepnoĭ dubravy. [Seasonal dynamics of chlorophyll content
in leaves of ephemeroids in forest-steppe oak groves.] - Tr. petergof. biol.
Inst. *27* (Voprosy Ėkologicheskoĭ Anatomii i Fiziologii Rasteniĭ) : 83 - 91,
1978. [In R, ab : E.]

36535 - ZACHHUBER, K., LARCHER, W. : Energy contents of different alpine species
of *Saxifraga* and *Primula* depending on their altitudinal distribution. -
Photosynthetica *12* : 436 - 439, 1978.

36536 - ZAJIC, J.E., KOSARIC, N., BROSSEAU, J.D. : Microbial production of hydrogen.
- In : GHOSE, T.K., FIECHTER, A., BLAKEBROUGH, N. (ed.) : Advances in Biochemical Engineering. Vol. 9. Pp. 57 - 109. Springer-Verlag, Berlin - Heidelberg - New York 1978.

36537 - **ZAKHARIEVA, T.** : Fe-khloroza pri tsarevichni rasteniya, otglezhdani na karbonaten chernozem. II. Vliyanie na formata na vnesenoto v pochvata Fe v"rkhu stepenite na Fe-khloroza i kontsentratsiyata na khlorofil v tsarevichniya khibrid SK-4. [Iron-chlorosis in maize plants grown on carbonate chernozem soils. II. Effect of the iron form applied to the soil on the degree of iron-chlorosis and chlorophyll concentration in the SK-4 maize hybrid.] - Pochvozn. Agrokhim. *13* (4) : 56 - 63, 1978. [In Bulg., ab : E, R.]

36538 - **ZAKHAROV, S.D., SYTNIK, S.K., MAL'YAN, A.N., MAKAROV, A.D.** : Vydelenie i svoĭstva $CF_1$-ATPazy s izmenennoĭ submolekulyarnoĭ strukturoĭ. [Isolation and properties of $CF_1$-ATPase from chloroplasts with changed submolecular structure.] - Biokhimiya *43* : 887 - 891, 1978. [In R, ab : E.]

36539 - **ZAKHAROVA, N.I., CHIBISOV, A.K.** : Issledovanie protsessov perenosa ĕlektrona v pigment-belkovykh kompleksakh fotosistemy 1. [Electron transfer in chlorophyll-protein complexes of photosystem 1.] - Mol. Biol. (Moskva) *12* : 1075 - 1084, 1978. [In R, ab : E.]

36540 - **ZAKRZHEVSKIĬ, D.A., ANAN'EV, G.M., GERTS, S.M.** : O kinetike vydeleniya kisloroda kletkami khlorelly v nachal'nyĭ period osveshcheniya. [The kinetics of oxygen evolution by *Chlorella* cells at the initial period of illumination.] - Fiziol. Rast. *25* : 829 - 835, 1978. [In R, ab : E.]

*36541 - **ZAKRZHEVSKIĬ, D.A., KALASHNIKOV, Yu.E., ROZONOVA, L.N., SINYAKOVA, R.S., NIKOLAEV, E.A., KUTYURIN, V.M.** : Izuchenie protsessa razlozheniya vody zelenymi rasteniyami v anaĕrobnykh usloviyakh. [Process of water-splitting by green plants in anaerobic conditions.] - In : Itogi Issledovaniya Mekhanizma Fotosinteza. Pp. 115 - 130. Pushchino 1974. [In R.]

*36542 - **ZAKRZHEVSKIĬ, D.A., ROZONOVA, L.N., KALASHNIKOV, Yu.E.** : O dvukh putyakh vydeleniya fotosinteticheskogo kisloroda. [On two ways of photosynthetic oxygen evolution.] - Dokl. Akad. Nauk SSSR *234* : 713 - 716, 1977. [In R.]

36543 - **ZÁLETOVÁ, E., PAULECH, C.** : Seasonal dynamics of photosynthesis intensity of the species *Prunus laurocerasus* L. - Biológia (Bratislava) *33* : 73 - 78, 1978.

36544 - **ZAMSKI, E., UMIEL, N.** : Streptomycin resistance in tobacco : II. Effects of the drug on the ultrastructure of plastids and mitochondria in callus cultures. - Z. Pflanzenphysiol. *88* : 317 - 325, 1978.

36545 - **ZANGERL, A.R.** : Energy exchange phenomena, physiological rates and leaf size variation. - Oecologia *34* : 107 - 112, 1978. [Ps.]

36546 - **ZANKEL, K.L.** : Energy transfer between antenna components and reaction centers. - In : CLAYTON, R.K., SISTROM, W.R. (ed.) : The Photosynthetic Bacteria. Pp. 341 - 347. Plenum Press, New York - London 1978.

36547 - **ZANNONI, D., JASPER, P., MAARS, B.** : Light-induced oxygen reduction as a probe of electron transport between respiratory and photosynthetic components in membranes of *Rhodopseudomonas capsulata*. - Arch. Biochem. Biophys. *191* : 625 - 631, 1978.

36548 - **ZAPOROZHCHENKO, V.A.** : Fotosinteticheskiĭ potentsial gibridov raznoĭ skorospelosti v TsChP. [Photosynthetic potential of hybrids of different earliness.] - Byull. vsesoyuz. nauch.-issl. Inst. Kukuruzy *1978* (2-3) : 36 - 38, 1978. [In R.]

36549 - **ZARIN', V.Ė., BOĬCHENKO, E.A.** : Uchastie metallov v temnovoĭ assimilyatsii uglekisloty. [Participation of metals in dark assimilation of $CO_2$.] - Tr. biogeokhim. Lab. Akad. Nauk SSSR *15* (Biogeokhimicheskoe Raĭonirovanie - Metod Izucheniya Ėkologicheskogo Stroeniya Biosfery) : 187 - 191, 1978. [In R.]

36550 - **ZAVODNIK, N.** : Environmental influences on the day and night rhythm of photosynthesis in some littoral marine algae. - Ekologija *13* (1) : 45 - 52, 1978.

36551 - **ZAV'YALOVA, N.S.** : Funktsional'naya kharakteristika fotosinteticheskogo apparata podrosta sosny, eli, pikhty i kedra v sosnyake travyanom v podzone yuzhnoĭ taĭgi Zaural'ya. [Functional characteristics of photosynthetic apparatus of undergrowth of *Pinus silvestris, Picea, Abies* and *Pinus sibi-rica* in pine forest in subzone of southern Transural taiga.] - Tr. Inst. Ėkol. Rast. Zhiv. (Sverdlovsk) *100* (Ėkologo-fiziologicheskie Issledovaniya Khvoĭnykh Drevesnykh Vidov na Urale) : 3 - 13, 96, 1976. [In R.]

*36552 - **ZAV'YALOVA, N.S.** : Svetovye krivye fotosinteza podrosta sosny i eli v pod-zone yuzhnoĭ taĭgi Zaural'ya. [Light curves of photosynthesis in undergrowth of pine and spruce in subzone of southern Transural taiga.] - Tr. Inst. Ėkol. Rast. Zhiv. (Sverdlovsk) *100* (Ėkologo-fiziologicheskie Issledovaniya Khvoĭ-nykh Drevesnykh Vidov na Urale) : 24 - 40, 97, 1976. [In R.]

36553 - **ZDANOWSKI, B., BNIŃSKA, M., KORYCKA, A., SOSNOWSKA, J., RADZIEJ, J., ZACH-WIEJA, J.** : The influence of mineral fertilization on primary productivity of lakes. - Ekol. pol. *26* : 153 - 192, 1978.

36554 - **ZEĬNALOV, Yu.** : Kolko fotosistemi uchastvuvat pri fotosintezata v zelenite rasteniya. [How many photosystems take part in photosynthesis of green plants.] - Fiziol. Rast. (Sofia) *4* (2) : 90 - 97, 1978. [In Bulg., ab : E, R.]

36555 - **ZEINALOV, Yu., PETKOVA, R.A.** : Kinetics of the $O_2$ evolution in isolated chloroplasts. - Dokl. bolg. Akad. Nauk *31* : 105 - 108, 1978.

36556 - **ZELENSKIĬ, M.I., MOGILEVA, G., SHITOVA, I., FATTAKHOVA, F.** : Hill reaction of chloroplasts from some species, varieties and cultivars of wheat. - Photosynthetica *12* : 428 - 435, 1978.

36557 - **ZELENSKIĬ, M.I., MOGILEVA, G.A.** : Fiziko-khimicheskie kharakteristiki vesh-chestv, ispol'zuemykh v issledovaniyakh po fotosintezu. [Physico-chemical data for substances used in photosynthetic research.] - Tr. priklad. Bot. Genet. Selektsii *61* (3) : 153 - 170, 1978. [In R, ab : E.]

36558 - **ZELENSKIĬ, M.I., MOGILEVA, G.A., SAKHAROVA, O.V.** : Fosfataktseptornyĭ kont-rol'i stekhiometriya fosforilirovaniya v izolirovannykh khloroplastakh. [Phosphate acceptor control and stoichiometry of phosphorylation in isola-ted chloroplasts.] - Uspekhi sovrem. Biol. *85* : 33 - 49, 1978. [In R.]

36559 - **ZELENSKIĬ, M.I., MOGILEVA, G.A., SAKHAROVA, O.V.** : Opredelenie parametrov protsessa fotofosforilirovaniya po izmereniyam skorosti vydeleniya kislo-roda. [Determination of photophosphorylation parameters based on oxygen evolution measurement.] - Tr. priklad. Bot. Genet. Selektsii *61* (3) : 86 - 95, 1978. [In R, ab : E.]

36560 - **ZELEPUKHIN, I.D., ZELEPUKHIN, V.D.** : Regulyatsiya vodnogo obmena yabloni biologicheski aktivnoĭ vodoĭ v razlichnykh ėkologicheskikh usloviyakh. [Water regime regulation in apple tree by biologically active water under different ecological conditions.] - In : Vodnyĭ Rezhim Rasteniĭ v Svyazi s Raznymi Ėkologicheskimi Usloviyami. Pp. 133 - 136. Izdatel'stvo Kazan-skogo Universiteta, Kazan' 1978. [Ps; in R.]

36561 - **ZELITCH, I.** : Effect of glycidate, an inhibitor of glycolate synthesis in leaves, on the activity of some enzymes of the glycolate pathway. - Plant Physiol. *61* : 236 - 241, 1978.

*36562 - **ZENCHENKO, V.A., CHURINA, M.B., SHAKHOV, A.A.** : Chuvstvitel'nost' izoliro-vannykh peroksisom semyan kleshcheviny k svetu. [Sensibility to light of peroxisomes isolated from *Ricinus* seeds.] - Fiziol. Rast. *19* : 752 - 758, 1972. [In R, ab : E.]

*36563 - **ZENCHENKO, V.A., CHURINA, M.B., SHAKHOV, A.A.** : Vliyanie belogo, sinego i krasnogo sveta na glikolatoksidaznuyu aktivnost' mikrotelets ėtiolirovan-nykh prorostkov pshenitsy. [Effect of white, blue, and red light on the activity of glycolate oxidase of microbodies from etiolated wheat seedlings.] - Fiziol. Rast. *24* : 725 - 731, 1977. [Proplastids; in R, ab : E.]

36564 - **ZEN'KEVICH, Ė.I., KOCHUBEEV, G.A., LOSEV, A.P., GURINOVICH, G.P.** : Spekt-ral'no-lyuminestsentnye svoĭstva otdel'nykh agregirovannykh form pigmentov v rastvorakh. [The spectral-luminescent properties of separate aggregated forms of pigments in solutions.] - Mol. Biol. (Moskva) *12* : 1002 - 1011, 1978. [In R, ab : E.]

36565 - ZENTNER, R.P., SONNTAG, B.H., LEE, G.E. : Simulation model for dryland crop production in the Canadian prairies. - Agr. Systems *3* : 241 - 251, 1978.

36566 - ZEYNALOV, Y. : On the possibility of the functioning of one sole photosystem at the photosynthesis of green plants. - Dokl. bolg. Akad. Nauk *31* : 731 - 734, 1978.

36567 - ZEYNALOV, Y. : An electric diagram for modelling the fluorescence induction phenomena in photosynthesis. - Dokl. bolg. Akad. Nauk *31* : 1185 - 1188, 1978.

36568 - ZEYNALOV, Y., MASLENKOVA, L. : On the Blinks effect and its explanation. - Dokl. bolg. Akad. Nauk *31* : 1189 - 1191, 1978.

36569 - ZEYNALOV, Y., MASLENKOVA, L. : On the light curves of photosynthesis. I. Theoretical analysis. - In : VI[th] National Conference of Plant Physiology. Pp. 138 - 143. Sofia 1978.

36570 - ZEYNALOV, Y., MASLENKOVA, L. : On the light curves of photosynthesis. II. Effect of suspension optical density. - In : VI[th] National Conference of Plant Physiology. Pp. 144 - 147. Sofia 1978.

36571 - ZHUK, L.I., PETRASH, V.G. : Produktivnost' listovogo apparata rassady tomatov pri razlichnykh usloviyakh vyrashchivaniya. [Productivity of the tomato seedling leaf apparatus under different growing conditions.] - Fiziol. Biokhim. kul't. Rast. *10* : 516 - 520, 560, 1978. [Chl, photosynthates; in R, ab : E.]

*36572 - ZIEGLER, I. : Sulfite action on ribulosediphosphate carboxylase in the lichen *Pseudevernia furfuracea*. - Oecologia *29* : 63 - 66, 1977.

36573 - ZIELINSKI, R., PRICE, C.A. : Some molecular properties of spinach chloroplast cytochrome *b*559. - Plant Physiol. *61* (Suppl.) : 103, 1978.

36574 - ZIELINSKI, R.E., PRICE, C.A. : Relative requirements for magnesium of protein and chlorophyll synthesis in *Euglena gracilis*. - Plant Physiol. *61* : 624 - 625, 1978.

36575 - ZILINSKAS, B.A., ZIMMERMAN, B.K., GANTT, E. : Allophycocyanin forms isolated from *Nostoc* sp. phycobilisomes. - Photochem. Photobiol. *27* : 587 - 595, 1978.

36576 - ZIMA, J., ŠESTÁK, Z. : Effect of abscisic acid on activities of photosystems 1 and 2 in French bean chloroplasts. - Biol. Plant. *20* : 232 - 233, 1978.

36577 - ZIMMERMANN, G., KELLY, G.J., LATZKO, E. : Purification and properties of spinach leaf cytoplasmic fructose-1,6-bisphosphatase. - J. biol. Chem. *253* : 5952 - 5956, 1978.

*36578 - ZORIN, N.A., GOGOTOV, I.N. : Aktivatsiya svetom gidrogenazy u fototrofnykh bakteriĭ *Thiocapsa roseopersicina*. [Light activation of hydrogenase in phototrophic bacterium *Thiocapsa roseopersicina*.] - In : Biologiya i Nauchno--Tekhnicheskiĭ Progress. Pp. 68 - 69. Pushchino 1974. [In R.]

36579 - ZUBRICKÝ, J. : Štúdium tenkovrstvovej chromatografie pigmentov zelenej silážnej kukurice a kukuričnej siláže z hľadiska krytia potreby karoténu vo výžive hovädzieho dobytka. [A study of thin-layer chromatography of pigments of green maize silage and maize silage from the viewpoint of satisfying the need of carotene in beef cattle nutrition.] - Poľnohospodárstvo *24* : 643 - 649, 1978. [In Slovak, ab : E, R.]

36580 - ZUBRICKÝ, J. : Vplyvy intenzifikačných faktorov na karoténovú hodnotu silážnej kukurice a kukuričnej siláže. [Influence of intensification factors on the carotene value of the silage maize and maize silage.] - Poľnohospodárska Veda, ser. C *1978* (1) : 1 - 97, 1978. [In Slovak, ab : E, R.]

36581 - ZUO BUO-YU, DUAN XU-CHUANG : [Ultrastructure and function of chloroplasts from the winter wheat leaves at different ranks of attachment to the main stem.] - Acta bot. sin. *20* : 223 - 228, 1978. [In Chin., ab : E.]

36582 - ZURZYCKI, J., GIERKA, A. : The effect of light intensity and spectral region on the structure of thylakoid membranes of *Phaseolus vulgaris*. - Acta Physiol. Plant. *1* : 27 - 34, 1978.

36583 - **ZVALINSKIĬ, V.I.** : Perenos élektrona poluprovodnikovogo tipa v biologiches-
kikh sistemakh. [The transfer of semiconductive electron in biological systems.] - Nauch. Soobshch. Inst. Biol. Morya, dal'nevostoch. nauch. Tsentr
Akad. Nauk SSSR *3* (Biologicheskie Issledovaniya Dal'nevostochnykh Moreĭ) :
35 - 38, 1978. [Ps; In R.]

36584 - **ZVALINSKIĬ, V.I.** : Kinetika i svetovye krivye fotosinteza zooksantell. [Ki-
netics and light curves of photosynthesis of *Zooxanthella*.] - Sb. Rabot
Akad. Nauk SSSR, dal'nevost. nauch. Tsentr, Inst. Biol. Morya (Vladivostok)
*12* [Biologiya Korallovykh Rifov (Fotosintez Zooksantell i Vodorosleĭ-Makrofitov)] : 53 - 64, 1978. [In R, ab : E.]

36585 - **ZVALINSKIĬ, V.I., IVANOV, N.A., CHERNOVA, S.I.** : Nativnoe sostoyanie khlorofilla *a* v morskikh vodoroslyakh v zavisimosti ot svetovykh usloviĭ obitaniya. [Native state of chlorophyll *a* in marine algae in dependence on
light conditions of their habitat.] - Sb. Rabot Akad. Nauk SSSR, dal'nevost.
nauch. Tsentr, Inst. Biol. Morya (Ékologicheskie Aspekty
Fotosinteza Morskikh Makrovodorosleĭ) : 88 - 101, 185, 1978. [In R, ab : E.]

36586 - **ZVALINSKIĬ, V.I., SEMEN'KOVA, E.A.** : Spektry vozbuzhdeniya sekundnogo kom-
ponenta poslesvecheniya zelenoĭ vodorosli *Ulva fenestrata*. [Excitation
spectra of the second component of delayed-light emission in the marine
alga *Ulva fenestrata*.] - Sb. Rabot Akad. Nauk SSSR, dal'nevost. nauch.
Tsentr, Inst. Biol. Morya (Vladivostok) *11* (Ékologicheskie Aspekty Fotosinteza Morskikh Makrovodorosleĭ) : 102 - 111, 185, 1978. [In R, ab : E.]

36587 - **ZVALINSKIĬ, V.I., TITLYANOV, É.A., LELĒTKIN, V.A., NOVOZHILOV, A.V.** :
Adaptatsiya korallov k svetu. [Adaptation of corals to light.] - Sb. Rabot.
Akad. Nauk SSSR, dal'nevost. nauch. Tsentr, Inst. Biol. Morya (Vladivostok)
*12* [Biologiya Korallovykh Rifov (Fotosintez Zooksantell i Vodorosleĭ-Makrofitov)] : 29 - 52, 1978. [In R, ab : E.]

Authors' names are presented in the form in which they appear in the respective publication. The names from papers published in Cyrillic characters are transcribed as
shown on p. III of this volume. Alternative spellings and forms of the name of the
same author are usually cross-indexed. The numbers in *italics* refer to publications
in which the respective author acts as an editor.

# A

AARNES, H. 32555
AASE, J.K. 32556
ABAD-ZAPATERO, C. 32557
ABDULAEV, N.G. 32558-9, 35199
ABDULLAEV, Kh.A. 32560-1
ABDULRAHMAN, F.S. 32562
ABDURAKHMANOV, A.A. 32865
ABDURAKHMANOV, I.A. 32563
ABE, A. 36517
ABO-EL LEL, G. 33449
ABRAHAMSSON, J. 33333
ABRAHAMSSON, S. *32733*
ABROS'KINA, L.S. 35191,36305
ABU-SHAKRA, S.S. 32832-3
ACEVEDO, E. 33527
ACKEFORS, H. 32564-5
ACKER, S. 32566, 35670
ACKERSON, R.C. 32567
ACOCK, B. 32568-9, 36000
ADAMS, M.S. 32570, 33441
ADAMS, M.W.W. 32571
ADAMS, S. 32570
ADAMSON, H. 32572
ADARI, H. 35569
ADEDIPE, N.O. 36080
ADELUSI, S.A. 34533
ADOMAKO, D. 35205
ADRIANO, D.C. *34885*
AÉROV, I.L. 32573
AFRIA, B.S. 34980
AFUSOAIE, I. 34876
AGAEV, M.G. 34709
AGAFODOROVA, M.N. 33647
AGALIDIS, I. 34686-7
AGARWALA, S.C. 32574
AGATA, W. 36069-70
AGHION, J. 32575
AGRAWAL, P.K. 32576-7
AGRIKOVA, I.M. 33468
AGRIS, P.F. *34753*, *36466*
AHARONI, N. 32578
AHL, P.L. 36380
AHLUWALIA, A.S. 36209
AHOKAS, H. 32579
AHRENS, E.H.Jr. 32580
AHRING, R.M. 34333
AIBA, S. 35138
AIGA, I. 33998
AÏRAPETYAN, A.L. 32581
AIRINEI, A. 34877
AKAO, S. 32582

AKAZAWA, T. 32583-5, 32683-5, 34103,
    34658, 35088
ÅKERLUND, H.-E. 32631, 33755
AKERS, C.P. 32586
AKETAGAWA, J. 36074
AKHMANOV, S.A. 32587-8
AKHUNDOV, Kh. 36085
AKIMOTO, Y. 32589
AKININA, D.K. 33545
AKIYAMA, T. 33998
AKOYUNOGLOU, G. *32566*, *32572*, *32590-1*,
    *32591*, *B32592*, *32593-5*, *32595*,
    *32650*, *32659*, *32660*, *32664*, *32811*,
    *32889*, *32902*, *32920*, *32932*, *32944*,
    *32958*, *32974*, *33026*, *33113*, *33202*,
    *33256*, *33298*, *33373*, *33389*, *33393*,
    *33419*, *33453*, *33481*, *33481*, *33499*,
    *33501*, *33517*, *33598*, *33603*, *33625*,
    *33639*, *33678*, *33783-4*, *33801*, *33810*,
    *33827*, *33841*, *33875*, *33942*, *33967*,
    *33983*, *34007*, *34026*, *34045*, *34054*,
    *34168*, *34174*, *34236*, *34340*, *34492*,
    *34526*, *34529*, *34578*, *34606*, *34619*,
    *34685*, *34692*, *34739*, *34739*, *34837*,
    *34846*, *34882*, *34922*, *35011*, *35035*,
    *35067*, *35072*, *35132*, *35152*, *35242*,
    *35383*, *35400*, *35479-80*, *35482*,
    *35498*, *35502*, *35519*, *35544*, *35602*,
    *35628*, *35655*, *35689*, *35703*, *35706*,
    *35730*, *35763*, *35835*, *35874*, *35967*,
    *35988*, *35991*, *36023*, *36031*, *36039*,
    *36101*, *36128*, *36193*, *36193*, *36233*,
    *36235*, *36327*, *36370*, *36409*, *36459*,
    *36465*
AKULOVA, E.A. 32596, 34321, 34734,
    35009, 35012, 35566
AKULOVICH, N.K. 32597
ALASAARELA, E. 32598-9
ALBERTE, R.S. 32600, 33195, 34478,
    35380, 36279, 36447
AL'BITSKAYA, O.N. 33148
ALBUZIO, A. 35248-9
ALEINIKOV, I.M. 32601
ALEKSANDROV, A.Yu. 32602
ALESHIN, E.P. 32603, 36493
ALESSIO, M. 34888
ALEXANDER, M. 36455
ALEXANDER, V. 34884
ALFANO, R.R. 35291
ALI, K.H. 32910
ALI, M.M. 35250-1

ALI, S.M.  35455
ALIEV, D.A.  32604-5
ALLAKHVERDIEV, S.I.  34349
ALLAKHVERDOV, B.L.  36054
ALLAN, R.J.  32606
ALLEN, E.J.  34091-2
ALLEN, J.F.  32607-9
ALLEN, L.H. Jr.  34633
ALLEN, M.J.  32610
ALLEN, R.J. Jr.  32611
ALLESSIO, M.L.  32612
ALLEWELDT, G.  34370
ALLMARAS, R.R.  35517
ALLOT-DERONNE, M.  32613
ALMASSY, R.J.  32614
ALMGREN, M.  32615
ALSCHER, R.  32616
ALVIN, R.  32758
AMANO, H.  32617
AMBASHT, R.S.  34498
AMEND, J.  32754
AMES, I.H.  32618
AMESZ, J.  32619-20, 35526
AMEZAGA, A. de  32621, 33722
AMMA, B.S.K.  see  SUMATHY KUTTY
     AMMA, B.
AMOUR, G.T.S.  32956
ANAN'EV, G.M.  32622, 36540
ANDERSEN, K.  32623
ANDERSEN, R.A.  32624
ANDERSON, J.E.  32625, B33281
ANDERSON, J.F.  *33936*
ANDERSON, J.M.  32626, 32889-90
ANDERSON, J.W.  32627, 34141
ANDERSON, L.E.  32628, 33809
ANDERSON, L.L.  33156-7
ANDERSON, W.K.  32629
ANDERSSON, B.  B32630, 32631-2
ANDRÉ, M.  32633
ANDREENKO, T.I.  33114
ANDREEV, N.G.  32634
ANDREEVA, A.N.  35202
ANDREEVA, N.E.  32635
ANDREO, C.S.  32636
ANDREU, J.M.  32637
ANDREWS, T.J.  32638
ANDRIANOV, V.K.  33329
ANGELOV, M.  32639, 35971
ANIKIEV, V.V.  32640
ANISIMOV, A.A.  34475
ANISTRATOVA, N.A.  33174
ANITOFF, O.E.  32641
ANTON, J.A.  32642
ANTONIELLI, M.  32643-7
ANTONIW, L.D.  32648
APASHEVA, L.M.  32649
APEL, K.  32650-1
APEL, P.  32652-4
APPLEBURRY, M.L.  32655
ap REES, T.  34556, 36005
APTE, S.K.  32656, 36002
ARAKI, S.  35178-9
ARAMBOURG, Y.  33573
ARATA, H.  32657

ARDITTI, J.  33892
ARENTS, J.C.  33949
ARESES, M.L.  32731
ARGYRAKIS, P.  32658
ARGYROUDI-AKOYUNOGLOU, J.H.  *32566,*
     *32572, 32591, B32592, 32593-5,*
     *32595, 32650, 32659-61, 32659,*
     *32664, 32811, 32889, 32902, 32920,*
     *32932, 32944, 32958, 32974, 33086,*
     *33113, 33202, 33256, 33298, 33373,*
     *33389, 33393, 33419, 33453, 33481,*
     *33499, 33501, 33517, 33598, 33603,*
     *33625, 33639, 33678, 33783-4, 33801,*
     *33810, 33827, 33841, 33875, 33942,*
     *33967, 33983, 34007, 34026, 34045,*
     *34054, 34168, 34174, 34236, 34340,*
     *34492, 34526, 34529, 34578, 24606,*
     *34619, 24685, 34692, 34739, 34837,*
     *34846, 34882, 34922, 35011, 35035,*
     *35067, 35072, 35132, 35152, 35242,*
     *35383, 35400, 35479-80, 35482,*
     *35498, 35502, 35519, 35544, 35602,*
     *35628, 35655, 35689, 35703, 35706,*
     *35730, 35763, 35835, 35874, 35967,*
     *35988, 35991, 36023, 36031, 36039,*
     *36101, 36128, 36193, 36233, 36235,*
     *36327, 36370, 36409, 36459, 36465*
ARIHARA, J.  32662
ARKEL, H. van  see  VAN ARKEL, H.
ARKHIPOV, V.N.  33429
ARKHIPOVA, N.D.  33177
ARKIN, G.F.  32663
ARMOND, P.A.  32664-8, 32866, 34729,
     35719
ARMSTRONG, J.E.  32669
ARNASON, T.  35879
ARNAUTOVA, A.I.  34318
ARNHEIM, K.  32670
ARNOLD, C.-G.  32671, 32872
ARNOLD, M.H.  *33928*
ARNOLD, W.  32672
ARNON, D.I.  32673-4, 36514
ARNTZEN, C.J.  32675, 32934, 33024-6,
     34609-10, 34990, 35947, 35987-8
ARO, E.-M.  32676, 36234
ARONOFF, S.  32677
ARTECA, R.N.  32678
ARYA, R.S.  36266
ASADA, K.  32679-82, 34229-30
ASADOV, A.A.  35846
ASAHI, T.  36168
ASAMI, S.  32585, 32683-5, 34658
ASANA, R.D.  32686
ASANOV, A.N.  32687
ASAY, K.H.  32912
ASENSI, A.  33312
ASGHAR, A.  32688
ASHCROFT, G.L.  33854
ASKAROV, M.  32816
ASLAM, M.  32689
ASLIN, R.G.  33984
ASTIER, C.  34199
ASTON, A.R.  33395
ASTON, M.J.  32690, 33790

GROSS, J.   35730
GROUT, B.W.W.   33790
GROUZIS, J.-P.   33791-2
GROVES, M.   33793
GROVES, M.R.   33794
GRUMBACH, K.H.   33795-6
GUBAR', G.D.   33797-8, 34439
GUCKERT, A.   34857
GUCSHA, N.I.   33360
GUDKOV, N.D.   33799-800
GUERRERO, M.G.   34745
GUGLIELMI, G.   33312
GUICHERD, R.   32721
GUICHON, G.   33829
GUIGNERY, G.   33801
GUIKEMA, J.A.   33802-3
GUILIZZONI, P.   32570, 33804, *33804*, *35586*
GUILLOT-SALOMON, T.   33502, 33805
GUINN, G.   34794
GULIEV, F.A.   33806
GULIEV, N.M.   33807
GULYAEV, B.I.   33808
GÜNTHER, K.   33141-2
GUPTA, D.S.   see  SEN GUPTA, D.
GUPTA, V.K.   33809
GUREVITZ, M.   33047, 33810, 35143
GURINOVICH, G.P.   36564
GUSEV, M.V.   32774-5, 33718, 33811, 35078, 36251
GUSEV, N.A.   33138, 33812, 35605
GUSS, J.M.   33220
GUTERSTAM, B.   33813
GUTEZEIT, B.   32695
GUTMAN, M.   33078 ·
GUTOWSKY, H.S.   34753, 36490
GUTSCHICK, V.P.   33814
GUZMAN, V.L.   33074
GVARDIYAN, V.N.   35777
GWYNN, G.R.   33815
GYSI, J.   33904
GYURJÁN, I.   33816-7
     see  DYURIAN, I.


H


HAAKSMA, C.   34698
HAAR, H. van de  see  VAN DE HAAR, H.
HABERMANN, H.M.   33818
HACHE, A.   33141-2, 33819
HACHTEL, W.   33820
HACKERT, M.L.   32557
HÄDER, M.   35120
HAEHNEL, W.   33822
HAFFNER, G.D.   35347
HAGAN, R.M.   B33281
HAGAR, W.G.   33823
HÄGELE, W.   33824-6
HAGEMAN, R.H.   35059
HAGEMANN, R.   33827

HAGEN, L.J.   35907
HAGGARD, S.S.   33225
HAGIHARA, B.   33828
HAGIMORI, M.   36495
HAGIWARA, H.   33906
HAJIBRAHIM, S.K.   33829
HÁLA, J.   33830
HALÁSZ, G.   35657-8
HALEVY, A.H.   33831
HALILOV, R.I.   33720
     see  KHALILOV, R.I.
HALL, A.J.   33832
HALL, D.O.   32571, *32583*, *32745*, *32779*, *32801*, *32826*, *32861*, *32874*, *32888*, *32898*, *32916*, *32953*, *32957*, *33037*, *33069*, *33116*, *33128*, *33190*, *33220*, *33250*, *33364*, *33381*, *33400*, *33406*, *33425*, *33454*, *33486*, *33591*, *33754*, *33821*, 33833, *33843*, *33872*, *33903-5*, *33919*, *33933*, *33943*, *33986*, *34106*, 34142, *34204*, *34292*, *34352*, *34378*, *34424*, *34645*, *34650*, *34659*, *34725*, *34741*, *34777*, *34804*, *34841*, *34938*, *35054*, *35064*, *35141*, *35143*, *35188*, *35220*, *35378*, 35456, *35475*, 35487, *35518*, *35669*, *35753*, *35799*, *35873*, *36002*, *36024*, *36118*, *36220*, *36264*, *36382*, *36401*, *36408*, *36512*
HALL, E.A.H.   33834
HALL, J.L.   33554-5
HALL, M.A.   35906
HALL, M.O.   34624
HALL, N.P.   33835-7, 34812, 36114
HALL, R.L.   33838
HALL, Z.   *34083*
HALLAIS, M.-F.   33805
HALLAM, N.D.   33839-40
HALLBERG, C.   33237
HALLDAL, P.   35635, 35944
HALLEGRAEFF, G.M.   34914
HALLER, W.T.   32953
HALLICK, R.B.   33841
HALLIER, U.W.   33842
HALLIWELL, B.   33843-4, 34142
HALLMAN, E.   33845
HALLMÉN, G.   34521
HALMER, P.   32837
HAM, G.E.   33846
HAMANAKA, T.   33993
HAMANN, H.K.   36381
HAMEEDI, M.J.   33847
HAMILTON, C.D.   33028
HAMMANS, J.W.K.  see  KLEINEN HAMMANS, J.W.
HAMMER, L.   33367
HAMMER, P.A.   33848, 36125
HAMMER, U.T.   33923, 34537
HAMMOND, L.C.   32923
HAMPP, R.   33849-50
HANCK, U.   33932
HAND, D.W.   32568-9
HANDA, N.   33851
HANDRO, W.   35310
HANEBUTH, W.F.   35460

This index contains a selection of primary items chosen according to their importance in photosynthesis research and to their relevance and occurrence. The word "Photosynthesis" is not regarded as a main theme, but partial processes, photosynthetic parameters and the factors affecting photosynthesis are listed. The processes and other characteristics are summarized into several main themes when presented in combination with individual factors, *e. g.* carbon fixation pathways, electron transport chain, chlorophyll, gas exchange, ecosystem and plant productivity (*including photosynthate distribution and translocation, and canopy organization and functioning*), photorespiration, resistances to $CO_2$ and water vapour transfer, *etc.*

Several items from branches related to photosynthesis research were also chosen for convenience, *e. g.* dealing with respiration, plant growth and development, water relations, anatomy, bioclimatology, *etc.* These items contain only references to papers within the scope of this bibliography.

# A

Abscisic acid   see   Growth regulators ...

Absorbance in canopy   see   Canopy, radiation profile

Accumulation of dry matter   see   Biomass distribution ...; Dry-matter production ...; Ecosystem production ...

Achlorophyllous cells and organs, respiration   see   Respiration of achlorophyllous tissues in light, light inhibition of respiration

Action spectra   see   Irradiance, spectral composition ...

Adenosine triphosphate   see   ATP

Age of algae, leaf, plant   see   Ontogeny ...; Canopy, leaf age

Agrotechnics and carotenoids   34707

Agrotechnics and chlorophyll   34707

Agrotechnics and ecosystem and plant productivity
        32713, 32738, 32761, 33532, 34500, 35101, 35514, 35620, 35871, 36086, 36388, 36433, 36497, 36400-1

Agrotechnics and electron transport chain   34316

Agrotechnics and gas exchange   32713, 33532, 33677, 33845, 34707, 35201, 36086, 36133, 36497

Agrotechnics and resistances to $CO_2$ and water vapour transfer   32713, 35610

Agrotechnics and respiration   36497

Air-conditioning in photosynthesis measurement   see   Gasometric system, conditioning of air

Air-flow rate   see   Wind ...

Albedo, canopy   see   Canopy, radiation distribution

Algae and photosynthetic bacteria, cultivation (*cf.* also Algae mass cultures productivity)   32623, 34143, 34191-2, 34363, 35109, 35193, 35195, 36001, 36248

Algae and secondary production of reservoirs
        32564, 32941, 33115, 33169, 33466, 33675, 33896, B34248, 34293, 34517, 34763,
        34818, 35298-9, 35436, 35462, 35552-3, 35597, 35715, 36141

Algae, blue-green, chromatophores in  see  Chromatophore ...

Algae carotenoids  see  Xanthophylls of algae

Algae chlorophylls  see  Chlorophylls $c$, $d$

Algae $CO_2$ and $O_2$ exchange  see  Gas exchange in algae

Algae, depth distribution in reservoirs
        32565, 32598, 32705, 32712, 32825, 32999, 33018, 33070-1, 33152, 33238-9,
        33274, 33365, 33367, 33379, 33466, 33531, 33586, 33656, 33676, 33721-2,
        33804, 33847, 33851, 33923, 33977, 34139, 34177, 34216, 34240, 34359, 34395,
        34407, 34474, 34519, 34528, 34537, 34567, 34571, 34592, 34793, 34878, 34994,
        35070, 35196, 35254, 35299, 35347, 35407, 35436, 35462, 35587, 35713, 35734,
        35799, 35841, 35864, 35880, 35896, 35908, 35946, 35965, 36061, 36079, 36123,
        36141, 36155, 36216, 36271, 36295, 36512

Algae in sediments  32998-9, 33320, 33531, 33594, 33877, 34184

Algae in sewage cleaning  34362, 35799

Algae life cycles  see  Ontogeny of algae ...

Algae mass cultures productivity (*cf.* also Algae and photosynthetic bacteria, culti-
        vation)  33728, 34143, 34363, 34404, 34538, 35109, 35195, 35995, 36197

Algae photosynthesis and production
        32564-5, 32570, 32598, 32669, 32694, 32705, 32712, 32814, 32818, 32825,
        32887, 33061, 33070-2, 33081, 33091, 33152, 33163, 33169, 33243, 33334,
        33340, 33365, 33367, 33422, 33428, 33511-2, 33586, 33588, 33619, 33635,
        33656, 33676, 33704, 33722, 33851, 33888, 33896, 33915, 33923, 33965, 33977,
        34016, 34030, 34052, 34058, 34143, 34177, 34184, 34222, 34359, 34395, 34407,
        34436, 34489, 34515, 34555, 34571, 34741, 34811, 34818, 34840, 34870, 34884,
        34951, 34985, 34991-2, 34994, 35070, 35196, 35217, 35299, 35313, 35462,
        35584, 35586, 35597, 35672, 35733, 35797-8, 35872, 35880, 35896, 35908,
        35946, 36061, 36079, 36088, 36117, 36123, 36141, 36153, 36271, 36298, 36477,
        36512

Algae, primary productivity in reservoirs (*cf.* also Chlorophyll and production of
        algae and water reservoirs)
        33852, 34016, 34099, 34387, 34436, 34571, 34818, 34840, 34955, 35127, 35227,
        35244, 35407, 35735, 35965, 36079, 36097, 36141, 36295, 36369

Algae, primary productivity, methods (*cf.* also $O_2$ electrode)
        32809, 32830, 32887, 32941, 32979, 33115, 33163, 33367, 33371, 33404, 33492,
        33535, 33721, 33813, 34247, 34386, 35216, 35546, 35635, 35672, 35797-8, 36149,
        36271, 36286, 36474

Algae synchronous cultures  see  Algae and photosynthetic bacteria, cultivation;
        Ontogeny of algae ...

Altitude  see  Pressure, altitude ...

Amino acids  see  Proteins, amino acids, nucleic acids ...

δ-Aminolaevulinic acid  see  Chlorophyll biosynthesis ...

Anaerobic atmosphere  see  $N_2$, anaerobic atmosphere ...

Antibiotics and carbon fixation pathways  33518, 34065, 35088, 36533

Biological clock   see   Diurnal changes ...

Biomass distribution and redistribution   in plant
        32612, 32629, 32633, 32648, 32652, 32678, 32752, 32762, 32832-3, 32836,
        32965, 32980, 33021, 33066, 33074, 33079, 33082, 33101, 33107, 33162, 33184,
        33196, 33209, 33222, 33226, 33229, 33232, 33255, 33260, 33262, 33279, 33293,
        33306, 33331, 33358-9, 33368-9, 33375, 33377, 33394, 33398, 33411, 33495,
        33507, 33513, 33523, 33536, 33547, 33570, 33611, 33614, 33624, 33655, 33668-
        -9, 33703, 33730, 33740, 33759, 33797, 33833, 33859-60, 33894, 33898-90,
        33921, 33925, 33951, 34037, 34076, 34091-2, 34095-6, 34129, 34137, 34176,
        34190, 34287, 34291, 34308, 34330, 34439, 34453, 34486-7, 34499, 34503,
        34507, 34611-2, 34621, 34635, 34721, 34735, 34754, 34787, 34810, B34849,
        34854, 34873-6, 34880, 34887, 34895, 34933, 34946, 35004, 35014, 35017,
        35021, 35025, 35034, 35039-40, 35047, 35079, 35091, 35117-8, 35154, 35169,
        35174, 35194, 35202-3, 35205, 35221, 35225, 35243, 35255, 35261, 35263,
        35268, 35293, 35301, 35314, 35321, 35335, 35350, 35394, 35448, 35458, 35463,
        35513-4, 35529, 35541-3, 35561, 35574, 35598, 35607, 35610, 35662, 35691,
        35721, 35781, 35785, 35787, 35808, 35815, 35870, 35882, 35884, 35907, 35928,
        35933, 35971, 36010, 36014, 36036, 36068, 36070, 36076, 36112, 36116, 36136,
        36163, 36165, 36201, 36226, 36234, 36275, 36281, 36290, 36315, 36319, 36332,
        36344, 36361, 36368, 36386-7, 36412, 36419, 36428, 36489, 36500, 36515

Biopotentials   see   Chloroplast and chromatophore biopotentials

Biosphere production   see   Ecosystem production ...

Blinks effect   see   Emerson effect, Blinks effect

Books on photosynthesis   see   General aspects ...

Boundary layer of air   see   Resistance, leaf boundary layer

Bundle sheaths   see   Carbon metabolism types ...; Carbon fixation pathways, compari-
        son in mesophyll and bundle sheath cells


C


$^{13}C/^{12}C$ ratio, $\delta^{13}C$   32803-4, 32950, 33473-4, 33587, 33595, 33857, 34131, B34354,
        35029, 35031, 35277, 35489, 35525, 35641, 35645, 35701, 35924, 36103, 36440,
        36444-5

$^{14}C$, $^{11}C$, $^{13}C$,   see   Carbon isotopes ...

$C_3$ pathway of carbon fixation
        32555, 32638, 32765-8, 32803, 32826-8, 32840, 32855, 32888, 32894, 32913,
        32950-1, 32953, 32996, 33002, 33053, 33062, 33069, 33116, 33158, 33235,
        33280, 33296, 33304-6, 33310, 33440, 33465, 33612, 33631, 33696, 33699,
        33759, 33945-6, 34013, 34095-6, 34122-3, 34173, 34251-4, 34295, 34429, 34525,
        34548-9, 34585, 34646-7, 34649, 34713, 34764, 34798, 34907, 34942, 34947,
        35029, 35034, 35140, 35167, 35186, 35267, 35285-6, 35307, 35341, 35425, 35428-
        -31, 35434, 35475, 35535, 35641, 35688, 35701, 35704, B35862, 35895, 35974-5,
        36121, 36146, 36148, 36274, 36303, 36330, 36443-5, 36473, 36480

$C_4$ pathway of carbon fixation
        32555, 32638, 32653, 32742, 32765-6, 32804, 32826-8, 32840, 32867, 32888,
        32894-5, 32913, 32950-1, 32953, 32987, 32996, 33053, 33062, 33080, 33130,
        33158, 33218, 33280, 33304-6, 33425, 33440, 33631, 33696, 33699, 33759,
        33843, 33913-4, 33946, 34013, 34044-5, 34052, 34095-6, 34122-3, 34193, 34252-
        -4, 34256, 34295, 34429, 34453, 34525, 34548-9, 34585, 34646-7, 34649, 34695,
        34709, 34836, 34907, 34916, 34942, 34947, 34958, 35031, 35034, 35139, 35186,
        35286, 35341, 35425-6, 35428-31, 35434, 35444, 35465-9, 35472, 35535, 35641,
                                                                           (continued)

Carotenoids chemical structure   33283, 33862, 34603-4, 34687, 35702, 35786

Carotenoids complexes *in vitro*   32575

Carotenoids complexes *in vivo*   33203-4, 33322, 34493, 34687, 35768, 36272, 36484

Carotenoids degradation   32731, 32998, 33476, 33610, 33796, 34380-1, 34941, 35443, 35949

Carotenoids determination   see also   Pigments determination, sampling and extraction

Carotenoids determination, column chromatography
        32966, 33263, 33303, 33538, 33829, 34114, 34169, 34381, 34425, 34603, 36022, 36405

Carotenoids determination, electrophoresis and other methods   33009, 36448

Carotenoids determination, paper chromatography, thin-layer chromatography
        32770, 33057, 33182, 33263, 33457, 33538, 34068, 34169, 34215, 34425, 34603, 34855, 35823, 36022, 36180, 36579-80

Carotenoids determination, spectral methods
        32770, 33538, 34169, 34212, 34425, 34624, 34855, 35015, 35522, 35652, 36405, 36579

Carotenoids energetic states *in vitro*   33580

Carotenoids energetic states *in vivo*   32947, 33139, 33203, 33337, 34435, 34797, 35745

Carotenoids, enzymes of synthetic and degradation processes   33827, 36499

Carotenoids fluorescence *in vitro*   35987

Carotenoids fluorescence *in vivo*   33203

Carotenoids in flowers   32740, 33457

Carotenoids in model systems   32575, 32991, 33580, 34115, 34435

Carotenoids in mutants   see   Mutants, carotenoids in

Carotenoids in photosynthesis mechanism   32685, 32733-4, 32744, 32800, 32947, 33139, 33186, 33190, 33203-4, 33252-3, 33266, 33337, 33400, 33402, 33484, 33610, 33746-7, 33779, 33838, 34020, 34115, 34179, 34204, 34215, 34344, 34400, 34435, 34607, 34686-7, 34773, 34789, 34797, 35058, 35386, 35653, 35699, 35702, 35728, 35739, 35745, 35763, 35915, 36055, 36193, 36272, 36484, 36503

Carotenoids in physiology of photosynthesis   33629, 36051, 36561

Carotenoids in seeds and fruits   33010, 33057, 34163, 34212, 34403, 34581, 34624, 34879, 35251, 35379

Carotenoids precursors   see   Carotenoids biosynthesis and precursors

Chamber, assimilation   see   Assimilation chamber

Chemiosmotic hypothesis, proton transport in chloroplast
        32596, 32631, 32636, 32657, 32669, 32697, 32703, 32732, 32744, 32792, 32807, 32834, 32935-6, 33142, 33160-1, 33207, 33251-3, 33350, 33400, 33402, 33448, 33452, 33553, 33560, 33606, 33687, 33691, 33743, 33754, 33762, 33778, 33819, 33850, 33864, 33882, 33919, 33932, 33945, 33989, 33994-5, 34025-6, 34032-3, 34039, 34093, 34127, 34186, 34202-3, 34205, 34282, 34300, 34355, 34431, 34605, 34723, 34762, 34769, 34774, 34803-5, 34841, 34868, 34975, 35012, 35054, 35068, 35111, 35188, 35209, 35356, 35392-3, 35402, 35487, 35538,
                                                                              (continued)

Chemiosmotic hypothesis ... (continued)
        35650, 35716, 35728, 35853, 35926, 36131, 36265, 36282, 36382, 36398, 36404,
        36449, 36484, 36490, 36496, 36502-3

*Chlorobium* chlorophyll   see   Chlorophylls, *Chlorobium*

Chlorophyll absorption spectra *in vitro*
        32677, 32744, 32754, 32790, 32999, 33240-1, 33274, 33303, 33308, 33575, 33609,
        33782, 33816, 33830, 33863, 33865, 33952, 33988, 33996-7, 34001, 34014, 34056,
        34086, 34214, 34335, 34381, 34564, 34662, 34722, 34770, 34795, 34860, 34877,
        34981, 35050, 35156, 35167, 35193, 35300, 35449, 35613, 35617, 35668, 35683,
        35741, 35757, 35763, 35833, 35890, 36033, 36209, 36228, 36260, 36470-1,
        36520, 36564

Chlorophyll absorption spectra *in vivo*
        32563, 32597, 32619, 32626, 32676, 32849, 32889-90, 32978, 33024, 33026,
        33031, 33035, 33048, 33064, 33138, 33186, 33189, 33204, 33237, 33298, 33322,
        33374, 33381, 33390, 33392-3, 33403, 33429, 33468, 33552, 33569, 33579,
        33618, 33702, 33706, 33746-7, 33782, 33788, 33874, 33956, 33958, 34007,
        34014, 34024, 34111, 34174, 34244, B34248, 34305, 34327, 34357, 34371, 34431,
        34490-1, 34582, 34594, 34627-8, 34761, 34845, 34957, 35000, 35015, 35066,
        35107, 35166-7, 35193, 35245, 35376, 35482, 35502, 35526, 35556, 35583,
        35613, 35658, 35665, 35668, 35670, 35725, 35757-8, 35762-3, 35835, 35847,
        35897, 35899, 35921-2, 36029-30, 36058, 36118-9, 36151, 36155, 36177, 36239,
        36305, 36355, 36377, 36399, 36411, 36460, 36519, 36585

Chlorophyll and its products determination   see also   Pigments determination, samp-
        ling and extraction

Chlorophyll and its products determination, column  chromatography
        32966, 33303, 33829, 33996-7, 34086, 34113-4,   34144, 34381-2, 35123, 35612,
        35663, 35760, 35837, 35855, 36022, 36525

Chlorophyll and its products determination, electrophoresis and other methods
        32626, 33455, 34758, 35496

Chlorophyll and its products determination, *in vivo*
        32600, 32675-6, 32890, 32988, 33024, 33063, 33131, 33515-6, 33806, 33880,
        34021, 34582, 34729, 34758, 34778, 34890, 35015, 35236, 35291, 35376, 35422,
        35576, 35599, 35665, 36399

Chlorophyll and its products determination, paper chromatography, thin-layer chro-
        matography
        32643, 32770, 32999, 33095, 33182, 33274, 33676, 33863, 33997, 34021, 34113,
        34144, 34169, 34502, 34765, 34770, 34855, 34914, 35156, 35520, 35612, 35663,
        35803, 35823, 35855, 35857, 36022, 36180, 36211, 36471, 36579

Chlorophyll and its products determination, spectral methods  32626, 32748, 32940,
        32988, 33018, 33131, 33577, 33706, 33733,   33880, 33927, 34056, 34069, 34082,
        34086, 34107-8, 34144, 34169, 34335, 34594, B34706, 34722, 34914, 35156,
        35206, 35435, 35520-2, 35556, 35612, 35652, 35756, 35839, 36082, 36211,
        36293, 36350

Chlorophyll and production of algae and water reservoirs
        32564-5, 32599, 32606, 32643, 32645, 32815, 32844, 32910, 32941, 32961,
        32998-9, 33018, 33044-5, 33070-1, 33083, 33238-9, 33274, 33297, 33333, 33340,
        33364-7, 33422, 33514, 33531, 33594, 33634-5, 33656, 33676, 33782-3, 33847,
        33851, 33866, 33896, 33923, 33927, 33965, 33977, 34002, 34030-1, 34034,
        34058, 34069, 34099, 34139, 34177, 34184, 34191-2, 34197, 34216, 34222,
        34240, B34248, 34272-3, 34387, 34394-5, 34489, 34514-5, 34537-8, 34567,
        34592, 34733, 34763, 34768, 34793, 34811, 34870, 34878, 34884, 34985, 34991,
        34994, 35015, 35020, 35027, 35063, 35070, 35127, 35196, 35287, 35360, 35407-
        -8, 35436, 35462, 35493, 35506, 35523, 35533, 35552-3, 35584, 35587, 35597,
                                                                        (continued)

Chlorophyll in mutants  see  Mutants, chlorophyll in

Chlorophyll in photosynthesis mechanism
        32587, 32619-20, 32650, 32658, 32661, 32667, 32672, 32675, 32714, 32733,
        32744, 32792, 32880, 32884, 32890, 32925-6, 32934, 32992, 32994, 33024,
        33037, 33063, 33100, 33120, 33139, 33141, 33186, 33188, 33190, 33204, 33258,
        33311, 33316, 33337, 33381, 33389-90, 33403, 33486, 33515-6, 33563-4, 33605,
        33609, 33626, 33708-9, 33745-8, 33750, 33778-9, 33788, 33814, 33864, 33885,
        33927, 33956, 33964, 33982, 34013, 34020, 34042, 34115, 34186, 34204-5, 34237,
        34277, 34313, 34338, 34344, 34349-50, 34357, 34367, 34419, 34441, 34461,
        34566, 34569, 34572, 34584, 34601, 34607, 34610, 34688, 34756, 34760, 34767,
        34775-6, 34795-6, 34843, 34845, 34860, 34893, 34965, 34968, 34990, 35058,
        35075, 35107-8, 35114, 35218, 35220, 35238, 35269, 35291, 35311, 35317,
        35366, 35386-7, 35422, 35428, 35432, 35526, 35665-6, 35670, 35719, 35725,
        35745, 35820, 35847, 35897, 35945, 35988, 36023-4, 36062, B36081, 36118,
        36193, 36246-7, 36256, 36272, 36276, 36311, 36352, 36355, 36377, 36411, 36430,
        36447, 36469, 36484, 36546, 36567, 36583

Chlorophyll in physiology of photosynthesis
        32574, 32666, 32769, 32899, 33034, 33070, 33076, 33118, 33379, 33499, 33630,
        33647, 33700, 33736, 33816, 33869, 33892, 34070, 34390, 34473, 34515, 34702,
        34729, 34782, 34870, 34932, 35062, 35191, 35270, 35339-40, 35475, 35609,
        35763, 35836, 35866, 35886, 36051, 36332, 36519, 36551

Chlorophyll in seeds and fruits
        32739, 33057, 33348, 34581, 34678, 35251, 35379, 35609, 35746, 35810, 36293

Chlorophyll luminescence  see  Chlorophyll, delayed light emission ...

Chlorophyll, methods  see  Chlorophyll and its products determination ...

Chlorophyll number  see  Chlorophyll in physiology of photosynthesis

Chlorophyll precursors  see  Chlorophyll biosynthesis ...

Chlorophyll unit  see  Photosynthetic (chlorophyll) unit

Chlorophyllase  32865, 33562, 33693, 34639, 34691, 34693, 34934-5, 34941, 35614,
        35810, 35814, 36032, 36098

Chlorophyllase and other enzymes of chlorophyll synthesis and degradation, methods
        36032, 36098

Chlorophylls $a,b$ and their ratio
        32566, 32572, 32590-1, 32593-5, 32601, 32695, 32626, B32630, 32632, 32639,
        32643-7, 32660, 32665-7, 32691, 32706, 32716, 32731, 32739-40, 32742, 32749,
        32769-70, 32819, 32829, 32844, 32865, 32889, 32892, 32910, 32929, 32931,
        32940, 32981, 33002, 33012-4, 33026, 33030, 33064, 33079, 33084, 33122,
        33173, 33177, 33219, 33245, 33268-9, 33280, 33305, 33331, 33336, 33339,
        33348-9, 33351, 33379, 33387, 33410, 33423, 33435, 33472, 33481, 33489,
        33499, 33502, 33513, 33523, 33540, 33543, 33589, 33608, 33655, 33670, 33676,
        33733, 33736, 33761, 33790, 33795-6, 33805, 33816, 33825, 33835, 33842,
        33892, 33909, 33952, 33956, 33987, 33997, 34008, 34016, 34026, 34030, 34042,
        34045, 34098, 34111, 34137, 34153, 34175, 34198, 34214, 34219, 34312, 34318-
        -9, 34437-8, 34441, 34457, 34464, 34468, 34480, 34482-3, 34490-1, 34493,
        34502, 34549, 34567, 34572, 34581, 34588, 34594-6, 34606-7, 34620, 34630,
        34670, 34693, 34702, B34706, 34707, 34722, 34739, 34765-6, 34779, 34790,
        34845, 34869, 34894, 34897, 34917, 34922, 34924-5, 34934, 34937, 34940,
        34945, 34948, 34956, 34959, 34990, 35021, 35035, 35044-5, 35048-9, 35051,
        35059, 35074, 35081, 35092, 35096, 35123, 35132, 35158, 35161, 35167, 35181,
        35232, 35241, 35248, 35250-1, 35261, 35270, 35309, 35321, 35342, 35348,
        35363-4, 35379, 35385, 35408, 35417, 35427, 35432, 35434, 35437, 35440,
        35449, 35454, 35471, 35477, 35496, 35583, 35609, 35617, 35619, 35638, 35657-
        -8, 35763, 35835-6, 35844, B35862, 35867, 35921, 35956, 35970, 35988, 36004,
                                                              (continued)

Compensation irradiance
        33496, 33565, 33736, 33759, 34061, 34445, 34792, 35616, 35762, 36051, 36259

Compensation point, $CO_2$  see  $CO_2$ compensation concentration

Compensation point, light  see  Compensation irradiance

Competition in ecosystem  33226, B 33343, 34541, 35102, 35508

Conductance for transfer of gases  see  Resistance ...

"Contribution" of individual organs  see  Biomass distribution and redistribution;
        Photosynthate translocation ...

Cosmic radiation  see  Ionizing radiation ...

Coupling factor 1  see  ATPase ...

Cover, vegetative  see  Canopy ...; Ecosystem ...

Crassulacean Acid Metabolism  see  CAM

Cultivar differences, carbon fixation pathways  33633, 33658, 35183-4

Cultivar differences, carotenoids  32722, 34214, 34403, 34425, 34493, 34879, 35363-4,
        36262

Cultivar differences, chlorophyll
        32605, 32647, 32729, 32881, 33002, 33034, 33513, 33761, 33815, 34153, 34214,
        34242, 34463-4, 34493, 34782, 34815, 34839, 35184, 35280, 35339, 35363-4,
        35389, 35437, 35454, 35512, 36213, 36262, 36528

Cultivar difference, chloroplast
        32893, 33761, 35296, 35482, 35875, 36047, 36262

Cultivar differences, ecosystem and plant productivity
        32567, 32737, 32965, 32973, 33059, 33066, 33092, 33101, 33222, 33513, 33698,
        33846, 33870, 33894, 34074, 34091-2, 34152, 34309, 34403, 34503, 34666,
        34781-2, 34834, 35184, 35202-3, 35389, 35394, 35454, 35512-4, 35534, 35541,
        35549, 35795, 35977, 36053, 36077, 36178, 36213, 36381, 36528, 36548

Cultivar differences, electron transport chain
        32604, 32786, 33761, 34214, 36300-1, 36556

Cultivar differences, gas exchange
        32689, 32751, 32810, 32965, 32973, 33002, 33110, 33112, 33604, 33698, 33761,
        33894, 33984, 34153-4, 34289, 34463, 34665, 34782-3, 34835, 35184, 35210,
        35222, 35295-7, 35339, 35622, B35862, 35920, 36010, 36064, 36213, 36222,
        36417

Cultivar differences, photorespiration
        32786, 32965, 33633, 34154, 34524, 35339, 36417

Cultivar differences, resistances to $CO_2$ and water vapour transfer
        32810, 32973, 33984, 34109, 35222, 35340, 35563, 35907, 36417

Cultivar differences, respiration  32965, 36178, 36213, 36299

Cultivation of algae and photosynthetic bacteria  see  Algae and photosynthetic
        bacteria, cultivation; Algae mass cultures productivity

Cuticular $CO_2$ and $O_2$ exchange  33355

Cuticular resistance  see  Resistance, cuticular

Cytochromes
        32602, 32614, 32675, 32680, 32701-3, 32711, 32733, 32756-7, 32764, 32799,
        32801, 32842, 32860-1, 32889, 32898, 32906-9, 32955, 32992, 33032, 33036,
        33114, 33119, 33142, 33189-90, 33251, 33253, 33294, 33307, 33309, 33400,
        33402-3, 33406, 33502, 33551-2, 33568-9, 33597-8, 33688, 33708, 33710, 33901,
        33919, 33964, 33983, 34004, 34042, 34138, 34205, 34224, 34277, 34321, 34357,
        34377-8, 34393, 34483, 34566, 34717, 34727, 34762, 34789, 34848, 34858,
        34891, 34974, 34990, 35058, 35066, 35104, 35177, 35236, 35245, 35322, 35420,
        35427, 35479, 35526, 35531, 35566, 35573, 35596, 35736, 35739, 35811-2,
        35818, 35863, 35897, 35915, 35917, 35969, 36118, 36129, 36342, 36355, 36377,
        36385, 36389, 36397-8, 36409, 36411, 36476, 36484-5, 36505, 36547, 36573

Cytochromes, methods
        32702, 32764, 32907, 34004, 34224, 35811-2, 36505, 36573


D


Dark $CO_2$ fixation
        32633, 32742, 32912, 33665, 34369, 34385, 34427, 34429, 34496, 34585, 35029,
        35087, 35097, 35281, 35293, 35527, 35631, 35678, 36044, 36068, 36241, 36512,
        36549

Data recording and processing  33039, 36000

Decapitation  see  Defoliation, decapitation ...

Defoliation, decapitation, ear and root removal, effect on biliproteins  33535

Defoliation, decapitation, ear and root removal, effect on carbon fixation pathways
        32717, 34937

Defoliation, decapitation, ear and root removal, effect on chlorophyll
        32863, 33301, 34508, 36213

Defoliation, decapitation, ear and root removal, effect on chloroplast  32717

Defoliation, decapitation, ear and root removal, effect on ecosystem and plant pro-
        ductivity
        32567, 32713, 33019, 33079, 33181, 33193, 33226, 33291, 33832, 33859, 33936,
        33950, 34037, 34176, 34541, 34936-7, 35056, 35073, 35455, 35549, 35712,
        35888, 35966, 36112, 36213, 36275, 36406, 36458

Defoliation, decapitation, ear and root removal, effect on gas exchange
        32689, 32713, 32717, 33079, 33291, 33832, 33859, 33879, 34132, 34176, 34936-
        -7, 35283, 35673, 36077, 36213, 36406, 36457, 36530

Defoliation, decapitation, ear and root removal, effect on photorespiration  33831

Defoliation, decapitation, ear and root removal, effect on resistances to $CO_2$ and
        water vapour transfer
        32713, 34384, 35283, 36457

Defoliation, decapitation, ear and root removal, effect on respiration
        33859, 34508, 36077, 36213, 36406

Desiccation of tissue  see  Water saturation deficit

Deuterium oxide, tritium oxide  35452

Development, leaf, plant  see  Leaf (and plant) development and ageing

Drought and gas exchange
    32698, 32986, 33440, 33660, 34295, 34368, 34632, 34677, 35594, 35632, 35636-
    -7, 35713, 36213, 36343

Drought and resistances to $CO_2$ and water vapour transfer  34651, 35594

Drought and respiration  33660, 35636, 36213

Dry-matter production, gravimetric determination  36543


E


Ear removal  see  Defoliation, decapitation, ear and root removal ...

Ecosystem production, primary productivity (terrestrial) (cf. also Biomass ...)
    32567, 32574, 32629, 32634, 32640, 32648, 32652, 32662, 32686, 32699, 32737,
    32761, 32783, B32852, 32868, 32882, 32888, 32918, 32950, 32973, 32980,
    33002, 33023, 33043, 33054-5, 33073-4, 33085, 33097, 33123, 33128, 33166,
    33175, 33184, 33196, 33209, 33222, 33255, 33260, 33293, 33331, 33394, 33412,
    33488, 33495, 33528, 33541, 33547, 33567, 33620, 33647, 33668, 33730, 33756,
    33870, 33872-3, 33894, 33951, 33961-3, 33976, 34000, 34012, 34035, 34037,
    34048, 34074, 34076, 34090, 34092, 34110, 34129, 34136, 34190, 34195, 34200,
    34279, 34289-90, 34330, 34351, 34365-6, 34387, 34444, 34471, 34487, 34545,
    34612-3, 34623, 34650, 34720, 34735, 34785, 34794, 34810, 34835, 34873-4,
    34876-7, 34880, 34888, 34919, 34940, 34942, 34948, 34970, 35001-2, 35004,
    35014, 35025, 35039-40, 35055-6, 35062, 35064, 35109, 35129, 35141-2, 35158,
    35221, 35258, 35273, 35283, 35301, 35341, 35346, 35351-2, 35363-4, 35389,
    35445, 35453, 35458, 35463, 35484, 35488, 35500, 35505, 35508, 35512-4,
    35518, 35598, 35608, 35610, 35619, 35662, 35677, 35691, 35740, 35781,
    35787-8, 35808, 35815, 35849, 35870, 35882-5, 35888, 35892-3, 35903, 35928,
    35932, 35973, 35979, 36010, 36014, 36027, 36037, 36076-7, 36107,
    36116, 36136, 36163, 36213, 36267, 36304, 36319, 36344, 36361, 36386, 36401,
    36408, 36417, 36419, 36427, 36458, 36477, 36489, 36565

Ecotypes, geographical types, and carbon fixation pathways
    33149, 34701, 36020, 36092-3, 36445, 36480

Ecotypes, geographical types, and chlorophyll  33435, 34673, 35045, 35947

Ecotypes, geographical types, and chloroplast  32895, 34481, 34926, 35045

Ecotypes, geographical types, and ecosystem and plant productivity
    33377, 33669, 34264, 34963, 35273, 35792, 35954, 36388, 36477

Ecotypes, geographical types, and electron transport chain  33248, 35947, 36556

Ecotypes, geographical types, and gas exchange
    32895, 33149, 33377, 33435, 34547, 34701, 34926, 35045, 35115, 35126, 35224,
    35886, 36480

Ecotypes, geographical types, and photorespiration  34701

Ecotypes, geographical types, and resistances to $CO_2$ and water vapour transfer
    33435, 35355

Efficiency, photochemical (cf. also Irradiance and gas exchange, analysis of light
        curves)
    33433, 33833, 34436, 34452, 34495, 34646-7, 34942, 35003, 35561, 35564,
    36037, 36195-6

Electron paramagnetic resonance  see  EPR, NMR

                                                        (continued)

Gas exchange in photosynthetic bacteria   see   Photosynthetic bacteria, gas exchange
        in

Gas exchange, model   see   Model ...

Gas exchange of organs other than leaf
        32965, 33132, 33566, 34153, 34523, 35030, 35074, 35121, 35161, 35349, 35527,
        36135

Gases, organic and algae productivity  35929

Gases, organic, and carotenoids  33010

Gases, organic, and chlorophyll  33012, 35810, 35814

Gases, organic, and chloroplast  35813

Gases, organic, and ecosystem and plant productivity  33846

Gases, organic, and gas exchange  36064, 36384

Gasometric methods, generally  33500, 33823, 34558, 35281

Gasometric system, closed and semiclosed  32821, 33038, 33278, 33796, 35029, 36406

Gasometric system, conditioning of air  32695, 35275, 36000

Gasometric system, open  33068, 33703, 33845, 34057, 34786, B36191, 36424

General aspects on carbon fixation pathways and electron transport chain; books
        B32592, B32630, B32723, B33140, B33191, B33295, B33447, B33583, B33628,
        B33685, 33844, B34248, B34257, B34354, B34706, B34849, 35013, B35575,
        B35639-40, B35862, B36081, B36191, B36297, 36330, 36566

General aspects on $CO_2$ exchange, photorespiration and productivity; books
        B32723, B32852, B33065, B33191, B33236, B33281, B33343, B33413, B33447,
        B33479, B33510, B33583, B33628, B33662, B33926, B34075, B34354, B34865,
        B35165, B35575, B35826, B35862, B36072, 36090, B36134, B36191

Genetics *cf.* also Mutagens ...; Mutants ...

Genetics and ecosystem and plant productivity
        32652, 32718, 33085, 33254, 33416, 33868, 33898, 33921, 33951, 34137, 34294,
        34308, 34437, 34782, 34876-7, 35025, 35034, 35052, 35268, 35534, 35660,
        35662, 35775, 35781, 35892-3, 35933, 36419

Genetics of carbon fixation pathways
        32902, 32912, 33053, 33076, 33135-6, 33172, 33571, 33686, 33767, 33992,
        34019, 34618-20, 34817, 35034, 35142, 35544, 35877, 35892-3, 35986, 36412

Genetics of carotenoids  32579, 32706, 32865, 33684, 33952, 34761, 35689, 35964

Genetics of chlorophyll
        32579, 32581, 32650, 32706, 32865, 33002, 33076, 33684, 33952, 34137, 34702,
        34761, 34782, 34877, 34897, 35034, 35398, 35440, 35446, 35680, 35689, 35893,
        36202, 36342, 36438

Genetics of chloroplast
        32579, 32650, 32856, 32902, 33180, 33202, 33282, 33766, 33841, 33942, 34762,
        34881, 35034, 35241, 35498, 35544, 35689, 35755, 35991, 36031

Genetics of electron transport chain
        32706, 32709, 33248, 33539, 34066, 34488, 34897, 35440, 35507, 35892-3  ·

Genetics of gas exchange
        32576, 32652, 32718, 32751, 32810, 32912, 33002, 33254, 33539, 33668, 33898,
        33921, 34137, 34310, 34702, 34782, 34792, 34881, 35034, 35634, 35775, 35807,
        35892, 36064, 36067, 36222, 36418-9, 36521, 36523

Genetics of photorespiration   32718, 32822

Genetics of resistances to $CO_2$ and water vapour transfer
        32810, 33106, 33571, 34792, 36418, 36523

Genetics of respiration   34137, 35807

Glycollate metabolism   see   Photorespiration ...

Glyoxysome   see   Peroxisome ...

Granum   see   Thylakoid ...

Gravimetric determination of photosynthesis   see   Dry-matter production ...

Gross photosynthetic rate
        32882, 32923, 32967, 33052, 33118, 33123, 33215, 33226, 33243, 33260, 33262,
        33379, 33428, 33496, 33561, 33566, 33741, 33915, 34037, 34101, 34485, 34535-
        -6, 34625, 34799, 34840, 35001, 35074, 35092, 35154, 35339, 35449, 35543,
        35611, 35664, 35673, 35761, 35791, 36067-8, 36133, 36198

Growth analysis, methods
        32891, 33232, 33432, B34075, 34090, 34784-5, 34913, 35174, 35902-3, 36391,
        36393

Growth analysis, net assimilation rate, leaf area ratio, relative growth rate
        32336, 32648, 32696, 32737-8, 32758, 32808, 32868, 32986, 33002, 33020,
        33028, 33050, 33074, 33096-7, 33167, 33184, 33195, 33278, 33306, 33330,
        B33343, 33368, 33375, 33394, 33416, 33449, 33508, B33510, 33547, 33559,
        33641, 33740, 33756, 33859, 33868, 33934, 33951, 33962, 33976, 34037, 34048,
        34095-6, 34122, 34134-6, 34152, 34176, 34206, 34269, 34291, 34308-9, 34325,
        34348, 34459, 34486-7, 34498, 34500, 34545, 34666, 34676-7, 34720, 34732,
        34781, 34794, 34808, 34830, 34834, 34939-40, 34946, 34959, 34972, 35003,
        35008, 35025, 35047, 35080, 35101, 35117-8, 35129, 35154, 35158-9, 35170,
        35192, 35244, 35261-3, 35272-3, 35341, 35448, 35500, 35505, 35512, 35514,
        35543, 35558, 35595, 35608, 35620, 35660, 35662, 35791, 35870-1, 35882,
        35884, 35892-3, 35903, 35905, 35914, 35931, 35933, 35971, 35980, 36010,
        36036, 36044, 36070, 36076-7, 36136, 36252, 36304, 36348, 36360, 36419,
        36428, 36501, 36528, 36571

Growth analysis, specific leaf area, leaf area index, leaf area duration
        32556, 32629, 32640, 32662, 32737-8, 32758, 32761, 32868-9, 32939, 32965,
        32973, 32986, 33019, 33022, 33096-7, 33123, 33184, 33195, 33222, 33242,
        33255, 33293, 33306, 33330-1, 33358-9, 33368, 33377, 33379, 33395, 33398,
        34433, 33435, 33449, 33495, 33508, B33510, 33542, 33547, 33565, 33572, 33669,
        33725, 33728, 33740, 33808, 33868, 33894, 33928, 33934, 33951, 33961-2,
        34000, 34037, 34048, 34076, 34091, 34135-6, 34151-2, 34181, 34206, 34228,
        34243, 34262, 34269, 34294, 34308-9, 34326, 34330, 34373, 34442, 34486,
        34500, 34622, 34676-7, 34690, 34720, 34732, 34782, 34794, 34808, 34829,
        34834, 34937, 34940, 34946, 34972, 35003, 35025, 35039, 35047, 35092, 35101,
        35103, 35154, 35158-9, 35169-71, 35202-3, 35262-3, 35273, 35303, 35340,
        35349, 35448, 35484, 35505, 35512, 35514, 35517, 35529, 35550, 35595,
        35607-8, 35632, 35660, 35662, 35691, 35775, 35790-1, 35808, 35870-1, 35882,
        35884, 35892-3, 35902, 35904-5, 35905, 35914, 35953-4, 35977, 36012, 36014,
        36027, 36068, 36075, 36076-7, 36120, 36133, 36136, 36225, 36242, 36252,
        36304, 36315, 36361, 36419, 36421, 36428, 36436, 36489, 36500-1, 36515,
        36521, 36528, 36548

Growth regulators and algae productivity   34474

Growth regulators and carbon fixation pathways  32930, 33344, 33518, 33835, 33978,
     34320, 34576, 34710, 34922

Growth regulators and carotenoids  32741, 33463, 33518, 33629, 34869, 36331

Growth regulators and chlorophyll
     32578, 32847, 32863, 32929-31, 33031, 33302, 33405, 33463, 33489, 33775,
     34209, 34225, 34267, 34320, 34322, 34869, 34922, 35021, 35023, 35065, 35133,
     35248, 35310, 35464, 35614, 35695, 35806, 36039, 36196, 36215, 36296, 36493,
     36520

Growth regulators and chloroplast (chromatophore)
     32717, 32929-31, 33405, 33463, 33518, 33692, 34249, 34320, 34579, 34606,
     34922, 35065, 35384, 35602, 36296, 36370, 36488

Growth regulators and ecosystem and plant productivity
     32567, 32847, 33079, 33270, 33318, 33449, 33831, 33935, 33970, 34266, 34275,
     34936, 35021, 35117, 35785, 35980

Growth regulators and electron transport chain
     32862, 32930, 34093, 34693, 35009, 35134, 35843, 36576

Growth regulators and gas exchange
     32929-31, 33079, 33096, 33288, 33388, 33489, 33835, 33969, 33978, 34125,
     34267, 34497, 34632, 34710, 34923, 34936, 35057, 35074, 35117, 35184, 35248,
     35504, 35759, 35895, 35939, 35980, 36196

Growth regulators and photorespiration  35248

Growth regulators and resistances to $CO_2$ and water vapour transfer  33286, 33985,
     34267

Growth regulators and respiration  33978, 35117

"Growth" respiration  see  Respiration, "growth"...


H


$H_2$ evolution, photoreduction
     32789, 32860, 32897, 32937, 33027, 33325, 33442, 33459, 33505, 33563, 33710-
     -1, 33734, 33833, 34041, 34130, 34161-2, 34217, 34303, 34328, 34343, 34405,
     34417, 34466, 34476, 34801, 34858, 34863, 34902-4, 35075, 35191, 35207,
     35304-7, 35315, 35456, 35649, 35856, 36002, 36025, 36096, 36189, 36359, 36492,
     36513, 36536, 36541, 36578

$H_2$ isotopes  see  Deuterium ...

$H^+$ transport in chloroplast  see  Chemiosmotic hypothesis

*Halobacterium* photosynthesis
     32558-9, 32655, 32727, 32735, 32781, 32793, 32800, 32878-9, 32901, 32903,
     33046, 33063, 33075, 33099, 33228, 33267, 33275-7, 33313, 33326-8, 33376,
     33421, 33437, 33443-4, 33636-8, 33650, 33681-2, 33752, 33833, 33903, 33924,
     33947-9, 33966, 33972-3, 33993, 34015, 34028, 34077-8, 34083-5, 34112, 34221,
     34263, 34268, 34270, 34345, 34391-2, 34398, 34449, 34516, 34590-1, 34662,
     34667-8, 34748, 34900, 34977, 35120, 35135, 35185, 35189, 35199, 35208,
     35228-9, 35253, 35312, 35486, 35551, 35565, 35568, 35718, 35722-3, 35782,
     35805, 35817, 35829-31, 35913, 35926, 35951, 36006-8, 36042, 36131, 36160-1,
     36317-8, 36321-2, 36349, 36374-5, 36380

Hatch-Slack cycle   see   $C_4$ pathway ...

Herbicides   see   Pesticides, herbicides ...

Heterogeneity of leaf blade (organ) and carbon fixation pathways   34473, 34618, 34749,
        35427-8, 35433, 35467

Heterogeneity of leaf blade (organ) and carotenoids   35400, 35768

Heterogeneity of leaf blade (organ) and chlorophyll   33108, 33523, 33630, 34043,
        34045, 35400, 35427, 35658, B36191

Heterogeneity of leaf blade (organ) and chloroplast   32859, 33170, 33498, 33502,
        33578, 34044, 34155, 34705, 34956, 35604, 35658, B35862, B36191

Heterogeneity of leaf blade (organ) and electron transport chain   33170, 33869, 34043,
        35059, 35427, 35433

Heterogeneity of leaf blade (organ) and gas exchange   32826, 33630, 33869, 34473,
        34835, 35059, 35504

Heterogeneity of leaf blade (organ) and photorespiration   34256

Heterogeneity of leaf blade (organ) and resistances to $CO_2$ and water vapour transfer
        34334

Heterotrophy   see   Carbon metabolism types ...

Hill reaction   see   Photosystem 2 activity ...

Hill reaction,methods   see   Photosystem·2 activity, methods

Humidity of air and carbon fixation pathways   33130

Humidity of air and chlorophyll   33014, 36039, 36235

Humidity of air and chloroplast 36487

Humidity of air and ecosystem and plant productivity   33020, 33060, 33073, 33330,
        35007, 36388

Humidity of air and electron transport chain   36235

Humidity of air and gas exchange   32810, 32824, 33060, 33441, 33984, 34260, 34646,
        34966, 35395, 35624, 35886, 36069, 36358

Humidity of air and resistances to $CO_2$ and water vapour transfer
        33000, 33020, 33050, 33451, 33547, 33984-5, 34636, 34646, 34829, 35537, 35594,
        36069, 36075, 36358

Humidity of air, methods  (*cf.* also Infra-red analyser for water vapour)   32877

Hydration level of leaf and carbon fixation pathways   33654, 34045, 34368, 34535,
        34949, 35326

Hydration level of leaf and carotenoids   34480, 34483, 35962

Hydration level of leaf and chlorophyll   32805, 33187, 33292, 33405, 33654, 34045,
        34464-5, 34480, 35962, B36081, 36439

Hydration level of leaf and chloroplast   32838, 33138, 33405, 33524, 33659, 33812,
        33839-40, 34129, 34440, 34480-3, 34663, 35094, 35603, 35605, 36526

Hydration level of leaf and ecosystem and plant productivity   33020, 33222, 33928,
        34181, 34831, 35194, 35904-5, 36205, 36325

Hydration level of leaf and electron transport chain  33187, 33659, 34663, 36439, 36526

Hydration level of leaf and gas exchange  32777, 32837, 32997, 33002, 33029, 33050, 33249, 33292, 33441, 33659, 33812, 33839, 34129, 34181, 34288, 34295, 34302, 34497, 34518, 34535, 34547, 34649, 34792, 34886, 35370, 35476, 35623, 35895, 36050, 36060, 36069, B36081, 36090, 36135, 36205, 36207

Hydration level of leaf and photorespiration  34535, 35324, 35370

Hydration level of leaf and resistances to $CO_2$ and water vapour transfer 32777, 33002, 33022, 33286, 33451, 33527, 33659, 33928, 33984-5, 34109, 34181, 34535, 34541, 34651, 34792, 34851, 34886, 35354, 35369-70, 35476, 35492, 35529, 35594, 35963, 36050, 36069, 36206, 36431

Hydration level of leaf and respiration  32837, 33925, 34440, 35370, 36205

Hydrogen  see  $H_2$ ...

Hydrogenase  see  $O_2$ evolution mechanism and kinetics; $H_2$ evolution ...

Hygrometer  see  Humidity of air, methods; Infra-red analyser for water vapour

I

Ideotype  see  Model ...

Immobilization of chloroplasts and photosynthetic systems  see  Photosystems stabilization ...

Induction phenomena  see  Transient phenomena ...

Infra-red analyser for $CO_2$
33068, 33192, 33278, 33361, 34786, 35235, 35275, B36191

Infra-red analyser for water vapour  35271

Infra-red radiation, effect on photosynthetic parameters  see  Irradiance, spectral composition ...; Temperature, high ...

Inhibitors of electron transport chain (*cf*. also Pesticides ...; Antibiotics ...)
32590, 32601, 32609, 32631, 32636, 32684, 32715, 32719, 32734, 32747, 32755, 32767-8, 32789, 32791, 32798, 32820, 32822, 32834, 32848, 32853, 32860, 32876, 32900, 32942, 32954, 32971, 33137, 33144, 33154, 33157, 33273, 33309, 33323, 33329, 33350, 33460, 33462, 33482, 33496, 33499, 33509, 33606, 33655, 33716--7, 33743-4, 33749, 33802, 33850, 33883, 33886, 33895, 33906, 33935, 33980, 33990-1, 34025, 34027, 34053, 34103, 34106, 34117, 34127, 34173, 34259, 34266, 34275-6, 34280, 34286, 34299, 34305, 34344, 34355, 34372, 34377, 34400, 34405, 34411, 34509, 34520-1, 34531, 34539, 34560-1, 34585, 34643, 34657, 34665, 34671, 34675, 34703, 34728-9, 34766, 34773, 34817, 34843-5, 34848, 34866, 34867, 34891, 34964, 34967, 34984, 34998, 35009, 35024, 35058, 35068, 35072, 35104, 35136, 35144, 35187, 35204, 35212-3, 35245, 35305, 35307, 35327-9, 35343, 35362, 35367-8, 35384, 35391, 35401-2, 35404, 35413, 35418, 35444, 35447, 35468, 35471-2, 35485, 35495, 35511, 35516, 35519, 35524, 35555, 35557, 35566, 35591, 35606, 35635, 35669, 35681-2, 35684-5, 35692, 35697, 35699, 35719, 35736, 35748, 35756, 35858, 35894, 35915-6, 35918, 35935, 35955, 35982, 35985, 36023, 36035, 36043, 36099, 36122, 36173-5, 36185, 36200, 36214, 36236, 36250, 36276, 36316, 36366, 36389, 36395-7, 36422, 36439, 36453, 36460-2, 36488, 36495, 36496, 36502-3, 36508, 36511, 36541-2, 36557, 36561

Irradiance (PhAR) and chlorophyll
        32572, 32595, 32645-6, 32714, 32749, 32763, 32805, 32814-5, 32820, 32929,
        32975, 32977-8, 33002, 33118, 33147, 33177, 33205, 33339, 33379, 33389-90,
        33418, 33482, 33551, 33596, 33602, 33608, 33761, 33901, 33952-3, 34174-5,
        34219, 34246, 34342, 34426, 34438, 34441, 34596, 34609, 34670, 34702, 34728,
        34736, 34766, 34774, 34786, 34791, 34934, 35044, 35062-3, 35137, 35143,
        35184, 35191, 35237, 35241, 35261, 35269, 35333, 35381, 35483, 35596, 35617,
        35619, 35689, 35762, 35773, 35777-8, 35919, 35923, 36097, 36104, 36119,
        36139, 36151, 36164, B36191, 36233-4, 36310, 36332, 36341-2, 36394, 36411,
        36494-5, 36534, 36585, 36587

Irradiance (PhAR) and chloroplast
        32595, 32661, 32763, 32774, 32929-31, 32974, 32978, 33602, 33606, 33737,
        33805, 34129, 34160, 34175, 34237, 34606, 34700, 34928, 35066, 35071-2,
        35157, 35241, 35391, 35482, 35585, 35689, 35714, 35749, 35755, 35861, B35862,
        B36191, 36233, 36291, 36309-10, 36326, 36488, 36582

Irradiance (PhAR) and ecosystem and plant productivity
        32629, 32648, 32794, 32976, 33019, 33073-4, 33107, 33330, 33368, 33375,
        B33413, 33513, 33698, 33728, 33797, 33808, 33831, 33897, 34005, 34036, 34269,
        34275, 34279, 34289-90, 34387, 34439, 34503, 34541, 34719, 34810, 34867,
        34959, 35052, 35062, 35102, 35151, 35170, 35184, 35261, 35263, 35301, 35331,
        35351-2, 35416, 35455, 35619, 35849, 35882, 36070, 36097, 36112, 36143,
        36163, B36191, 36288, 36290, 36304, 36348, 36388, 36394, 36494

Irradiance (PhAR) and electron transport chain
        32591, 32595, 32604, 32673, 32733, 32834, 32867, 32889, 33004-5, 33323,
        33389, 33482, 33502, 33513, 33551, 33582, 33714, 33749, 33761, 33869, 33901,
        34219, 34234, 34300, 34432, 34441, 34521, 34588, 34589, 34607, 34746, 34769,
        34774, 34805, 34841, 34868, 35143, 35157, 35187, 35240-1, 35269, 35273,
        35410, 35413, 35530, 35617-8, 35773, 35780, 35819, 35879, 36122, 36139,
        36234, 36244, 36250, 36310, 36397, 36569

Irradiance (PhAR) and gas exchange
        32568-70, 32573, 32613, 32648, 32665-6, 32685, 32777, 32787, 32804, 32814,
        32824, 32830, 32836, 32867, 32876, 32965, 32986, 33002, 33028-9, 33039,
        33052, 33070, 33076, 33084, 33107, 33118, 33121, 33153, 33175, 33212-3,
        33231, 33244, 33262, 33280, B33281, 33292, 33305, 33324, 33335, 33368, 33375,
        33377, 33379, 33384, 33431, 33435-6, 33465, 33482, 33496, 33508, B33510,
        33513, 33551, 33565, 33588, 33611, 33621, 33677, 33694, 33696, 33698-9,
        33703, 33725, 33736, 33740, 33742, 33761, 33798, 33808, 33813, 33823, 33831,
        33847, 33878-9, 33901, 33915, 33976, 33984, 33999, 34005, 34036, 34100,
        34143, 34162, 34174, 34193, 34218-9, 34260, 34288-90, 34302, 34374-6, 34402,
        34404-5, 34417, 34423, 34429, 34445, 34451, 34471, 34484, 34495, 34497,
        34518, 34535, 34547, 34562, 34570, 34625, 34632, 34635, 34644, 34646-7,
        34651, 34664, 34676-7, 34689, 34702, 34718-9, 34755, 34786-7, 34792, 34800,
        34810, 34832-3, 34853, 34860, 34886, 34888, 34899, 34944, 34961-2, 34972,
        35028, 35044, 35062, 35072, 35074, 35096, 35129, 35131, 35140, 35153, 35171-
        -3, 35184, 35224, 35243, 35261, 35269, 35273, 35293, 35297, 35314, 35349,
        35373-4, 35381, 35395, 35412, 35430, 35470, 35476, 35543, 35559, 35594,
        35616, 35622-4, 35643, 35755, 35761-2, 35762, 35765, 35773, 35780, 35802,
        35895, 35946, 35954, 36049, 36051, 36058-61, 36064, 36069, 36077, 36085,
        36133, 36135, 36137, 36150, 36152, 36190, B36191, 36198-9, 36207, 36234,
        36270, 36332, 36348, 36381, 36388, 36410, 36411, 36417, 36424, 36434, 36457,
        36472, 36480, 36497, 36509, 36550-2, 36570, 36584

Irradiance (PhAR) and gas exchange, analysis of light curves
        32687, 33124, 33175, 33244, 33513, 33694, 33741, 33759, 33798, 33879, 34070,
        34101, 34123, 34219, 34330, 34373, 34467, 34792, 34899, 35092, 35105, 35129,
        35140, 35158, 35184, 35200, 35281, 35314, 35664, 35674, 35773, 35807, 36067-
        -8, 36077, 36132, B36191, 36195, 36203, 36417, 36569-70

Irradiance (PhAR) and photorespiration
        33324, 33785, 34375, 34718, 35044, 35334, 36280, 36562

Irrigation and gas exchange  33857, 34512, 35192, 35624, 35912, 36207, 36560

Irrigation and resistances to $CO_2$ and water vapour transfer  33857, 35815, 36017,
        36207

Irrigation and respiration  32980, 36560

## K

Kok effect  33231, 33385, 34123

## L

Leaf anatomy (*cf*. also Leaf thickness)
        32568, 32572, 32653, 32717, 32929, 32931, 33080, 33093, 33158, 33168, 33280,
        33429, 33500, 33513, 33524, 33534, 33549, 33602, 33695, 33759, 33937, 33974,
        34052, 34260, 34310, 34416, 34471, 34510, 34548, 34597, 34648, 34730, 34827,
        34924-6, 35031, 35045, 35051, 35091, 35096, 35122, 35200, 35261, 35429-30,
        35450-1, 35460, 35604, 35659, 35708, 35842, 35924, 35931, 35936, 36057, 36190,
        B36191, 36217, 36432, 36521, 36552

Leaf and plant development and ageing, morphology (*cf*. also Ontogeny ...)
        32696, 33835, 34176, 34243, 34279, 34318, 34552, 34937, 35021, 35106, 36109,
        36196, 36345, 36478-9, 36509

Leaf area duration  see  Growth analysis, specific leaf area ...

Leaf area index  see  Growth analysis, specific leaf area ...

Leaf area measurement  32846, 33051, 33573, 34088, 34737, 35707, 36218, 36221, 36237,
        36313

Leaf area ratio  see  Growth analysis, net assimilation rate ...

Leaf chamber  see  Assimilation chamber

Leaf dimensions  see  Extension growth, leaf dimensions

Leaf epidermis, anatomy
        32586, 32610, 32690, 32696, 32895, 33288, 33434-6, 33506, 33916, 33937,
        34132, 34334, 34384, 34416, 34472, 34636, B34865, 35016, 35096, 35122, 35234,
        35283, 35460, 35491, 35594, 35708, 35773, 35920, 36080, 36206, 36432, 36524

Leaf epidermis, stomata (*cf*. also Amfistomatous leaf, gas exchange in; Resistance,
        stomatal...)  35886

Leaf life span, plastochron index
        32648, 32696, 33290, 33346, 33669, 33797, 33899, 33976, 34036, 34506-7,
        34794, 35122, 35578, 35664, 35867, 36386, 36479

Leaf morphology
        32648, 32696, 32717, 32870, 32929, 32939, 33106, 33222, 33278, 33330, 33368,
        33375, 33411, 33416, 33495, 33513, 33559, 33669, 33703, 33728, 33737, 33740,
        33790, 33797, 33808, 33860, 33894, 33898-9, 33951, 34095-6, 34132, 34176,
        34243, 34294, 34319, 34326, 34486, 34523, 34754, 34794, 34829, 34924, 34960,
        34963, 35021, 35045, 35051, 35065, 35073, 35154, 35169, 35183-4, 35247,
        35255, 35261, 35263, 35350, 35389, 35459, 35492, 35528, 35542-3, 35558, 35632,
                                                                          (continued)

Leaf morphology (continued)
        35664, 35744, 35787, 35867, 35882, 35884, 35892-3, 35904-5, 35907, 35933,
        35936, 35971, 36010, 36044, 36070, 36076, 36116, 36195-6, 36213, 36254, 36313,
        36346, 36393, 36428, 36521, 36528, 36545

Leaf movements  33627, 34551-2

Leaf optical properties (*cf.* also Carotenoids absorption spectra *in vivo* ; Chloro-
        phyll absorption *in vivo*)
        32665, 32956, 33174, 33432, 33434, 33436, 33471, 33540, 33998, 34550, 34784,
        35449, 35491, 35842, 36011, 36058, 36138, B36191, 36203

Leaf resistance  see  Resistances to water vapour ...; Resistance, stomatal ...

Leaf, sun- and shade leaf  see  Leaf anatomy

Leaf temperature (methods and results)  33436, 33438, 33695, 33984, 34219, 34333-4,
        34451, 34485, 34511, 34747, 34755, 34814, 35096, 35344, 35529, 35595, 35930,.
        36050, 36069, 36348, 36403, 36545

Leaf temperature measurement  see  Leaf temperature (methods and results)

Leaf thickness  32652, 33107, 33158, 33368, 33513, 33694-5, 34265, 34310, 34597,
        34730, 34783, 34827, 35045, 35065, 35091, 35096, 35122, 35261, 35350, 35514,
        35708, 35842, 35931, B36191

Light  see  Irradiance ...; Canopy, radiation ...

Lighting system  see  Irradiation, illumination equipment and systems

Linear dichroism  see  Dichroisms ...

Lipids, fatty acids, and carbon fixation pathways  33796, 33939, 34182, 34820,
        35005, 35851

Lipids, fatty acids, and carotenoids  34941

Lipids, fatty acids, and chlorophyll  32876, 32890, 33308, 35284, 35754

Lipids, fatty acids, and chloroplast (chromatophore)
        32632, 32668, 32858-9, 32876, 32893, 32917, 33113, 33373, 33387, 33481,
        33939-40, 33964, 33994-5, 34180, 34297, 34307, 34426, 34446, 34820, 34856,
        34928-9, 34931, 35054, 35066, 35296, 35400, 35438, 35749, 36101, 36177,
        36238, 36262

Lipids, fatty acids, and electron transport chain  34232-3, 36240

Lipids, fatty acids, and gas exchange  33521, 36557

Lutein  see  Carotenoids ...; Xanthophylls ...


M

"Maintenance" respiration  see  Respiration, "growth" and "maintenance"

Malate dehydrogenase, methods  see  Malic enzyme, malate dehydrogenase, methods

Malic enzyme, malate dehydrogenase
        32867, 32912, 33099, 33148, 33554-5, 33612, 33849, 33912-3, 34147, 34253,
        34256, 34259, 34353, 34525, 34549, 34684, 34950, 34952, 35192, 35430, 35434,
        35442, 35467, 35471, 35675, 36019, 36121, 36241, 36366, 36426, 36443, 36533

                                                                   (continued)

Mutants, chlorophyll in (continued)
  34770-1, 34779-80, 34858, 34883, 34959, 35035, 35191, 35376, 35502, 35560,
  35596, 35617-8, 35633, 35687, 35706, 35763, 35835, B35862, 35887, 35901,
  35922, 35970, 36029, 36119, 36156-7, 36187, 36230, 36258, 36266, 36268,
  36409, 36411

Mutants, chloroplast (chromatophore) in
  32651, 32811, 32958, 32975, 33598, 33625, 33816-7, 33827, 33861, 33942,
  34026, 34491, 34642, 34692, 34762, 34771, 34883, 35026, 35035, 35067, 35071,
  35128, 35700, 35703, 35873-4, 35922,  35967, 35988, 36257

Mutants, ecosystem and plant productivity of  33899, 34959

Mutants, electron transport chain in
  32811, 32860, 32889, 33186, 33190, 33300, 33311, 33597-8, 33842, 33853,
  33901, 34026, 34098, 34185, 34214, 34303, 34478, 34490, 34529, 34607, 34726-
  -7, 34762, 34801, 34858, 34882-3, 35376, 35617-8, 35700, 35763, 35780,
  35922, 36257-8, 36409, 36411

Mutants, gas exchange in
  32892, 33816, 33842, 33899, 33901, 34404, 34456, 34779, 34801, 34858, 34905,
  34958-9, 35035, 35121, 35191, 35856, 36411

Mutants, photorespiration in  34379, 34905, 35152

Mutants, photosynthetic, isolation and selection  32623, 35780

Mutants, resistances to $CO_2$ and water vapour transfer in  34959

Mutants, respiration in  34490, 34905


N


$N_2$, anaerobic atmosphere and chlorophyll  36292

NAD  see  NADP ...

NADP, NAD  32656, 32674, 32848, 33402, 33462, 33710, 33811, 33911, 34013, 34080,
  34133, 34166-7, 34250, 34255, 34281, 34299, 34321, 34356, 34400, 34446,
  34573-5, 34746, 34767, 34866, 34952, 34982, 35012, 35036, 35075, 35077,
  35131, 35136, 35305, 35387, 35427, 35434, 35566, 35581, 35591, 35676, 35818,
  35969, 36043, 36240, 36301

NADP, NAD, methods  33456, 35978

Net assimilation rate  see  Growth analysis, net assimilation rate ...

Net photosynthetic rate  see  Gas exchange ...

Nitrogen  see  $N_2$ ...; Mineral elements (N, P, K) ...

NMR  see  EPR, NMR

Nuclear magnetic resonance  see  EPR, NMR

Nucleic acids  see  Proteins, amino acids, nucleic acids ...

# O

Ontogeny of algae and chloroplast (chromatophore)  32854, 33225, 33810, 35241

Ontogeny of algae and ecosystem and plant productivity  34638, 35109

Ontogeny of algae and electron transport chain
        33001, 33047, 33810, 34589, 34845, 35241, 36214, 36409, 36411

Ontogeny of algae and gas exchange
        32622, 32787, 32813, 33001, 33423, 34070, 34473, 34589, 35241, 35712, 36195-
        -6, 36214

Ontogeny of canopy and ecosystem and plant productivity
        33073, 33199, 34365, 35787, 36138, 32681

Ontogeny of chloroplast and carbon fixation pathways  35241

Ontogeny of chloroplast and carotenoids  35071, B35862

Ontogeny of chloroplast and chlorophyll
        32778, 32890, 33108, 33389-90, 33598, 33781, 34922, 35071, 35602, B35862

Ontogeny of chloroplast and electron transport chain
        32594, 32799, 33390, 34042, 35071, 35241, 35731, 36177, 36487

Ontogeny of chloroplast and gas exchange  32799, 33390, 33916

Ontogeny of leaf, insertion level, and carbon fixation pathways
        33273, 33654, 33731, 33835, 33893, 34683, 34987, 34937, 34949, 34972, 35023,
        35051, 35087, 35247-8, 35294, 35877, 35970, 36111, 36114, 36132, 36327,
        36426, 36440, 36442, 36450-1, 36527

Ontogeny of leaf, insertion level, and carotenoids
        33015, 33174, 33269, 33303, B33628, 34111, 34457, 34707, 34752, 35237,
        35768, 36551

Ontogeny of leaf, insertion level, and chlorophyll
        32605, 32639, 32643, 32646-7, 32739, 32863, 33015, 33173-4, 33268-9, 33301,
        33303, 33348, 33523, 33589, B33628, 33630, 33654-5, 33731, 33815, 33835,
        33889, 33909, 33953-4, 34111, 34319, 34348, 34457, 34463, 34480, 34508,
        B34706, 34707, 34937, 34956, 34959, 34972, 35021-3, 35035, 35042, 35051,
        35081, 35137, 35237, 35247-8, 35257, 35280, 35294, 35613, 35774, 35777,
        35866, 35906, 35970, 36111, 36137, 36195-6, 36235, 36378, 36450, 36527-8,
        36551

Ontogeny of leaf, insertion level, and chloroplast
        32895, B33628, 33889, 34319, 34457, 34705, 34956, 35051, 35603-5, 36190,
        36262

Ontogeny of leaf, insertion level, and ecosystem and plant productivity
        32612, 33040, 33043, 33050, 33181, 33368, 33467, 33542, 33655, 33671, 33697,
        33703, 33950, 34036, 34040, 34348, 34498, 34577, 34808, 34831, 34876, 34959,
        35021, 35039, 35255, 35455, 35529, 35542, 35712, 35744, 35866-7, 35998,
        36075, 36136-8, 36308, 36571

Ontogeny of leaf, insertion level, and electron transport chain
        32706, 32739, 34233-4, 34319, 34462, 34588, 35764, 35767, 35774, 36262,
        36300, 36581

Ontogeny of leaf, insertion level, and gas exchange
        32613, 32639, 32654, 32751, 32777, 32895, 32923, 33002, 33015, 33029, 33050,
        33052, 33068, 33162, 33290, 33409, 33489, 33523, B33628, 33630, 33671,
        33697, 33703, 33731, 33835, 33893, 33899, 33916, 33950, 33976, 34036, 34061,
        34284, 34295, 34348, 34463, 34471, 34485, 34523, 34639, 34683, 34707, 34712,
        34730-1, 34835, 34933, 34937, 34940, 34956, 34961-2, 34997, 35022-3, 35030,
                                                                          (continued)

Pesticides, herbicides and carotenoids
      32763, 33265, 33338-9, 33463, 33519, 33607, 34468, 34995, 35519, 35964,
      36142, 36580

Pesticides, herbicides and chlorophyll
      32708, 32728, 32763, 32780, 32899, 33159, 33338-9, 33391, 33463, 33519,
      33603, 33607, 33641, 33651, 33670, 33867, 33869, 34140, 34238, 34468, 34897,
      34995, 35343, 35947, 35957, 36142, 36250, 36527

Pesticides, herbicides and chloroplast (chromatophore)
      32763, 32883, 32899, 33463, 33603, 33917-8, 35555, 35948, 36488

Pesticides, herbicides and ecosystem and plant productivity  33007, 33641, 33867,
      34468

Pesticides, herbicides and electron transport chain
      32834, 32853, 32900, 33008, 33048, 33159, 33200, 33309, 33347, 33391, 33458,
      33502, 33530, 33582, 33687, 34032, 34231, 34328, 34468, 34801, 34866, 34897,
      35396, 35539, 35555, 35947, 36043, 36171, 36192, 36250-1

Pesticides, herbicides and gas exchange
      32831, 33003, 33007, 33159, 33530, 33532, 33652-3, 33867, 33917-8, 34411-3,
      34417, 34468, 34801, 34807, 35343, 355'9, 35784, 35957, 36192, 36255, 36527

Pesticides, herbicides and photorespiration  33519

Pesticides, herbicides and respiration  33917-8, 34807, 34995

Petiole  see  Stem, petiole, morphology, structure and physiological activity in

pH, effect on algae productivity  33357, 33704

pH, effect on biliproteins  34740

pH, effect on carbon fixation pathways
      32628, 32720, 32894, 32953, 32968, 33155, 33555, 33777, 33887, 33914, 33943,
      33944, 33981, 34573, 34658, 34950, 35141-2, 35214, 35294, 35375, 35428,
      35468, 35726, 35925, 36019, 36409-10, 36441, 36451

pH, effect on chlorophyll  32992, 33268, 33577, 33782, 33875, 34188, 34838, 35093,
      35284, 35809, 35858, 36062, 36160, 36278

pH, effect on chloroplast (chromatophore)
      33341, 33791, 34033, 34563, 34780, 35026, 35111, 35391, 35414, 35606, 35651,
      35853

pH, effect on electron transport chain
      32631, 32657, 32697, 32702-3, 32801, 32807, 32886, 32935, 32954, 33016,
      33117, 33142, 33160, 33253, 33289, 33299, 33356, 33448, 33552-3, 33647,
      33691, 33718, 33778, 33787, 33864, 33929, 33945, 33989, 34025, 34032, 34126-
      -7, 34138, 34203, 34277, 34286, 34300, 34355, 34424, 34723, 34734, 34764,
      34803, 34805, 34868, 34918, 35068, 35076, 35125, 35177, 35209, 35252, 35322,
      35433, 35568, 35591, 35649, 35685, 35696-7, 35774, 35819, 36130, 36282,
      36397, 36482, 36484, 36506, 36538, 36558

pH, effect on gas exchange
      32570, 32830, 32953, 33146, 33384, 33426, 33759, 33929, 34162, 34188, 34558,
      34671, 35402, 35802, 36362, 36455, 36509

pH, effect on photorespiration  32841

PhAR  see  Irradiance ...; Canopy, radiation ...

Phosphoenolpyruvate carboxylase
        32691, 32826, 32912, 32931, 32953, 32996, 33148-9, 33344, 33474, 33555,
        33571, 33612, 33654, 33777, 33843, 33913-4, 33981, 34017, 34052, 34196,
        34253, 34256, 34312, 34429, 34452, 34456, 34496, 34525, 34549, 34556, 34716,
        34738, 34800, 34834, 34905-6, 34949-50, 34952, 34978-9, 35029, 35034-5,
        35051, 35087, 35191-2, 35197-8, 35248-9, 35266-7, 35302, 35425, 35427,
        35431, 35466-9, 35471-2, 35489, 35593, 35631, 35701, 35892-3, 35924-5,
        35970, 35986, 36019, 36121, 36241, 36366, 36426, 36440, 36442, 36491, 36533

Phosphoenolpyruvate carboxylase, methods    33914, 34196, 34979, 36442

Phosphorus   see   Mineral elements (N, P, K) ...

Photoperiod and carbon fixation pathways   35192

Photoperiod and chlorophyll   32594, 33481, 34607, 35035

Photoperiod and chloroplast (chromatophore)   33481

Photoperiod and ecosystem and plant productivity   33019, 34545, 35052, 35458, 36219,
        36348

Photoperiod and electron transport chain   32591, 32594

Photoperiod and gas exchange   32613, 32953, 34368, 35115

Photophosphorylation, cyclic
        32636, 32673, 32706, 32769, 32792, 32834, 32848, 32853, 32862, 32908, 33207,
        33253, 33323, 33425, 33687, 33716, 33353, 33932, 33986, 33994-5, 34026-7,
        34059, 34093, 34372, 34431, 34441, 34447, 34598, 34805, 34866, 34891, 35009,
        35124, 35167, 35177, 35396, 35404, 35427, 35432, 35434, 35440, 35539, 35698,
        35752, 35764, 35892-3, 35917, 35926, 36002, 36175, 36210, 36316, 36382

Photophosphorylation in photosynthetic bacteria   see   Photosynthetic bacteria, pho-
        tophosphorylation

Photophosphorylation, methods   32697, 34680, 34918, 35650, 35889, 36559

Photophosphorylation, model   see   Model ...

Photophosphorylation, non-cyclic
        32636, 32664, 32673, 32709, 32795, 32819, 32826, 32834, 32839, 32848, 32862,
        32866, 32886, 32908, 33008, 33032, 33041, 33108, 33207, 33323, 33425, 33530,
        33647, 33687, 33714-5, 33743, 33748-9, 33853, 33882, 33886, 33932, 33939-40,
        34025, 34059, 34205, 34256, 34372, 34422, 34441, 34475, 34521, 34575-6,
        34589, 34661, 34674, 34769, 34803-4, 34850, 34866, 34918, 34922, 34963,
        34984, 35009, 35012, 35054, 35131, 35134, 35177, 35187-8, 35209, 35306,
        35396, 35404, 35427, 35432, 35434, 35487, 35618, 35676, 35681, 35698, 35700,
        35716-7, 35764, 35843, 35868, 35893, 35915, 35926, 35981, B36081, 36173-4,
        36200, 36223-4, 36236, 36265, 36283, 36316, 36363, 36382, 36398, 36558,
        36581

Photophosphorylation, pseudo-cyclic
        33425, 34250, 34255, 34447, 34694, 34764, 36175

Photoreduction   see   $H_2$ evolution ...

Photorespiration enzymes
        32585, 32841-2, 32952, 33125, 33156, 33235, 33705, 33837, 33887, 33960,
        34073, 34254, 34256, 34424, 34509, 34658-9, 34683, 34905, 35061, 35152,
        35213, 35247-9, 35324, 35431, 35444, 35627, 35694, 35766, 36104, 36162,
        36410, 36415, 36561

Photorespiration metabolic cycles
        32577, 32585, 32638, 32683-5, 32822-3, 32826, 32828, 32855, 32953, 33069,
        33125, 33217, 33235, 33324, 33543, 33561, 33631, 33633, 33696, 33785, 33843-
        -4, 33960, 34103, 34251, 34254-6, 34379, 34383, 34410, 34424, 34524-5,
        34539, 34652, 34658-9, 34695, 34701, 34718, 34905, 35061, 35162-4, 35167,
        35186, 35265, 35324, 35375, 35468, 35535, 35601, 35675, 35765-6, 35944,
        36116, 36162, 36169, 36185, 36280, 36330, 36410, 36432, 36475, 36480, 36561

Photorespiration metabolic cycles enzymes, methods  see  Enzymes of glycollate cycle,
        methods

Photorespiration rate
        32652, 32773, 32804, 32823, 32965, 32967, 32996, 33062, 33123-4, 33153,
        33231, 33363, B33510, 33543, 33561, 33631, 33633, 33785, 34052-3, 34097,
        34122, 34154, 34379, 34524, 34535, 34549, 34689, 35044, 35081, 35138, 35140,
        35161-2, 35164, 35222, 35286, 35324-5, 35334, 35341-2, 35368, 35370, 35423,
        35465, 35471, 35535, 35574, 35678, 35714, 35765, 35842, 36114-6, 36162,
        36280, 36344, 36406, 36417

Photosynthate translocation and distribution
        32576, 32582, 32612, 32686, B32723, 32742, 32808, 32850, 32882, 32971, 33002,
        33011, 33021, 33040, 33043, 33111, 33159, 33162, 33168-9, 33181, 33226,
        33232, 33260-1, 33270, 33279, 33318, 33324, 33358-9, 33369, 33396-7, 33415,
        33467, 33500, 33507, 33521, 33525, 33534, 33537, 33547-9, 33599, 33614,
        33624, 33658, 33671, 33813, 33831-2, 33846, 33859, 33867, 33871-2, 33893,
        33928, 33950, 33970, 33974-5, 34005-6, 34040, 34156, 34258, 34264, 34304,
        34312, B34354, 34408, 34429-30, 34437, 34458, 34510, 34525, 34536, 34555,
        34614, 34622, 34637-40, 34712, 34719, 34721, 34754, 34831, 34834, 34873,
        34896, 34923, 34932, 34937, 35033, 35110-1, 35169, 35180, 35205, 35225-6,
        35243, 35255-6, 35279, 35295, 35330-2, 35373, 35394, 35416, 35451, 35481,
        35504, 35524, 35656, 35712, 35721, 35744, 35759, 35766, 35790, 35824, 35866-
        -7, 35888, 35938, 35952-4, 35966, 35972, 35974, 35979-80, 35997-8, 36009,
        36014, 36021, 36044, 36116, 36126, 36136, 36144, 36172, 36194, 36204-5,
        36227, 36267, 36288, 36308, 36329, 36333, 36343-4, 36381, 36422-3, 36531

Photosynthates and intermediates of carbon fixation pathways
        32584, 32795, 32803, 32855, 32894, 33054-5, 33069, 33116, 33148, 33151,
        33154, 33198, 33491, 33553, 33696, 33777, 33842, 33887, 33943-4, 34199,
        34250, 34292, 34427-8, 34496, 34559-61, 34614, 34657, 34659, 34742, 34800,
        34813, 34824, 34906, 34978, 35005, 35064, 35087, 35192, 35368, 35378, 35390,
        35428, 35469, 35535, 35925, 35940, 35976, 36002, 36019, 36204, 36308, 36331,
        36365'

Photosynthates and intermediates of carbon fixation pathways and chloroplast (chro-
        matophore)  33146, 35747

Photosynthates and intermediates of carbon fixation pathways and gas exchange
        34923, 35162-3, 35368, 36557

Photosynthates and intermediates of carbon fixation pathways and respiration  35471

Photosynthetic bacteria carbon fixation pathways
        32623, 32683, 33401, 33612, 33707, 33807, 34658, 35323, 35693, 35937, 35975,
        36189

Photosynthetic bacteria carotenoids  see  Xanthophylls of photosynthetic bacteria

Photosynthetic bacteria chlorophylls  see  Bacteriochlorophylls; Chlorophylls,
        *Chlorobium*

Photosynthetic bacteria chromatophores  see  Chloroplast and chromatophore ...;
        Chromatophore ...

                                                                        (continued)

Photosystem 2 (continued)
          33802, 33806, 33810, 33827, 33850, 33869, 33880-1, 33883, 33901, 33927,
          33932, 33956-7, 33982, 33990-1, 33995, 34027, 34042-3, 34045, 34093, 34106,
          34126-7, 34155, 34165-8, 34202-3, 34208, 34215, 34223, 34234, 34286, 34349-
          -50, 34357, 34367, 34390, 34417, 34433, 34462, 34490, 34527, 34529, 34531,
          34569, 34572, 34600, 34607, 34610, 34628, 34669, 34700, 34703, B34706, 34723-
          -4, 34728, 34756, 34759, 34766-7, 34775, 34777, 34799, 34801, 34832, 34843-5,
          34848, 34858-9, 34882-3, 34891, 34893, 34919, 35058-9, 35068, 35114, 35124-5,
          35143, 35157, 35177, 35218-9, 35230, 35264, 35269, 35291, 35306-7, 35318,
          35327-9, 35362, 35366-7, 35392-3, 35413, 35415, 35434, 35447, 35502, 35526,
          35548, 35557, 35571-2, 35576, 35606, 35615, 35644, 35649, 35653, 35665,
          35669, 35685, 35696, 35698-9, 35719-20, 35731, 35736, 35739, 35741, 35763-4,
          35774, 35780, 35804, 35818-9, 35844, 35878-9, 35894, 35935, 35944-5, 35947,
          35994, 36023-4, 36043, 36122, 36140, 36155, 36173-5, 36177, 36200, 36234,
          36235, 36239, 36250-1, 36256, 36258, 36264-5, 36276, 36307, 36310, 36312,
          36314, 36316, 36377, 36397-8, 36411, 36439, 36460, 36462, 36469, 36490,
          36502-3, 36511, 36540, 36542, 36554

Photosystem 2 activity measurement
          32591, 32594-5, 32604, 32610, 32667, 32706, 32725-6, 32739, 32765, 32769,
          32786, 32819, 32834, 32861-2, 32864, 32867, 32889, 32934, 33001, 33030,
          33047, 33058, 33108, 33200-1, 33248, 33310, 33347, 33353, 33391, 33502,
          33513, 33530, 33539, 33551, 33582, 33663, 33665, 33687, 33719, 33748-9,
          33761, 33790, 33802-3, 33842, 33901, 33982, 33995, 34043, 34045, 34093,
          34100, 34102, 34126-7, 34165, 34180, 34231-4, 34286, 34316, 34319, 34372,
          34441, 34475, 34490, 34527, 34576, 34588, 34606-7, 34609, 34630, 34643,
          34663, 34693, 34703, B34706, 34766-7, 34799, 34845, 34848, 34866, 34883,
          34897, 34918, 35058, 35068, 35076, 35124-5, 35157, 35177, 35207, 35320,
          35328-9, 35358, 35427, 35432, 35434, 35440, 35507, 35547, 35555, 35637-8,
          35685, 35697-9, 35739, 35763-4, 35767, 35773, 35780, 35804, 35818-9, 35840,
          35894, 35921-2, 35935, 35947, 35987, 36003, 36101, 36122, 36171, 36175,
          36192, 36210, 36214, 36250-1, 36257, 36300-1, 36310, 36312, 36316, 36355,
          36409, 36411, 36506-7, 36511, 36541, 36556, 36576, 36581

Photosystem 2 activity measurement, methods  33464, 34406, 34918

Photosystem 2, primary acceptor
          32755, 33253, 33720, 33748-9, 34127, 34205, 34357, 34601-2, 34724, 34775,
          35240, 35362, 35526, 35548, 35945, 36173, 36462

Photosystem 2 reaction centre  see  P680

Photosystems stabilization, chloroplast immobilization, methods
          32934, 33789, 33995, 34217, 34956, 35124-5, 35230, 35456, 35992, 36003,
          36402

Phycobilins  see  Biliproteins

Phycobilisome  32927, 33233, 33757-8, 33883, 34029, 35366, 35738, 36179, 36504,
          36575

Phycocyanins
          32617, 32785, 32981, 33134, 33234, 33258, 33420, 33535, 33569, 33576, 33584,
          33657, 33757-8, 33776, 33811, 34324, 34339, 34394, 34629, 34696-8, 34846,
          34976, 35015, 35058, 35147, 35308, 35366-7, 35452, 35478, 35621, 35684,
          35738, 35804, 35935, 36180, 36186, 36294, 36429, 36504, 36575

Phycoerythrins
          32557, 32617, 32981, 33420, 33535, 33576, 33613, 33657, 33757-8, 33811,
          34339, 34596, 34629, 34696-7, 34740, 34846, 34976, 35147, 35366-7, 35738,
          36071, 36294

Phylogeny of biliproteins  34382, 35167

Proteins, amino acids, nucleic acids and carbon fixation pathways
        32970, 33491, 33842, 34253, 34315, 34428, 34554, 34620, 35359, 35592, 36146,
        36365, 36426

Proteins, amino acids, nucleic acids and chlorophyll   33094, 33600

Proteins, amino acids, nucleic acids and chloroplast (chromatophore)
        32593, 32650-1, 32661, 32671, 32783, 32811, 32854, 32858, 32890, 32902,
        32920, 32932, 32944, 32975, 33053, 33090, 33098, 33170, 33172, 33180, 33202,
        33256, 33381, 33383, 33419, 33452-4, 33517, 33592, 33597, 33625, 33648,
        33678, 33686, 33766, 33773, 33801, 33810, 33820, 33827, 33838, 33841, 33942,
        33957, 34054, 34060, 34164, 34223, 34237, 34360, 34578, 34692, 34704, 34762,
        34816, 34871-2, 34892, 34908, 34927, 34930, 34956, 35034-5, 35054, 35066,
        35072, 35114, 35128, 35130, 35143, 35211, 35241-2, 35382-3, 35479, 35497-8,
        35515, 35544, 35590, 35655, 35689, 35698, 35730, 35755, 35804, 35859, 35873-
        -4, 35967, 35991, 36031, 36127-8, 36177, 36210, 36261, 36284, 36327, 36459,
        36461, 36529

Proteins, amino acids, nucleic acids and electron transport chain
        33170, 33906, 35100, 36476

Proteins, amino acids, nucleic acids and gas exchange   36407, 36557

Proteins, amino acids, nucleic acids and photorespiration   32822, 33593, 35601

Protochlorophyll(ide)   see   Chlorophyll biosynthesis ...

Proton transport in chloroplast   see   Chemiosmotic hypothesis ...

Protoplasts, isolated   see   Tissue culture ...

Pteridines   see   Ferredoxin-NADP reductase ...

Pyrenoid   33211, 33765, 35626


Q


Quantum yield and requirement
        32587, 32615, 32664, 32744, 32792, 32812, 32826, 32866, 32897, 32914, 32916,
        33024, 33047, 33056, 33062-3, 33187, 33323, 33433, 33436, 33626, 33707,
        33714, 33758, 33779, 33788, 33799, 34023, 34099, 34208, 34277, 34393, 34447,
        34451, 34462, 34572, 34588, 34859, 34893, 35096, 35138, 35285, 35320, 35374,
        35762, 35819, 35945, 35955, 36150, 36250, 36256, 36569

Quantum yield  and requirement, methods   see   Quantum yield and requirement

Quinones in photosynthesis
        32703, 32755, 32769, 32834, 32845, 32860-1, 32915, 32955, 33030, 33142,
        33187, 33190, 33253, 33300, 33350, 33400, 33402, 33460, 33515, 33551, 33717-
        -8, 33748-9, 33802-3, 33864, 33901, 33919, 34032, 34126, 34138, 34202,
        34205, 34277, 34357, 34377, 34389, 34606-8, 34727, 34743, 34767, 34774-5,
        34789, 34845, 34965, 35066, 35167, 35239, 35282, 35292, 35366, 35615, 35649,
        35741, 35812, 35881, 35917, 35945, 36122, 36164, 36173, 36175, 36189, 36246,
        36250, 36256, 36262, 36265, 36389, 36449, 36483-4, 36502, 36511, 36540

Quinones, methods   35239, 35812


R


Radiation in canopy   see   Canopy, radiation ...

Radiation,light  see  Irradiance

Reaction centres  see  *P680*; *P700* ...

Recycling of $CO_2$ inside the cell and leaf  36185

Relative growth rate  see  Growth analysis, net assimilation rate ...

Relative water content  see  Water saturation deficit

Resistance, carboxylation and excitation  32573, 33002, 33879, 34625, 34648, 35634,
    36190, B36191, 36521

Resistance, cuticular  35355

Resistance, intracellular (mesophyll)
    32573, 32772, 32810, 32913, 32967, 32973, 32997, 33002, 33231, 33244,
    33304, 33306, 33358-9, 33506, 33703, 33879, 33976, 34038, 34097, 34170,
    34375, 34453, 34625, 34646-8, 34683, 34689, 34886, 35097-8, 35234, 35261,
    35370, 35610, 35634, 35712, 35910, 36190, B36191, 36358, 36424, 36457,
    36521-2

Resistance, leaf boundary layer
    32810, 32973, 32997, 33223, 33358, 33385, 33436, 33451, B33510, 33547, 33985,
    34038, 34713, 34743, 34886, 35129, 35234, 35278, 35303, 35529, 35610, 35827,
    36013, 36545

Resistance, stomatal (and intercellular) (*cf*. also Resistance for water vapour ...)
    32562, 32573, 32690, 32698, 32713, 32772, 32777, 32810, 32866, 32913, 32967,
    32973, 32997, 33000, 33002, 33050, 33068, 33105-6, 33195, 33244, 33280,
    33285-6, 33358-9, 33388, 33396, 33435, 33451, 33506, 33527, 33547, 33558-9,
    33703, 33742, 33818, 33857-8, 33858, 33879, 33916, 33976, 33985, 34008,
    34038, 34097, 34132, 34135, 34170, 34181, 34193, 34267, 34295, 34332-4,
    34352, 34368, 34375, 34384, 34453, 34494, 34497, 34504-5, 34511, 34535,
    34541, 34625, 34632, 34636, 34646-7, 34649, 34683, 34743, 34814, 34829,
    B34865, 34959, 35010, 35095, 35097, 35192, 35222, 35234, 35260-1, 35273,
    35283, 35295, 35303, 35340, 35355, 35369-70, 35374, 35460, 35476, 35492,
    35529, 35536-7, 35563, 35775, 35845, 35884, 35904, 35907, 35930, 35963,
    36017, 36050, 36069, 36075, 36087, 36120, B36191, 36205, 36207, 36358, 36372,
    36418, 36424, 36431, 36457, 36521, 36523

Resistances to $CO_2$ and water vapour transfer at canopy level
    32693, 33022, 33223, 33916, 34690, 34713, 35151, 35529, 35845, 36011

Resistances to $CO_2$ transfer
    32568, 32573, 32653, 32716, 32810, 32913, 32973, 33002, 33050, 33215, 33231,
    33290, 33358-9, 33436, 33614, 33703, 33879, 34193, 34352, 34416, 34504,
    34648, 34689, 34743, 35155, 35222, 35234, 35273, 35286, 35303, 35460, 35910,
    35939, B36191

Resistances to water vapour transfer,"leaf resistance"
    32690, 32871, 33020, 33022, 33167, 33287-8, 33305, 33388, 33395, 33398,
    33435-6, 33505, 33508, 33527, 33694, 33703, 33984-5, 34109-10, 34135, 34170,
    34193, 34239, 34504, 34690, 34747, 34792, 34851, 34886, 34945, 34954, 35096-
    -8, 35129, 35194, 35234, 35335, 35354, 35369, 35460, 35594-5, 35610, 35712,
    35773, 35815, 35827, 35905, 35931, 35939, 36013, 36069, 36075, 36132, 36132,
    36417-8, 36472

Respiration and photosynthesis
    32656, 32694, 32710-1, 32732, 32744, 32798, 32802, 32806, 32868, 32899,
    32941, 32953, 32967, 32969, 32974-5, 32980, 33001, 33070, 33124, 33162,
    33201, 33210, 33231, 33243, 33260, 33278, 33292, 33294, 33340, 33378-80,
    33384, 33409, 33496, 33514, 33652, 33675, 33687, 33731, 33759, 33764, 33832,
    33896, 33898, 33912, 33915, 33999, 34012, 34052, 34100, 34158, 34213, 34216,
                                                                    (continued)

Respiration and photosynthesis (continued)
        34252, 34281, 34293, 34370, 34373, 34404, 34409, 34437, 34490, 34518, 34520,
        34528, 34538, 34562, 34650, 34654-5, 34661, 34674, B34706, 34741, 34807,
        34809-10, 34853, 34867, 34905, 34912, 34915, 35138, 35140, 35167, 35171-2,
        35175, 35179, 35319, 35333, 35368, 35381, 35413, 35431, 35454, 35574, 35622,
        35643, 35676, 35688, 35762, 35802, 35868, 35872, 35885, 35920, 35927, 35957,
        36077, 36135, B36191, 36198-9, 36373, 36424, 36434, 36445, 36497, 36512,
        36547

Respiration, dark $CO_2$ efflux
        32562, 32568, 32573, 32633, 32654, 32742, 32773, 32804, 32837, 32882, 32965,
        32969, 33013, 33084, 33124, 33192, 33215, 33245, 33260, 33262, 33335, 33358,
        33385, 33465, B33510, 33523, 33566, 33660, 33703, 33725, 33740, 33859,
        33891, 33917-8, 33925, 34005-6, 34012, 34017, 34096, 34137, 34219, 34269,
        34302, 34326, 34330, 34471, 34485, 34536, 34613, 34677, 34689, 34754, 34792,
        34886, 34888, 34940, 34945, 34997, 35019, 35040, 35074, 35092, 35106, 35117,
        35129, 35154, 35161, 35169, 35173, 35233, 35243, 35255, 35261, 35286, 35370,
        35423-4, 35449, 35559, 35636, 35664, 35674-5, 35761, 35789, 35807, 35932,
        36037, 36049, 36113, 36116, 36132, 36135, 36203, 36213, 36299, 36303, 36329,
        36406, 36427, 36435, 36515-6

Respiration, "growth" and "maintenance"
        32753, 33260, 33808, 33859, 33999, 34325-6, 34650, 34809-10, 36344, 36516

Respiration of achlorophyllous tissues in light, light inhibition of respiration
        34301, 35974

Ribosome of chloroplast
        32616, 32671, 32763, 32902, 32949, 32958, 33170, 33282, 33518, 33520, 33686,
        33817, 33827, 33861, 34164, 34616, 34750, 34929, 34934, 35382-3, 35689,
        35755, 35936, 35967, 36532

Ribulose 1,5-bisphosphate carboxylase
        32583-5, 32623, 32661, 32685, 32691, 32717, 32719-20, 32759, 32767-8, 32783,
        32803, 32826, 32840-2, 32862, 32866, 32902, 32929-31, 32944, 32953, 32962-3,
        32996, 33002, 33042, 33076, 33090, 33110, 33135-6, 33148-9, 33154-7, 33171-2,
        33180, 33198, 33213, 33271, 33273, 33296, 33310, 33387, 33445, 33453-4,
        33473-4, 33513, 33518-20, 33571, 33597, 33612, 33625, 33633, 33642-5, 33648,
        33672-3, 33686, 33731, 33767-8, 33770-2, 33817, 33827, 33835-7, 33842,
        33849, 33873, 33887, 33892, 33895, 33913, 33944, 33979, 33992, 34003, 34017-
        -9, 34045, 34054-5, 34065, 34071-3, 34173, 34178, 34199, 34219, 34253,
        34270, 34276, 34312, 34315, 34320, 34379, 34388, 34429, 34437, 34452, 34456,
        34470, 34473, 34475, 34509, 34525, 34534, 34549, 34556, 34574, 34618-20,
        34656-8, 34683, 34692, 34710, 34716, 34738, 34749, 34786, 34800, 34812-3,
        34817, 34834, 34911, 34922, 34937, 34949, 34972, 34978-9, 35006, 35023,
        35034-5, 35051, 35087-90, 35141-2, 35152, 35183-4, 35186, 35191-2, 35212-5,
        35226, 35241, 35247-9, 35294, 35372, 35382, 35411, 35420, 35425, 35427,
        35431, 35444, 35467-8, 35471, 35475, 35479, 35490, 35497-9, 35509, 35544,
        35569-70, 35596, 35626-7, 35646-7, 35692-4, 35701, 35703-4, 35755, 35762,
        35794, 35877, 35892-3, 35970, 35986, 36002, 36026, 36114, 36116, 36121,
        36132, 36159, 36162, 36234, 36327, 36366, 36379, 36395-6, 36410-2, 36414-6,
        36426, 36440, 36450-2, 36473, 36486, 36491, 36495, 36561, 36572

Ribulose 1,5-bisphosphate carboxylase, methods
        32962-3, 33155, 33171, 33271, 33273, 33473-4, 33767, 33837, 33842, 33887,
        33895, 34178, 34388, 34812, 34817, 34925, 34979, 35006, 35510, 35569-70,
        36412, 36414, 36451, 36491

Ribulose 1,5-bisphosphate oxygenase  see  Ribulose 1,5-bisphosphate carboxylase ...;
        Photorespiration enzymes

Root removal  see  Defoliation ...

Root, underground part, and chlorophyll  33034

Tissue cultures, gas exchange in  32823, 33219, 33663, 33665, 33790, 34710, 35057, 36051

Tissue cultures, growth of  35932, 36494-5

Tissue cultures, photorespiration in  32822

Tissue cultures, respiration in  34800, 35163, 35265, 35267, 35471, 35932, 35950

Transient phenomena in gas exchange  32799, 33068, 33430, 34570, 35482, 36540, 36555, 36568, 36584

Transpiration and photosynthesis
        32573, 32633, 32871, 32973, 32996-7, 33050, 33280, 33304-6, 33388, 33441,
        33547, 33694-5, 33857, 33920, 34295, B34354, 35096, 35512, 35561, 35773,
        35886, 35920, 36480

Tritium oxide  see  Deuterium oxide, tritium oxide ...


U

Ubiquinones  see  Quinones ...

Ultraviolet radiation  see  Irradiance, spectral composition ...

Uncouplers of electron transport chain (*cf.* also Antibiotics and electron transport
        chain)
        32636, 32725, 32734, 32755, 32807, 32908, 32936, 33086, 33114, 33402, 33687,
        33744, 33748, 33753, 33886, 33929, 33986, 33990-1, 34026-7, 34102, 34138,
        34250, 34405, 34422, 34424, 34431, 34575-6, 34601, 34769, 34897, 35068,
        35076, 35305, 35307, 35573, 35681, 35915, 35955, 36094, 36397, 36496


V

Virus diseases  see  Phytopathological effects ...

Vitamin $K_3$  see  Quinones ...

Volume changes in chloroplast (chromatophore)  see  Chloroplast and chromatophore
        volume changes

Volume changes in leaf and other organs  33111, 33524, 33985, 35604


W

Warburg effect  see  $O_2$ and gas exchange

Water, heavy  see  Deuterium oxide, tritium oxide ...

Water saturation deficit
        32713, 33022, 33084, 33261, 33287, 33396, 33398, 33527, 33680, 33840, 33928,
        34120, 34129, 34170. 34181, 34239, 34314, 34747, 35354-5, 35512, 35537,
(continued)

Water saturation deficit (continued)
    35607, 35645, 35815, 35906, 35952, 36013, 36056-7, 36205, 36207, 36431

Water splitting mechanism  see  $O_2$ evolution mechanism and kinetics

Wind (air-flow rate) and ecosystem and plant productivity  33546, 33559, 34714,
    35594-5, 36343, 36388

Wind (air-flow rate) and gas exchange  36069, 36343

Wind (air-flow rate) and resistances to $CO_2$ and water vapour transfer  33287, 33559,
    34038, 34886, 35303, 35529, 35594-5


X


Xanthophylls  32595, 32770, 32966, 33030, 33057, 33263-6, 33284, 33336, 33351, 33472,
    33589, 33684, 33795, 33862, 34030, 34163, 34175, 34214, 34399-400, 34457,
    34533, 34581, 34583, 34595-6, 34707, 34752, 34779, 34845, 34869, 34948,
    35048, 35358, 35379, 35421, 35427, 35432, 35449, 35519, 35653, 35696, 35763,
    35844, 35854, 36004, 36154, 36180, 36338, 36447, 36456

Xanthophylls of algae
    32643, 32966, 33006, 33263, 33266, 33336, 33535, 33780, 33795-6, 33811,
    33834, 33862, 34215, 34380, 34595-6, 34604, 34607, 35380, 35763, 35768,
    36154-5, 36447-8, 36587

Xanthophylls of photosynthetic bacteria
    32685, 32924, 33203, 33326, 33707, 33862, 34363-4, 34603-4, 34687, 34744,
    34791, 35066, 35236, 35323, 35386, 35899, 35901, 36119, 36189, 36484

Xerophytes  see  Drought ...; Temperature, high ...

X-rays  see  Ionizing radiation ...


Y


Yield formation  see  Biomass distribution ...; Photosynthate translocation ...


Z


Zeaxanthin  see  Carotenoids ...; Xanthophylls

This index contains a selection of plant genera and types interesting as experimental material for physiological, ecological and agricultural studies. Latin scientific names of plant genera and English names of plant groups and types are the main items which present the reference numbers.

# A

*Abies*    32698, 33422, 34284, 34747, 34806, 35808, B35862, 36049, B36191, 36201, 36551

*Acacia*   35008, 36050, 36445

*Acer*     33451, 33617-8, 33936, 33984, 34480, 34504-5, 34924, 35269, 35525, 35691,
           35808, 35949, 36057, 36190, B36191, 36241, 36403

*Acetabularia* 32920, 33762, 33774, 34396, 34578, 34917, 34948, 35211, 35356, 35400,
           36045

*Aesculus* 33595, 34246, 35006, 35269, 36139, 36190, B36191

*Agave*    33440-1, 34649, 35097, 35442

Alder  see  *Alnus*

Alfalfa  see  *Medicago*

Algae (*cf.* also *Acetabularia*, A. blue-green, A. brown, A. green, A. red, *Anabaena*,
       *Anacystis*, *Ankistrodesmus*, *Chlamydomonas*, *Chlorella*, *Chrysophyta*, Diatoms,
       *Dinoflagellatae*, *Dunaliella*, *Euglena*, *Nostoc*, *Porphyridium*, *Scenedesmus*,
       *Ulva*)
       32564-5, 32598-9, 32606, 32612, 32617, 32621, 32681, 32705, 32809, 32818,
       32824-5, 32907, 32938, 32941, 32961, 32988, 33006, 33017-8, 33045, 33061,
       B33065, 33070-2, 33081, 33098, 33115, 33152, 33210, 33239, 33266, 33292,
       33294, 33297, 33320, 33329, 33334, 33336, 33340, 33357, 33365, 33367, 33385,
       33466, 33496, 33509, 33512, 33514, 33544-5, 33586, 33588, 33619, 33622,
       33634, 33651, 33674, 33676, 33722-3, 33804, 33847, 33866, 33888, 33896,
       33971, 33977, 33996-7, 34011, 34034, 34069, 34099, 34139, 34144, 34157,
       34177, 34191-2, 34200, 34240, 34247, 34280, 34293, 34330, 34386, 34407,
       34436, 34473-4, 34489, 34514-5, 34517, 34528, 34538, 34555, 34558, 34567,
       34569, 34571, 34593, 34599, 34601, 34604, 34628, 34659, 34669, 34671-2,
       34697, 34715, 34733, 34763, 34768, 34811, 34818, 34840, 34859, 34870, 34878,
       34885, 34901, 34912, 34964, 34971, 34985, 34991-4, 35020, 35043, 35063,
       35070, 35127, 35217, 35227, 35254, 35298-9, 35356-7, 35360, 35365, 35381,
       35405, 35407-8, 35439, 35456, 35488, 35506, 35520, 35523, 35552-3, 35556,
       35586-7, 35597, 35654, 35672, 35733-5, 35799, 35841, 35852, B35862, 35864,
       35868, 35872, 35896, 35908, 35911, 35941-2, 36028, 36045, 36061, 36079,
       36088-9, 36100, 36117, 36123, 36141, 36149-50, 36179-82, 36208, 36216, 36271,
       36286, 36295, 36298, 36320, 36390, 36420, 36448, 36474, 36550

Algae, blue-green (*cf.* also *Anabaena*, *Anacystis*, *Nostoc*)
       32564-5, 32599, 32608, 32614, 32621, 32669, 32679-81, 32767, 32775, 32789,
       32794, 32803, 32911, 32937-8, 32979, 32993, 33070, 33083, 33133-4, 33148,
       33163, 33169, 33183, 33218, 33266, 33294, 33333, 33399, 33420, 33422, 33442,
       33446, 33464, 33482-3, 33521, 33535, 33568-9, 33584, 33650, 33656-7, 33749,
       33757-8, 33780, 33811, 33904, 33923, 33927, 34004, 34021, 34030-1, 34081,
       34197, 34199, 34216, 34222, B34248, 34329, 34382, 34394-5, 34418, 34430,
       34434, 34466, 34555, 34604, 34626, 34629, 34634, 34660, 34669, 34696, 34698,
       34717, 34726, 34768, 34780, 34798, 34832-3, 34881, 34885, 34893, 34902,
       34904, 34971, 34976, 34991-4, 35015, 35043, 35058, 35075, 35078, 35138,
       35147, 35167, 35182, 35195, 35206, 35209, 35244, 35306, 35323, 35328-9,
                                                                    (continued)

Algae, blue-green (continued)
        35378, 35452, 35456, 35502, 35584, 35587, 35654, 35663, 35684, 35688, 35696,
        35699, 35706, 35763, 35811, 35935, 35965, 35974, 36002, 36028, 36054, 36071,
        36110, 36143, 36179-82, 36248, 36251, 36279, 36294, 36336-7, 36369, 36375,
        36446-7, 36455, 36466, 36490, 36505-7

Algae, brown
        32679-81, 32859, 32911, 33011, 33072, 33121, 33128, 33147, 33243, 33266,
        33312, 33384-5, 33813, 33915, 33922, 34016, 34058, 34143, 34251, 34258,
        34468, 34473, 34594, 34596, 34604, 34626, 34629, 34637, 34638, 34669, 34741,
        34824, 34903, 34912, 34914, 34955, 34979, 35069, 35109, 35119, 35585, 35713,
        35729, 35735, 35756, 35804, B35862, 36098, 36149, 36152, 36334, 36585

Algae, green (*cf.* also *Acetabularia, Ankistrodesmus, Chlamydomonas, Chlorella, Duna-
        liella, Scenedesmus, Ulva*)
        32584, 32599, 32621, 32645, 32679-81, 32687, 32712, 32770, 32803, 32859,
        32873, 32910-1, 32938, 32948, 32995, 33070, 33083, 33147, 33163, 33183,
        33197-8, 33201, 33243, 33246, 33263, 33266, 33292, 33333, 33385, 33446,
        33496, 33545, 33606, 33679, 33813, 33852, 33922-3, 33927, 34030-1, 34069,
        34183, 34197, 34215, 34224, 34251, 34303, 34385, 34387, 34418, 34595-6,
        34626, 34629, 34669, 34768, 34817, 34848, 34885, 34903, 34912, 34917, 34955,
        34971, 34985, 34991-4, 35027, 35043, 35078, 35153, 35175, 35206, 35244,
        35264, 35378, 35449, 35587, 35597, 35603, 35606, 35626, 35654, 35715, 35735,
        35756, 35799, 35802, 35822, 35860, B35862, 35880, 35995, 36028, 36149,
        36151-2, 36154, 36182, 36214, 36323, 36334, 36369, 36439, 36446-7, 36455,
        36474, 36486, 36491, 36496, 36585

Algae, red (*cf.* also *Porphyridium*)
        32679-81, 32859, 32911, 32927-8, 33035, 33147, 33211, 33243, 33266, 33294,
        33399, 33550, 33875, 33884, 33905, 33922, 34029, 34113, 34251, 34418, 34427-
        -8, 34478, 34562, 34594, 34596, 34604, 34626, 34629, 34669, 34819, 34853,
        34884, 34893, 34903, 34912, 34985, 35167, 35178-9, 35293, 35663, 35715,
        35735, 35756, 35793, 35798, B35862, 36071, 36122, 36149, 36152, 36187, 36204,
        36309, 36334, 36365, 36439, 36510, 36585

*Allium*  33730, 35006, 35426, 35840, 35912, 36054, 36308

*Alnus*  33422, 35006

*Aloe*   34480, 34482, 35442

Alpine plants  33547, 34747, 35129, 35403, 36333, 36535

*Amaranthus*  32766, 32987, 32996, 33080, 33388, 33914, 34254, 34256, 34513, 35186,
        35263, 35403, 35427-8, 35430-3

*Anabaena*  32599-600, 32656, 32681, 32715-6, 32774, 32859, 32948, 33027, 33062,
        33070, 33083, 33133, 33169, 33200-1, 33218, 33422, 33442, 33535, 33610,
        33656, 33718, 33749, 33833, 34051, 34161-2, 34197, 34277, 34405, 34418,
        34555, 34629, 34798, 34967, 34992-4,35015, 35043, 35078, 35148, 35216,
        35315, 35477-8, 35587, 35761, 35780, B35862, 36033, 36096, 36099, 36182,
        36209, 36455

*Anacystis*  32668, 32714, 32792, 32859, 32988, 33218, 33233-4, 33257-8, 33360, 33442,
        33483, 33679, 33880, 33883, 34344, 34559-61, 34628-9, 34745, 34998-9, 35015,
        35177, 35242, 35307-8, 35755, 35890, 35944, 36065, 36273, 36279, 36365,
        36429, 36455, 36504

*Ananas*  33547, 34649, 34950

*Ankistrodesmus*  32814, 33070, 33721, 34555, 34994, 35043, 35216, 35587, 36654,
        35819, 36182, 36455

*Antirrhinum*  32948, 33967, 34372, 34850, 34882, 35420-1, 35698-700

*Apium*  36255

Apple  see  *Malus*

Apricot  see  *Armeniaca*

Aquatic macrophytes (*cf.* also *Elodea, Phragmites, Typha*)
    32570, 32643, 32645, 32752, 32803-4, 32830-1, 32953, 33364, 33379, 33412,
    33422, 33428, 33570, 33624, 33675, 33694, 33703, 33759, 33804, 33852, 33893,
    34017, 34030-1, 34035, 34057-3, 34213, 34216, 34304, 34551, 34587, 34421,
    34472, 34486-7, 34621, 34672-3, 34715, 34915-6, 34924, 35048, 35099, 35113,
    35160, 35168-9, 35171-4, 35244, 35298, 35334, 35401, 35425, 35597, 35607,
    35643, 35727, 35797, 36103, 36113, 36756-7, 36441, 36553

*Arachis* 32686, 32765, 32967, 33394, 33871-3, 34847, 35103, 35272, 35286, 35744

Arbor vitae  see  *Thuja*

*Armeniaca*  33547, 34274, 34549, 35481

*Armoracia*  32948

Ash  see  *Fraxinus*

*Asparagus*  35426, 36445

Aspen  see  *Populus*

*Atriplex*  32766, 32828, 32867, 32889, 32996, 33062, 33304-6, 33440, 33547, 33551,
    33696, 33843, 33914, 34013, 34219, 34256, 34659, 34717, 34926, 35186, 35273,
    35286, 35425, 36445

*Avena*  32765, 32769, 32847, 32956, 32973, 33007, 33062, 33158-9, B33343, 33375,
    33388, 33849-50, 33871-2, 33951, 33981, 34129, 34140, 34225, 34295, 34685,
    34815, 34924, 35142, 35410, 35555, 35590, 35614, 35711, 35925, 36019, 36067-
    -8, 36224, 36299, 36570, 36440, 36534

Avocado  see  *Persea*


# B

Bacteria, photosynthetic (*cf.* also *Chlorobium, Chromatium, Halobacterium, Rhodopseu-*
    *domonas, Rhodospirillum*)
    32602, 32614, 32619, 32623, 32679-81, 32760, 32764, 32779, 32803, 32859,
    33063, 33114, 33185, 33190, B33191, 33203, 33253, 33294, 33300, 33355,
    33380, 33403, 33406, 33473, 33494, 33509, 33514, 33552, 33612, 33688, 33702,
    33710-1, 33807, 33833, 33863, 34009, 34020, 34023, 34131, 34133, 34237,
    34270, 34281, 34296-7, 34362-4, 34393, 34435, 34460, 34469, 34583, 34603,
    34659, 34669, 34686-7, 34726, 34762, 34772, 34795, 34817, 34841, 34863,
    34902, 34957, 34968, 35078, 35166-7, 35238-9, 35274, 35306, 35311, 35333,
    35378, 35386-7, 35456, 35475, 35501, 35531, 35663, 35686, 35707, 35729,
    35855, 35868, 35900-1, 35937, 36001-2, 36118-9, 36188-9, 36247, 36399-400,
    36471, 36515, 36536, 36540, 36578

Banana  see  *Musa*

Barley  see  *Hordeum*

Bean  see  *Phaseolus*

Beech  see  *Fagus*

Bermuda grass  see  *Cynodon*

*Beta*     32573, 32699, 32701, 32865, 32874, 32888, 32948, 32987, 33062, 33310, 33398,
           33562, 33590-1, 33650, 33658, 33671, 33808, 34012, 34182, 34412, 34506,
           34545, 34625, 34650, 34670, 34735, 34857, 34875, 34940-1, 35006, 35104,
           35155, 35159, 35341, 35488, 35505, 35528, B35862, 35871, 35933, 36032, 36058,
           36224, 36357, 36375, 36431

*Betula*   32891, 33264, 33559, 34101, 34471, 34747, 35006, 35014, 35194, 35269, 35358,
           36190, B36191, 36408

Birch  see  *Betula*

Blackberry  see  *Rubus*

Blueberry  see  *Vaccinium*

Bluegrass  see  *Poa*

*Brassica*  32765, 32948, 32956, 32965, 33059, 33184, 33348, B33510, 33565, 33604,
            33643, 33790, 33808, 33958, 34159, 34312, 34448, 34712, 34835, 34924, 35006,
            35029, 35268, 35512-4, 35545, 35659-60, 35849, B35862, 35948, 36033, 36058,
            36137-8, 36527

Brinjal  see  *Solanum*

Bristlegrass see  *Setaria*

Broadbean  see  *Vicia*

Broccoli  see  *Brassica*

Bromegrass  see  *Bromus*

*Bromus*   32950, 36076, 36203

Brussels sprouts  see  *Brassica*

*Bryophyllum*  33890-1, 33960, 34196, 34585, 34952, 35087, 35403, 35442, 36146,
               36253-4

Buckwheat  see  *Fagopyrum*


C


Cabbage  see  *Brassica*

Cacti    33062, 33440, 33547, 33663-5, 33857, 33858, 35095, 35930, 36093

*Cajanus*  33934

*Calluna*  see  Heath plants and communities

CAM plants (*cf.* also *Agave, Ananas, Bryophyllum,* Cacti, Succulents)
            32840, 33547, 33777, 33809, 33857, 34298, 34352-3, B34354, 34368-9, 34585,
            34684, 34950, 35097, 35192, 35442, 35625, 35950, 36442-5

*Camellia*  see  *Thea*

Canarygrass  see  *Phalaris*

Caper-bush  see  *Capparis*

*Capparis*  36445

*Capsicum*  33057, 33832, 34403, 34624, 34678, 35006, 35033, 35446, 35776, 36124

Caraway  see  *Carum*

*Carex*  33050, 33331, 33728, 33756, 33804, B33926, 34052, 34351, 34825, 34924,
36037, 36132, 36136, 36201, 36205, 36408

*Carpinus*  33262, 33451

Carrot  see  *Daucus*

*Carum*  36255

*Carya*  33632, 33984, 34639-40, 35282, 36124

Cassava  see  *Manihot*

*Castanea*  33477, 33567

Castor bean  see  *Ricinus*

Cat's tail  see  *Typha*

Cattail flag  see  *Typha*

Cauliflower  see  *Brassica*

Cedar  see  *Cedrus*

*Cedrus*  32698

Celery  see  *Apium*

Cereals  see  *Avena, Hordeum, Oryza, Panicum, Secale, Sorgum, Triticum, Zea*

*Chenopodium*  32766, 32948, 32996, 33080, 33310, 33391, 33399, 34506, 34924, 35102,
35456, 35947, 36165

Chestnut  see  *Castanea*

Chick pea  see  *Cicer*

Chinese cabbage  see  *Brassica*

*Chlamydomonas*  32599, 32681, 32716, 32811, 32827, 32872, 32902, 32932-3, 32948,
33026, 33070, 33083, 33126-7, 33163, 33170-1, 33212-3, 33218, 33265, 33282,
33311, 33453, 33545, 33686, 33783, 33816, 33861, 34054-5, 34066-7, 34185,
34303, 34305, 34328, 34343, 34490-1, 34531, 34618, 34750, 34801, 34817,
34883-4, 34917, 34994, 35027, 35043, 35143, 35191, 35211, 35241, 35244,
35544, 35755, 35763, B35862, 35873, 35967, 36182, 36341-2, 36455, 36476,
36541-2

*Chlorella*  32622, 32649, 32669, 32681, 32687, 32728, 32748, 32767-8, 32792,
32794, 32798, 32803, 32813-4, 32816, 32822, 32859, 32945-8, 32966, 33062-3,
33068, 33070, 33072, 33083, 33131, 33146, 33217-8, 33265, 33314-6, 33323-4,
33351-2, 33355, 33360, 33430, 33475, 33511, 33545, 33625, 33652, 33660,
33721, 33749, 33765, 33784, 33795-6, 33814, 33856, 33875, 33880, 33911,
33912, 33987, 34007, 34025, 34082, 34185, 34197, 34303, 34328, 34361, 34363-
-4, 34404, 34417-8, 34447, 34468, 34529-31, 34589, 34607, 34628-9, 34634,
34658-9, 34664, 34691, 34717-9, 34759, 34773, 34775, 34797, 34817,

(continued)

*Chlorella* (continued)
> 34837-8, 34846, 34893, 34905-6, 34914, 34967, 34992, 34994, 35024, 35043,
> 35126, 35144-5, 35191, 35195, 35206, 35250, 35252, 35304-5, 35319-20, 35336,
> 35348, 35366, 35378, 35475, 35485, 35494, 35524, 35635, 35729, 35755, 35761,
> 35763, 35836, 35878-9, 35881, 35916, 35940, 35945, 35957, 36043, 36081,
> 36097, 36105-6, 36145, 36176, 36197, 36239, 36276, 36342, 36365, 36409,
> 36411, 36462, 36540, 36569-70

*Chlorobium*  32619, 32673, 32681, 32764, 32915, 32977-8, 33294, 33402-3, 33612,
> 33688, 33902, 34194, 34229, 34296-7, 34356, 34522, 34537, 34817, 34852,
> 35167, 35239, 35323, 35333, 35387, 35411, 35501, 35702, 35855, 35968, 36188-
> -9, 36513

*Chromatium*  32563, 32585, 32657, 32673, 32681, 32683-5, 32764, 32915, 32925, 33119,
> 33188, 33294, 33325, 33402, 33468-9, 33515, 33552, 33612, 33661, 33688,
> 33833, 34130, 34230, 34277, 34281, 34297, 34355-6, 34522, 34566, 34658,
> 34817, 34852, 34968, 35066, 35107, 35193, 35239, 35323, 35387, 35397, 35501,
> 35573, 35702, 35855, 35937, 36025, 36119, 36129, 36357, 36513

*Chrysanthemum*  32569, 33082, 34332

*Chrysophyta*  32599, 32621, 32624, 32898-9, 32907-8, 32921, 32993, 33070, 33371,
> 33446, 33545, 33856, 34016, 34555, 34604, 34884-5, 34991-4, 35043, 35216,
> 35587, 35654, 35715, 35763, B35862, 36182, 36420, 36455, 36474

*Cicer*  32686, 34834, 35225-6, 35603, 35891

*Citrullus*  36328

*Citrus*  32797, 32877, 33463, 33589, 34940, 35073, 35140, 35810, 35813-4, 36198,
> 36231-2

Clover  see  *Trifolium*

Cocksfoot  see  *Dactylis*

Cocoa  see  *Theobroma*

Coconut palm  see  *Cocos*

*Cocos*  33409, 35426

*Coffea*  34743

Coffee tree  see  *Coffea*

Coniferous plants (*cf.* also *Abies, Cedrus, Juniperus, Larix, Metasequoia, Picea,*
> *Pinus, Pseudotsuga, Taxus, Thuja, Tsuga*)
> 32681, 33296, 33646, 35115, 35156, 35808, 36346, 36515

*Corchorus*  34050, 34151

Corn  see  *Zea*

Cornelian cherry  see  *Cornus*

*Cornus*  33559

*Corylus*  33062

Cotton  see  *Gossypium*

Cottonwood  see  *Populus*

Cowberry  see  *Vaccinium*

Cowpea  see  *Vigna*

Crabgrass  see  *Digitaria*

*Crataegus  32948, 33602*

Cucumber  see  *Cucumis*

*Cucumis*  32581, 32776, 32783, 32948, 33094-5, 33205, 33301-2, 33432, 33503, 33589-91,
     33593, 33775, 33797-8, 33875, 34036, 34098, 34232-3, 34250, 34256, 34266,
     34275-6, 34426, 34438-9, 34450, 34523, 34666, 34784, 34790, 34924, 34940,
     35029, 35604, 35620, 35749-50, 35781, 35787, 35800, 35809, 35923, 35943,
     36060, 36158, B36191, 36199, 36221, 36288-90, 36327, 36367

*Cucurbita*  32948, 33470-1, 34320, 34506, 34556, 34871-2, 34924, 35006, 35180, 35604,
     35729, 36215, 36293, 36437

Currant  see  *Ribes*

Cyanobacteria  see  Algae, blue-green

*Cycas*  32681, 32948

*Cynodon*  32766, 33130, 33696, 34100, 34249, 34907, 34926, 35344

*Cyperus*  32752, 32766, 34193, 34279, 35101-3, 35425

Cypress  see  *Cupressus*


# D

*Dactylis*  32686, 32766, 32868, 33488, 34458, B35862, 36165, 36183, 36205, 36456

Dallis grass  see  *Paspalum*

Date palm  see  *Phoenix*

*Daucus*  32722, 32808, 33089, 33572, 33590-1, 34477, 35687, 35914, 35920, 36255

Deciduous trees and shrubs (*cf.* also *Acer, Aesculus, Alnus, Armeniaca, Betula, Car-
     pinus, Carya, Castanea, Citrus, Cornus, Corylus, Crataegus, Eucalyptus, Fa-
     gus, Fraxinus, Hevea, Hibiscus, Juglans, Malus, Mangifera, Morus, Olea, Per-
     sea, Persica, Pirus, Populus, Prunus, Quercus, Ribes, Robinia, Rubus, Salix,
     Sambucus, Sorbus, Syringa, Tamarix, Tilia, Ulmus, Vitis*)
     32646, 32667, 32681, 32948, 33167, 33260, 33294, 33346, 33440, 33559, 33602,
     B33628, 33669, 33696, 34101, 34278, 34677, 34690, 34826-7, 34924, 34945,
     35006, 35092, 35167, 35269, 35525, 35719, 35808, 35885, 35930-1, 36037,
     B36191, 36201, 36319, 36445, 36509, 36515

Desert plants and ecosystems
     32665-7, 32824, 33434-6, 33547, 33696, 33699, 33839-40, 34239, 34511-2,
     34613, 34945, 35096, 35098, 35258, 35885, 35912, 35930-1, 36050

Dewberry  see  *Rubus*

Diatoms 32564-5, 32599, 32669, 32679, 32681, 32787, 32887, 32910, 32993, 33062,
     33070, 33083, 33118, 33127, 33163, 33169, 33183, 33224, 33266, 33333, 33366,
     33422, 33428, 33480, 33496, 33531, 33544-5, 33594, 33721, 33867, 33923,
                                                                    (continued)

Diatoms (continued)
    33927, 34002, 34016, 34021, 34030-1, 34058, 34069-70, 34160, 34191-2, 34197,
    34216, 34273, 34359, 34429, 34484, 34515, 34538, 34555, 34557, 34604, 34628-
    -9, 34634, 34669, 34768, 34793, 34885, 34914, 34953, 34991-5, 35043, 35204,
    35216, 35244, 35343, 35493, 35522, 35533, 35562, 35584, 35587, 35597, 35654,
    35811-2, 35880, 35946, 36061, 36182, 36369, 36420, 36474

*Digitaria*  32653, 32766, 32867, 33130, 33167, 33425, 33914, 34063, 34541, 34618,
    35101-3, 35186, 35466, 35468

*Dinoglagellatae, Dinophyceae*
    32564-5, 32599, 32859, 32887, 33006, 33070, 33118, 33927, 34021, 34216,
    34272-3, 34380, 34519, 34604, 34991-4, 35043, 35380, B35862, 35880, 36061,
    36117, 36474

*Dioscorea*  35426

Dogwood  see  *Cornus*

Douglas fir  see  *Pseudotsuga*

*Dunaliella*  32669, 32795, 32988, 33001, 33371, 33545, 33687, 33669, 33721, 34157-8,
    34223, 34538, 34978, 34988-9, 35390, 36291, 36364-5, 36474


E


Egg plant  see  *Solanum*

*Elaeis*  32888, 33229, 33868

Elder  see  *Sambucus*

Elm  see  *Ulmus*

*Elodea*  32831, 33103, 33675, 33804, 34052, 34216, 34421, 34629, 34715, 34859, 35048,
    35099, 35113, 35653, 35727, 36453-4, 36553

*Equisetum*  32681, 32948, 33804, 33904, 34715, 34917, 34924

*Ericaceae*  see  Heath plants and communities

*Eucalyptus*  32888, 33368, 34101, 34124, 34946, 34954, 35488, 35909-10, 36472

*Euglena* 32565, 32580, 32599, 32621, 32679-81, 32701-2, 32854, 32876, 33003, 33047,
    33070, 33113, 33225, 33235, 33256, 33266, 33294, 33389-90, 33417-8, 33423,
    33461-2, 33475, 33596-8, 33648, 33674, 33678, 33810, 33841, 33875, 33942,
    34058, 34164, 34188, 34327, 34360, 34557, 34604, 34629, 34644, 34652, 34669,
    34688, 34692, 34694, 34817, 34867, 34885, 34917, 35043, 35071-2, 35143,
    35167, 35190, 35206, 35241, 35244, 35289-90, 35382-3, 35479, 35567, 35596,
    35628-30, 35655, 35689, 35729, 35731, 35755, 35799, B35862, 35992, 36028,
    36031, 36056, 36182, 36249, 36414, 36455, 36459, 36574

*Euphorbia*  33054-5, 33440, 33653, 34980, 35160, 35186, 35425, 35442, 35906, 36201

Evergreen plants  see  Sempervirent plants


F


*Fagopyrum*  32948, 33655, 34410, 35245, 35403

*Fagus*    33700, 34227, 34285, 34471, 34607, 34860, 35200, 35808, B36191, 36201, 36345

Fern-palm  see  *Cycas*

Ferns    32679-81, 32948, 33103, 33823, 33879, 34485, 34669, 35096, 35098, 35477-8,
         35746, 36345

Fescue  see  *Festuca*

*Festuca*  32766, 32867, 32912-3, 32997, 33022, 33130, 33216, 33416, 33756, 34100,
           34181, 35014, 35040, 35052, 35142, 35280, 35301, 35595, 35744, 36156-7,
           36165, 36417-9, 36456

*Ficus*    33589, 35006

Fig  see  *Ficus*

Fir  see  *Abies*

Flax  see  *Linum*

Forage crops (*cf.* also *Brassica*, Grases, Leguminous plants, *Lupinus*, *Medicago*, *Tri-
          folium, Vicia, Vigna etc.*)
          33658, 34074, 34541, 35118, 35158

Forest (including undergrowth) plants and ecosystems (*cf.* also Coniferous plants,
          Deciduous trees and shrubs, Ferns, *Fragaria*, Grasses, Heath plants and com-
          munities, Lichens, Liverworts, Medicinal plants, Mosses, *Sphagnum, Vacci-
          nium,etc.*)
          32749, 32891, 33209, 33358-9, 33608, 33736-8, 33984-5, 34061, 34290-1, 34330,
          34613, 34677, 34987, 35014, 35165, 35445, 35488, 35808, 35885, 35919, 36201,
          36319, 36360, 36534

Fountain-grass  see  *Pennisetum*

Foxtail millet  see  *Setaria*

*Fragaria*  33107, 33415, 35202, 35450, 35776, 36201, 36213

*Fraxinus*  33062, 33089, 33440, 33559, B33628, 34246, 34924, 35006, 35269, 36139,
            36190, B36191

*Fungi* (parasitic)
          32850, 32869, 33372, 33477, 33599, 33909, 34146, 34413, 34443, 34501-2,
          34847, 34851, 34894, 34898, 35210, 36047, 36053, 36184


G

Garlic  see  *Allium*

Gherkin  see  *Cucumis*

Ginger  see  *Zingiber*

*Glycine*  32686, 32765, 32871, 32879, 32888, 32923, 32956, 33002, 33019-21, 33060,
           33068, 33076, 33092, 33103, 33181, 33222, B33281, 33369, 33394, 33453,
           33472, 33525, 33548-9, 33703, 33846, 33871-3, 33875, 33916-8, 33952, 33999-
           -4000, 34003, 34083, 34096, 34098, 34136, 34195, 34228, 34384, 34416, 34467,
           34506, 34524, 34544, 34549, 34609-10, 34794, 34800, 34831, 34936-7, 34940,
           35061, 35103, 35141-2, 35184, 35340-1, 35403, 35458, 35476, 35564, 35611,
           35632, 35744, 35765, 35809, B35862, 35866-7, 35902-5, 35977, 35998, 36004,
                                                                          (continued)

*Glycine* (continued)
        36058, 36066, 36069, 36077, B36191, 36207, 36243, 36303, 36325, 36348, 36372,
        36378, 36388, 36402, 36412, 36421, 36425, 36497

*Gossypium*  32706, 32803, 33019-20, 33096, 33125, 33261, 33344, 33388, 33500, 33506,
        33547, 33684, 33827, 33871-2, 33928, 34040, 34319, 34416, 34450, 34508,
        34648, 34794, 34856, 34898, 34954, 35016, 35079-80, 35146, 35221, 35260-1,
        35263, 35302, 35324-5, 35416, 35440, 35893, 36063, 36124, B36191, 36202,
        36528, 36532

Gourd  see  *Cucurbita*

Gram chick pea  see  *Vigna*

Grape fruit  see  *Citrus*

Grape vine  see  *Vitis*

Grasses (*cf.* also *Avena, Bromus, Carex, Cynodon, Cyperus, Dactylis, Digitaria, Fes-
        tuca, Hordeum, Lolium, Oryza, Panicum, Paspalum, Pennisetum, Phalaris, Phleum,
        Poa, Saccharum, Secale, Setaria, Sorgum, Triticum, Zea*)
        32576, 32625, 32653, 32867, 32882, 32888, 32895, 32950, 33014, 33050, 33130,
        33158-9, 33167, 33176, 33178-9, 33209, 33216, 33232, 33331, 33335, 33378-9,
        33395, 33412, 33433, 33440, 33537, 33547, 33607, 33620, 33730, 33742, 33756,
        33843, B33926, 34098, 34100, 34134-5, 34137, 34269, 34308, 34351, 34365-6,
        34387, 34442, 34444, 34452, 34458, 34499, 34536, 34541, 34613, 34646-7,
        34676, 34825, 34907, 34924, 34926, 35031, 35037, 35040, 35056, 35118, 35168,
        35174, 35176, 35186, 35243, 35272, 35301, 35403, 35426, 35441, 35468-9,
        35484, 35488, 35529, 35550, 35607-8, 35674, 35678, 35711, 35744, 35885,
        35891, 35924, 36014, 36027, 36037, 36070, 36092, 36103, 36115, 36132-3,
        36135, 36165, 36252, 36333, 36340, 36386-7, 36424, 36445, 36457

Groundnut  see  *Arachis*


H


*Halobacterium*  32558-9, 32655, 32727, 32734, 32781, 32793, 32800, 32878-9, 32901,
        32903, 33046, 33048, 33075, 33099, 33228, 33267, 33275-7, 33313, 33326,
        33328, 33376, 33421, 33437, 33443-4, 33636-8, 33650, 33681-2, 33752, 33833,
        33862, 33903, 33924, 33947-9, 33966, 33972-3, 33993, 34015, 34028, 34077-8,
        34084-5, 34221, 34263, 34270, 34345, 34391-2, 34398, 34449, 34516, 34590,
        34604, 34748, 34900, 34977, 35120, 35135, 35185, 35189, 35199, 35208, 35228-
        -9, 35253, 35312, 35486, 35551, 35565, 35568, 35718, 35722-3, 35782, 35805,
        35817, 35829-31, 35913, 35926, 35951, 36006-8, 36131, 36160-1, 36317-8,
        36321-2, 36349, 36373-5, 36380

Halophilous plants (*cf.* also Salt marsh and strand plants)
        32562, 32681, 32895, 32996, 33554-5, 33696, 33900, 35186, 35401, 36445

Hawthorn  see  *Crataegus*

Hazel  see  *Corylus*

Heath plants and communities
        33925, B33926, 34190, 34880, 35792, 35928, 36037

*Hedera*  32772-3, 34260, 35006, 36018, 36345

*Helianthus*  32577, 32629, 32690, 32788, 32828, 32867-8, 32948, 32969, 33068-9,
        33087, 33089, 33291, 33310, 33424, 33426, 33524, 33561, 33742, 33808, 33818,
        33827, 34037, 34271, 34325-6, 34585, 34659, 34794, 34877, 34940, 34954,
                                                                    (continued)

# K

*Kalanchoë*   34585, 34684, 34952, 35326, 35466

Kale   see  *Brassica*

Kenaf   see  *Hibiscus*

Kohlrabi   see  *Brassica*

# L

*Lactuca*   32578, 32736, 32829, 32896, 32948, 32996, 33048, 33060, 33078, 33089,
            33347, 33399, 33650, 33670, 33748-51, 33956-7, 34530, 34629, 34924, 35054,
            35291, 35456, 35511, 35752, 35838, 35859, 35982-4, 36126, 36167, 36175,
            36375, 36499

Larch   see  *Larix*

*Larix*   33221, 33268, 33920, 34654, 35808, 35949, 36049

*Lathyrus*   32739, 32843, 34924, 35605

Leguminous plants (*cf.* also *Arachis, Cajanus, Cicer, Glycine, Lathyrus, Lens, Lupi-
            nus, Medicago, Phaseolus, Pisum, Trifolium, Vicia, Vigna*)
            32739, 32762, 33547, 34049, 34825, 35464, 35828

*Lemna*   see  *Lemnaceae*

*Lemnaceae*   32571, 32645, 32681, 32780, 33419, 33600-1, 33739, 34052, 34060, 34145,
            34726, 35110-1, 35426, 35497-500, 36083-4, 36219

Lemon   see  *Citrus*

*Lens*   36438

Lentil   see  *Lens*

Lettuce   see  *Lactuca*

Lichens   32990, 33084, 33215, 33249, 33264, 33476-7, 33491, 33507, 33660, 33756,
            33974, 34239, 34302, 34518, 34547, 34966, 35281, 35691, 36361, 36408, 36427,
            36572

Lilac   see  *Syringa*

Linden   see  *Tilia*

Linseed   see  *Linum*

*Linum*   *34087, 35221, 35379*

Liverworts   32681, 32989, 34700, 35018, 35463, 36051, 36057, 36060

Locust   see  *Robinia*

*Lolium*   32686, 32867, 32980, 33130, 33193, 33242, 33488, 33633, 33859, 34100, 34176,
            34181, 34408, 34458-9, 34541, 34649, 35011, 35158, 35176, 35373, 35508,
            35542-3, 35594-5, 36060, 36111-2, 36165, B36191, 36205, 36348, 36433, 36457

Lucerne  see  *Medicago*

Lupine  see  *Lupinus*

*Lupinus*  32665-6, 32691, 32948, 33214, 33216, 33791-2, 33961-3, 34506, 35029, 35255-6, 35338, 35464, 35554, 36057

*Lycopersicon*  32568, 32639, 32741, 32759, 32796, 32948, 33010, 33093, 33219, 33642, 33644, 33761, 33827, 33967, 33976, 34005-6, 34098, 34163, 34212, 34232-4, 34241-2, 34250, 34252-6, 34271, 34425, 34448, 34506, 34649, 34683, 34851, 34939, 34972, 35004, 35006, 35029, 35060, 35128, 35142, 35259, 35335, 35601, 35613, 35633, 35776, 35971, 36060, 36213, 36329, 36335, 36571

*Lycopodium*  32680-1, 33264, 34917, 35014


# M

Macereed  see  *Typha*

Maize  see  *Zea*

*Malus*    32708, 32729, 32751, 32846, 33089, 33104, 33285-6, 33322, 33504, 33522, 33532, 33697-8, 33733, 33860, 34463-4, 34480, 34483, 34494-5, 34497, 34665, 34742, 34924, 35201, 35395, 35577-8, 35740, 35784, B35826, B35862, 36078, 36086, 36120, 36371, 36560

*Mangifera*  33129, 35251

Mango  see  *Mangifera*

Mangold  see  *Beta*

Mangrove communities  33166, 33292, 34887

*Manihot*  32758, 32867, 32888, 35064, 35297

Manioc  see  *Manihot*

Maple  see  *Acer*

Marrow  see  *Cucurbita*

Meadowgrass  see  *Poa*

*Medicago*  32582, 32686, 32762, 32888, 33223, 33294, 33338, 33488, 33547, 33894, 33904, 34052, 34100, 34142, 34301, 34541, 35100, 35158, 35295-6, 35341, 35359, 35488, 35729, 35775, 35885, 36124, 36166, 36391, 36445, 36458

Medicinal plants (*cf.* also *Cynodon, Hibiscus, Papaver, Ricinus, etc.*)
        32665-6, 32766, 32819, 32948, 33054, 33089, 33440, 34174, 34924, 35006, 35687, 35930, 35932

Melon  see  *Cucumis*

*Metasequoia*  32681

Millet  see  *Panicum*

*Morus*    34940, 34997, 35149, 36060, 36225, 36509

Mosses (*cf.* also *Sphagnum*)
        32676, 32679-81, 32694, 32803, 32837-8, 33028-9, 33052, 33410, 33660, 33756,
        33804, 33875, 34171, 34288-90, 34421, 34440, 34445, 34669, 34715, 34765,
        34886, 34917, 34924, 35014, 35094, 35129, 35244, 35463, 35525, 35928, 36013,
        36172, 36201, 36233-4

Mulberry  see *Morus*

Mung bean  see *Vigna*

*Musa*    35426

Musk-melon  see *Cucumis*

Mustard  see *Sinapis*


N


Napier grass  see *Pennisetum*

*Nicotiana*  32586, 32589, 32822-3, 32890, 32948, 33062, 33089, 33103, 33135-6, 33150,
        33153-7, 33271, 33273, 33303, 33445, 33489, 33629, 33642, 33672-3, 33686,
        33701, 33767-8, 33798, 33815, 33836, 33919, 33950, 33958, 33967, 34007,
        34026-7, 34066-7, 34072, 34125, 34205, 34267, 34286, 34346, 34372, 34379,
        34388, 34438-9, 34470, 34488, 34506, 34513, 34539, 34572, 34579, 34608,
        34618, 34670, 34676, 34683, 34716, 34758, 34776, 34781-3, 34817, 34882,
        34940, 35006, 35035-6, 35051, 35100, 35142, 35152, 35162, 35164, 35341,
        35359, 35378, 35404, 35456, 35459, 35502, 35507, 35509, 35687, 35697-700,
        35704, 35752, 35794, 35851, 35898, 35959-60, 35983, 35996, 36032, 36146,
        36280, 36288-9, 36344, 36351, 36381, 36461, 36494-5, 36529, 36544, 36561

*Nostoc*  32674, 32681, 32943, 33083, 33201, 33294, 33399, 33442, 33679, 33757-8,
        33776, 33903-5, 34081, 35696, 35699, 35955, 36096, 36279, 36575


O


Oak  see *Quercus*

Oat  see *Avena*

Oil palm  see *Elaeis*

*Olea*    33247, 33573, B33628, 34107-8

Olive  see *Olea*

Onion  see *Allium*

Orange  see *Citrus*

Orchardgrass  see *Dactylis*

Orchids 33892, 35426

Ornamental plants (*cf.* also *Agave, Antirrhinum, Asparagus,* Coniferous plants, *Cype-
        rus,* Deciduous trees and shrubs, *Eucalyptus, Euphorbia, Ficus, Hedera, Hi-
        biscus, Ilex, Lathyrus, Lupinus, Pelargonium, Perilla, Tradescantia, Tuli-
        pa,* etc.)

                                                        (continued)

Ornamental plants (continued)
            32560, 32664, 32766, 32866, 32948, 32996, 33089, 33248, 33349, 33440, 33565,
            33590-1, 33642, 33669, 33756, 33804-5, 33831, 33842, 33858, 33968, 34018-9,
            34067-8, 34101, 34175, 34507, 34549, 34608, 34788, 34814, 34817, 34924,
            35006, 35140, 35260-1, 35263, 35314, 35330-2, 35426, 35438, 35687, 35691,
            35703, 35776-8, 35785, B35862, 36063, 36217, 36257-8, 36379, 36386, 36407,
            36437, 36445, 36478, 36535, 36549

*Oryza*     32574, 32603, 32681, 32686, 32737-8, 32863-4, 32874, 32948, 33109-10, 33130,
            33158, 33162, 33607, 33871-2, 33918, 33921, 33999-4000, 34038, 34095-8,
            34118-21, 34123, 34152-4, 34200, 34261, 34397, 34615, 34738, 34786-7, 34907,
            34940, 34942, 34956, 35022-3, 35044, 35064, 35091, 35103, 35183, 35210,
            35232, 35257, 35270, 35341, 35363-4, 35428, 35430-4, 35437, 35647, 35759,
            36056-8, 36060, 36064, 36066, 36069-70, 36077, 36185, 36212, 36227, 36267,
            36303, 36412, 36497, 36500-1, 36524


P


Paddy  see  *Oryza*

Palms  see  *Cocos, Elaeis*

*Panicum*  32653, 32867, 32888, 32913, 32950, 32997, 33053, 33130, 33167, 33537, 33705,
            33843, 33914, 34095-7, 34134-5, 34252-4, 34256, 34365-6, 34452-3, 34541,
            34549, 34618, 34676, 34924, 34943, 35186, 35286, 35425-6, 35466, 35468,
            35471-2, 35704, 35766, 36056, 36146, 36205, 36252, 36303, 36386, 36436

*Papaver*  32768, 33264, 35265-7

Paprika  see  *Capsicum*

Para-rubber tree  see  *Hevea*

Parasitic plants   33230, B35862, 36194

Parsley  see  *Petroselinum*

Parsnip  see  *Pastinaca*

*Paspalum*  32653, 32766, 32976, 33914, 34301, 34365-6, 35186, 35403, 36348

*Pastinaca*  32948, 36255

Pasture plants  see  Forage plants

Pea  see  *Pisum*

Peach  see  *Persica*

Peanut  see  *Arachis*

Pear  see  *Pirus*

Peavine  see  *Lathyrus*

Pecan  see  *Carya*

*Pelargonium*  33015, 33827, 33967, 36422

*Pennisetum*  32653, 33073, 33130, 33614, 33855, 33914, 34134-5, 34942, 35007, 35186,
            35257, 35427-8, 35430-3, 35455, 35560, 35888, 36205, 36252

Pepper  see  *Capsicum*

*Perilla*  32613

*Persea*  35691

*Persica*  33111,  34098,  34480,  34483,  34742

*Petroselinum*  32948,  32962,  33556,  34717,  34812,  34938,  35743,  36255

*Phalaris*  32950,  33703,  33728,  34181,  34351,  36386

*Phaseolus*  32590-1,  32593-5,  32648,  32660,  32680,  32731,  32835,  32868-9,  32889-90,
          32929-31,  32948,  32986,  33035,  33062-3,  33079,  33150,  33205,  33219,  33244,
          33272,  B33281,  33290,  33298-9,  33392,  33429,  33481,  33526,  33542,  33561,
          33590-1,  33639,  33742,  33778,  33805,  33808,  33870-3,  33875,  34066-7,  34111,
          34147,  34170,  34198,  34232-3,  34277,  34294,  34306,  34409,  34418,  34420,
          34480,  34482,  34493,  34501,  34627-9,  34703,  34742,  34817,  34821,  34927,
          34929,  34934-5,  34940,  35019,  35021,  35161,  35241,  35278,  35283,  35321,
          35359,  35368-70,  35374,  35385,  35423-4,  35541,  35545,  35613,  35641,  35656,
          35677,  35710,  35729,  35767,  35773-4,  35891,  35893,  35897,  35979-80,  36023-4,
          36044,  36140,  36146,  36168,  36193,  36195-6,  36270,  36274,  36280,  36384-5,
          36431,  36435,  36478-9,  36487,  36516,  36519,  36576,  36582

*Phleum*  34458,  34743,  34924,  36165,  36333

*Phoenix*  35426

Photosynthetic bacteria  see  Bacteria, photosynthetic

*Phragmites*  33364,  33411-2,  33536,  33703,  33804,  34035,  34052,  34216,  34442,  34487,
          34621,  34715,  34907,  35168-9,  35171-4,  35607,  35674,  36386,  36553

*Picea*  32698,  32777,  32891,  32960,  33199,  33278,  33422,  33677,  33783,  33878,  33920,
          34260,  34283-4,  34318,  34323,  34402,  34532,  34588,  34689,  34747,  35081,
          35115,  35156-7,  35491,  35536,  35764,  35808,  35936,  35949,  36049,  B36191,
          36201,  36358,  36551-2

Pigeon pea  see  *Cajanus*

Pine  see  *Pinus*

Pineapple  see  *Ananas*

*Pinus*  32644,  32681,  32698,  32713,  32802,  32806,  32810,  32891,  33043,  33066,  33077,
          33103,  33209,  33221,  33269,  33378,  33440,  33467,  33508,  33528,  33587,  33595,
          33693,  33725,  33845,  33878,  33920,  34174,  34287,  34307,  34318,  34402,  34416,
          34441,  34471,  34651,  34654-5,  34752,  34766-7,  35014,  35115,  35156,  35181,
          35224,  35358,  35453,  35559,  35808,  B35862,  35885,  35949,  36049,  B36191,  36310,
          36392-3,  36465,  36478,  36530-1,  36551-2

*Pirus*  34480,  34483,  35203,  35282,  35800

*Pistacia*  36445

*Pisum*  32561,  32594,  32601,  32610,  32616,  32627-8,  32660-1,  32675,  32679,  32687-8,
          32700,  32739,  32746-7,  32801,  32832-3,  32836,  32848-9,  32855,  32859,  32889,
          32892,  32934-6,  32948-9,  33024,  33063,  33090,  33137,  33149,  33160-1,  33171-2,
          33206-7,  33219,  33310,  33317,  33323,  33370,  33399,  33450,  33453-4,  33461-2,
          33550,  33579,  33684,  33727,  33731,  33750,  33770-2,  33806,  33809,  33827,
          33871-3,  33898-9,  33914,  33959,  33979,  33990-1,  34026-7,  34042,  34141,
          34207-8,  34214,  34250,  34253-6,  34277,  34312,  34317,  34321-2,  34349-50,
          34371,  34400,  34418,  34431-2,  34475,  34527,  34586,  34610,  34629,  34717,
          34723,  34753,  34779-80,  34817,  34855-6,  34869,  34892,  34924,  34982-3,  35009,
                                                                    (continued)

*Pisum*   (continued)
         35012, 35026, 35065, 35076, 35104, 35142, 35264, 35356, 35361-2, 35378,
         35456, 35464, 35515, 35530, 35566, 35569-70, 35572, 35576, 35581, 35591,
         35685, 35755, 35796, 35800, 35809, 35844, B35862, 35976, 35988, 35997,
         36005, 36054, 36094, 36192, 36200, 36223, 36233, 36250, 36268, 36302-3,
         36306, 36331, 36375, 36423, 36469, 36486, 36558-9

Plum   see   *Prunus*

*Poa*      32567, 32766, 33130, 33331, 33391, 33728, 34021, 34100, 35176, 35521, 35795,
         36445, 36456

Poplar   see   *Populus*

Poppy   see   *Papaver*

*Populus*   32618, 32692, 33013, 33089, 33105-6, 33220-1, 33230, B33281, 33377, 33422,
         33558-9, 33566, 33590-1, 34150, 34246, 34506, 34827, 34866, 34924, 35154,
         35269, 35807, 35827, B35862, 35949, 36060, 36190, B36191, 36445, 36509

*Porphyridium*   32557, 32744, 32988, 33613, 33758, 33884, 34007, 34696, 35366-7,
         35738, 36279, 36455

*Portulaca*   32766, 32867, 32996, 33080, 33914, 34368, 34980, 35101-3, 35186, 35961

Potato   see   *Solanum*

Prune   see   *Prunus*

*Prunus* (*cf.* also *Armeniaca*)
         32948, 33504, 33602, 34098, 34239, 34511-2, 34581, 34742, 36478, 36543

*Pseudotsuga*   32698, 33422, 33595, 34287, 34471, 35721, 35808, 36049, 36075, 36332

Pumpkin   see   *Cucurbita*

Purslane   see   *Portulaca*


# Q

*Quercus*   32693, 33262, 33345, 33440, 33451, 33477, 33559, 33595, 33632, 33736-7,
         33936, 33984, 34577, 34730, 34924, 35006, 35194, 35233, 35808, 35949,
         35962, 36144, B36191, 36403


# R

Radish   see   *Raphanus*

Rape   see   *Brassica*

*Raphanus*   32681, 33031, 33062, 33643, 34606-7, 35140, 35310, 35752, 35849, 35983,
         36060, 36137-8, 36142

Raspberry   see   *Rubus*

Redwood   see   *Metasequoia*

Reed   see   *Phragmites*

Sempervirent plants (*cf.* also *Coffea*, Coniferous plants, *Hedera*, *Ilex*, *etc.*)
     34260, 35354, 35537, 35808, 36543

Service-tree  see  *Sorbus*

*Sesamum* 33089, 33504, 35033

*Setaria* 32653, 32766, 32950, 33130, 34365-6, 34676, 35427-8, 35430-4, 36252

Shrubs  see  Deciduous trees and shrubs; Sempervirent plants

*Sinapis* 32959, 33062, 33102, 33231, 33551, 33603, 33655, 33692, 34218, 34699, 34920-
     -2, 35132-4, 36410

Sisal  see  *Agave*

*Solanum* (*cf.* also *Lycopersicon*)
     32678, 32717, 32750, 32874, 32929-31, 32948, 33008, 33034, 33103, 33195-6,
     33291, B33510, 33513, 33645, 33729, 33875, 33897, 33908, 34091-3, 34132,
     34597, 34737, 34808, 34923-5, 34932-3, 34940, 34942, 35006, 35051, 35064,
     35447, 35547, 35704, 35729, 35776-7, 35789, B35862, 35979, 36142, B36191,
     36226, 36262, 36489

*Sorbus* 32891, 33264

Sorghum  see  *Sorgum*

*Sorgum* 32653, 32663, 32765-6, 32803, 32881, 32888, 32918, 32969, 33074, 33150,
     33162, 33192, 33222, 33527, 33547, 33705, 33843, 33914, 34048, 34109-10,
     34147, 34228, 34256, 34295, 34312, 34480, 34506, 34649, 34713, 34743, 34794,
     34942-3, 35186, 35257, 35302, 35341, 35378, 35476, 35893, 35907, 35966,
     36016-7, 36205, 36206-7, 36237, 36267, 36274, 36303, 36325, 36348

Soybean  see  *Glycine*

*Sphagnum* 33264, B33926, 34190, 35014, 35244, 35463, 35928

Spinach  see  *Spinacia*

Spinach beet  see  *Beta*

*Spinacia* 32594, 32608-9, 32626, B32630, 32631-2, 32636-7, 32641, 32659-60, 32673,
     32675, 32679-82, 32687, 32697, 32720, 32725-6, 32744, 32755, 32767, 32769,
     32782, 32791, 32828, 32834, 32839, 32841-2, 32845, 32851, 32853, 32889-90,
     32894, 32898, 32905-7, 32922, 32932-3, 32944, 32948, 32954, 32968, 32992,
     32994, 33005, 33008, 33035, 33037, 33058, 33063-4, 33068, 33100, 33116-7,
     33137, 33151, 33171-2, 33212-4, 33256, 33294, 33310, 33341, 33350, 33353-6,
     33383, 33387, 33452, 33458-9, 33461-2, 33473-4, 33533, 33553, 33560, 33649,
     33690-1, 33712-3, 33743-4, 33748, 33751, 33755, 33772, 33779, 33785, 33801-2,
     33805, 33821-2, 33836-7, 33865, 33869, 33875, 33880-1, 33886-7, 33895,
     33904, 33910, 33919, 33929-33, 33938-9, 33944-5, 33960, 33986, 33994-5,
     33997, 34004, 34032, 34062-3, 34067, 34089, 34093, 34102-6, 34114, 34117,
     34126-7, 34138, 34142, 34147, 34165-6, 34168, 34173, 34178, 34180, 34205,
     34226, 34231-3, 34277, 34286, 34292, 34328, 34335, 34357-8, 34364, 34367,
     34377-8, 34417, 34422-4, 34446, 34488, 34509, 34526, 34530, 34563, 34573-6,
     34598, 34659, 34694, 34726, 34728, 34734, 34749, 34760, 34769, 34775, 34802,
     34804-5, 34816-7, 34820, 34882, 34889-91, 34956, 34975, 34984, 34996, 35005,
     35026, 35086, 35088, 35093, 35100, 35114, 35125, 35140, 35187-8, 35206,
     35212, 35214-5, 35231, 35286, 35291, 35345, 35353, 35375, 35378, 35382,
     35391, 35402, 35414-5, 35438, 35444, 35456, 35466, 35470, 35479, 35495,
     35503, 35510, 35516, 35526, 35540, 35545, 35547-8, 35600, 35615, 35646,
     35649-50, 35663, 35665-6, 35669-70, 35681, 35692, 35694, 35716, 35728-9,
     35736-7, 35739, 35743, 35752-3, 35816, 35818, 35822-3, 35848, 35853-4,
     35858, B35862, 35879, 35889, 35894, 35917-8, 35981, 35983, 35991, 36003,
                                                            (continued)

*Spinacia* (continued)
        36022-3, 36025-6, 36035, 36046, 36055, 36063, 36074, 36095, 36102, 36127-8,
        36130, 36175-6, 36210, 36239-40, 36244-5, 36263, 36277-8, 36282-3, 36314,
        36331, 36355, 36375, 36377, 36389, 36397-8, 36404-5, 36413, 36416, 36430,
        36460, 36462-3, 36475, 36488, 36502-3, 36511, 36526, 36555, 36573, 36577

*Spirodela*  see  *Lemnaceae*

Spruce  see  *Picea*

Squash  see  *Cucurbita*

Strawberry  see  *Fragaria*

Submersed plants  see  Aquatic macrophytes

Succulents (*cf.* also *Agave, Aloe, Bryophyllum,* Cacti, CAM plants,*etc.*)
        32555, 33055, 33270, 33858, 33960, 34368-9, 34585, 34924, 35097, 35258,
        35442, 35625, 35930, 35958, 36093

Sugar beet  see  *Beta*

Sugar cane  see  *Saccharum*

Sunflower  see  *Helianthus*

Sweet potato  see  *Ipomoea*

*Synechococcus*  see  *Anacystic*

*Syringa* 34924


T

Tamarisk  see  *Tamarix*

*Tamarix* B33281, 36445

Tapioca  see  *Manihot*

*Taxus*    32948, 34457, 35006, 35808

Tea  see· *Thea*

*Thea*    34047, 35017, 35123, 35535, 35619, 35622, 35963

*Theobroma*  35205

*Thuja*    33422, 34318, 35949

*Tilia*    34924, 35949, B36191

Timothy  see  *Phleum*

Tobacco  see  *Nicotiana*

Tomato  see  *Lycopersicon*

*Tradescantia*  32572, 33615, 34629, 35426, 35571, 35652, 36326, 36541-2

*Trifolium*   32948, 33022, 33331, 33338, 33488, 33611, 34086, 34100, 34181, 34541,
    34701-2, 34809-10, 34924, 35158, 35508, B36191, 36434, 36549

*Triticum*   32556, 32576, 32604-5, 32634, 32679, 32681, 32689, 32704, 32718, 32761,
    32763, 32786, 32805, 32855, 32893, 32900, 32919, 32948, 32956, 32971, 33000,
    33040, 33085, 33090, 33116, 33128, 33132, 33158, 33162, 33164-5, 33173-5,
    33226-7, 33279, 33339, 33363, 33372, 33374-5, 33405, 33424, 33426-7, 33438,
    33478, 33488, 33519, 33523, 33540, 33547, 33592, 33647, 33655, 33660, 33680,
    33760, 33827, 33835, 33871-2, 33914, 33970, 34014, 34056, 34059, 34066,
    34088, 34090, 34098, 34137, 34228, 34253-4, 34256, 34262, 34308-10, 34312,
    34314, 34316, 34333-4, 34348, 34373-6, 34416, 34437-9, 34443, 34451-2,
    34465, 34467, 34479, 34483, 34499, 34513, 34540, 34542, 34549, 34551-2,
    34554, 34582, 34721, 34817, 34830, 34839, 34879, 34918, 34940, 35029, 35047,
    35064, 35121, 35140, 35142, 35237, 35249, 35257, 35279, 35285, 35294, 35321,
    35327, 35337, 35339, 35341, 35349-52, 35377, 35389, 35394, 35441, 35448,
    35474, 35482-3, 35505, 35517, 35550, 35610, 35616, 35623, 35673, 35676,
    35701, 35704, 35759, 35790-1, 35815, B35862, 35883-4, 35886, 35891, 35893,
    35938, 35952-4, 36010, 36038-9, 36053, 36056, 36058, 36067-8, 36077, 36114,
    36116, 36137-8, 36205, 36281, 36284, 36289, 36296, 36299-301, 36303, 36315-
    -6, 36343, 36406, 36412, 36440, 36450-2, 36556, 36563, 36581

*Tsuga*   33422, 35808, 36049

Tulip   see   *Tulipa*

*Tulipa*   33506

Tundra plants and ecosystems

    33050, 33052, 33245, 34536, 34553, 34613, 34888, 35407, 35885, 36014,
    36132-3, B36134, 36361

Turnip   see   *Brassica*

Turpentine tree   see   *Pistacia*

*Typha*   32867, 33412, 33422, 33536, 33804, 34052, 34351, 34387, 34442, 34486, 34631,
    34942, 35006, 35168, 35174, 35244, B35862, 36386-7, 36553


U


*Ulmus*   33013, 34285

*Ulva*   32681, 33922, 34251, 34385, 34568, 34570, 34626, 34629, 34899, 34912, 35449,
    35557, 35735, 36153-4, 36586


V


*Vaccinium*   33264, B33926, 34291, 35014, 35525, 35691, 35808, 36037, 36201

Vegetables (*cf.* also *Allium, Asparagus, Beta, Brassica, Capsicum, Cucumis, Cucurbita,*
    *Daucus, Lactuca, Lycopersicon, Pastinaca, Petroselinum, Phaseolus, Pisum,*
    *Portulaca, Raphanus, Solanum, Spinacia*)
    32765, 32948, 33790, 33958, 36058, 36137, 36226, 36255

Vetch   see   *Vicia*

(continued)

*Zea* (continued)
          35186, 35249-50, 35257, 35286, 35303, 35321, 35341, 35346, 35398, 35410,
          35444, 35454, 35460, 35489, 35492, 35504, 35558, 35564, 35582, 35603-5,
          35631, 35636-8, 35641, 35652, 35656-8, 35662, 35701, 35704, 35730, 35806,
          35835, B35862, 35870, 35877, 35891-3, 35964, 35970, 35986, 35993, 35996,
          36004, 36019, 36021, 36056-7, 36060, 36069, 36077, 36146, 36148, 36163,
          36178, B36191, 36205, 36212, 36225, 36230, 36237, 36269, 36275, 36303,
          36348, 36412, 36425-6, 36440, 36497, 36520, 36533, 36537, 36548, 36579-80

*Zebrina*  see  *Tradescantia*

*Zingiber*  35426